Prioritizing Environmental Sustainability in Local Industrial Systems

Syed Abdul Rehman Khan
Business Research and Consultancy Limited, Sheffield, UK

Zhang Yu
Yunnan University of Finance and Economics, Kunming, China

Muhammad Tanveer
https://orcid.org/0000-0002-2019-6544
Mohammad Ibn Saud Islamic University, Riyadh, Saudi Arabia

IGI Global
Scientific Publishing
Publishing Tomorrow's Research Today

Vice President of Editorial	Melissa Wagner
Director of Acquisitions	Mikaela Felty
Director of Book Development	Jocelynn Hessler
Production Manager	Mike Brehm
Cover Design	Jose Rosado

Published in the United States of America by
IGI Global Scientific Publishing
701 East Chocolate Avenue
Hershey, PA, 17033, USA
Tel: 717-533-8845 | Fax: 717-533-7115
Website: https://www.igi-global.com E-mail: cust@igi-global.com

Library of Congress Cataloging-in-Publication Data

Names: Khan, Syed Abdul Rehman, 1987- editor. | Yu, Zhang, 1990- editor. | Tanveer, Muhammad editor.
Title: Prioritizing environmental sustainability in local industrial systems / [edited by] Syed Abdul Rehman Khan, Zhang Yu, Muhammad Tanveer.
Description: Hershey, PA : IGI Global Scientific Publishing, [2026] | Includes bibliographical references and index. | Summary: "This book explores existing circular economy approaches, green ideologies, and sustainability practices within industries, while also emphasizing the role of digital technologies in prioritizing environmental considerations in local industrial systems"-- Provided by publisher.
Identifiers: LCCN 2025036617 (print) | LCCN 2025036618 (ebook) | ISBN 9798337364353 hardcover | ISBN 9798337364360 paperback | ISBN 9798337364377 ebook
Subjects: LCSH: Industrial management--Environmental aspects. | Social responsibility of business.
Classification: LCC HD30.255 .P754 2026 (print) | LCC HD30.255 (ebook) | DDC 658.4/083--dc23/eng/20260122
LC record available at https://lccn.loc.gov/2025036617
LC ebook record available at https://lccn.loc.gov/2025036618

British Cataloguing in Publication Data
A Cataloguing in Publication record for this book is available from the British Library.

Table of Contents

Detailed Table of Contents

Plasma-based technologies are emerging as cutting-edge solutions in fostering eco-friendly production and sustainable industrial practices, offering exceptional abilities for circular economic solutions and sustainable eco-friendly pollution control. Using evidence-based assessment, it examines practical applications of waste minimization or waste-to-energy conversion techniques while comparing current performance with traditional approaches in terms of technological feasibility, ecological impact, and economic viability. The chapter also emphasizes the knowledge gap and future research directions by focusing on advanced manufacturing, sustainable energy, and science technology. It concludes by highlighting the importance of interdisciplinary research and strategic policy frameworks to accelerate the integration of plasma-based technologies. Bringing together scientific, technological, and policy perspectives, this chapter demonstrates how plasma approaches can significantly contribute to the United Nations Sustainable Development Goals (SDGs) and broader environmental targets.

This paper formulates and substantiates a pragmatic integration of enterprise risk management and the circular economy (ERM–CE), facilitating resource-limited organisations in diminishing their environmental impact while enhancing operational and financial outcomes. The study uses a field-tested ERM–CE framework that was used in eleven small pottery businesses in West Bengal, India, as a starting point. It then adapts and translates the framework for use in other industries, such as textiles, food processing, electronics, and logistics. The baseline shows that aligning risk processes with life-cycle stages according to ISO 14040/14001 and using CE levers

a Sustainability-Oriented GMP Framework that embeds circular economy principles, carbon reduction, and responsible resource use into local production systems, particularly for SMEs. Grounded in industrial ecology, knowledge-based theory, and stakeholder responsibility, the framework redefines GMP as a pathway to achieve the UN Sustainable Development Goals (notably SDG 12 and 13). By integrating energy efficiency, recycling, and knowledge-driven innovation with digital tools, industries can shift from compliance-based models to sustainable ecosystems. The chapter offers conceptual insights and practical directions for policymakers and managers, positioning GMP as a transformative tool for advancing sustainability in local industries.

The paper lingers on the critical role that the Management and Information System (MIS) play in enhancing the activities of the local industrial systems in the environmental sustainability. It submits the significance of the MIS and other online platforms to the ecological accountability by incorporating environmental observation, compliance system and business sustainability. The research article will talk about the use of Environmental Management Systems (EMS) like, ISO 14001, Environmental Management Information System (EMIS) and Sustainability Information System (SIS) in the attainment of resource efficiency, reduction of emissions and responsibility of an organization. The paper identifies the increased importance of the new technologies of cloud-based platforms, the Internet of Things (IoT), digital twins, predictive analytics, and the tools of sustainability planning. It also shows that these technologies will be useful in making real-time decisions, enhancing the ability to assess a risk, and making more informed and sustainable decision-making.

This chapter investigates the convergence of event management operations and sustainable practices to guide their function of stimulating innovation and development within the event management sector. With increasing necessity, sustainability becomes central and unavoidable to the world economy, and event professionals are increasingly embracing green practices in order to minimize environmental footprints while enhancing overall experiences for attendees. This chapter gives an extensive overview of important sustainability concepts, energy efficiency, sustainable procurement and encompassing waste minimization and outlines their day-to-day applicability in the planning, staging and post-event assessment of events It also centers the role of stakeholder engagement, ranging from suppliers through to guests, in influencing sustainable action. The chapter ends with suggesting future research directions for embedding sustainability into event management, focusing on the imperative of innovation, cooperation, and constant industry-wide dedication to protecting the planet.

This chapter introduces the Ecoleviant Governance Model, a novel conceptual framework developed through doctoral research, designed to address the pressing need for environmentally sustainable governance in terms of crisis governance. The model can be applied to local industrial systems as well. Based on interdisciplinary inquiry the model emphasizes participatory, adaptive, and multi-scalar mechanisms of environmental decision-making. It responds to the complex ecological interdependencies and socio-economic challenges faced by industrial zones, especially in emerging economies. The chapter will first outline the model's key components. It then discusses its empirical grounding in case studies from China and Italy. The chapter concludes with practical recommendations for policymakers and industry leaders seeking to localize sustainability efforts through integrated governance strategies.

Neoliberal globalization, driven by U.S. economic policies, trade liberalization, and technology, has transformed production, consumption, and environmental outcomes. Deregulation and weakened environmental oversight enabled resource-intensive industries and offshoring of polluting production to countries such as China, Vietnam, and Indonesia. Trade agreements like NAFTA and WTO frameworks reinforced this trend, externalizing carbon emissions and social impacts. Industries such as fast fashion contribute roughly four percent of global emissions. Addressing these structural harms requires enforceable supply chain standards, sustainable production, and transformative consumption to protect both the planet and vulnerable populations.

The study investigates assessment and monitoring strategies for Sampalok Lake in San Pablo City to promote sustainable practices. Qualitative research involving key informants reveals knowledge disparities and fragmented monitoring systems, as well as emerging threats such as microplastic contamination. Community initiatives like clean-up campaigns and youth involvement show potential in addressing these challenges. The research highlights the need for a collaborative, multidisciplinary approach incorporating scientific advancements, indigenous knowledge, and policy coordination. The study discusses various suggested measures to safeguard the ecological health of lakes in the Philippines. Among these are setting up a collaborative lake monitoring group, promoting community-based scientific ventures, launching a university-driven extension program, utilizing real-time monitoring technology, and strengthening public outreach campaigns. These approaches underscore the significance of employing diverse and adaptable methods to guarantee the protection of lakes across the nation.

Companies require formulating marketing strategies which can promote sustainable manufacturing and green products. In today's world, growing environmental concerns and consumer expectations are important. The objective of the study is to analyze how companies can sort their marketing efforts with sustainability goals. A qualitative methodology using secondary data and case studies is adopted. The analysis focuses on key elements such as transparency, green branding, consumer education, and digital engagement. Case studies of prominent companies viz. Unilever, Apple, and Tesla demonstrate how companies incorporate sustainability into their operations and brand image. The study also identifies challenges such as the attitude-behaviour gap and greenwashing. A strategic model is suggested to connect marketing with sustainability. Practitioners and companies may apply the analysis to adopt sustainability initiatives. The findings provide practical understandings for businesses and how companies should incorporate sustainability in their business actions.

CSR, known as dimensions. At its core, CSR is about operating. This concept has evolved significantly over the years, moving from being a peripheral issue to one that is integral to business strategies. The modern understanding of CSR recognizes that businesses operate closely with companies. The foundation of CSR lies in the belief that corporations have a responsibility that goes beyond generating profit for shareholders construct society by integrating. The traditional view sole responsibility a business to increase profits, as famously advocated by economist Milton Friedman, has gradually shifted towards a broader view where corporations are seen as key

Preface

INTRODUCTION

Environmental sustainability has emerged as one of the most pressing challenges of our time, particularly in the context of emerging economies where rapid industrialization and climate change have intensified environmental degradation and pollution-related health concerns. Local industrial systems, while critical for economic growth and social development, are increasingly called upon to balance productivity with environmental responsibility. In this context, sustainability is no longer a peripheral consideration but a central imperative for industries, policymakers, and society at large.

This edited reference book, *Prioritizing Environmental Sustainability in Local Industrial Systems*, is conceived against this backdrop. The volume brings together contemporary theoretical perspectives and the latest empirical research to examine how industries can address environmental challenges without compromising economic performance. In recent years, policymakers and regulatory authorities have strongly encouraged the adoption of circular economy models, green principles, and sustainability-oriented practices. When these approaches are strategically combined with digital technologies, they offer significant potential to transform local industrial systems and deliver optimal environmental and economic outcomes.

The primary objective of this book is to provide a comprehensive and integrative framework that captures the multifaceted nature of environmental sustainability in industrial contexts. The chapters collectively explore circular economy approaches, green ideologies, and sustainability practices across diverse industrial settings, while highlighting the enabling role of digital technologies in monitoring, managing, and enhancing environmental performance. By presenting diverse methodological approaches and real-world insights, the book seeks to bridge the gap between academic research and practical application, fostering meaningful dialogue and knowledge exchange between scholars, practitioners, and decision-makers.

This volume is intended as a valuable reference for researchers, academics, policymakers, and professionals engaged in sustainability and environmental management. Beyond conceptual understanding, it offers actionable strategies and innovative approaches for managing expertise, information, and knowledge, as well as for supporting organizational learning and development in varied industrial and institutional environments. We hope that this book will inspire further research, inform policy formulation, and support industries in their transition toward more sustainable, resilient, and environmentally responsible local industrial systems.

ORGANIZATION OF THE BOOK

Chapter 1 sets the foundation for the volume by examining how digital technologies enable circular economy transitions in localized industrial systems within emerging economies. Drawing on case evidence from India, Bangladesh, and East Africa, the chapter introduces an integrated framework that connects digital readiness, innovation intensity, institutional support, organizational culture, and environmental commitment to digitally enabled circular economy outcomes. By highlighting the transformative role of IoT and circular business model innovation—while also acknowledging structural, financial, and policy constraints—the chapter offers both conceptual advancement and actionable insights for policymakers and industry leaders seeking inclusive and sustainable industrial growth.

Chapter 2 explores plasma-based technologies as advanced solutions for sustainable industrial production and environmental protection. Through an evidence-based assessment, the chapter evaluates plasma applications in waste minimization, pollution control, and waste-to-energy systems, comparing their ecological and economic performance with conventional technologies. By integrating scientific, technological, and policy perspectives, the chapter underscores the need for interdisciplinary research and enabling policy frameworks to scale plasma innovations, positioning them as powerful contributors to circular economy objectives and the United Nations Sustainable Development Goals.

Chapter 3 presents a pragmatic integration of enterprise risk management and circular economy principles, demonstrating how resource-constrained organizations can achieve measurable environmental and financial improvements. Anchored in a baseline study of pottery SMEs in India and extended to other sectors, the chapter shows how aligning risk management with life-cycle thinking and ISO environmental standards can generate significant operational gains without additional resource inputs. This chapter offers compelling evidence that circular practices, when embedded in risk governance, can enhance resilience, competitiveness, and sustainability across diverse industries.

Chapter 4 provides a conceptual grounding of the circular economy as a transformative alternative to the linear "take–make–dispose" model. Emphasizing regenerative design, closed-loop systems, and resource efficiency, the chapter articulates how technology-enabled circular strategies decouple economic growth from finite resource consumption. By framing circularity as both an economic and ecological imperative, the chapter reinforces the foundational principles that inform many of the empirical and applied discussions throughout the book.

Chapter 5 repositions Good Manufacturing Practices as a strategic driver of sustainability in local industrial systems. Moving beyond compliance-focused approaches, the chapter proposes a sustainability-oriented GMP framework that integrates circular economy principles, carbon reduction, and knowledge-driven innovation. Grounded in industrial ecology and stakeholder theory, the chapter highlights how SMEs can leverage digital tools and knowledge-centric practices to transition from linear production models to sustainable industrial ecosystems aligned with global development goals.

Chapter 6 emphasizes the critical role of Management Information Systems in advancing environmental sustainability within local industries. The chapter discusses how systems such as EMS, EMIS, and sustainability information platforms enable real-time monitoring, compliance, and data-driven decision-making. By examining emerging technologies including cloud computing, IoT, digital twins, and predictive analytics, the chapter demonstrates how integrated information systems enhance risk assessment, resource efficiency, and organizational accountability in sustainability-oriented industrial management.

Chapter 7 addresses environmental sustainability within the event management industry, highlighting its growing relevance as a driver of innovation and economic growth. The chapter examines how green practices—ranging from energy efficiency and sustainable procurement to waste minimization—can be embedded across the event lifecycle. By emphasizing stakeholder engagement and collaborative responsibility, the chapter positions sustainable event management as both an operational necessity and a catalyst for broader industry transformation.

Chapter 8 introduces the Ecoleviant Governance Model, a novel framework designed to strengthen environmentally sustainable governance in local industrial systems. Drawing on interdisciplinary insights and empirical cases from China and Italy, the chapter highlights participatory, adaptive, and multi-scalar governance mechanisms capable of addressing complex ecological and socio-economic challenges. The chapter concludes with policy-relevant recommendations for embedding localized sustainability through integrated and crisis-responsive governance structures.

Chapter 9 examines the role of Green Human Resource Management in cultivating environmentally responsible organizational cultures. Based on empirical evidence from Indian IT firms, the chapter demonstrates how green training, performance appraisal, and employee participation enhance engagement and pro-environmental behavior. By positioning HR functions as catalysts for sustainability, the chapter underscores the human dimension of environmental transformation within organizations.

Chapter 10 focuses on the role of green technologies in reshaping local industrial systems toward sustainability. Through examples such as renewable energy integration, cleaner production, and waste-to-energy systems, the chapter illustrates how technological innovation reduces environmental footprints while enhancing efficiency and competitiveness. Case studies from diverse economic contexts demonstrate the strategic value of green technologies in advancing circular economy goals.

Chapter 11 critically examines the environmental consequences of neoliberal globalization, with particular emphasis on U.S.-driven trade policies and industrial offshoring. The chapter highlights how deregulation and global supply chains have externalized environmental and social costs, contributing to carbon-intensive production patterns. It calls for enforceable sustainability standards, responsible consumption, and structural reforms to address the global dimensions of industrial environmental degradation.

Chapter 12 presents a qualitative assessment of monitoring and governance challenges facing Sampalok Lake in the Philippines. By examining institutional gaps, community initiatives, and emerging threats such as microplastics, the chapter underscores the importance of collaborative, multidisciplinary approaches to ecosystem protection. The proposed strategies highlight the role of community engagement, technological monitoring, and policy coordination in sustaining local aquatic ecosystems.

Chapter 13 explores sustainability-driven marketing strategies through cross sectoral case studies of leading global firms. The chapter examines how transparency, green branding, consumer education, and digital engagement can align marketing practices with sustainability goals while addressing challenges such as greenwashing. By proposing a strategic model, the chapter offers practical insights for organizations seeking to integrate sustainability into their brand and market positioning.

Chapter 14 traces the evolution of Corporate Social Responsibility from a profit-centered view to a values-driven strategic imperative. The chapter discusses CSR as an integrated approach encompassing ethical labor practices, environmental stewardship, and social accountability. By framing CSR as a core organizational mission rather than peripheral philanthropy, the chapter reinforces its relevance to sustainable industrial development.

Chapter 15 provides an overview of the origins, principles, and evolution of the Sustainable Development Goals. Highlighting their inclusive and integrated design, the chapter explains how the SDGs address economic, social, and environmental dimensions simultaneously. The discussion emphasizes the principle of "leaving no one behind" and situates the SDGs as a unifying framework guiding global and local sustainability initiatives.

Chapter 16 examines waste management and resource optimization strategies within local industrial systems, with particular attention to circular economy practices. Through case studies and policy analysis, the chapter demonstrates how waste-to-resource approaches can reduce environmental impact while generating economic value. The chapter highlights the importance of technology adoption, regulatory frameworks, and stakeholder engagement in institutionalizing sustainable waste management.

Chapter 17 concludes the volume by exploring biofortification as a sustainable pathway to improved nutrition and food security. Aligning with multiple SDGs, the chapter demonstrates how nutrient-enriched crops contribute to health, poverty reduction, and climate resilience. By emphasizing enabling factors such as policy support, farmer training, and gender inclusion, the chapter positions biofortification as a critical, scalable strategy for achieving healthy and sustainable futures.

CONCLUSION

As this edited reference book draws to a close, it becomes evident that environmental sustainability in local industrial systems is not a singular goal but a multidimensional, evolving process that demands coordinated action across technology, governance, policy, organizational practice, and society. The chapters assembled in this volume collectively demonstrate that sustainable industrial transformation is both achievable and necessary, particularly in emerging economies where environmental pressures and development imperatives intersect most sharply.

Across the contributions, a consistent message emerges: meaningful progress toward sustainability requires integrated approaches. Digital technologies, circular economy principles, green innovations, responsible governance, and human-centered management practices must operate in synergy rather than in isolation. Whether through digitally enabled circular models, advanced manufacturing technologies, knowledge-driven practices, or community-based environmental stewardship, the chapters highlight practical pathways for reducing environmental impact while strengthening economic resilience and social well-being.

This volume also underscores the critical role of institutions, leadership, and collaboration. Policymakers, industry leaders, researchers, and communities share responsibility in shaping enabling ecosystems that support sustainable decision-making and long-term environmental accountability. Equally important is the recognition that local contexts matter—solutions must be adaptable, inclusive, and responsive to regional realities, cultural norms, and resource constraints.

As Editors, we hope this book serves not only as a scholarly reference but also as a catalyst for action. By bridging theory and practice, and by fostering dialogue across disciplines and sectors, this volume aims to inform policy formulation, inspire innovation, and guide industries toward more sustainable and responsible pathways. Ultimately, prioritizing environmental sustainability in local industrial systems is central to achieving global sustainability goals and ensuring a resilient, equitable future for generations to come.

Syed Abdul Rehman Khan
Business Research and Consultancy Limited, Sheffield, UK

Zhang Yu
Yunnan University of Finance and Economics, Kunming, China

Muhammad Tanveer
Mohammad Ibn Saud Islamic University, Riyadh, Saudi Arabia

Chapter 1
Digitally-Enabled Circular Economy Framework for Localized Industrial Sustainability in Emerging Economies

Shivani Patil
School of Humanities and Social Sciences, Jain University, India

MV Satya Soumya
School of Humanities and Social Sciences, Jain University, India

Patcha Bhujanga Rao
https://orcid.org/0000-0003-4736-8497
Jain University, India

P. Nandeeswara Rao
Jain University, India

ABSTRACT

This chapter investigates how digital technologies—particularly the Internet of Things (IoT) and Circular Business Model Innovation (CBMI)—drive circular-economy (CE) adoption in emerging-economy industries. It proposes a framework linking Digital Infrastructure Readiness (DIR), Digital Innovation Intensity (DII), Policy and Institutional Support (PIS), Circular Economy Culture (CEC), and Environmental Commitment and Investment (ECI) to Digitally-Enabled Circular Economy (DE-CE) outcomes. Case evidence from India, Bangladesh, and East Africa

DOI: 10.4018/979-8-3373-6435-3.ch001

shows that IoT adoption (M_1) enhances traceability and compliance, while CBMI (M_2) fosters innovation but remains limited by capital and ecosystem constraints. Socio-institutional readiness and policy coherence determine digital effectiveness. The study contributes an integrated model combining technology, governance, and culture to advance sustainable industrial transformation and provides actionable insights for policymakers and managers promoting inclusive, digitally enabled circular growth.

INTRODUCTION

The transition towards a Circular Economy (CE) is now coming to be a significant global priority, with industries being forced to react to growing pressures of resource shortage, waste, and pollution. The linear take–make–dispose economy is the opposite of the CE, which in turn is based on reducing, reusing, recycling, and regenerating to build sustainable and restorative industrial ecosystems. In spite of the fact that the principles of the CE are well explained, the issue of their effective implementation is critical, especially in the case of emerging economies experiencing rapid industrialization under the shifting regulatory system.

Digital technologies are becoming identified as one of the major enablers of the shift towards the CE. The aid in operationalizing CE principles can consist of such technologies as artificial intelligence (AI), the Internet of Things (IoT), cloud computing, and blockchain. These tools enable industries to trace resources across their life cycles, increase their efficiency in energy consumption, anticipate and prevent malfunctions, and permit recycling or remanufacturing. According to evidence-based practice, IoT-enabled systems can extend the life and usage of products, and data-driven intelligent manufacturing can cut down costs and emissions in energy-intensive sectors. Digital technologies are also related to innovation because they encourage resource optimization and the creation of new business models, which contribute to a higher level of operational efficiency.

The transformation is supported by business model innovation. The conversion of product ownership to service models like product–service systems or reverse logistics systems extends the financial lifespan of products. Digitalization contributes to these changes by facilitating real-time data flow, predictive analytics, intelligent supply-chain coordination, and technology-based business models. Localized industries in up-and-coming economies, which are usually cash-strapped and in need of innovative yet low-cost solutions, can benefit from such digitally enabled models.

However, there are still a number of major problems. The barriers to the diffusion of digitally enabled CE practices include the high cost of adoption, a lack of interoperability, poor institutional arrangements, and disconnected data systems.

Moreover, digitally enabled CE is mostly researched and evidenced through case studies in highly developed economies, constraining the insights that can be obtained on the sociocultural and institutional realities of developing nations.

This study fills that gap and suggests a digitally facilitated circular-economy model of localized industrial sustainability in emerging economies. The framework incorporates the knowledge behind digitalization, circular-economy models, and sustainable manufacturing in order to assist industries in developing settings to avoid linear models and instead enable the development of economically viable, environmentally regenerative, and socially inclusive systems.

EMERGING ECONOMIES SOCIO-INSTITUTIONAL AND CULTURAL DRIVERS

Circular-economy (CE) transitions in emerging economies cannot be considered technological phenomena that occur in perfect isolation, but they are highly predetermined by social institutions, cultural norms, and regulatory capacity. Systems with digital capabilities, including the Internet of Things (IoT), blockchain, and data-analytics tools, operate in institutional settings that are commonly starkly different from those of developed industrial countries (Parida et al., 2022). The knowledge of such contextual drivers is hence key to the design of inclusive and scalable CE pathways.

Weaknesses in Institutions and Gaps in Enforcement

The industrial ecosystems of developing countries are often typified by disjointed regulatory control, multiple jurisdictions, and inadequate resources to respond to compliance (World Bank, 2023). In India, Kenya, and Indonesia, environmental enforcement departments, for example, have a difficult time tracking waste streams or imposing extended producer responsibility (EPR) (Chaturvedi & Gupta, 2024). Reporting can be automated with digital tools, and auditable data trails can be created; however, in the absence of effective coordination between institutions, these tools are not going to be transformative but symbolic (Singh et al., 2023). Therefore, effective CE digitalization requires institutional capacity building, i.e., training of inspectors, aligning digital standards, and integrating data systems.

Role of the Informal Sector

In South Asia and Sub-Saharan Africa, 50–80% of the informal economy deals with recyclable material flows (Wilson et al., 2021). Waste pickers, refurbishers, and micro-dismantlers are the invisible workforce of complete circular activity and are not associated with formal compliance and taxation systems (Kumar & Bhatnagar, 2024). These actors can be linked to formal value chains with the help of mobile-based material tracking, micro-identity tokens, and incentive wallets (Gupta & Sarkar, 2023). However, digital inclusion requires trust and mutual advantage: the structures have to protect privacy and provide equitable payment and market access (Kassim & Alemu, 2022).

Trust, Traceability, and Cultural Norms

Trust in most areas is based on interpersonal networks and not on institutions. Formal certification can be replaced by long-established supplier relationships (Dasgupta & Mukherjee, 2021). Traceability projects that are implemented based on blockchains and are not sensitive to these relational norms may meet disapproval. On the other hand, cooperatives and producer associations may be used as trust anchors and prove data and resolve conflicts. Traceability as a cultural aspect is therefore reliant on participatory governance, informality, and overt illustration of local value creation (Kumar & Bhatnagar, 2024).

Linguistic, Digital-Literacy, and Connectivity Constraints

The structural barriers to CE digitalization are language diversity and uneven digital literacy. Dashboards written in English are usually incomprehensible to factory operators as well as small recyclers (Munyua & Wanjohi, 2022). In low-bandwidth areas, delays in data transmission disrupt real-time tracking (UNCTAD, 2024). Voice-based reporting applications, multilingual interfaces, and data caching available offline have proven to fill these gaps (Patel & Raj, 2023). Contextual training and maintenance are offered through partnerships with local technical institutes.

Affordability and Economic Incentives

IoT sensors and blockchain nodes require high initial costs that are not always justified for small and medium enterprises (SMEs) unless they can pay off in the short term (Mordor Intelligence, 2025). Entry barriers are reduced through shared service platforms and subscription models, and demand-side pull can be created through public procurement and green labeling (OECD, 2023). Therefore, CE digitalization will have to proceed through economic and socio-institutional preparedness to ensure the transition beyond the pilot phase.

Consequences of the Digital-Circular Nexus

These socio-institutional forces alter the channels by which digital technologies can affect CE performance. Lax enforcement undermines the regulatory worth of data; engagement of informal sectors distorts network topology; cultures of trust establish new rules of governance; and cost dictates the magnitude of technology. The digitally enabled CE in developing economies can thus be explained as a co-evolutionary technology-institution-culture system as opposed to a progressive diffusion process (Parida et al., 2022). These realities can be incorporated at the beginning of framework design to make digital circular strategies inclusive, context-specific, and stable.

CONCEPTUAL LITERATURE

Its current direction in the literature of the circular economy (CE) and digitalization shows that digital tools are catalysts of CE policies, although fewer empirical and contextualized frameworks are available that could be applied to emerging economies, especially. In that regard, the literature reviews contribute to mapping the minimum possible spectrum of digital technologies (such as IoT, AI, and blockchain) that have so far been utilized in the practices of CE. Although in this aspect, most focus in terms of studies has been on the setting of developed contexts, thus leaving an empty void around the less analytic and socially inclusive or institutionally viable localized frameworks the emerging economies need (Parida, Sjödin, & Reim, 2022).

At the firm level, digitalization exploits CE opportunities primarily through the augmentation of understanding of resource flows, and later by apprehending and redefining the value creation and capture action. Both business-model innovation evidence have indicated that circularity strategies emerge in small steps (e.g., resource efficiency) when data capture is primitive (e.g., just accounting). Nevertheless, the strategies of circularity that encompass the goals of radicalism (like the loop closing of resources) demand more sophisticated processes of data integration and analytics. According to recent studies, the frameworks for the involvement of emerging economies should be lined up on a spectrum of potential impact, and sensory and monitoring routes are engaged with integrated analytics towards the formalization of more aggressive ambitions of new types of business models (Ranta, Aarikka-Stenroos, & Väisänen, 2021; Gupta & Sarkar, 2023).

The Internet of Things (IoT) is the basis of CE strategies, which involve product use, maintenance, reuse, and recycling. The capabilities of the IoT are separated into tracking, monitoring, control, optimization, and design-evolution functions by the capability of dematerialized space studies. Tracking enhances the life of product usage, monitoring aids predictive maintenance, control enhances energy and material consumption, and design evolution ensures feedback of real changes in the design into real time. Practically, IoT simplifies efficiency and durability of products used. IoT applications in reuse, remanufacturing, and recycling are less developed. This implies that the emerging economies could first use IoT to achieve better efficiency and maintenance and then proceed to full-loop systems of reuse, remanufacturing, and recycling (Ingemarsdotter, Jamsin, Kortuem, & Balkenende, 2019; Patel & Raj, 2023).

According to case studies, tracking and monitoring through IoT enhance efficiency in the management of resources by being able to locate the products over a broader area as well as communicate their overall status. Control functionality enables remote repair and updating of products, optimization tools improve preventative maintenance schedules and energy usage, and design evolution enables feedback loops to allow product adaptation and redesign. IoT has numerous digital capabilities that are tailor-made and can be easily applied in settings that do not require any serious infrastructural capacity, thereby producing essential building blocks of localized sustainability (Parida et al., 2022).

One of the additional pieces of evidence of how digital CE can be practiced refers to energy-intensive sectors. IoT-based data-driven manufacturing systems and those powered by cloud computing, cyber-physical systems, and digital twins have demonstrated a range of energy cost reductions, ranging in real-life cases up to 20% (Ma, Zhang, Ding, & Li, 2020). These savings reveal how digital technologies can be used to follow the principles of CE by minimizing the intensity of materials and energy and creating a foundation for the increased life cycle of products and the reuse of resources. In the case of emerging economies, where resource efficiency and energy costs are paramount, the applications can be of immediate benefit with tangible results. Digital CE demands more than technology to exist at the business-model level; it relies on the way companies create and distribute value. Service-oriented models, performance-based contracts, and data sharing at the ecosystem level are often accompanied by advanced circularity. This applies particularly to the case of emerging economies, where small firms and informal actors are significant. A localized framework should thus comprise information-sharing norms, governing institutions, and tools of equitable value allocation so that each industrial participant can be involved (Chaturvedi & Gupta, 2024; Wilson, Rodic, & Velis, 2021).

On the system level, CE and Industry 4.0 will have enablers and barriers. The enablers consist of AI and other digital tools, supportive policy, and CE strategies themselves. These obstacles comprise poor interface and interoperability design, among others. It implies that in the case of emerging economies, a new digital CE framework should not only concern technology but also policy, standards, and human-oriented design. Interfaces should be simple, multilingual, and low-bandwidth to make sure a wide audience adopts them, whereas policies, standards, and incentives should provide conducive environments for the adoption of digital CE (OECD, 2023; UNCTAD, 2024; Chaturvedi & Gupta, 2024).

Combined, the literature gives a four-layer conceptual framework. The first one is the technology layer, and it is used to denote technologies of IoT, edge computing, analytics, and the capability to scale an IoT or edge-computing solution to the degree and functionality of local capabilities. The second level is the circular-strategy level, which offers digital solutions to organizations to narrow (improve efficiency), slow (engage product lifespan), and loop-close (recycle and reuse). The third level is the business-model level, which assists in making the shift between the model of product sales to a service-based model that is facilitated by data sharing among the supply chains. The fourth and last layer is the layer of governance and inclusion, which introduces governmental support of usability, policy support, and fair participation of local industries (Bag, Pretorius, Gupta, & Dwivedi, 2021; Parida et al., 2022; Ranta et al., 2021).

Overall, the potential of digital technologies to facilitate the implementation of the circular economy is enormous, yet their effectiveness will be determined by the extent to which they are introduced in strategies, business models, and systems of governance. Recent research is united in the view that emerging economies must be approached in stages and locally, in ways that combine technical feasibility, policy backing, and social inclusion. By doing so, these industries in such settings will be able to start to transition to sustainable, circular, and digitally enabled futures (Gupta & Sarkar, 2023; Chaturvedi & Gupta, 2024).

THEORETICAL FRAMEWORK IN LITERATURE

The growing body of scholarship on circular economy (CE) and digitization has produced many theoretical and analytical structures that make it clear how these domains intersect into practice. These framework provides ideological equipment to understand how digital technologies enable circular strategies that affect the adoption of ambitions and obstacles, and how the business models change in this process. Three of the most impressive framework in this domain are IoT -Circular Strategies Matrix, CE -Industry 4.0 Enabler -Barrier model and circular business model innovation (CBMI) framework in digital technologies.

IoT–Circular Strategies Matrix

Ingemarsdotter, Jamsin, Kortuem, and Balkenende, 2019 developed a two-dimensional matrix that maps IoT capabilities - including tracking, monitoring, control, optimization, and design evolution - against circular strategies, namely efficiency-in-use, increased utilization, lifetime extension, reuse, remanufacturing, and recycling. This matrix establishes a structured way of identifying how specific IoT functionalities can enable particular CE practices.

Through the analysis of 40 industry cases, Ingemarsdotter et al found that most current IoT applications cluster around efficiency-in-use (e.g., energy monitoring systems) and lifetime extension (e.g., predictive maintenance). By contrast, strategies involving reuse, remanufacturing, and recycling were found to be underdeveloped, largely because these require more complex cross-stakeholder coordination and robust reverse logistics systems.

This framework highlights two important insights: (A) IoT is most effective when applied to "slow" strategies, such as expansion of product usage, and (B) strategies exist a difference between potential and real implementation that exists "closure the loop", such as recycling or remanufacturing. For emerging economies, where infrastructure for reverse logistics and formal recycling is often weak, matrix provides a roadmap to prioritize digital investments that can first win in efficiency and product life expansion, slowly before moving towards more complex circular ends

CE–Industry 4.0 Enabler–Barrier Model

Rajput and Singh (2019) used the principal component analysis (PCA) and decision -making trial and evaluation laboratory (DEMATEL) technology to conduct 161 Indian manufacturing enterprises by the intersection of CE and Industry 4.0, which then mapped 26 enablers and 15 obstacles that he found to detect the cause interrelationships.

The CE–Industry 4.0 integration is the most important factor that facilitates integration, according to their model, AI applications, which encourage supportive policy frameworks, and is a strong organizational culture focused on CE. By offering the institutional validity and technical capacity required for systemic changes, these components promote adoption. On the other hand, poor interface design and system compatibility became the biggest obstacle, which describes how significant integration and purpose are in digitally enabled CE systems.

This model suggests that the existence of technology alone may not be responsible for successful digital-circular integration. Rather, it asks for systemic preparations, including socio-technical integration, organizational commitment and accessories. The approach emphasizes the need to address organizational and institutional obstacles, besides adopting technical to emerging economies, which is characterized by fragmented ecosystem and gaps in the infrastructure.

Digital Technologies in Circular Business Model Innovation

1. Ranta, Aarikka-Stenroos, Ritala, and Mäkinen, 202,) focused on the role of digital technologies in catalyzing circular business model innovation (CBMI). The study identified four separate CBMI itineraries based on the depth of CE reform and the range of business model change:
2. Incremental CE improvements with incremental business model change, such as using IoT monitoring to optimize energy efficiency.
3. Incremental CE improvements with radical business model change, such as transitioning from product sales to service-based models using digital platforms.
4. Radical CE improvements with incremental business model change, such as integrating remanufacturing without fundamentally altering revenue models.
5. Radical CE improvements with radical business model change, such as platform-based ecosystems that facilitate large-scale resource sharing, reuse, and closed-loop systems.

This framework emphasizes the strategic spectrum of digital-enabled innovation, where digital technologies act not only as operational enablers but also as drivers of systemic reconfiguration in value creation, delivery, and capture. Importantly, it reveals that firms can engage with digital–circular integration at different levels of ambition-from small-scale optimizations to radical rethinking of business models.

This framework is particularly valuable in the emerging economies, because it illustrates multiple entry points into CE. Firms constrained by infrastructure or capital may embark on with incremental improvements, while more digitally mature firms or sectors may pursue radical CBMIs to capture competitive advantage.

Data-Driven Intelligent Manufacturing for Sustainability

The convergence of a circular economy (CE) principles with data-operated intelligent manufacturing represents one of the most promising paths to achieve the tangible, measurable sustainability outcomes. As industries become digitalized, the ability to capture, process and analyze large amounts of operating data has transformed how firms manage resources, energy and production cycles. Intelligent manufacturing systems- machine learning, real-time optimization, and cyber-physical integrations such as lapping tools- CE strategies not only ideologically, but capable of advancing in tangible, quantitative ways.

Ma, Zhang, Ding, and Li (2020) provide one of the most compelling empirical illustrations of this ability through its development of a multi-level demand response (DR) structure. An energy-intensive ball-mill system is applied to the three hierarchical levels: site-level coordination, workshop-level scheduling and equipment-level adaptation. By integrating real -time data flow with advanced analytics, the system gained 19.33% in energy consumption. This case is important because it indicates how digital analytics directly translates to industrial sustainable outcomes, validating the practical effects of data-operated approaches.

Ma et al. (2020) The widespread implication of work is that intelligent manufacturing can conduct CE principles by narrowing and slowing the resource flow. Narrow flow is obtained through increased efficiency in energy and use of material, while slow flow is supported by future predictive maintenance and equipment adaptation, which expands the asset lifespan. In this sense, data-driven approaches serve as a technical originator of CE strategies and a measurement mechanism, providing industries with evidence base to justify further circular investments.

For emerging economies, where industries such as cement, steel, textiles and chemicals account for an important part of GDP and environmental impact, Data-driven intelligent manufacturing is particularly salient. These sectors are often energy-intensive and resource-arms, making them prominent candidates for interventions that provide immediate efficiency benefits. Importantly, the return (ROI) on investment in such contexts is not only environment, but also economic: energy cost reduction and improvement in efficiency directly increase competition in global markets.

In addition, adopting intelligent manufacturing is a step towards wider CE adoption. By demonstrating the average performance results - such as Ma et al.'s (2020) finding of 19.33% energy savings - show that firms in emerging economies can instill confidence in digital -circular strategies. This incremental passage is important in contexts where capital deficiency, infrastructural weaknesses, and skill deficiency often prevents large -scale changes. Starting with data-powered efficiency benefits allows industries to cultivate capabilities, governance structures, and the belief necessary to move towards more complex circular strategies

Thus, data-driven intelligent manufacturing illustrates how digitalization serves as a catalyst for localized industrial sustainability. It offers industries in emerging economies a pragmatic entry point: immediate efficiency benefits that reduce costs and emissions, while simultaneously laying the groundwork for systemic circularity. In this sense, it bridges the gap between short-term industrial competitiveness and long-term circular transformation.

Thus, data-powered intelligent manufacturing demonstrates how digitization serves as a catalyst for localized industrial stability. It provides a practical entry point to industries in emerging economies: immediate efficiency benefits that reduce costs and emissions, as well as laying groundwork for systemic circularity. In this sense, it bridges the gap between short -term industrial competitiveness and long -term circular transformation.

Emerging Economies: Contextual Challenges and Opportunities

Integration of digital technologies with circular economy (CE) principles presents a clear -different landscape in emerging economies compared to developed contexts. While advanced economies often work within the environment of strong infrastructure, mature digital ecosystems, and rigorous regulatory structure, emerging economies are characterized by institutional weaknesses, infrastructural boundaries and socio-economic duality. Nevertheless, these provide unique opportunities for similar reference innovation, leapfrogging and localized stability solutions.

Barriers in Emerging Contexts

Many hindrances spread over technical, institutional and socio -economic sectors are disrupting the change of emerging economies for digitally capable circular economy.

1. **Limited infrastructure for digitization**. Stable internet access, sensors with appropriate value and reliable data systems are required to adopt IoT devices, advanced analytics and blockchain. Scalability of digital-circular solutions is forced by issues with broadband penetration with cloud infrastructure, cyber security preparations and many emerging countries (Chauhan, Parida, and Dhir, 2022).
2. **Fragmented industrial ecosystems** Industrial ecosystems in developing economies are often fragmented and controlled by spacious informal regions and small & medium -sized firms (SMEs), unlike integrated value chains in industrial contexts. Reverse logistics, traceability and cooperative CE practices are made more difficult by this fragmentation (Rajput & Singh, 2019). For example, informal recycles, already required for material recovery, may be rarely digital or combined with functionary supply chains.
3. **Weak policy enforcement.** There are still a lot of implementation gaps, even though many 3. emerging economies have implemented policies that encourage digital use, environmental control and CE. Policies that can otherwise encourage digital-circular activities are less effective due to institutional capacity, corruption and deficiency of conflicting development agenda (Chauhan et al., 2022).
4. **Skill shortages.** Adopting digital technologies in CE depends a lot on technical expertise in areas such as Analytics, AI, IoT system integration and advanced manufacturing. Emerging economies often face lack of efficient labor and lack making technology availability and its practical deployment difficult.(Rajput & Singh, 2019).

Collectively, these obstacles create a challenging environment for digital -rich transitions, suggesting that the global answer model is unlikely to be successful. Instead, analogous structures accommodating local limitations.

OPPORTUNITIES FOR LEAPFROGGING

Despite these obstacles, emerging economies are specifically deployed to leapfrog and pursue digital-enabled circularity to traditional linear industrial models. This is due to the absence of entangled heritage systems and urgency of addressing resource and sustainability challenges.

1. **Linear model leapfrogging.** Emerging economies can bypass aged infections experienced in developed contexts by directly adopting advanced digital -rich circular practices. For example, embedded decentralized renewable energy system with IoT monitoring can provide both energy access and stability, without the requirement of a comprehensive centralized infrastructure.
2. **Localized circular solution.** Industrial sectors in emerging economies often offer field-specific opportunities for CE.. For example, in agriculture and agro-processing, digital equipment can support waste valorization, convert crop residues into bioenergy or bioproducts. Similarly, digital platforms can enable localized recycling markets, which can link waste generators with recycles via mobile-based solutions (Chauhan et al., 2022).
3. **Employment and social stability**. CE strategies are labor-intensive in areas such as reverse logistics, repair, refurbishing, and remanufacturing, which create adequate opportunities for employment. In emerging economies, where unemployment and informalization are consistent issues, digitization can formalize these activities, increase productivity, and align industrial strategies with inclusive social stability goals (Rajput & Singh, 2019).

These opportunities underline that emerging economies, face significant structural challenges, they also have specific competitive advantages in furthering digitally capable sphere of influence. By focusing on relevant adaptation and inclusive innovation, such economies can craft the routes that not only meet stability goals, but also contribute to industrial modernization and social development.

Case Vignette 1: Informal E-Waste Refurbishers and IoT Traceability (India)

The discarded computers and smartphones are processed by clusters of small refurbishers in the Bengaluru industrial belt of Peenya. One of the Indian start-ups installed low-cost IoT tags and QR-coded receipts into the workflow of refurbishers. Every device that had been fixed was registered in a cloud registry, which was available to recyclers and bulk purchasers. Initially, the survey was met with suspicion, but by working with a local cooperative and learning Kannada, the cooperation improved, and within six months, the rate of recovery of reusable parts increased by 22 percent, and the cooperative obtained a municipal recycling contract. The case demonstrates that mediation of trust and linguistic localization is just as vital as technology design for traceable circular practices.

Case Vignette 2: Textile and Leather Clusters in Bangladesh and East Africa

The small leather-goods workshops of Dhaka and Nairobi have artisans who recycle off-cuts into belts and wallets. These products do not reach export markets because they are not certified. A collaborative provenance platform based on blockchain was applied through a partnership between local chambers of commerce and an EU development agency and executed by cooperative "data stewards." The system captured waste minimization and fair-labor observation, showing simplified measures on a mobile dashboard. Buyer confidence and unit prices grew by 18 percent, which proves that models of collective governance and affordability - not technological sophistication itself - lead to CE digitalization in culturally embedded craft industries (UNIDO, 2024).

CONCEPTUAL GAPS AND RESEARCH DIRECTIONS

There are various competing definitions and arguments of the Circular Economy.

However, although the Circular Economy (CE) is a popular term, the scope and operationalization of the concept as a sustainable alternative to the notorious linear model, i.e., take–make–dispose, does not have a consensus. Narrowly, some authors perceive CE as material recovery and recycling (Kirchherr et al., 2017), but the broader and more systemic concept refers to resource efficiency, social equity, and innovation (Geissdoerfer et al., 2017).

As per Bocken et al. (2016), the strategy of CE can be narrow, slow, and close resource loops: narrowing reduces resource usage by efficiency, slowing extends the life of products by reuse and repair, and closing promotes recycling. Nevertheless, critics report that these strategies can come back to bite them - when efficiency upsurges lead to new consumption, undermining overall sustainability results (Zink & Geyer, 2017).

Another controversy is related to downcycling and upcycling. Recycled products are usually of lower quality, which can lead to additional degradation of materials rather than regeneration (Korhonen et al., 2018). Downcycling is likely to prevail in CE practices in emerging economies that have poor material purity and collection networks.

Arguments and Critiques of Digitalization

Digital technologies (IoT, blockchain, and AI) are likely to be promoted as the cause of CE transitions, but through their application, new risks, dependencies, and inequalities are presented.

Firstly, cloud computing and blockchain are highly energy-consuming and data-intensive technologies that partially negate the environmental advantages they are expected to deliver (Pihlajamaa et al., 2023). Second, decentralized supply chains have a higher number of players that transmit sensitive information, exacerbating the problem of cybersecurity and data privacy (Radanliev et al., 2022). Third, small businesses frequently cannot innovate locally or make their own decisions as vendor lock-in is established by proprietary platforms operated by multinational technology companies (Rosenlund et al., 2021).

Moreover, there exists a promise of digital traceability, which is linked to occasional circularity washing when companies state that they are sustainable because of digital dashboards but achieve very little material recovery (Hobson & Lynch, 2016). In such situations of resource scarcity, technological zeal may cloud labor, management, and culture that are crucial in the long-term application.

Prospects and Lenses and Contextual Gaps in Emerging Economies

The current models of CE are mostly concentrated on developed economies in which institutions, logistics, and markets are formalized. On the other hand, emerging markets have institutional gaps, dominance of the informal sector, and fragmented supply chains (Khanna et al., 2021; Pansera & Owen, 2022). The local segments of waste segregation, safeguarding, and product-traceability criteria are of low consistency, and imitation of Western CE models can easily be troublesome (Agyemang et al., 2019).

Besides, much of the literature fails to sufficiently portray socio-cultural and governance heterogeneity - such as community-based repair networks or cooperative recycling models that are common in South Asia and Africa. These systems support livelihoods and material recovery but have no access to capital and digital infrastructure.

According to new evidence, this divide can be bridged by digital inclusion, namely, by low-cost IoT and mobile data sharing, but only under the condition of being adjusted to local realities (Parida et al., 2022). Therefore, the discourse in CE needs to change focus from imitating global templates to collaborative, bottom-up digitalization that aligns with institutional and cultural logics.

Despite the increased scholarship on the topic of integrating the concept of the Circular Economy (CE) with digital technologies in recent years, there are still gaps in ideological and empirical knowledge. These shortcomings restrict the ability to adapt existing structures to the realities of emerging economies, and this is the reason why new theoretical and practical approaches are needed.

To begin with, there is a technocentric preconception still found in the literature. The available literature is mostly dedicated to the capability of devices in the Digital–Circular Nexus, predominantly through the technological adoption lens, emphasizing IoT, AI, blockchain, and big data. Although these technologies are certainly significant, these strategies tend to minimize the significance of socio-institutional enablers such as governance frameworks, cultural orientation, labor-market forces, and the contribution of the informal sector. For example, Rajput and Singh (2019) have focused on the element of organizational culture as an enabling environment, but most frameworks continue to focus on the "what" of technology and not on the "how" of socio-technical integration. This narrow focus risks missing the conditions needed to ensure successful digital–circular implementation in situations where institutional weaknesses and informal economies prevail.

Second, there is a geographical imbalance in the literature. Most empirical studies, as Chauhan, Parida, and Dhir (2022) and Ranta, Aarikka-Stenroos, Ritala, and Mäkinen (2021) noted, are based in Europe and North America, where industrial ecosystems enjoy good infrastructure, high digital maturity, and regulatory conditions. Conversely, research in Asia, Africa, and Latin America is scarce, even though these regions are experiencing severe threats to industrial sustainability and resource depletion. This asymmetry not only limits the generalizability of theory but also creates a practical gap, as strategies that have proven successful in the Global North cannot be directly applied to emerging markets with disaggregated supply chains, poor enforcement, and unique socio-economic contexts.

RESEARCH GAP

Although earlier studies on the digital–circular economy (DE-CE) spotlight the transformative potential of Industry 4.0 technology in permitting useful resource performance and closed-loop practices, the literature remains largely technocentric, with limited attention to the socio-institutionalal and ecosystem-stage enablers required for achievement in rising economies (Moktadir et al., 2020; Rajput & Singh, 2019). Much of the empirical evidence stems from advanced-economic system contexts, where regulatory clarity, infrastructure, and corporation-degree capabilities are greater mature, thereby leaving localized and inclusive fashions-especially the ones integrating informal zone actors-underexplored (Nobre & Tavares, 2021). Moreover, even as conceptual frameworks regularly posit that digital capabilities translate into circular results, the mechanisms thru which such capabilities are operationalized-especially via IoT-enabled visibility and traceability or circular business model innovation - are seldom empirically unpacked (Centobelli et al., 2022; Garcia-Muiña et al., 2019). Finally, despite growing calls requires quantitative checking out, few tested, multi-construct survey instruments exist which might be tailored to the limitations of rising economies, consisting of low bandwidth, variable enforcement of coverage, and uneven virtual skills (Bag et al., 2021). These gaps collectively motivate the present study, which develops and validates a measurement version and tests the multi-path mechanisms" through which virtual and institutional competencies power DE-CE implementation in emerging-economy.

CONCEPTUAL FRAMEWORK DIAGRAM

Building on the identified research gaps, this study proposes a conceptual framework to examine how digital and institutional capabilities influence the implementation of a digitally-enabled circular economy, mediated by IoT-enabled visibility and traceability as well as circular business model innovation (Figure 1).

Figure 1. Conceptual framework of the digitally-enabled circular economy model

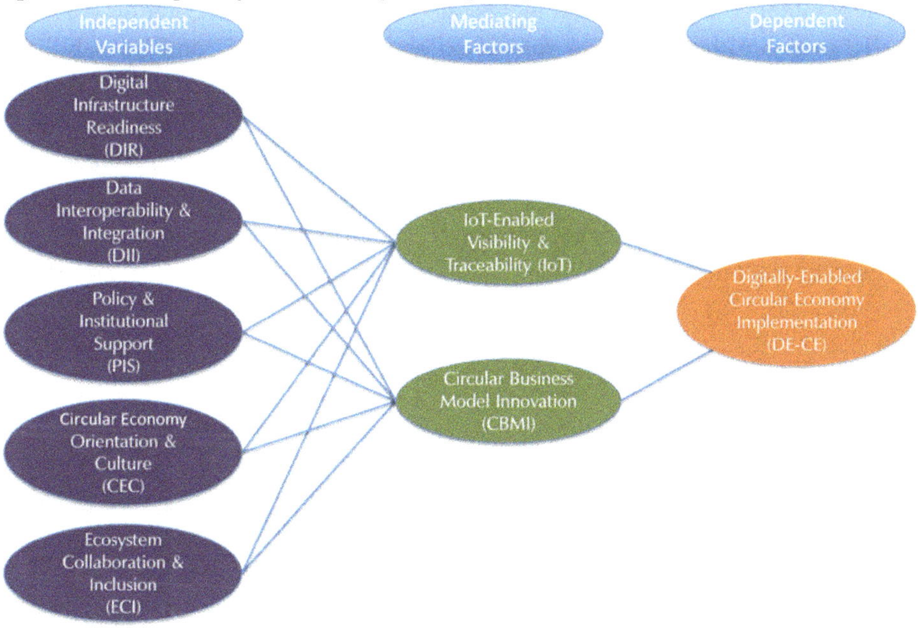

As shown in Figure 1, the model positions the five capability and contextual factors as independent variables influencing Digitally-Enabled Circular Economy DE-CE implementation. Their effects are transmitted through two mediators: (i) IoT-enabled visibility and traceability, which enhances operational transparency and resource monitoring; and (ii) circular business model innovation, which captures organizational transformation towards circular practices. This framework thus integrates both technological and institutional pathways, providing the basis for developing the study's hypotheses.

RESEARCH OBJECTIVES

Building on the identified research gaps, this study sets out the following objectives:

1. To develop and validate a multi-construct measurement model that captures the dimensions of digitally-enabled circular economy (DE-CE) implementation in emerging-economy manufacturing contexts.
2. To analyze the direct effects of five critical capability and contextual factors - digital infrastructure readiness, data interoperability and integration quality,

policy and institutional support, circular economy orientation and culture, and ecosystem collaboration and inclusion - on DE-CE implementation.

3. To examine the mediating role of IoT-enabled visibility and traceability in linking capability/contextual factors with DE-CE outcomes.
4. To investigate whether circular business model innovation (CBMI) mediates the relationship between capability/contextual enablers and DE-CE implementation.
5. To assess the moderating influence of firm-level characteristics (such as size, age, sector, export intensity, and digital maturity) and, on this basis, provide managerial and policy insights for scaling DE-CE in emerging economies.

HYPOTHESES

Based on the research framework and extant literature, the following hypotheses are formulated:

Direct Effects

H1: Digital infrastructure readiness (DIR) positively influences DECE implementation.
H2: Data interoperability and integration quality (DII) positively influences DECE implementation.
H3: Policy and institutional support (PIS) positively influences DECE implementation.
H4: Circular economy orientation and culture (CEC) positively influences DECE implementation.
H5: Ecosystem collaboration and inclusion (ECI) positively influences DECE implementation.

Mediation via IoT-Enabled Visibility & Traceability (M1)

H6a: IoT-enabled visibility and traceability mediates the relationship between DIR and DECE implementation.
H6b: IoT-enabled visibility and traceability mediates the relationship between DII and DECE implementation.
H6c: IoT-enabled visibility and traceability mediates the relationship between PIS and DECE implementation.
H6d: IoT-enabled visibility and traceability mediates the relationship between CEC and DECE implementation.

H6e: IoT-enabled visibility and traceability mediates the relationship between ECI and DECE implementation.

Mediation via Circular Business Model Innovation (M2)

H7a: Circular business model innovation mediates the relationship between DIR and DECE implementation.
H7b: Circular business model innovation mediates the relationship between DII and DECE implementation.
H7c: Circular business model innovation mediates the relationship between PIS and DECE implementation.
H7d: Circular business model innovation mediates the relationship between CEC and DECE implementation.
H7e: Circular business model innovation mediates the relationship between ECI and DECE implementation.

EMPIRICAL ANALYSIS

Table 1. Demographic profile of the respondents

Sl. No.	Particulars	Classification	No. of Respondents	%
1	Gender	Male	111	62.7%
		Female	66	37.3%
2	Age Group	<25	21	11.9%
		25–34	48	27.1%
		35–44	45	25.4%
		45–54	34	19.2%
		55+	29	16.4%
3	Highest Education	Diploma	26	14.7%
		Bachelor	70	39.5%
		Master	53	29.9%
		PhD/Doctorate	28	15.8%

continued on following page

Table 1. Continued

Sl. No.	Particulars	Classification	No. of Respondents	%
4	**Job Role**	CEO/Owner/MD	6	3.4%
		Sr. Management	38	21.5%
		Middle Management	56	31.6%
		Engineer/Technical	54	30.5%
		Supervisor	23	13.0%
5	**Industry Sector**	Manufacturing	106	59.9%
		Production	33	18.6%
		Services	18	10.2%
		Other / Mixed	20	11.3%
6	**Product Type**	Discrete	37	20.9%
		Continuous	40	22.6%
		Process/Batch	30	16.9%
		Assembly	38	21.5%
		Services	32	18.1%
7	**Ownership**	Private/Family	90	50.8%
		Public/Listed	24	13.6%
		State-owned	18	10.2%
		JV/Foreign	27	15.3%
		Cooperative/Other	18	10.2%
8	**Region**	North	24	13.6%
		South	26	14.7%
		East	30	16.9%
		West	32	18.1%
		Central	32	18.1%
		International	33	18.6%
9	**Sector**	Urban	110	62.1%
		Peri-urban	39	22.0%
		Other	28	15.8%

The 177 respondents represented in the survey with a wide range of organizational and individual characteristics, offering a foundation for researching and the adoption of the digitally enabled circular economy in emerging economies. Although the gender distribution was inclusive (37.3% female) and male (62.7%), reflects the industrial workforce and showed encouraging female representation. The majority of respondents were mid-career professionals, well-positioned for the shifts to digital innovation and sustainability, with ages ranging from 25 to 44. The credibility of the responses was reinforced by the high level of educational attainment, with 15.8% having doctorates and nearly 70% having bachelor's or master's degrees.

The distribution of job roles included supervisors (14.0%), top leaders (3.4%), engineers/technical staff (30.5%), middle managers (31.6%), and senior managers (21.5%). There was a good balance between long-tenured employees and new hires, combining institutional knowledge with flexibility. The different firm sizes being micro (11.3%), small (22.0%), medium (36.2%), and large (30.5%). The age of the firm was ranging from young businesses (less than five years, 18.1%) to established businesses (more than twenty years, 23.2%).

In addition to a wide range of product types (discrete, continuous, batch, assembly, and services), the sectors represented were electronics, automotive, chemicals, textiles, machinery, plastics, food, construction, and services. The ownership structures included state-owned (10.2%), private/family businesses (50.8%), cooperatives/others (10.2%), public-listed (13.6%), and joint ventures/foreign-owned (15.3%). Geographically, respondents were 18.6% residing abroad, 62.1% based in cities, and the remaining respondents in rural or peri-urban settings.

Table 2. Reliability summary (Cronbach's Alpha)

No. of Items	Factor (Construct)	No. of Items	Cronbach's Alpha
1.	Digital Infrastructure Readiness (DIR)	6	**0.748**
2.	Data Interoperability & Integration (DII)	6	**0.725**
3.	Policy & Institutional Support (PIS)	6	**0.788**
4.	CE Orientation & Culture (CEC)	6	**0.804**
5.	Ecosystem Collaboration & Inclusion (ECI)	6	**0.790**
6.	IoT-Enabled Visibility & Traceability (IoT)	6	**0.800**
7.	Circular Business Model Innovation (CBMI)	6	**0.732**
8.	Digitally-Enabled CE Implementation (DECE)	6	**0.750**

The reliability of measurement constructions was evaluated using Cronbach's alpha. The results suggest that all eight constructions achieved the acceptable credibility level, with the values from 0.725 to 0.804. In particular, digital infrastructure readiness ($\alpha = 0.748$), data interoperability and integration ($\alpha = 0.725$), policy and institutional support ($\alpha = 0.788$), CE orientation and culture ($\alpha = 0.804$), ecosystem cooperation and synergy ($\alpha = 0.790$), IOT-CE ($\alpha = 0.790$), visibility and trading ($\alpha = 0.790$), ($\alpha = 0.732$), and digitally enabled CE implementation ($\alpha = 0.750$) all were above the recommended limit of 0.70, as suggested by Nunnally (1978). Of these, CE Orientation and Culture and IoT-Enabled Visibility and Traceability recorded the highest reliability, which demonstrates strong internal consistency within their measurement items. Although data interoperability and integration and circular business model innovation reported slightly lower alpha values, they remain within the acceptable limits.

DESCRIPTIVE STATISTICS OF THE CONSTRUCTS

Table 3. Descriptive statistics of the constructs

Factors	N	Minimum	Maximum	Mean	Std. Deviation
Digital Infrastructure Readiness (DIR)	177	12.00	30.00	22.10	4.41
Data Interoperability & Integration (DII)	177	8.00	30.00	19.73	4.72
Policy & Institutional Support (PIS)	177	7.00	30.00	22.62	4.55
Circular Economy Orientation & Culture (CEC)	177	8.00	30.00	22.98	4.49
Ecosystem Collaboration & Inclusion (ECI)	177	11.00	30.00	22.76	4.40
IoT-Enabled Visibility & Traceability (IoT)	177	7.00	30.00	22.79	4.56
Circular Business Model Innovation (CBMI)	177	9.00	30.00	21.68	4.68
Digitally-Enabled Circular Economy Implementation (DE-CE)	177	11.00	30.00	21.62	4.67

Descriptive figures of constructions offer substantial information about the overall position of the digital and circular economy between sample firms. Digital Infrastructure Readiness (DIR) recorded an average score of 22.10 (SD = 4.41) within the range of 12 to 30, indicating that the firms are well prepared with digital technologies, although some variability is present in cases. Data Interoperability and Integration (DII) reported a relatively low meaning of 19.73 (SD = 4.72), with a wide range (8 to 30), showing that many organizations are still in early to medium stages of achieving uninterrupted data flow and system integration. Policy and Institutional Support (PIS) obtained a meaning of 22.62 (SD = 4.55), suggesting that the respondents experience a fairly supportive regulator and institutional environment, although stability varies. Circular economy orientation and culture (CEC) showed the highest medium of 22.98 (SD = 4.49), followed by ecosystem at 22.76 (SD = 4.40), closely followed by cooperation and inclusion (ECI). These results suggest that most firms are firmly aligned with stability-oriented values, cultural orientation, and stakeholder cooperation. The IoT-Competent Visibility and Traceability score was 22.79 (SD = 4.56), indicating that firms are quickly adopting IoT-Competent Tracking and Monitoring Systems to improve permanent practices. In contrast, Circular Business Model Innovation (CBMI) was introduced; the level of maturity is modest compared to the product-service system, take-back-back model, culture, and weaving. Finally, the dependent construction-digital-enabled circular economy implementation (DE-CE) scored 21.62 (SD = 4.67), indicating that although firms show readiness and competence, the implementation of digital ownership practices is still in development and not fully realized. Overall, descriptive figures suggest that the firms in the sample are equipped with digital and circular economy capabilities, showing strong growth with cultural orientation and IOT adoption, while business model innovation and systemic implementation remain in their early stages.

Exploratory Factor Analysis Results (Principal Component Analysis)

To examine the underlying factor structure of the proposed constructs, an Exploratory Factor Analysis (EFA) was conducted using Principal Component Analysis (PCA). The analysis employed the eigenvalue-greater-than-one rule as the extraction criterion.

Table 4. Exploratory factor analysis results (principal component analysis)

Construct / Variable	Communality	Factor 1	Factor 2	Eigenvalue	% Variance Explained
Digital Infrastructure Readiness (DIR)	.750	**.854**	.148	4.124	51.545
Data Interoperability & Integration (DII)	.842	.259	**.880**	1.039	12.983
Policy & Institutional Support (PIS)	.680	**.808**	.168	.902	-
Circular Economy Orientation & Culture (CEC)	.672	**.812**	–.111	.649	-
Ecosystem Collaboration & Inclusion (ECI)	.775	**.854**	–.214	.529	-
IoT-Enabled Visibility & Traceability (IoT)	.755	**.846**	–.198	.373	-
Circular Business Model Innovation (CBMI)	.221	.397	.251	.213	-
Digitally-Enabled CE Implementation (DECE)	.466	**.642**	–.232	.171	-

Note: Variance Explained: 64.53% (Factor 1 = 51.55%, Factor 2 = 12.98%).
Factor loadings ≥ .50 are bolded.

A two-factor solution that together explicated 64.53% of the variance was found using exploratory factor analysis (PCA extraction).

Factor 1: Readiness of Institutions and Infrastructure: This factor included IoT-enabled visibility and traceability (IoT), policy and institutional support (PIS), ecosystem collaboration and inclusion (ECI), circular economy orientation and culture (CEC), digital infrastructure readiness (DIR), and the dependent construct - digitally-enabled circular economy implementation (DE-CE). It was responsible for

51.55% of the variance, with an eigenvalue greater than 4.0. The robust and steady loadings of these variables suggest that the main force behind CE implementation is enterprises' readiness in terms of digital infrastructure, institutional frameworks, cultural alignment, and IoT uptake. This implies that the foundation for successful CE adoption in emerging market environments is institutional support and structural preparedness.

Factor 2: Ability to Integrate Data: Data Interoperability and Integration (DII), which had an eigenvalue over 1.0 and explained 12.98% of the variance, was the only factor representing this aspect. Data sharing, interoperability, and system integration form a distinct capability dimension that is important for the expansion of CE but has not yet been fully embedded within the broader structure of institutional preparedness, as shown by the specific clustering of DII.

Innovation in Circular Business Models (CBMI): Remarkably, CBMI displayed modest loadings for both factors (below the 0.50 criterion). Despite being a major component of CE theory, the empirical analysis indicates that business model innovation is still in its early stages among enterprises. This result is in line with the reality of emerging economies, where businesses prioritize institutional alignment and infrastructure preparedness over rigorous business model transformations.

Overall, EFA results suggest that data integration is a separate but underdeveloped factor, while institutional–infrastructure preparedness is the fundamental basis for DE-CE in developing economies. Although the CE's idea is widely accepted, business model transformation trails behind institutional and infrastructure readiness, with CBMI remaining an emerging construct.

Mediation Estimates

Table 5. IoT as the mediator, DECE as the dependent variable

Predictor	Effect	Estimate	SE	95% Confidence Interval		Z	p	% Mediation
				Lower	Upper			
DIR	Indirect	0.2326	0.0749	0.0834	0.3810	3.10	.002	52.2%
	Direct	0.2126	0.0949	0.0253	0.4000	2.24	.026	47.8%
	Total	0.4452	-	-	-	-	<.001	100%
DII	Indirect	0.0309	0.0390	-0.0418	0.1125	0.79	.428	33.7%
	Direct	0.0609	0.0666	-0.0705	0.1924	0.92	.362	66.3%
	Total	0.0918	-	-	-	-	NS	100%

continued on following page

Table 5. Continued

Predictor	Effect	Estimate	SE	95% Confidence Interval		Z	p	% Mediation
				Lower	Upper			
PIS	Indirect	0.2016	0.0648	0.0751	0.3314	3.11	.002	46.2%
	Direct	0.2344	0.0853	0.0660	0.4028	2.75	.007	53.8%
	Total	0.4360	-	-	-	-	<.001	100%
CEC	Indirect	0.1439	0.0623	0.0266	0.2758	2.31	.021	26.5%
	Direct	0.3993	0.0818	0.2379	0.5606	4.88	<.001	73.5%
	Total	0.5432	-	-	-	-	<.001	100%
ECI	Indirect	0.1676	0.0924	-0.0090	0.3500	1.81	.070	32.2%
	Direct	0.3523	0.1107	0.1339	0.5707	3.18	.002	67.8%
	Total	0.5199	-	-	-	-	<.001	100%

Table 6. Path estimates

Predictor	Path	Estimate	SE	95% Confidence Interval		Z/t	p
				Lower	Upper		
DIR	IV → IoT (a)	0.6958	0.0580	0.5813	0.8102	12.00	<.001
	IoT → DECE (b)	0.3343	0.0917	0.1533	0.5153	3.65	.0004
	IV → DECE (c′)	0.2126	0.0949	0.0253	0.4000	2.24	.026
DII	IV → IoT (a)	0.0661	0.0729	−0.0779	0.2100	0.91	.366
	IoT → DECE (b)	0.4680	0.0689	0.3320	0.6039	6.79	<.001
	IV → DECE (c′)	0.0609	0.0666	−0.0705	0.1924	0.92	.362
PIS	IV → IoT (a)	0.6115	0.0601	0.4930	0.7301	10.18	<.001
	IoT → DECE (b)	0.3297	0.0851	0.1618	0.4977	3.88	.0002
	IV → DECE (c′)	0.2344	0.0853	0.0660	0.4028	2.75	.007
CEC	IV → IoT (a)	0.6064	0.0615	0.4849	0.7278	9.85	<.001
	IoT → DECE (b)	0.2373	0.0805	0.0784	0.3963	2.95	.0037
	IV → DECE (c′)	0.3993	0.0818	0.2379	0.5606	4.88	<.001
ECI	IV → IoT (a)	0.8073	0.0491	0.7103	0.9043	16.43	<.001
	IoT → DECE (b)	0.2075	0.1068	−0.0032	0.4182	1.94	.054
	IV → DECE (c′)	0.3523	0.1107	0.1339	0.5707	3.18	.002

When IoT was recorded as a variable dependent as a mediator with DECE, the results gave clear evidence of partial mediation for many predictors. In particular, DIR, PIS, and CEC each demonstrated important indirect effects through IoT, while maintaining important direct effects on DECE, suggesting that IoT explains a meaningful part of their impact on effectiveness, but not completely. For example, the effect of DIR on DECE was almost equally divided between the indirect path through IoT ($\beta = 0.233$, $p = .002$) and the direct path ($\beta = 0.213$, $p = .026$), reflecting a balanced mediation. A similar pattern emerged for PIS and CEC, in which IoT transmitted between a quarter and about half of their total effects. ECI also showed evidence of partial mediation, although the indirect path did not reach conventional significance, which only reflects a trend-level effect. Conversely, DII neither significantly affected IoT nor had a direct impact on DECE, which does not indicate any mediation. Overall, these conclusions confirm that IoT acts as a meaningful mediator through which some predictors (DIR, PIS, CEC, and to some extent ECI) increase their impact on decision effectiveness. predictors (DIR, PIS, CEC) and DECE, highlighting its role as a pathway through which these factors influence decision effectiveness.

Mediation Estimates

Table 7. CBMI as the mediator, DECE as the dependent variable

Predictor	Effect	Estimate	SE	95% Confidence Interval		Z/t	p	% Mediation
				Lower	Upper			
DIR	Indirect (via CBMI)	0.0061	0.0181	-0.0267	0.0454	–	–	–
	Direct	0.4392	0.0750	0.2912	0.5872	5.8576	0.0000	–
DII	Indirect (via CBMI)	0.0117	0.0140	-0.0066	0.0476	–	–	–
	Direct	0.0802	0.0747	-0.0672	0.2276	1.0738	0.2844	–
PIS	Indirect (via CBMI)	-0.0059	0.0234	-0.0507	0.0435	–	–	–
	Direct	0.4420	0.0744	0.2950	0.5889	5.9377	0.0000	–
CEC	Indirect (via CBMI)	-0.0026	0.0184	-0.0419	0.0334	–	–	–
	Direct	0.5458	0.0694	0.4088	0.6827	7.8677	0.0000	–
ECI	Indirect (via CBMI)	-0.0052	0.0205	-0.0478	0.0360	–	–	–
	Direct	0.5251	0.0730	0.3809	0.6692	7.1877	0.0000	

Table 8. Path estimates (with CBMI as mediator, DECE as dependent variable)

Predictor (IV)	Path	Estimate	SE	95% Confidence Interval		Z/t	p
				Lower	Upper		
DIR	DIR → CBMI	0.2468	0.0781	0.0926	0.4010	3.1592	0.0019
	CBMI → DECE	0.0245	0.0706	-0.1147	0.1638	0.3477	0.7285
	DIR → DECE (direct)	0.4392	0.0750	0.2912	0.5872	5.8576	0.0000
DII	DII → CBMI	0.1041	0.0746	-0.0431	0.2513	1.3953	0.1647
	CBMI → DECE	0.1121	0.0753	-0.0365	0.2606	1.4890	0.1383
	DII → DECE (Direct)	0.0802	0.0747	-0.0672	0.2276	1.0738	0.2844
PIS	PIS → CBMI	0.3315	0.0736	0.1862	0.4769	4.5022	0.0000
	CBMI → DECE	-0.0178	0.0723	-0.1606	0.1249	- 0.2466	0.8055
	PIS → DECE (Direct)	0.4420	0.0744	0.2950	0.5889	5.9377	0.0000
CEC	CEC → CBMI	0.2595	0.0763	0.1090	0.4100	3.4025	0.0008
	CBMI → DECE	-0.0100	0.0666	-0.1414	0.1215	- 0.1497	0.8811
	CEC → DECE (Direct)	0.5458	0.0694	0.4088	0.6827	7.8677	0.0000
ECI	ECI → CBMI	0.2972	0.0772	0.1448	0.4496	3.8493	0.0002
	CBMI → DECE	-0.0174	0.0687	-0.1529	0.1182	- 0.2530	0.8006
	ECI → DECE (Direct)	0.5251	0.0730	0.3809	0.6692	7.1877	0.0000

In comparison, when CBMI was specified as an intermediary, the pattern of results was particularly different. Across nearly all predictors, indirect effects through CBMI were consistently non-significant, as the confidence intervals included zero, indicating that CBMI did not act as an intermediary between the predictors and DECE. At the same time, strong and highly significant direct effects of DIR (β = 0.439, $p < .001$), PIS (β = 0.442, $p < .001$), CEC (β = 0.546, $p < .001$), and ECI (β = 0.525, $p < .001$) were evident. Only DII failed to show a significant impact in either pathway. Thus, unlike IoT, which emerged as an important mediation path, CBMI was not responsible for any significant indirect impact, underlining that the relationship between major predictors and decision effectiveness is direct in nature when CBMI is considered.

FINDINGS

Empirical results provide many important insights into the digitally-capable circular economy (DE-CE) in emerging-economy manufacturing contexts.

First, the analysis confirmed the four capability-related and important direct effects of relevant factors on DE-CE implementation. In particular, Digital Infrastructure Readiness (DIR) had a positive impact on DE-CE ($\beta = 0.213$, $p = 0.026$), supporting H1. Similarly, Policy and Institutional Support (PIS) ($\beta = 0.442$, $p < 0.001$), Circular Economy Orientation and Culture (CEC) ($\beta = 0.546$, $p < 0.001$), and Ecosystem Collaboration and Inclusion (ECI) ($\beta = 0.525$, $p < 0.001$) also showed significant positive effects. In contrast, Data Interoperability and Integration (DII) did not significantly affect DE-CE ($\beta = 0.081$, $p = 0.284$), leading to the rejection of H2. Collectively, these conclusions highlight that while infrastructure, institutional support, cultural alignment, and ecosystem collaboration are the central drivers of digital–circular changes, integration and interoperability challenges remain unsolved.

Second, IoT-enabled visibility and mediation analysis through traceability gave clear evidence of partial mediation in many routes. DIR had a significant indirect impact through IoT ($\beta = 0.233$, $p = 0.002$) alongside its direct effect ($\beta = 0.213$, $p = 0.026$), supporting H6a and indicating balanced mediation. Similarly, PIS demonstrated an important indirect effect through IoT ($\beta = 0.202$, $p = 0.002$) and a direct effect ($\beta = 0.234$, $p = 0.007$), supporting H6c. CEC also displayed partial mediation, confirming a strong direct effect ($\beta = 0.399$, $p < 0.001$), with an indirect effect ($\beta = 0.144$, $p = 0.021$), supporting H6d. ECI showed a trend-level indirect effect ($\beta = 0.168$, $p = 0.070$), which suggests only partial support for H6e. Conversely, DII neither predicted IoT nor directly impacted DE-CE ($\beta = 0.061$, $p = 0.362$; $\beta = 0.031$, $p = 0.428$), resulting in the rejection of H6b. These results outline the role of IoT as an important path through which the digital readiness of firms, institutional support, and cultural orientation are translated into effective DE-CE implementation.

Third, when Circular Business Model Innovation (CBMI) was tested as an intermediary, the pattern of the results was different. Although DIR, PIS, CEC, and ECI significantly predicted CBMI (DIR → CBMI: $\beta = 0.247$, $p = 0.002$; PIS → CBMI: $\beta = 0.332$, $p < 0.001$; CEC → CBMI: $\beta = 0.260$, $p = 0.001$; ECI → CBMI: $\beta = 0.297$, $p < 0.001$), the subsequent paths from CBMI to DE-CE were not significant (from $\beta = -0.018$ to 0.112, all $p > 0.13$). Transformative business models appear to be less influential at this stage, with firms prioritizing IoT adoption over business model redesign.

SUGGESTIONS

Strengthen digital infrastructure and IoT adoption: Firms should prioritize investment in IoT-enabled systems to increase the traceability and visibility of resources, as these tools greatly enhance the effectiveness of existing capabilities.

Address interoperability gaps: Industry consortia and policymakers must develop common digital standards to improve system integration, which remains the weakest link in the transition towards DE-CE.

Foster institutional and policy support: Governments and industry associations must continue to strengthen the positive impact of Policy and Institutional Support (PIS) and Ecosystem Collaboration and Inclusion (ECI).

Promote gradual CBMI adoption: Firms should be encouraged to experiment with product–service systems, closed-loop logistics, and other models that gradually position CBMI as a strategic driver, even if its current mediating role is limited.

Target capacity building: Training middle-level managers and engineers - who make up the majority of respondents - can accelerate the adoption of IoT technologies and circular economy practices.

THEORETICAL IMPLICATIONS

This study contributes to the digital sustainability and circular economy literature in three important ways. First, it validates a multi-construct measurement model for DE-CE in emerging-economy manufacturing, expanding the existing framework that is often more relevant to developed markets. Second, the results highlight that IoT plays an important mediating role, providing empirical evidence that digital technologies serve as a bridge between organizational capabilities and sustainability outcomes. Third, the findings suggest that CBMI is not yet an effective intermediary in emerging economies, offering a nuanced understanding that challenges prevailing perceptions in CE theory that business model innovation universally drives circular adoption.

MANAGERIAL IMPLICATIONS

For managers, the study underlines the need to take advantage of IoT technologies to strengthen the link between organizational readiness and circular practices. Investment in IoT-enabled traceability can yield immediate benefits in efficiency, compliance, and resource optimization. In addition, managers should not underestimate the role of institutional alignment - the integration with industry networks and regulators can significantly reduce barriers to adoption. Finally, while CBMI is not yet a major driver, managers must consider incremental innovations in business models to remain competitive in the long term.

POLICY AND MANAGERIAL IMPLICATIONS

The findings of this study emphasize that the success of digitally enabled circular-economy (DE-CE) transitions in emerging economies depends not only on technological availability but also on the alignment of policy, governance, and managerial practices. A multilevel policy and managerial response is therefore necessary.

Policy Implications

National and sub-national governments must strengthen institutional capacity for monitoring and enforcing circular-economy norms through digital platforms. Policies that invest in digital infrastructure readiness (DIR) and data-governance frameworks can ensure that IoT-enabled traceability systems translate into actual environmental compliance.

Governments should adopt open data standards and interoperable IoT protocols to prevent vendor lock-in and promote inclusion of small firms. Targeted financial incentives - such as tax rebates, soft loans, and performance-linked subsidies - can encourage Circular Business Model Innovation (CBMI) among micro-, small-, and medium-sized enterprises (MSMEs). Moreover, integration of the informal sector through digital identity and cooperative registration programs will enhance traceability without displacement.

Managerial Implications

For managers, IoT adoption (M_1) represents an immediate lever for operational visibility, real-time decision support, and compliance assurance. Organizations should therefore prioritize IoT-based data capture and analytics as a foundation for broader circular transformation. However, CBMI (M_2) requires longer-term strategic commitment - redesigning value chains, service models, and partnerships - which may initially appear risk-intensive.

Managers can mitigate these barriers by fostering a circular-economy culture (CEC) that rewards innovation and continuous learning, backed by environmental commitment and investment (ECI) from senior leadership. Cross-functional teams that integrate IT, production, and sustainability units can create synergies between digital transformation and circular practices.

Finally, companies should engage in public–private partnerships to co-develop sectoral digital infrastructure (such as e-waste traceability portals or materials passports) that distribute costs and accelerate collective learning.

CONCLUSION

This research shows that digitally-enabled circular economy implementation in emerging economies is mainly driven by infrastructure readiness, institutional and cultural alignment, and ecosystem cooperation, with IoT acting as an important pathway. Conversely, data interoperability and CBMI remain underdeveloped dimensions, limiting their current effects. The study confirms that while firms are adopting digital technologies and CE-oriented practices, system integration and transformative business model innovation are still in the early stages. Overall, the conclusions reinforce the idea that technological enablers such as IoT and strong institutional structures form the backbone of DE-CE adoption in emerging markets.

FUTURE RESEARCH DIRECTIONS

Building on these findings, future studies can explore several avenues. First, longitudinal research designs can capture the evolving role of CBMI as firms progress in their CE adoption journey. Second, cross-comparative studies can reveal institutional and cultural contingencies that shape DE-CE adoption across different contexts. Third, future research should integrate moderating variables such as firm size, age, and digital maturity to unpack heterogeneous adoption pathways. Fourth, sector-specific case studies can provide rich insights into how IoT and business model innovation manifest in industries such as automotive, textiles, or food processing. Finally, incorporating environmental performance outcomes into the framework (e.g., carbon footprint, resource efficiency) can strengthen the empirical link between digitization, circular economy practices, and sustainability performance.

REFERENCES

Bag, S., Pretorius, J. H. C., Gupta, S., & Dwivedi, Y. K. (2021). Role of institutional pressures and resources in the adoption of big data analytics powered artificial intelligence for enhancing circular economy performance in organizations. *Technological Forecasting and Social Change*, *163*, 120420. DOI: 10.1016/j.techfore.2020.120420

Bocken, N. M. P., de Pauw, I., Bakker, C., & van der Grinten, B. (2016). Product design and business model strategies for a circular economy. *Journal of Industrial and Production Engineering*, *33*(5), 308–320. DOI: 10.1080/21681015.2016.1172124

Centobelli, P., Cerchione, R., Chiaroni, D., Del Vecchio, P., & Urbinati, A. (2022). Designing business models in circular economy: A systematic literature review and research agenda. *Business Strategy and the Environment*, *31*(2), 585–607. DOI: 10.1002/bse.2905

Chaturvedi, S., & Gupta, N. (2024). Institutional capacity and enforcement challenges in India's circular-economy transition. *Environmental Policy and Governance*, *34*(1), 45–59.

Chauhan, C., Parida, V., & Dhir, A. (2022). Linking circular economy and digitalisation technologies: A systematic literature review and future research agenda. *Technological Forecasting and Social Change*, *177*, 121508. DOI: 10.1016/j.techfore.2022.121508

Dasgupta, P., & Mukherjee, S. (2021). Trust, culture, and sustainability transitions in South Asia. *Development and Change*, *52*(4), 879–901.

Geissdoerfer, M., Savaget, P., Bocken, N. M. P., & Hultink, E. J. (2017). The circular economy—A new sustainability paradigm? *Journal of Cleaner Production*, *143*, 757–768. DOI: 10.1016/j.jclepro.2016.12.048

Gupta, R., & Sarkar, A. (2023). Integrating informal waste workers into formal recycling through digital inclusion. *Resources, Conservation and Recycling*, *195*, 107333.

Ingemarsdotter, E., Jamsin, E., Kortuem, G., & Balkenende, R. (2019). Circular strategies enabled by the Internet of Things: A framework and analysis of current practice. *Sustainability (Basel)*, *11*(20), 5689. DOI: 10.3390/su11205689

Kirchherr, J., Reike, D., & Hekkert, M. (2017). Conceptualizing the circular economy: An analysis of 114 definitions. *Resources, Conservation and Recycling*, *127*, 221–232. DOI: 10.1016/j.resconrec.2017.09.005

Kumar, S., & Bhatnagar, M. R. (2024, April). Differential Modulation-Based Non-Orthogonal Multiple Access. In *2024 IEEE Wireless Communications and Networking Conference (WCNC)* (pp. 1-6). IEEE. DOI: 10.1109/WCNC57260.2024.10570917

Ma, Z., Zhang, D., Ding, S., & Li, Z. (2020). Data-driven sustainable intelligent manufacturing based on demand response for energy-intensive industries. *Journal of Cleaner Production, 274*, 122934. DOI: 10.1016/j.jclepro.2020.123155

Mordor Intelligence. (2025). *Emerging-economy circular-economy market outlook 2025–2030*. Mordor Intelligence LLP.

Nunnally, J. C. (1978). An overview of psychological measurement. *Clinical diagnosis of mental disorders: A handbook*, 97-146.

OECD. (2023). *Policy instruments for circular-economy transitions in developing economies*. OECD Publishing.

Parida, V., Sjödin, D., & Reim, W. (2022). Digital transformation and circular-economy implementation: Contextual challenges and pathways. *Journal of Business Research, 150*, 556–567. DOI: 10.1016/j.jbusres.2022.06.045

Patel, R., & Raj, S. (2023). Voice-enabled IoT systems for low-literacy manufacturing contexts. *Computers in Industry, 148*, 103874. DOI: 10.1016/j.compind.2023.103874

Ranta, V., Aarikka-Stenroos, L., & Väisänen, J.-M. (2021). Digital technologies catalysing business model innovation for circular economy: Multiple case study. *Resources, Conservation and Recycling, 164*, 105155. DOI: 10.1016/j.resconrec.2020.105155

Stahel, W. R. (2016). The circular economy. *Nature, 531*(7595), 435–438. DOI: 10.1038/531435a PMID: 27008952

UNCTAD. (2024). Digital connectivity and sustainability in developing economies: 2024 report. *United Nations Conference on Trade and Development.*

UNIDO. (2024). *Inclusive industrial digitalization casebook: Craft sectors in Asia and Africa*. United Nations Industrial Development Organization.

Wilson, D., Rodic, L., & Velis, C. (2021). Waste management in the global south: Informality, innovation, and inclusion. *Waste Management & Research, 39*(8), 915–928.

World Bank. (2023). *Strengthening environmental governance for circular-economy transitions*. World Bank Group.

Chapter 2
Advancing Industrial Sustainability Through Plasma Technologies by Exploring Environmental Applications, Circular Economy, and Policy Pathways

Javaria Razzaq
https://orcid.org/0000-0001-5985-7752
Quaid-i-Azam University, Islamabad, Pakistan

Adnan Mehmood Bhatti
Center of Excellence in Science and Applied Technologies, Islamabad, Pakistan

Arshad M. Mirza
Physics Department, University of Management and Technology, Lahore, Pakistan

ABSTRACT

Plasma-based technologies are emerging as cutting-edge solutions in fostering eco-friendly production and sustainable industrial practices, offering exceptional abilities for circular economic solutions and sustainable eco-friendly pollution control. Using evidence-based assessment, it examines practical applications of waste minimization or waste-to-energy conversion techniques while comparing current performance

DOI: 10.4018/979-8-3373-6435-3.ch002

with traditional approaches in terms of technological feasibility, ecological impact, and economic viability. The chapter also emphasizes the knowledge gap and future research directions by focusing on advanced manufacturing, sustainable energy, and science technology. It concludes by highlighting the importance of interdisciplinary research and strategic policy frameworks to accelerate the integration of plasma-based technologies. Bringing together scientific, technological, and policy perspectives, this chapter demonstrates how plasma approaches can significantly contribute to the United Nations Sustainable Development Goals (SDGs) and broader environmental targets.

INTRODUCTION

Nowadays, plasma-based technologies provide potential pathways for eco-efficient industrial operations, coupled with other innovative climate change mitigation methods. Industries are facing pressure to reduce emissions and adopt clean technologies. Plasma-based technologies provide smart and distinctive solutions to complex environmental challenges. For example, non-thermal plasma has demonstrated near-complete removal of persistent pollutants under ambient conditions (Tamang et al., 2024), while magnetohydrodynamic (MHD) separators have achieved recovery rates of up to 99% for high-purity metals from e-waste, offering an efficient route for resource recycling (Wang & Liu, 2023).

The advantage of the technologies used on the basis of plasma is the unusual physical characteristics. Examples include selective dissociation with extremely high-energy electrons typically in the 10–20 eV range (Bogaerts & Neyts, 2024), advancements in nanofabrication and atomic manipulation (Chiappim et al., 2022). Other advancements in technology have been used in optimization of plasma reactors, including the high-power pulse systems and the AI-optimized reactor systems. Such innovations can reduce the carbon dioxide emissions by 35-50 percent as compared to traditional robust thermal systems (Nauri et al., 2024).

Despite these advances, industrial adoption remains limited. The current literature demonstrates major gaps in the adaptability, durability, and optimization of energy consumption versus performance of plasma-based technologies. Plasma systems are deployed in fewer than 5% of major sectors, constrained by technical complexity, higher capital costs (30–50% above traditional systems), and insufficient policy integration (World Bank, 2024; OECD, 2024). This gap represents a missed opportunity, as the IPCC (2023) estimates that large-scale deployment could contribute up to 12% of industrial decarbonization. Encouragingly, economic studies indicate that firms adopting plasma technologies report cost reductions of approximately 22% over three years, demonstrating long-term benefits when supported by appropriate policies and regional adaptation strategies. This chapter not only addresses the unrealized potential of plasma technologies in accelerating sustainable industrial transitions but also critically evaluates current progress and financial constraints. Basically, the industrial sector is actively pursuing sustainable production and decarbonization through the adoption of new technologies. These technologies provide various techniques for handling waste management and mitigating pollution by offering other paths, for example, electrolysis, catalysis, and carbon dioxide capture. The deployment of plasma-based technologies into various industrial sectors can facilitate eco-friendly production and key UN sustainability targets. This chapter also provides policy-based future research, multi-domain and cross-cutting ends of plasma, and addresses gaps in plasma-based technologies across various sectors. It adopts a cross-disciplinary perspective to analyze both technical innovations and policy frameworks, guided by the following research questions:

- **RQ1:** Which plasma technologies (catalytic, thermal, or electromagnetic) demonstrate the greatest adaptability in terms of performance and cost–benefit?
- **RQ2:** How can plasma reactor designs be optimized to reduce barriers for small and medium enterprises (SMEs)?
- **RQ3:** What policy initiatives (e.g., tax incentives, R&D grants) are most effective across industrial sectors?
- **RQ4:** How do regional factors (e.g., energy costs, workforce skills) shape deployment strategies for plasma-based technologies?

2 LITERATURE REVIEW

2.1 Historical Background

The development of plasma-based technologies spans over a century, evolving from fundamental scientific discoveries into a wide range of industrial applications. In the late 19th century, Sir William Crookes identified plasma as "radiant matter" during experiments with electrical discharges in low-pressure gases. Plasma results from ionization, a process in which neutral atoms or molecules lose or gain electrons. In 1928, Irving Langmuir introduced the term *"plasma"* to describe the collective behavior of ionized gases, laying the foundation for plasma physics as a distinct discipline (Chen, 1984).

By the mid-20th century, advances in vacuum technology, power electronics, and materials engineering enabled the development of low-temperature plasma devices for experimental and industrial applications (Lieberman & Lichtenberg, 2005). In the

1970s, NASA introduced plasma gasification processes for waste treatment (Huang et al., 2006). Various plasma torches, including direct current (DC), alternating current (AC), microwave, and radio frequency (RF), became central to the energy sector. In the 1990s, increasing demand for environmentally sound waste management shifted attention toward eco-friendly practices, where non-thermal plasma (NTP) catalysis emerged as a promising technology for sustainable chemical manufacturing. NTP systems provide rapid response, low heat input, and atmospheric-pressure operation, making them suitable for integration with renewable energy sources.

By the early 21st century, plasma technologies were applied in waste recycling and water purification, supporting progress toward the Sustainable Development Goals (SDGs). Today, leading suppliers such as Westinghouse, Euro Plasma, Tectonics, and Phoenix Solutions provide high-power plasma torches, including portable DC models with water-cooled electrodes. Based upon these advancements, recent plasma-based technologies have attained great progress in various commercial applications, specifically in plasma gasification, resource recovery, and waste-to-energy volorization.

Plasma gasification exposes waste to extreme temperatures (1,500–14,000 °C), with heat transferred through conduction, convection, and radiation (Messerle et al., 2024; Unnisa & Hassanpour, 2017). Modeling studies using COMSOL (Azhagu et al., 2025) have advanced the understanding of arc behavior and heat transfer in torches, supporting improved designs for plasma reactors. Integrated three-stage plasma recycling now provides a cost-effective approach for converting mixed waste into energy without pre-treatment, contributing to circular economy models (Arena, 2012). Plasma-based systems provide significant benefits for waste management and resource recovery, but they constitute a single dimension of a broader set of solutions. Catalytic process, transition to electric power, and decarbonization are complementary to plasma-based methods and hold equal importance.

Recently, magnetohydrodynamic (MHD) separators for atomic-scale metal recovery from e-waste (Chen et al., 2023) and AI-assisted plasma reactors for energy management have moved plasma-based applications toward digitally optimized, low-carbon plasma systems that are technically viable and environmentally friendly, and will substitute the old methods. Despite these new advancements, adoption remains minimal owing to many reasons, like technical complexity, high costs, and lack of policy support. Plasma technologies are still characterized by considerable energy losses, which restrict their use on large-scale projects. Even current research (2020–2025) still highlights drawbacks and limitations in industrial-scale validation and sustained performance metrics.

2.2 Scope and Objectives

This chapter examines the latest developments and breakthroughs in plasma-based technologies, which advance industrial sustainability, promote circular economic practices, and focus on resource recovery for energy production and reprocessing waste materials. The review comprises both thermal and non-thermal plasma systems, analyzing their underlying concepts, functional approach, and implementation strategies.

The contents of this chapter encompass four fundamental aspects:

- Economic and operational aspects: such as financial viability and stability, and compatibility with existing workflows in the industry.
- Environmental impacts: including waste product purification, environmental protection, and integration with renewable and sustainable energy.
- Policy and governance strategies: considering the needs of both developed and emerging economies.
- Comparative assessment: contrasting alternative plasma-based treatments with traditional approaches.

The objectives of this chapter are fourfold. First, to explain the operational principles of plasma systems and their relevance to industrial sustainability. Second, to identify current industrial and environmental challenges where plasma-based methods can provide effective solutions. Third, to compare plasma technologies with traditional methods across technical, environmental, and economic dimensions. Finally, the chapter highlights policy-oriented strategies and presents real-world industrial applications to demonstrate the practical integration of plasma-based solutions into sustainable development frameworks.

It is vital to incorporate the latest findings and current studies (2020-2025) here, which have significantly increased the productivity, efficiency, and industrial applicability of technologies based on plasma. The latest studies highlight remarkable progress in both thermal and non-thermal plasma-based systems applicable in industrial waste management. Thermal plasma waste treatment has revealed better quality syngas, enhanced hydrogen production, and minimized coal tar production, as indicated in recent research findings. These efficiency gains improve the economic viability of plasma gasification and apply to varied organic residues, industrial waste materials, and chemical sludge.

In contrast, NTP (non-thermal plasma) based technologies have made substantial progress in the elimination of pollutants. The hybrid NTP-catalysis system, as mentioned by Subash et al. (2025), proved removal of high destruction efficiencies exceeding 95% for VOCs (Volatile Organic Compounds), plastic pellets, and low-temperature industrial discharges, emphasizing the commercial viability.

Plasma gasification and pyrolysis have also developed into an effective approach for plastic trash, biomedical waste, and harmful chemical pollutants. (Oliveria et.al., 2022) indicates enhanced reactor stability, emission reduction, and Higher Heating Value (HHV) of syngas generated via pyrolytic and thermochemical decomposition. Installations of commercial plasma-arc furnaces in the United Arab Emirates, South Korea, and Japan have shared positive outcomes. (Samal & Blanco, 2022) showed an improved rate of metal recovery of more than 98% through thermal plasma treatment of e-waste, while (Olaoye & Amlogu, 2024) indicated the asbestos encapsulation via safe vitrification, incineration residue, and biomedical waste. Environmental impact and Life Cycle Assessment (LCA) assessments validate plasma's ecological advantage (Mallick & Vairakannu, 2025). These studies show mitigation in carbon emissions and toxic byproducts as compared to traditional incineration. Most recent studies, for example, (Messerle et al. 2024) indicated that coupling of solar-PV or biomass with plasma gasifier can decrease dependence on external energy sources while enhancing syngas stabilization. Similarly, (Hu et al., 2023) have shown extended electrode durability and superior thermal performance using high-frequency plasma torches, fostering equitable industrial development for both emerging and developed economies.

(Panwar et. al., 2025) His recent Life Cycle Thinking (LCT) based analysis shows that plasma gasification mitigates greenhouse gas output, sanitary landfill reliance, and hazardous waste production, but still encounters economic obstacles due to costly infrastructure development. Similarly, (Wang et. al., 2025) analysis on carbon dioxide and thermal plasma treatment of medical waste shows optimal carbon yield and reduction of tar content, assessing plasma's viability for toxic waste materials, and indicated that plasma arc melting has attained REE (Rare-Earth-Element) extraction yield of more than 95%. Moreover, (Pachecho et. al., 2024) demonstrated plasma-based mineralization technologies and facilitated the conversion of hazardous industrial waste material into non-reactive materials, applicable for construction sections. (Bashir et al., 2025) highlights that plasma-based waste-to-energy gasification acts as an effective approach for low-emission hydrogen, integrating waste management into climate change mitigation strategies. Plasma technologies demonstrate considerable promise; numerous applications, specifically plasma catalytic reactions and assisted conversion, are not yet commercially viable. Life Cycle Assessment (LCA) findings indicate that the benefits of plasma are heavily dependent on the availability of low-carbon power, positioning it as only one option among many for emission reduction, not a universally optimal strategy. NTP (Non-thermal plasma) catalysis, plasma-based waste treatment, and combustion exhibit potential at lab and large-scale, yet these technologies largely remain at an early stage and need more development. So, these research studies primarily address technical aspects instead of economic sustainability and environmental analysis. Hence, this chapter intends to address this gap by providing a more comprehensive and analytical perspective of plasma waste management technologies.

2.3 Conceptual Framework

The conceptual framework of this chapter is based on the interaction between technological innovation, sustainability goals, and policy integration. Plasma technologies are positioned as catalysts that can simultaneously improve environmental performance, address industrial inefficiencies, and support transitions toward a circular economy. The framework is prioritizing sustainable developmental goals via the circular economy and is structured around three interconnected pillars. The first one is technological infrastructure: covering the fundamental principles of plasma (ionization, energy transfer mechanisms, electron and ion temperatures), process types (thermal and non-thermal), and system configurations, including hybrid systems for waste treatment, gas cleaning, and energy recovery. The second one describes sustainability integration, focusing on outcomes such as waste-to-energy conversion, pollutant destruction, carbon footprint reduction, recovery of valuable metals, chemical synthesis from waste gases, and the use of Life Cycle Assessment (LCA) to evaluate impacts across the process chain. The third one describes policy and deployment strategies, encompassing government incentives, public–private partnerships, and mechanisms for transferring scientific knowledge into industrial practice, particularly in emerging economies, while ensuring alignment with the United Nations Sustainable Development Goals (SDGs). This three-pillar model demonstrates how plasma technologies can act both as technical enablers and as policy-driven instruments for advancing industrial sustainability at a global scale. Instead of just a basic description, the framework analyses the interaction of these three pillars, for instance, the impact of technology adoption on environmental, social, and economic outcomes, and how inadequate regulation or a financial model can cause deployment delays. By analyzing these relationships, this conceptual framework gives insight that emphasizes limitations and practical constraints, and the essential criteria for real-world deployment, aligning with the Technology-Organization-Environment framework.

3. RESEARCH METHODOLOGY

This study adopts a theoretical framework and narrative literature review to evaluate plasma's contribution to industrial sustainability. Rather than empirical investigation, the chapter gives a comprehensive review of current advancements, peer-reviewed literature, cross-disciplinary examination, and policy directives that are both reliable and beneficial in real-world scenarios. Pertinent literature was systematically identified through major scientific databases (Web of Science, Scopus, Google Scholar), focusing on peer-reviewed articles, authoritative reports, and key reviews. The review focuses on research articles that largely fall into the publication period of 2000 to 2024, prioritizing influential literature reviews and applied industry examples. Initially, a comprehensive review of literature has established the basis of plasma's role in waste treatment, pollution management, and assessing technology readiness levels. Following that, evidence was gathered from pilot projects and industrial case studies, which were then benchmarked against conventional methods to assess their technical, ecological, and economic performance. After that, a comparative method was used to compare Plasma-based technologies against established alternatives like incineration, pyrolysis, and conventional gasification, evaluating their strengths and weaknesses regarding efficiency, emissions, and cost-effectiveness. Additionally, the study incorporated policy, international collaboration, and governance to provide a broader view of technology adoption

In the end, the chapter adopts a future-based analytical approach that exceeds description, unveiling new frontiers and advancements of plasma-based technologies in energy systems, advanced manufacturing, and medical sciences. The methodology integrates a theoretical framework with evidence-based discussion, which acts as a bridge between high-tech innovations, global sustainability goals, and strategic approaches. This approach analyzes critically and compares plasma technologies against traditional methods by evaluating environmental sustainability and performance metrics. It also combines insight from various sectors and recent TRL projects to identify emerging patterns, trends, and limitations. Throughout the chapter, plasma-based technologies are evaluated and differentiated by common evaluation criteria such as technical efficiency, environmental effect, TRLs, financial viability, and policy implications.

4. APPLICATION-BASED EVIDENCE

This section provides an overview of current applications, but also provides a critical analysis of systems based on plasma in real working conditions. The analysis demonstrates the differences and gap between small and large-scale implementation, pinpoints technical obstacles such as energy consumption and reactor reliability, and compares results with traditional thermochemical methods. These findings highlight the applicability, restrictions, and future directions for plasma-based technologies.

4.1 Industrial Waste Treatment

Plasma technology offers diverse applications in materials processing. Leveraging the unique physical properties of ionized gases, three main types of processing can be achieved: (1) destruction of hazardous substances, (2) surface modification of materials, and (3) creation of new functional materials (Bonizzoni & Vassallo, 2002).

Thermal plasmas (in which electrons and heavy particles are nearly at the same very high temperature) generated by direct current (DC), alternating current (AC), radio frequency (RF), or microwave sources at pressures above 10 kPa, these plasmas exhibit electron and ion temperatures of 1–2 eV with relatively low degrees of ionization. Such systems, commonly referred to as plasma torches, are widely used for the destruction of toxic materials and for coating applications such as plasma spraying. Among emerging methods, plasma arc processes have demonstrated especially high destruction efficiency for halogenated wastes. Their ability to operate at extremely high temperatures allows flexible, high-throughput reactions that contribute to environmental integrity and, in some cases, resource recovery. Non-thermal plasmas (NTPs), or cold plasmas (has highly energetic electrons), differ in that their electron temperature greatly exceeds ion temperature. Produced at low pressure using low-power RF, microwave, or DC discharges, NTPs are particularly suited to delicate applications such as textiles, polymers, and even biological tissues (Meichsner et al., 2012). By enabling surface modifications at near-room temperatures, they expand the possibilities for sustainable and energy-efficient material processing.

Plasma-based treatments also play a critical role in industrial manufacturing. Approximately one-third of the processes involved in microelectronics fabrication rely on plasma techniques (d'Agostino et al., 2005). Everyday technologies such as fluorescent lamps, energy-efficient bulbs, and plasma televisions emerged from such advances (Shahzad 2022; Khan & Abas, 2011). Although light-emitting diodes (LEDs) have transformed lighting markets, plasma remains essential in producing coatings, plastics, and optical materials, including lenses and reading glasses (Liston et al., 1993; Eliasson et al., 1987).

Overall, both thermal and non-thermal plasma technologies present significant opportunities for hazardous waste destruction, sustainable manufacturing, and advanced material development. Their broader adoption, however, will depend on continued collaboration between academia and industry to translate laboratory advances into scalable industrial solutions. Although both types of plasma show advanced and effective waste management abilities, their industrial operation and output vary significantly regarding their operational metrics. Thermal plasma ensures that all the molecules are effectively dissociated and ionized, but it is associated with huge power consumption and electrode erosion, restricting cost-effectiveness in ongoing operations. Nonthermal plasma operates effectively at low temperatures with less energy consumption, but the depth of operation and consistency fail when handling dense industrial waste management. Specialized plasma installation demands qualified personnel and safety guidelines. Effective management of workplace safety and continuous staff training are crucial for enduring industrial practices. An in-depth and critical evaluation shows that individually, both systems are insufficient for overcoming industrial limitations, underscoring the importance of hybrid thermal–nonthermal designs and advanced reactor development and deployment.

4.2 Plasma Technology in Pollution Control

Conventional pollution control systems, such as incineration, adsorption, catalysis, and landfilling, are already commercialized but face challenges related to cost, energy intensity, and secondary waste generation. Plasma-based technologies have gained prominence as alternatives due to their higher energy efficiency, smaller footprint, and potential to generate reusable by-products. Large-scale commercialization, however, still requires improvements in treatment rates, energy efficiency, reactor design, and integrated power supply systems. Pollution control strategies are generally grouped into three approaches:

- Dilution: Collecting solid waste or diluting liquid and gaseous waste to acceptable thresholds before disposal. This method is primarily used for liquid and gas emissions; it minimizes pollution by contamination up to the maximum allowable concentration level, and after that, it can be discharged. Although it fails to reduce pollutants, it only mitigates their effect via controlled scattering.
- Neutralization: converting toxic pollutants into non-hazardous waste products, and ensures that pollutants lose their toxic nature before discharge, such as water, before safe release.
- Valorization: converting waste into reusable materials such as construction aggregates, fertilizers, or syngas. It basically converts garbage into profitable goods and re-engages them back into the economic phase. These methods may encompass gasification or nutrient reclamation. It promotes sustainability practices by waste minimization and resource recovery.

Plasma technologies bridge approaches 2 and 3 by enabling both detoxification and resource recovery, offering near-zero-emission solutions. For example, plasma oxidation and catalytic reduction systems are increasingly used in automotive exhaust treatment, while integrated plasma systems are being developed to manage mixed pollutants in advanced clean rooms, bacterial control units, and VOC treatment devices.

In local industries such as iron and steel, plasma systems primarily target combustion-derived gases, including greenhouse and acid gases. In high-tech sectors such as semiconductors, ceramics, and glass fibers, emissions are more complex, consisting of toxic gases, liquid effluents, and solid residues. Plasma's ability to degrade volatile organic compounds (VOCs), ozone-depleting substances, radioactive gases, and even biological contaminants makes it a versatile platform for industrial environmental management.

For liquid waste treatment, three Plasma-based methods are currently in use: remote plasma, indirect plasma/radiation, and direct plasma discharge (Chang et al., 2002). Laboratory-scale studies suggest that hybrid approaches, combining two or more of these methods, can outperform single systems in efficiency and scalability.

Solid waste treatment relies mainly on thermal plasma processes, including plasma oxidation/incineration, plasma pyrolysis, and pulsed arc shockwave methods. In many cases, non-thermal plasma purification is integrated with thermal plasma systems to achieve comprehensive pollutant removal. In summary, plasma technologies provide flexible, high-efficiency solutions for pollution control across multiple industrial sectors, enabling pollutant destruction, resource recovery, and alignment with circular economy principles. While Plasma-based systems strengthen pollution reduction through the above-mentioned strategies, their performance varies for solid and liquid waste material. NTP (Nonthermal plasma) offers energy-saving solutions for fluids and less density materials, but may fall short of expectations for challenging waste streams. In contrast, thermal plasma almost decomposes entirely at higher temperatures and capital expenditures. This assessment underscores the necessity of optimizing plasma type and operational strategies of waste characterization for industrial applications.

4.3 Technology Readiness Levels (TRLs) and Demonstration Projects

The Plasma-driven waste technologies can be assessed using the framework of Technology Readiness Levels (TRLs), which trace the progression from fundamental research to large-scale deployment. Early-stage applications, such as laboratory-scale Plasma-enhanced pyrolysis and plasma-activated catalysis, typically fall within TRL 2–4, where basic principles are investigated and small prototypes are tested. Intermediate readiness levels (TRL 5–7) correspond to pilot-scale systems, including Plasma-enhanced gasification and plasma arc melting, which validate technical performance under semi-industrial conditions. At the highest levels (TRL 8–9), commercial demonstration remains limited but significant. Industrial plasma vitrification and arc gasification plants in Japan and Europe have achieved sustained operation, particularly for hazardous waste treatment. Overall, TRL analysis indicates that while plasma technologies demonstrate strong technical promise, most remain at mid-range levels and require further validation before widespread adoption.

Large-scale demonstration projects are essential in bridging laboratory research and commercialization by proving feasibility, adaptability, and long-term efficiency under real operating conditions. Examples include plasma arc facilities for ash vitrification in Japan, advanced plasma gasification plants in China, and European initiatives for hazardous waste disposal. These projects have confirmed technical viability but also revealed persistent barriers, particularly those related to system optimization, durability, and economic feasibility. Documented outcomes suggest that broad industrial adoption will depend on policy incentives, cost reductions, and advances in plasma reactor design (RQ2).

Even though multiple plasma technologies have reached mid-level TRLs, numerous studies illustrate differences or gaps between lab performance and long-term industrial stability, operating costs, and cost-effectiveness. TRLs show that mostly major projects demonstrate potential for the reduction of pollution, but are hindered by system integration challenges for energy, biomedicine, etc., with the established industrial framework, which indicates the necessity of large-scale validation and direct influence on policy relevance. The capability and efficiency of the aforementioned plasma technologies determine their potential for broad implementation. As a result, these technical features ensure the basis for the regulatory frameworks, management systems, and reward/incentive schemes as discussed in the following section.

5. RELATIVE PERFORMANCE ANALYSIS AND IMPACT ASSESSMENT

Plasma technologies, despite their significant technical capabilities, have performance benefits that differ considerably when compared to conventional established practices like thermochemical and biological methods. This section assesses these multifaceted directions and impacts to elucidate where Plasma-based systems provide industrial benefits and where certain limitations remain.

5.1 Technical Performance vs. Conventional Methods

Plasma-assisted waste treatment demonstrates clear advantages and competence over traditional waste-to-energy treatment and thermochemical technologies. Traditional high-temperature treatment and incineration processes are generally operated under aerobic conditions and produce secondary pollutants. On the other hand, plasma reactors need to operate at extremely high temperatures (5,000–10,000 K), enabling near-complete disintegration of complex waste streams. Studies indicate that plasma systems can achieve pollutant destruction efficiencies surpassing legal requirements while exceeding regulatory standards and suppressing combustion byproduct generation. (Zhovtyansky & Valinčius, 2018). Another major advantage lies in high-standard product quality and obtaining energy from waste streams. Plasma gasification produces high-purity synthesis gas ($CO + H_2$), suitable for both energy generation and chemical synthesis, whereas conventional systems typically yield lower-quality gas with higher contaminants. In addition, plasma furnaces generate vitrified slag, an inert and recyclable by-product, unlike the hazardous fly ash commonly produced by incineration. Performance shows that plasma gasification increases energy output and improves mass conversion efficiency, specifically for different types of municipal solid waste (Munir et al., 2019). Even though plasma systems commonly result in accelerated reaction rates and enhanced decomposition as compared to conventional methods, their performance strongly depends upon the stability of the power system and reactor design. Practical challenges often diminish the value of technological advances, such as energy losses and electrode erosion, while a comparative evaluation indicates that plasma technologies are particularly effective in specialized fields; they need additional engineering refinements before widespread adoption/commercialization. Overall, plasma technologies give greater technical feasibility and environmental performance as compared to conventional methods.

5.2 Environmental Impact Analysis

Plasma-based waste treatment can also serve as an environmentally sound substitute to traditional approaches. In addition to waste minimization, these methods also incorporate advanced waste management and sustainable techniques, such as resource recovery, waste reclamation, and generating inert byproducts suitable for recycling (Evangelisti et al., 2014). The study also highlights the numerous environmental benefits of Plasma-based technologies. For example, as compared to traditional incineration methods, plasma gasification reliably achieves lower pollutant emissions, less ash formation, high gas conversion efficiency, and better quality syngas. (Ramos et al., 2019). Plasma-based systems substantially minimize hazardous pollutants and waste byproducts; the sustainability of the operation is conditional on its energy input and comprehensive life cycle assessment. Large electricity requirements can undermine the potential for emission reductions, especially where the power supply is not sourced from clean, renewable resources. Plasma-based technologies maximize efficiency when combined with sustainable conditions with low-carbon sources. These findings mark a paradigm shift from conventional waste-to-energy processes toward systems designed for both emission minimization and resource valorization. Hence, these results show that Plasma-based technologies offer a sustainable and energy-efficient alternative for waste management. By reducing dangerous emissions and maximizing resource recovery, they can contribute directly to long-term sustainability goals and align with global environmental targets.

5.3 Economic Feasibility

Economic outcome is dependent on the site's context; however, a recurring pattern is evident. Initial investment is typically greater than that of conventional systems, while active benefits come from enhanced devastation, refined syngas, and income from avoided tipping fees or sales of associated goods. Reported case studies show strong outcomes when feedstock quality, power prices, and gate-fee structures align in plasma's favor. For example, one plastic waste scenario reported a return on investment of nearly 80%, a payback period of about 1.5 years, and a gross margin of around 130%. These results should be treated as scenario-specific rather than universal expectations. At the municipal scale, economic conditions are shaped by several key factors: (i) feedstock composition and moisture content, (ii) electricity and gas takeout pricing, (iii) capital cost of the torch and power supply, and (iv) policy instruments such as tax credits, R&D grants, or performance-based incentives. Studies on residential-complex applications reach the same conclusion: the business case depends on local waste generation rates, energy prices, and optimized capital expenditure. So, in spite of the benefits offered by advanced plasma technology, their financial sustainability faces challenges due to high investment costs, energy consumption concerns, and scarce data on widespread application. Compared to standard, traditional approaches, plasma technology proves financially advantageous only within specific, limited applications. This analysis highlights the importance of cost management, standardized designs, and financial feasibility analysis, as shown in Table 1.

Table 1. Comparative analysis of plasma-based versus conventional waste management techniques across key metrics, based on literature review and technical reports

Parameter	Plasma-based technologies	Conventional methods (incineration / standard gasification / landfill)	Implication
Energy efficiency & process severity	Very high process temperatures (≈5,000–10,000 K); reports of up to 35–50% CO_2 reduction in certain industrial uses	Lower temperature, incomplete combustion, and secondary pollutants are more likely	Better emissions profile when energy and off-gases are well utilized
Products & residues	Clean syngas ($CO + H_2$); vitrified, inert slag	Flue gas with higher contaminants; ash requiring special disposal	Supports resource recovery and circular uses
Environmental performance	Lower pollutant formation including dioxins/furans; reduced landfill	Higher pollutant generation and ash volumes	Easier compliance with strict standards
Economics	Scenario-dependent. Attractive where feedstock and policy align; ROI up to ~80% reported for plastics	Typically lower CapEx; longer payback when dealing with heterogeneous waste	Policy and offtake contracts are decisive
Adoption barriers	Higher CapEx (often 30–50% above alternatives); technical complexity; low current penetration	Mature supply chains and skills-based	Incentives, skilled operators, and design standardization help close the gap

6. POLICY AND GOVERNANCE PATHWAYS

6.1 Regulatory Drivers and Obstacles

The adoption of Plasma-based technologies depends not only on technical feasibility but also on stable incentives and effective regulatory frameworks. Policies promoting renewable energy deployment, financial support for biomass and waste-to-energy projects, and strict emission reduction targets have proven instrumental in stimulating investment in gasification and Plasma-based systems. By aligning with environmental and sustainability policies, plasma technologies gain momentum as part of broader decarbonization and circular economy strategies. This section is consistent with our conceptual model by demonstrating the combination of plasma technologies with technological, sustainable, and policy elements.

Significant barriers remain; however, the high upfront capital required for plasma reactors, particularly cold plasma systems, limits accessibility for small and medium-sized enterprises (SMEs) and regions with restricted financing options. Inconsistent policy support and limited awareness among regulators also hinder large-scale deployment, and mostly, regulatory systems are centered on established conventional methods, prolonging the approval timeline and creating uncertainty for Plasma-based systems. Market analyses suggest that while plasma applications are expanding across industries such as textiles, polymers, semiconductors, agriculture, and healthcare, adoption continues to be constrained by financial and institutional obstacles (Ghosh, 2025).

Ultimately, the effectiveness of regulatory drivers will depend on balancing incentives and safeguards with targeted support for SMEs and emerging economies. Without dedicated financial mechanisms, the high initial costs of plasma technologies risk slowing their integration into industrial sustainability frameworks.

6.2 Industrial Adoption Incentives

Scaling up Plasma-based technologies depends on a targeted financial framework and incentive mechanisms that essentially reduce early capital risks while ensuring profitability over time. Established tools, such as feed-in tariffs, tax allowances, and direct subsidies, play a key role in promoting renewable energy and waste-to-energy systems by reducing operational expenditure and boosting revenues (Nchofoung, Fotio, and Miamo, 2023). When extended to Plasma-based systems, such policies can mitigate high capital cost while allowing firms to benefit from efficient waste-to-resource transformation.

Global standards indicate that purpose-driven policy structures are essential for the effective scaling of Plasma-based technologies. For example, the EU's Waste Framework Directive and Circular Economy Action Plan provide regulatory incentives to encourage industries to shift toward circular economic practices and employ advanced technologies (European Commission, 2020). When integrated with tools like R and D grants, preferential tariffs, or green procurement schemes, these frameworks enable the conquest of the obstacle between small-scale innovations and industrial-scale deployment. Even economic inducements, tax reliefs, and ecological grants can increase the uptake of Plasma-based technologies, industries are reluctant due to unpredictable investment risk and a lack of performance record. Aligning Plasma-based technologies with the national sustainable development framework and the UN Sustainable Development Goals (SDGs) can offer integrated solutions by enhancing renewable energy and restoring the natural ecosystem to build a prosperous and environmentally sound future. Enabling Frameworks Accelerate Plasma Uptake: Japan capitalized on energy recovery rewards while Germany leveraged EU mandates. Given the resource constraints and governance weaknesses, policies must be specifically adapted to their context. In this manner, incentive-based policy strategies, targeted subsidies, and funding can simultaneously overcome adoption barriers but also position plasma technologies at the core of a future low-carbon industrial framework.

6.3 Worldwide Collaboration and Knowledge Transfer

International collaboration serves as a core driver in the advancement of Plasma-based technology. Since plasma research requires a significant expenditure and large-scale infrastructure, international joint ventures enable countries and institutions to share their experience, economic resources, and facilities. These joint efforts advance progress in plasma applications while also facilitating competency building and information sharing simultaneously between developed and underdeveloped economies. One of the best examples is the International Plasma Nanotechnology Center of Excellence Network, which undertakes joint projects supported by national governments. Such initiatives illustrate how international corporations can be aligned with industrial requirements through government-sponsored R&D, technology implementation strategy for SMEs, and research partnership and knowledge-sharing agreement between academia, industry, and policymakers. These endeavors are highly worthwhile for SMEs, which are incapable of independently designing or implementing plasma systems. By integrating SMEs within global innovative networks, governments and research centers can ensure that plasma advancements benefit diverse sectors instead of concentrating them in a particular R&D group. Ensuring long-term and sustained international cooperation also requires strong and effective governance systems. Co-funded research projects and bilateral agreements foster mechanisms for long-term and enduring partnerships. This type of collaboration and polycentric model of governance would not only guarantee that the Plasma-based innovations are able to remain within the four walls of the research lab, but can also be applied to other subjects and industries.

Despite the growth of international partnerships and knowledge-sharing networks, the adoption of Plasma-based technologies does not persist equally owing to inequalities in research potential, financing, and framework. In many emerging economies, there are gaps in the provision of pilot-scale data, formalized methodologies, and training opportunities. Promoting international research cooperation and accessible technical infrastructure is essential for covering these gaps and boosting global deployment.

7.1 EMERGING APPLICATIONS OF PLASMA TECHNOLOGIES

Beyond its application in converting waste to power, plasma is finding substantial roles and innovative uses in sophisticated manufacturing, catalysis, and the production of low-carbon fuels. Plasma's diverse roles, from industrial scale to high-tech niche, are detailed in sections 7.1 and 7.2.

7.1.1 Sustainable Energy and Power Systems

Several solutions offered by plasma technologies can be described as emerging trends. Key research requirements include improving DC (Direct Current) switch performance in HVDC (High Voltage Direct Current) transmission systems and enhancing the durability of components in localized DC networks (Chemisana, 2011). Photovoltaic devices also need research into arc behavior and electrode–wall interactions. Other examples include plasma waste and biomass conversion and gas reforming, for which more efficient and economical plasma technologies need to be developed. Similarly, this technology is beneficial for producing new materials, energy valorization, and storage, which, in addition to offering unique material properties, also require improvements in efficiency and scalability. Plasma systems also offer potential in energy recovery, as well as in the development of lightweight components for additive manufacturing, which can improve industrial efficiency and resource conservation (Uriondo et al., 2015).

Although Plasma-based solutions offer innovative avenues for gas reforming and energy-saving transformations, their ability to be sustained is heavily influenced by how much power they consume and how long they last. Numerous small-scale experiments achieve significant performance levels, but become unreliable for large-scale industrial operation. This finding underscores the importance of enhanced energy efficiency, long-lasting, reliable elements, and renewable energy integration to attain sustainability benefits.

7.1.2 Plasma-Enabled Material Science and Optics

One of the most interesting applications of plasma lies in creating coatings of materials synthesized with new properties. Complex but controllable plasma chemistries have already been used to synthesize nanomaterials, such as graphene or gold nanoparticles, which provide better quality and higher yields as compared to conventional technologies (Martinu and Poitras, 2000). Smart windows, or switchable windows, which change their light transmittance (from transparent to translucent when potential, light, or heat are applied), will enable the design of climate-friendly building structures in the future (Khandelwal et al., 2017; Chemisana, 2011; Tavares et al., 2015). Another potential area of Plasma-based systems is in optics, where an optical element replaces standard lenses and mirrors. By tuning reflectivity or electron density, plasma lenses can operate across a wide range of frequency spectrum, from visible to terahertz radiation, without degradation risks which are inherent in conventional optics. These advances open opportunities in communications, sensing, and imaging technologies. Plasma offers non-destructive ways to fine-tune surface chemistry and morphology; however, several limitation still exists in achieving the required uniformity and scalability for industrial implications. An in-depth analysis indicates that optimizing production workflow is fundamental to successfully scaling up laboratory results for reliable manufacturing.

7.1.3 Transportation and Aerospace Innovation

Plasma applications in logistics focus on emission, reduction, and propulsion efficiency. Studies show that Plasma-assisted processes can significantly eliminate nitrogen oxides (NOx) and sulfur oxides (SOx) emissions when combined with catalytic converters (Hammer, 2014; Chen et al., 2009). Plasma may also assist and improve fuel combustion directly in internal combustion engines (Cathey et al., 2007). In aerospace, highly reactive and unstable chemical substances can enhance oxidation mechanisms, resulting in higher outputs and improved combustion efficiency, which would result in greater propulsion for both atmospheric flight and deep-space missions. For aircraft, plasma actuators may improve aerodynamic control by keeping airflows attached to wings during high-altitude and high-speed flight (Enloe et al., 2004; Corke et al., 2010). High-altitude, high-speed flight has increased interest in using plasma to increase maneuverability with plasma engines, extending mission ranges, and opening new opportunities for interplanetary exploration (Winglee et al., 2007).

Plasma-based technologies offer capabilities for producing lightweight surface coatings, enhancing the efficiency of fuel, and facilitating advanced material handling techniques, but large-scale industrial adoption is constrained by high cost and strict safety requirements. A detailed analysis shows that more research on the reliability of aerospace standards is required for practical execution.

7.1.4 Plasma for Surface Treatments of Polymers

Atmospheric Plasma-based systems offer economical prospects and practical solutions for surface treatment of polymers, textiles, and composites. Operating under room conditions with gases (e.g., CO_2, moisture, or water vapor), these systems virtually eliminate the need for ultra-high vacuum setups. Conventional corona treatment has been extensively studied for decades but is generally considered unfeasible and uneconomical for the textile industry (Ghazal et al., 2024). As an alternative, atmospheric pressure glow discharge (APGD) and diffuse coplanar surface barrier discharge (DCSBD) systems have been established. Specifically, DCSBD generates a high-density, low-temperature plasma in ambient air and has been shown to accelerate fabrics to speeds of up to 100 m/min, the highest speed reported to date (Samanta et al., 2006). These technologies represent alternative solutions for corona treatment, although their mechanisms and scalability remain under-researched, limiting wider adoption. Plasma surface modification enhances polymer bonding and surface chemistry alteration, but inequitable treatment and scalability problems are still the major limitations. An in-depth and critical analysis shows the need for highly advanced reactors to maximize the manufacturing outcomes.

7.1.5 Biomedical and Agriculture Applications

In the domains of biomedicine and agriculture, plasma technologies are recognised as revolutionary achievements that offer Plasma-based solutions for scenarios where standard approaches are intrinsically confined. In the medical field, one of the most important applications lies in plasma oncology, where reactive oxygen and nitrogen species (RONS) produced by thermal plasma can kill cancer cells without harming healthy tissue. This evades genomic resistance, offering complementary cancer therapies (Keidar et al., 2013; Ishaq et al., 2014). Plasma-enhanced therapy can also be studied for microbial treatments, with tangible outcomes against anti-infective therapies. Hydrogen peroxide gas plasma sterilization further indicates potential for sanitizing medical equipment, specifically by decontaminating pathogens such as prions, where traditional methods fail to meet the standard.

Plasma-based processes can also enhance nutrient extraction, pharmaceutical production, and biofuel generation from plant biomass. Their ability to inactivate pathogens makes them a sustainable alternative to chemical disinfectants, aligning with global efforts to promote eco-friendly farming practices.

Despite these advances, numerous research gaps remain:

- Standardization: limited clinical trials and agricultural field studies hinder regulatory approval and commercialization. The inconsistent testing standards make it unsuitable for comparison and validating the performance of plasma in practical situations. More collaborative and joint assessments are required to set trustworthy benchmarks and formal rules for broader implementation.
- Mechanistic understanding: biological pathways through which plasma interacts with cells, tissues, and plants require deeper study. Gaining insight into the formation and transport of reactive species and interconnection is beneficial for refining treatment variables and assuring reliable and reproducible results. Lacking this mechanistic transparency, Plasma-based applications may only be based on observations rather than evidence.
- Scalability: moving from laboratory demonstrations to large-scale deployment presents technical and economic challenges. Commercial plasma reactors need imperative and high stability, strong and durable designs of reactors, and economical management to work effectively. Moreover, combining these

operational setups into current industrial frameworks requires considerable financing and efficient process management and control.

- Safety and ethics: long-term impacts of plasma exposure on human health, food quality, and ecosystems remain insufficiently understood. Extensive toxicity evaluation and environmental observation are required to verify that species generated by plasma species do not create unforeseen issues. Moral and ethical guidelines must be responsible for the implementation of plasma-based technologies, particularly in farming, food manufacturing, and medical facilities.

Addressing these issues through interdisciplinary research that connects plasma physics, biomedical science, and agricultural engineering will determine the extent to which plasma becomes a mainstream technology in healthcare and sustainable agriculture. Plasma-based technologies are emerging solutions in disinfection, pest control, and soil steaming; however, issues endure in dose management, consistency, and broad implementation. To ensure safe and effective adoption in these sectors, a critical analysis confirms the need for standardized protocols and field-scale validation.

7.2 RESEARCH NEEDS AND KNOWLEDGE GAPS

Despite significant progress, Plasma-based technologies face scientific and technical limitations that hinder large-scale industrial adoption. These gaps span fundamental plasma physics, applied material processing, and plasma–chemistry interactions.

7.2.1 Fundamental Plasma Physics Challenges

Although thermal and non-thermal plasmas have been studied for decades, many processes remain poorly understood. Arc discharge behavior, molten metal–plasma interactions, and transient discharge phenomena such as arc quenching are not yet fully modeled or experimentally validated. Plasma–surface interactions and detailed molecular plasma chemistry also remain unresolved, particularly for non-equilibrium discharges. These limitations restrict the predictive capability of non-LTE (local thermodynamic equilibrium) plasma simulations. Advances in high-resolution diagnostics and atomic/molecular data generation are urgently needed to strengthen modeling accuracy and reproducibility.

7.2.2 Plasma-Based Optical Coatings and Surface Processing

Plasma coatings are already used commercially, but their performance and scalability remain constrained. Controlling ionization dynamics and excitation of gaseous species during deposition is essential for producing high-quality, stable coatings. Research priorities include:

- development of *in situ* diagnostics for real-time monitoring, which is beneficial for surface micro topography, deposition rate, and flux of film-forming species through plasma-based coating techniques. These methods help to maintain optical evenness and anticipate imperfections.
- advanced simulation of plasma surface chemistry, providing a profound understanding of sputtering and erosion mechanisms, annealing, and substrate-ion interactions that evaluate the system's optical efficiency. Such models increase the fine-tuning of coating conditions for strength and optical transparency.
- design of reliable, large-area plasma sources. Expanding plasma-based systems is essential for thin-film deposition on large-area optical instruments like mirrors, lenses, and display panels. Enhanced designs must certify minimized thermal stress across broad surface areas.

By combining experimental research with modeling, coating reliability and scalability can be improved, enabling broader industrial applications. Even though plasma techniques facilitate the production of highly accurate optical coatings, key difficulties exist in attaining stability and homogeneous film formation under different conditions. A critical analysis indicates that enhanced command over plasma power dynamics and extended operational assessment are required.

7.2.3 Automotive and Aerospace Applications

In safety-critical sectors such as automotive and aerospace, reliability and process control are essential. Plasma-enhanced combustion, surface modification, and material synthesis require advanced diagnostics and simulation tools that can replicate high-altitude, high-pressure, or chemically reactive environments. Particular attention should be given to atmospheric-pressure plasma (APP) sources, which offer lower costs but present unique challenges compared with low-pressure systems. Research into hybrid Plasma-assisted manufacturing, such as integration with PVD or PECVD, may provide transferable insights across industries. These Plasma-assisted systems provide advantageous surface finishes, but their adoption is hindered by mandatory compliance criteria and performance metrics. For the industry to formally endorse this, more evidence and a robust assessment framework are essential.

7.2.4 Polymer and Textile Surface Treatments

Plasma treatments for polymers and textiles continue to face challenges related to plasma uniformity and scalability. Filamentary discharges often produce inconsistent corona-based systems and are not optimized for large-area polymer finishing. Future research should focus on application-specific plasma sources that provide uniform treatment, use low-cost process gases, and meet industrial-scale requirements. Plasma treatments facilitate adhesion and improve the bonding strength of polymer and textile surfaces, but the results are highly variable, depending on the specific materials and process conditions used, as discussed in Table 2. Further validated research and broad-scale application guidelines are necessary to establish reliable operational outcomes.

7.2.5 Plasma–Chemistry Coupling and Interface Interactions

A comprehensive understanding of plasma–chemistry coupling is still lacking. Key parameters such as temperature, number density, and free electron energy must be correlated with plasma-induced chemical reactions and by-product formation. Improving energy efficiency, chemical selectivity, and reactor scalability are major research priorities. Comparative techno-economic assessments against advanced oxidation and water treatment methods are also critical for determining the true commercialization potential of plasma–chemical systems, also discussed in Table 2. Especially under complex conditions at large-scale setups, current knowledge regarding plasma chemistry interaction is still limited, and reactions are difficult to control. More in-depth experimental and modeling research is required to optimize interface performance and enhance process reliability.

Table 2. Current and emerging applications of plasma technology, under-researched areas, scale-up, and limitations

Area	Plasma contribution	Current limitation
Sustainable energy & power	Waste/biomass conversion, hydrogen routes, DC switching	Efficiency and scale; better arc modeling and component durability
Materials & optics	Nanomaterials, smart windows, plasma optical elements	Need miniaturized sources; atmospheric-pressure quality control
Transport & aerospace	NOx/SOx control, assisted combustion, propulsion, flow control	Real-condition efficiency and reliable diagnostics

continued on following page

Table 2. Continued

Area	Plasma contribution	Current limitation
Polymers & textiles	Open-air surface treatment (APGD, DCSBD)	Uniformity and durability; incomplete understanding of DCSBD
Hybrid systems	Plasma + catalysts/filters/CO_2 capture	Lab-scale maturity; energy optimization and LCA proof

8 RESEARCH IMPLICATIONS, CONTRIBUTIONS, LIMITATIONS, AND FUTURE DIRECTIONS

The analysis provided in this chapter holds significant research applications by highlighting how plasma technologies can assist in environmental and circular economic viability, sustainable industrialization through decarbonization, optimized energy output, waste valorization, resource recovery, and supportive governance frameworks. The findings highlight plasma's transformative potential to complement and, in many cases, surpass conventional approaches, positioning it as a critical enabler of the circular economy. Building on TRLs and comparative assessments, this chapter contributes significantly to existing literature by incorporating recent technological breakthroughs, including AI-powered plasma control systems, Hybrid Power Systems (HPS) with gasification, plasma-based mineralization, and advanced applications in optics and material surface engineering. Yet, there are still some drawbacks; as mentioned before, plasma techniques are still essentially valid in a laboratory setting, but their durability and financial factors have not yet been verified. This suggests a need for future research on mega projects, where plasma technologies are expected to play pivotal roles, for example, in energy transitions by supporting hydrogen production, power-to-gas systems, and renewable energy storage, and advanced manufacturing by enabling precision coating, surface engineering, MEMS/NEMS (Micro-Electro-Mechanical Systems/ Nano-electro-Mechanical systems) fabrication, and scalable additive manufacturing. It can also play an important role in transportation and aerospace engineering by providing flow control, lightweight materials processing, and next-generation propulsion systems. Further, advancements in the biomedical and agricultural sectors could help in delivering novel cancer therapies, pathogen control, and sustainable alternatives to pesticides and sterilization.

CONCLUSION

Plasma technologies are emerging as critical enablers of sustainable industrialization, offering advanced solutions for waste-to-energy conversion, high-performance material synthesis, environmental protection, and also play an important role in hard-to-treat waste streams. Plasma also offers high destruction efficiency as compared to other low-carbon treatments. Compared with conventional approaches, plasma systems achieve almost complete pollutant elimination, produce environmentally friendly by-products, and contribute directly to industrial decarbonization. Reviewed evidence confirms the growing momentum of plasma research, reflecting increasing potential from both academia and industry. The area of study mentions common benefits and shared drawbacks, specifically substantial energy demands; all of these findings also align with the three pillars of the conceptual framework

Despite this progress, large-scale adoption remains constrained by technical complexity, high investment costs, and insufficient policy support. Overcoming these challenges requires integrated strategies that connect technological innovation with financial incentives, regulatory alignment, and international collaboration. The ability to implement these technologies on a large scale is contingent upon improved energy efficiency, making electricity cost more affordable and a more robust regulatory, industrial supportive structure. Such efforts are particularly important for extending plasma adoption in regions where sustainable economic development is a priority. Future-oriented applications in advanced manufacturing, aerospace, biomedical engineering, and sustainable agriculture highlight plasma's transformative potential. Addressing fundamental challenges in plasma–chemistry coupling, plasma–surface interactions, and the operational stability of atmospheric-pressure systems will be essential to move from small-scale projects to large-scale deployment. Fair site location and civic management are also essential for the implementation of sustainable plasma-based technologies.

Plasma technologies represent more than incremental innovation; they embody a strategic shift toward cleaner, smarter, and more adaptive industrial systems. By embedding plasma innovations within circular economy frameworks and global sustainability agendas, industries can accelerate towards a resilient and low-carbon future. In conclusion, Plasma-based technologies stand at the frontier of sustainable industrial innovation. By combining advances in plasma science with supportive policy and international collaboration, they can become a milestone of future industries, reducing harmful emissions, conserving resources, and enabling cleaner, more resilient, and adaptive production systems worldwide. Plasma's industrial implementation is constrained by significant barriers, including capital expenditure and energy demands. Aligning plasma's drawbacks with its capabilities and mapping these barriers against plasma's capabilities helps identify its potential contribution to the sustainable industry.

REFERENCES

Arena, U. (2012). Process and technological aspects of municipal solid waste gasification. *Waste Management (New York, N.Y.)*, *32*(4), 625–639. DOI: 10.1016/j.wasman.2011.09.025

Bashir, M. A., Ji, T., Weidman, J., Soong, Y., Gray, M., Shi, F., & Wang, P. (2025). Plastic waste gasification for low-carbon hydrogen production: A comprehensive review. *Energy Advances*, *4*(3), 330–363. DOI: 10.1039/D4YA00292J

Bogaerts, A., & Neyts, E. (2024). Plasma-driven selective molecular dissociation and its role in sustainable industrial processes. *Plasma Sources Science & Technology*, *33*(1), 015004. DOI: 10.1088/0963-0252/33/1/015004

Bonizzoni, G., & Vassallo, E. (2002). Plasma physics and technology; industrial applications. *Vacuum*, *64*(3-4), 327–336. DOI: 10.1016/S0042-207X(01)00341-4

Cathey, C. K., Nunnally, W. C., & Kristiansen, M. (2007). Plasma-assisted combustion: Concepts and applications. *IEEE Transactions on Plasma Science*, *35*(6), 1638–1645. DOI: 10.1109/TPS.2007.910939

Chang, J. S., Lawless, P. A., & Yamamoto, T. (2002). Corona discharge processes. *IEEE Transactions on Plasma Science*, *19*(6), 1152–1166. DOI: 10.1109/27.125038

Chemisana, D. (2011). Building integrated concentrating photovoltaics: A review. *Renewable & Sustainable Energy Reviews*, *15*(1), 603–611. DOI: 10.1016/j.rser.2010.07.017

Chen, F., Cheung, M. C., Rempel, M., & Chintzoglou, G. (2023). Data-driven radiative magnetohydrodynamics simulations with the MURaM code. *The Astrophysical Journal*, *949*(2), 118. DOI: 10.3847/1538-4357/acc8c5

Chen, F. F. (1984). *Introduction to plasma physics and controlled fusion* (Vol. 1). Plenum press. DOI: 10.1007/978-1-4757-5595-4

Chen, L., Li, W., Liu, S., Tao, M., Long, Y., Duan, W., Zhang, C., Xiao, J., Qin, Q., Luo, K., Liu, J., & Liu, Y. (2009). Novel genetic markers derived from the DNA fragments of Sox genes. *Molecular and Cellular Probes*, *23*(3-4), 157–165. DOI: 10.1016/j.mcp.2009.03.001

Chiappim, W., Botan Neto, B., Shiotani, M., Karnopp, J., Gonçalves, L., Chaves, J. P., da Silva Sobrinho, A., Leitão, J. P., Fraga, M., & Pessoa, R. (2022). Plasma-assisted nanofabrication: The potential and challenges in atomic layer deposition and etching. *Nanomaterials (Basel, Switzerland)*, *12*(19), 3497. DOI: 10.3390/nano12193497

Corke, T. C., Enloe, C. L., & Wilkinson, S. P. (2010). Dielectric barrier discharge plasma actuators for flow control. *Annual Review of Fluid Mechanics*, *42*(1), 505–529. DOI: 10.1146/annurev-fluid-121108-145550

d'Agostino, R., Favia, P., Kawai, Y., Ikegami, H., Sato, N., & Arefi-Khonsari, F. (2005). *Plasma processing of materials: scientific and industrial applications*. Springer.

Eliasson, B., Hirth, M., & Kogelschatz, U. (1987). Ozone synthesis from oxygen in dielectric barrier discharges. *Journal of Physics. D, Applied Physics*, *20*(11), 1421–1437. DOI: 10.1088/0022-3727/20/11/010

Enloe, C. L., McLaughlin, T. E., VanDyken, R. D., Kachner, K. D., Jumper, E. J., Corke, T. C., Post, M., & Haddad, O. (2004). Mechanisms and responses of a single dielectric barrier plasma actuator: Geometric effects. *AIAA Journal*, *42*(3).

Evangelisti, S., Lettieri, P., Borello, D., & Clift, R. (2014). Life cycle assessment of energy from waste via anaerobic digestion: A UK case study. *Waste Management (New York, N.Y.)*, *34*(1), 226–237. DOI: 10.1016/j.wasman.2013.09.013

Ghazal, H., Allam, L. N., & Beltagy, Z. Sh. Abd Elaal, L. S., Elshamy, M. N., & Maraae, A. G. (2024). *The role of plasma technology in surface modification of textiles. Egyptian Journal of Chemistry, 67*(SI), 1–12. http://ejchem.journals.ekb.eg/

Ghosh, S. (2025, March 18). Cold plasma market: Growth, demand & forecast 2025 to 2035. [Market research report].

Hammer, T. (2014). Atmospheric pressure plasma application for pollution control in industrial processes. *Contributions to Plasma Physics*, *54*(2), 187–201. Advance online publication. DOI: 10.1002/ctpp.201310063

Hu, Y., Zhang, W., Han, J., Zhu, H., & Yang, Y. (2023). Design and Study of a Large-Scale Microwave Plasma Torch with Four Ports. *Processes (Basel, Switzerland)*, *11*(9), 2589. DOI: 10.3390/pr11092589

Huang, J., Guo, W., & Xu, P. (2006). *Thermodynamic study of water–steam plasma pyrolysis of medical waste for recovery of CO and* H_2. *Plasma Science and Technology, 7*(6), 3148. IPCC (2023) *AR6 synthesis report: climate change 2023*. Intergovernmental Panel on Climate Change, Geneva.DOI: 10.1088/1009-0630/7/6/018

Ishaq, M., Evans, M. D. M., & Ostrikov, K. K. (2014). Effect of atmospheric gas plasmas on cancer cell signaling. *International Journal of Cancer* 134(7):1517–1528. DOI: Shahzad, A. (Ed.). (2022). *Plasma science and technology*. IntechOpen. https://doi.org/DOI: 10.1002/ijc.28323

Keidar, M., Walk, R., Shashurin, A., Srinivasan, P., Sandler, A., Dasgupta, S., Ravi, R., Guerrero-Preston, R., & Trink, B. (2013). Cold plasma selectivity and the possibility of a paradigm shift in cancer therapy. *British Journal of Cancer*, *105*(9), 1295–1301. DOI: 10.1038/bjc.2011.386

Khan, N., & Abas, N. (2011). Comparative study of energy-saving lamps and plasma-based lighting technologies. *Renewable & Sustainable Energy Reviews*, *15*(1), 296–309. DOI: 10.1016/j.rser.2010.07.072

Khandelwal, H., Schenning, A. P. H. J., & Debije, M. G. (2017). Infrared regulating smart window based on organic materials. *Advanced Energy Materials*, *7*(14), 1602209. DOI: 10.1002/aenm.201602209

Lieberman, M. A., & Lichtenberg, A. J. (2005). *Principles of plasma discharges and materials processing* (2nd ed.). Wiley-Interscience., DOI: 10.1002/0471724254

Liston, E. M., Martinu, L., & Wertheimer, M. R. (1993). Plasma surface modification of polymers for improved adhesion: A critical review. *Journal of Adhesion Science and Technology*, *7*(10), 1091–1127. DOI: 10.1163/156856193X00600

Mallick, R., & Vairakannu, P. (2025). Life cycle assessment of plasma gasification integrated molten carbonate fuel cells and chemical-looping reforming using RDF feedstock. *Sustainable Energy & Fuels*, *9*(2), 627–639. DOI: 10.1039/D4SE01338G

Martinu, L., & Poitras, D. (2000). Plasma deposition of optical films and coatings. *Journal of Vacuum Science & Technology. A, Vacuum, Surfaces, and Films*, *18*(6), 2619–2645. DOI: 10.1116/1.1314395

Meichsner, J., Schmidt, M., Schneider, R., & Wagner, H. E. (2012). Nonthermal plasma chemical processes of general interest, Nonthermal plasma chemistry and physics.

Messerle, V. E., Ustimenko, A. B., Lavrichshev, O. A., & Nugman, M. K. (2024). The Gasification and Pyrolysis of Biomass Using a Plasma System. *Energies (19961073)*, *17*(22).

Munir, M. T., Mardon, I., Sulaiman, A.-Z., & Shawabkeh, A. (2019). Plasma gasification of municipal solid waste for waste-to-value processing. *Renewable & Sustainable Energy Reviews*, *116*, 109461. DOI: 10.1016/j.rser.2019.109461

Nauri, I. M., Andoko, A., Prasetya, R., Pasha, M. F., Akbar, M. R., Darmawan, M. W., & Puspitasari, P. (2024). Optimization of preparation parameters of palm oil-based nanofuel with multi wall carbon nanotube (MWCNT) for stability using Taguchi-grey relation analysis (GRA) combination. *Mechanical Engineering for Society and Industry*, *4*(2), 223–236. DOI: 10.31603/mesi.11945

Nchofoung, T., Fotio, H. K., & Miamo, C. W. (2023). Green taxation and renewable energy technologies adoption: A global evidence. *Renewable Energy Focus, 44*(4). OECD. (2024) *Eco-friendly manufacturing policies and technology integration.* OECD Publishing, ParisDOI: 10.1016/j.ref.2023.01.010

Olaoye, F., & Amlogu, R. (2024). Optimization of chemical reactor design using AI algorithms. *International Journal of Chemical Reactor Engineering*.

Panwar, N. L., Lanjekar, P. R., & Soni, K. (2025). Plasma gasification of municipal solid waste: A life cycle thinking perspective on energy, emissions, and economic feasibility. *Discover Sustainability*, *6*(1), 1164. DOI: 10.1007/s43621-025-01583-1

Ramos, A., Teixeira, C. A., & Roboa, J. P. (2019). Environmental assessment of municipal solid waste by two-stage plasma gasification. *Energies*, *12*(1), 137. DOI: 10.3390/en12010137

Samal, S., & Blanco, I. (2022). An Overview of Thermal Plasma Arc Systems for Treatment of Various Wastes in Recovery of Metals. *Materials (Basel)*, *15*(2), 683. DOI: 10.3390/ma15020683

Samanta, K., Jassal, M., & Agrawal, A. K. (2006). Atmospheric pressure glow discharge plasma and its applications in textiles. *Indian Journal of Fibre and Textile Research*, *31*(1), 83–98.

Subash, T. D., Babu, R. G., Josphine, C. V., & Subha, T. D. (2025). Optimization of absorber layer thicknesses for enhanced efficiency and stability in lead-free perovskite/silicon tandem solar cells. *Renewable Energy*, •••, 124642.

Tamang, S. G., Umlauf, G., & Barz, J. (2024). Degradation of PFOA solutions and PFAS-contaminated groundwater using atmospheric non-thermal plasma treatment. *Water Practice & Technology*, *19*(7), 2645–2654. DOI: 10.2166/wpt.2024.137

Tavares, P. F., Gaspar, A. R., Martins, A. G., & Frontini, F. (2015). The impact of electrochromic windows on the energy performance of buildings in Mediterranean climates: a case study. In *Eco-Efficient Materials for Mitigating Building Cooling Needs* (pp. 499–524). Woodhead Publishing. DOI: 10.1016/B978-1-78242-380-5.00018-2

Unnisa, S. A., & Hassanpour, M. (2017). Plasma technology and waste management. Resources. *Recycling & Waste Management*, *1*(1), 7.

Uriondo, A., Esperon-Miguez, M., & Perinpanayagam, S. (2015). The present and future of additive manufacturing in the aerospace sector: A review of important aspects. *Proceedings of the Institution of Mechanical Engineers. Part G, Journal of Aerospace Engineering*, *229*(11), 2132–2147. DOI: 10.1177/0954410014568797

Uthayakumar Azhagu, S., Kumaravel, S., & Arumugam, N. (2025). Computational modeling of arc plasma torches for welding and waste treatment applications. *Journal of Thermal Science and Engineering Applications*, *17*(3), 031001. DOI: 10.1115/1.4054210

Wang, L., & Liu, S. (2023). Metal Recovery and Separation from Wastes. *Metals*, *13*(8), 1411. DOI: 10.3390/met13081411

Wang, M., Yu, Y., Su, B., Zhang, W., & Yang, Q. (2025). Efficient Syngas Production from Medical Waste by CO2 Thermal Plasma Gasification. *Chinese Journal of Chemical Engineering*.

Winglee, R., Ziemba, T., Giersch, L., Prager, J., Roberson, B., & Carscadden, J. (2007). Advances in High Power Beamed Plasma Propulsion. In *38th AIAA Plasmadynamics and Lasers Conference In conjunction with the 16th International Conference on MHD Energy Conversion* (p. 4384). DOI: 10.2514/6.2007-4384

World Bank. (2024). *Industrial decarbonization pathways and financing mechanisms*. World Bank Publications.

Zhovtyansky, V., & Valinčius, V. (2018). Efficiency of plasma gasification technologies for hazardous waste treatment. In *Gasification for low-grade feedstock*. IntechOpen. DOI: 10.5772/intechopen.74485

Chapter 3
Integrating Circular Economy and Enterprise Risk Management for Diverse Industries:
Lessons From a Pottery SME Baseline

Souvik Banerjee
https://orcid.org/0000-0001-9988-8887
Management Development Institute, Murshidabad, India

Abhijit Pandit
https://orcid.org/0000-0003-2122-3468
Brainware University, Kolkata, India

Nabyendu Roy
https://orcid.org/0009-0002-6469-9685
Independent Researcher, India

Merina Banerjee
https://orcid.org/0009-0001-4974-5893
Management Development Institute, Murshidabad, India

ABSTRACT

This paper formulates and substantiates a pragmatic integration of enterprise risk management and the circular economy (ERM–CE), facilitating resource-limited organisations in diminishing their environmental impact while enhancing operational and financial outcomes. The study uses a field-tested ERM–CE framework that was

DOI: 10.4018/979-8-3373-6435-3.ch003

used in eleven small pottery businesses in West Bengal, India, as a starting point. It then adapts and translates the framework for use in other industries, such as textiles, food processing, electronics, and logistics. The baseline shows that aligning risk processes with life-cycle stages according to ISO 14040/14001 and using CE levers across procurement, production, distribution, consumption, and reverse logistics can lead to quantifiable improvements without needing more resources. In the pilot, 62–88 percent of pottery SMEs said their operations and finances got better at all stages of the business cycle, with an average improvement of 78 percent. Environmental sustainability got better for 61–100 percent of SMEs at all stages.

INTRODUCTION

Background Context

Climate change, global resource depletion, and increasing institutional accountability have transformed environmental sustainability into a strategic and financial need rather than a marginal social responsibility issue. Small and medium-sized firms, especially in emerging economies like India, face challenges in transitioning to sustainability due to restricted resources, disjointed governance frameworks, and established operational dependencies. In these contexts, the Circular Economy provides a means to alleviate environmental impacts via waste reduction, material circularity, and regenerative design (Geissdoerfer et al., 2017; Stahel, 2016). Nonetheless, circularity alone does not ensure continuity, scalability, or congruence with overarching organisational objectives. In numerous traditional clusters, such as pottery communities where artisans encounter compounded occupational, environmental, and health hazards, sustainability issues are interlinked with resource inefficiencies and exposure to airborne contaminants (Abo-el-Ata et al., 2023).

Enterprise Risk Management, in contrast, offers systematic governance methodologies for the identification, assessment, mitigation, and monitoring of risks that influence strategic decision-making and value generation (COSO, 2017; Nocco & Stulz, 2006). When integrated with circular economy principles, ERM may transform sustainability from a theoretical goal into a systematic operational practice ingrained throughout life-cycle phases, supply chains, and stakeholder interactions. ERM thus serves as a conduit between the normative expectations of regulators and markets and the practical reality of SME operations.

Rationale for Integration

Despite the increasing academic and policy focus on sustainability transitions, there exists a paucity of applied research illustrating the simultaneous operationalisation of ERM and CE, particularly in microenterprise settings characterised by constrained budgets, formal documentation, and digital capabilities (Brustbauer, 2014; Henschel & Lantzsch, 2022). The integration is notably pertinent as the adoption of a circular economy frequently presents new risks, such as uncertainties in reverse logistics, complications in supplier transitions, or increased compliance requirements, while concurrently alleviating other risks, including resource dependency, reputational exposure, and cost fluctuations (Rizos et al., 2016).

When SMEs endeavour to apply circular processes without a risk perspective, they often find it challenging to prioritise efforts, assess trade-offs, and maintain enhancements. The combination of ERM and CE is not only methodologically sound but also strategically imperative. It guarantees that environmental activities enhance corporate viability rather than undermine it. By integrating CE activities into organised ERM cycles, SMEs can cautiously experiment, expand successful interventions, and mitigate the risk of unanticipated adverse outcomes.

Research Gap

Current research often regards circular economy and enterprise risk management as parallel but unconnected domains. Research on circular economy focusses on modelling waste streams, reengineering materials and logistics, and assessing environmental consequences throughout product life cycles (Dey et al., 2020; Gonçalves et al., 2022; Ormazabal et al., 2018). The literature on Enterprise Risk Management (ERM) emphasises governance maturity, compliance, risk culture, and the performance implications of risk management frameworks, especially within financial and corporate settings (Fraser & Simkins, 2016; Hoyt & Liebenberg, 2015). Limited research investigates the interaction of these areas in real-world SMEs, and even fewer provide quantifiable results across several life-cycle stages in a resource-limited context.

Furthermore, much of the current research on Circular Economy (CE) and Enterprise Risk Management (ERM) implicitly presupposes the existence of extensive digital infrastructure, established governance frameworks, or robust institutional backing—conditions frequently lacking in conventional manufacturing clusters within developing nations (Chakraborti & Mishra, 2018; Ghazilla et al., 2015). This study fills existing gaps by creating and empirically validating a field-tested ERM–CE integration methodology, utilising baseline data from eleven ceramic SMEs, and subsequently translating the findings into cross-sectoral roadmaps.

Research Objectives

The primary objective of this study is to devise and implement a theoretically informed ERM–CE integration framework tailored for resource-limited SMEs. The study aims to illustrate that this integration can yield quantifiable enhancements in operational and environmental performance. It also seeks to convert this validated framework into sector-specific adaptation pathways for industries characterised by differing material intensity, regulatory scrutiny, and logistical intricacy, including textiles, food processing, electronics assembly, and logistics.

Research Questions

This study aims to address three primary questions. The initial focus pertains to the function of Enterprise Risk Management (ERM) in enhancing Circular Economy (CE) implementation and governance, enquiring how risk-based methodologies may delineate, prioritise, and oversee circular initiatives within Small and Medium-sized Enterprises (SMEs). The second examines which circular economy mechanisms produce quantifiable operational and environmental enhancements within resource limitations and identifies the life-cycle stages where the most substantial benefits arise. The third topic investigates the institutional, stakeholder, and policy mechanisms that facilitate long-term adoption and scalability, especially in contexts where ecosystem preparation and infrastructure are constrained.

Contribution of the Study

This research enhances academic knowledge and practical application in multiple aspects. This study presents one of the initial empirically validated ERM–CE frameworks evaluated within the context of Indian SMEs, illustrating that sustainability initiatives can yield concurrent financial and environmental benefits without significant capital investment. It offers policy-oriented and sector-specific frameworks that can be adapted across many industries and geographies, while acknowledging contextual variations. It positions ERM not merely as a compliance tool but as a catalyst for capability enhancement in circular transformation, linking risk governance to the development of dynamic capabilities and institutional flexibility (Teece, 2007, 2018).

Theoretical Framework

This section enhances the theoretical framework of the study by elucidating the rationale for the integration of ERM and CE and its implications for organisational decision-making, capability enhancement, and institutional responsibility.

Dynamic Capabilities Theory

Dynamic Capabilities Theory elucidates how businesses cultivate adaptable and learning-oriented behaviours to address uncertainty, resource limitations, and changing market situations (Teece, 2007, 2018). It highlights three interrelated mechanisms: sensing, seizing, and transforming. Within the framework of this study, sensing denotes the capacity of SMEs to discern risks, opportunities, and environmental indicators, including waste streams, supply vulnerabilities, and compliance obligations. Seizing refers to the ability to act on these insights by employing circular strategies such as material substitution, reuse, recovery, or process redesign. Transforming denotes the continual reorganisation of structures, procedures, and stakeholder relationships to institutionalise effective practices and eliminate unsuccessful ones.

In SMEs, characterised by brief learning cycles and direct leadership impact, dynamic skills dictate the success, stagnation, or failure of sustainability initiatives under operational challenges. ERM implements these skills by organising the sensing–seizing–transforming loop through methodical risk identification, analysis, treatment planning, and monitoring. The ERM–CE architecture presented in this research can be understood as a dynamic capability-building mechanism that directs circular efforts via a structured yet adaptable governance approach.

Institutional Theory

Institutional Theory elucidates organisational behaviour via coercive, normative, and mimetic influences that define legitimacy and appropriateness within a certain domain (DiMaggio & Powell, 1983). Coercive pressures stem from legal and regulatory mandates, encompassing waste management legislation, disclosure regulations, and environmental standards, such as those integrated under SEBI's BRSR framework or extended producer responsibility initiatives. Normative pressures arise from professional associations, academic frameworks, industry norms, and the progression of standards for ESG reporting and resource efficiency. Mimetic pressures arise when organisations imitate perceived leaders, including competitors or supply chain partners that implement circular models or practices aligned with ESG principles.

In emerging markets, institutional pressure frequently precedes technological preparedness. Companies are anticipated to adhere to new circular or ESG requirements despite the inadequacy of their data systems, infrastructure, and capabilities. ERM offers the operational rigour necessary to analyse these challenges, prioritise solutions, and incrementally develop competence. When combined with CE, ERM enables SMEs to perceive institutional requirements as defined risk categories instead of ambiguous obligations, thus converting compliance into concrete operational and reputational benefits.

Complementarity of ERM and CE

Despite their origins in distinct intellectual and practical domains, ERM and CE frameworks exhibit structural compatibility. The components of Enterprise Risk Management, including risk identification, risk analysis and prioritisation, treatment planning, and monitoring, can be aligned with Circular Economy enablers such as life-cycle assessment, material flow analysis, decision points for substitution and recovery, and circular performance metrics (ISO, 2006, 2015, 2018). Life-cycle assessment and hotspot mapping facilitate risk detection by highlighting concentrations of material usage, pollutants, or waste generation. Material flow analysis and circular economy decision points align with risk analysis and prioritisation, since they underscore trade-offs and chances for intervention. Treatment planning corresponds with certain circular techniques, including the substitution of hazardous products, reduction of consumption, reuse of components, and recovery of resources. Monitoring and governance relate to CE indicators, reporting systems, ESG dashboards, and feedback mechanisms that guarantee responsibility and facilitate course correction (Garza-Reyes, 2015; Tseng et al., 2018).

This alignment guarantees that CE is controlled, measurable, and scalable, rather than reliant on ad hoc activities or isolated projects. The ERM–CE integration paradigm provides a cohesive decision-making structure by superimposing circular interventions on a risk management foundation, thereby facilitating the achievement of both environmental and economic objectives.

Conceptual ERM–CE Integration Model

This study's conceptual approach establishes ERM as the foundational governance framework and CE as the transformative operational route. Enterprise Risk Management (ERM) delineates processes for risk identification, analysis, mitigation, and monitoring; Corporate Environmental (CE) integrates methodologies that tackle environmental and resource-related risks and opportunities within these processes. The integration is influenced by leadership culture, communication clarity, stakeholder alignment, trust networks, institutional and legal frameworks, as well as technology readiness and data accessibility. In the pottery cluster, artisan identity and intergenerational knowledge affected the speed of acceptance of new practices, while the accessibility of fundamental data on materials, energy consumption, and waste impacted the thoroughness of the analysis conducted (Galeazzo et al., 2014; Paul et al., 2014).

This model directs the methodological and empirical aspects of the investigation. It indicates that the integration of ERM–CE will be more successful when leadership transparently conveys objectives and trade-offs, stakeholders possess a shared sense of collective purpose, institutions deliver consistent signals via policy and regulation, and technological solutions align with the capabilities and limitations of the participating firms.

Theoretical Proposition

The study posits that when SMEs integrate Enterprise Risk Management with Circular Economy through intentional capability development and alignment with institutional expectations, they attain enhanced operational resilience, environmental performance, and regulatory compliance without necessitating substantial capital investment. The argument suggests that integration promotes a transition from reactive compliance to proactive circular transformation, allowing SMEs to regard sustainability not as an external need but as an internal strategic asset.

LITERATURE REVIEW

Evolution of Circular Economy Thinking

The circular economy has transitioned from a waste-reduction narrative to a regenerative and systemic industrial paradigm that prioritises resource efficiency, reverse logistics, and closed-loop product design (Geissdoerfer et al., 2017; Yong, 2007). Stahel (2016) emphasises that circular solutions, including reuse, repair, and remanufacturing, can prolong product lifespan and dissociate value generation from the exploitation of raw resources. Over time, circular economy (CE) thinking has become progressively integrated with comprehensive environmental policy frameworks and industrial strategies, such as the European Union's Circular Economy

Action Plan and national resource efficiency roadmaps (European Commission, 2024; Ministry of Environment, Forest and Climate Change, 2023).

Recent research demonstrates that circular business models can improve competitiveness, diminish supply chain vulnerability, and mitigate lifecycle consequences; yet, their implementation is inconsistent, particularly among SMEs (Ormazabal et al., 2018; Rizos et al., 2016). Technological disparities, legal uncertainty, restricted access to money, and organisational resistance often impede adoption. Corvellec et al. (2021) warn that the circular economy is not a universal solution and that conceptual limitations and implementation difficulties must be recognised. This study addresses these problems by integrating circular economy methods inside a pragmatic risk governance framework designed for the restrictions of small and medium-sized enterprises.

Circular Economy in SMEs: Opportunities and Barriers

Small and medium-sized enterprises can significantly gain from circular practices by enhancing resource efficiency, minimising production waste, and creating alternative revenue models, including product-service systems, reuse, repair services, and take-back initiatives (Dey et al., 2022; Gonçalves et al., 2022). Gandhi et al. (2018) and Garza-Reyes (2015) demonstrate that the integration of lean and green techniques can generate synergies in waste reduction and process efficiency, especially within manufacturing SMEs. Draghici and Iva cu (2022) contend that green manufacturing can serve as a precursor to more advanced circular solutions.

Notwithstanding these opportunities, SMEs encounter enduring obstacles. Research highlights restricted access to information and technical resources, the absence of standardised environmental accounting, lack of reverse logistics infrastructure, and resistance to change among leadership and staff as prevalent barriers (Chakraborti & Mishra, 2018; Ghazilla et al., 2015; Rizos et al., 2016). The implementation of circular economy practices in small and medium-sized enterprises generally occurs gradually. Preliminary actions often emphasise material replacement, enhanced process efficiency, and energy optimisation. Complex activities, such as building circular business models or engaging in industrial symbiosis networks, typically arise only after basic capabilities and relationships have been developed (Tseng et al., 2018).

Foundations of Enterprise Risk Management

Enterprise Risk Management frameworks, including ISO 31000 and COSO, underscore the methodical identification, evaluation, and surveillance of financial, operational, supply chain, regulatory, and reputational risks (Committee of Sponsoring Organisations of the Treadway Commission, 2017; ISO, 2018). Enterprise Risk Management (ERM) aims to include risk considerations into strategic planning and performance oversight, thereby diminishing the probability and consequences of negative occurrences while allowing organisations to capitalise on value-generating prospects (Fraser & Simkins, 2016; Nocco & Stulz, 2006).

Research demonstrates a significant correlation between the maturity of Enterprise Risk Management (ERM) and enhanced performance in small and medium-sized enterprises (SMEs), especially in volatile and dynamic contexts (Brustbauer, 2014; Henschel & Lantzsch, 2022). Hoyt and Liebenberg (2015) present empirical proof of the usefulness of Enterprise Risk Management (ERM) by correlating integrated risk management skills with enhanced market valuation. Fraser and Simkins (2016) and Furlan and Pacheco (2021) emphasise that implementation continues to be difficult, particularly when resource limitations, cultural obstacles, and insufficient skills impede formalisation. These issues are further exacerbated in conventional clusters and informal SME networks.

ERM and Sustainability Integration

There is increasing acknowledgement that risk management and sustainability must not be regarded as distinct areas. Banjo et al. (2022) illustrate that environmental risk factors are becoming increasingly pivotal to Enterprise Risk Management (ERM) in small and medium-sized enterprises (SMEs), whereas Dahlan and Nurhayati (2022) reveal that green innovation and ERM can collaboratively improve organisational performance. Lee et al. (2023) investigate the adoption of ESG in SMEs within developing economies, emphasising challenges associated with governance, disclosure, and capability deficiencies, and stressing the necessity for cohesive frameworks that connect risk management, sustainability, and reporting.

Nonetheless, the majority of research continues to examine ERM and CE independently, seldom considering them as an integrated operational model. Dey et al. (2019, 2020, 2022) and Garza-Reyes (2015) demonstrate that the integration of lean and green practices can enhance supply chain sustainability and triple-bottom-line performance. However, the explicit integration of circular economy techniques into a risk management framework remains insufficiently developed. The lack of such integration renders SMEs unclear regarding the prioritisation, sequencing, and scaling of sustainability initiatives without jeopardising operational efficiency. The ERM–CE paradigm established in this study rectifies this conceptual and practical deficiency.

Policy, ESG and Institutional Alignment

Recent changes in global policy frameworks have heightened the importance of circular economy and ESG integration in corporate governance. The EU Circular Economy Action Plan, India's National Resource Efficiency Policy, SEBI's Business Responsibility and Sustainability Report for listed entities, and ISO environmental standards collectively indicate heightened expectations regarding resource efficiency, lifecycle assessment, and waste accountability (European Commission, 2024; ISO, 2006, 2015; Ministry of Environment, Forest and Climate Change, 2023; Securities and Exchange Board of India, 2024). The OECD due diligence guidance underscores ethical corporate practices, supply chain transparency, and risk-based strategies on environmental and social concerns (OECD, 2018).

For SMEs, these institutional advancements signify both a challenge and an opportunity. On one side, they augment compliance responsibilities and reporting obligations. Conversely, they generate opportunities for market distinction, inclusion into ethical supply chains, and improved access to financing. Alignment with the United Nations Sustainable Development Goals, specifically SDG 9 (Industry, Innovation and Infrastructure), SDG 12 (Responsible Consumption and Production), and SDG 13 (Climate Action), offers an additional normative framework that connects ERM–CE integration to broader developmental outcomes (United Nations, 2015).

The Gap in Existing Literature

The literature identifies three interconnected limitations. There is a deficiency of integrated models that empirically operationalise Enterprise Risk Management (ERM) and Circular Economy (CE) concurrently, particularly in the context of small and medium-sized enterprises (SMEs). Secondly, several existing frameworks presuppose the presence of extensive digital, governance, and institutional infrastructure, which are frequently lacking in conventional clusters and low-capacity environments. Third, a limited number of studies measure both operational and environmental consequences resulting from such integration across various life-cycle stages.

This study fills these gaps by implementing a validated ERM–CE methodology in a ceramic SME cluster, utilising mixed-method data to assess changes in operational and environmental performance and to analyse behavioural and institutional dynamics. The findings are subsequently applied to other industrial sectors to evaluate transferability and necessary adaptations.

METHODOLOGY

Research Design

The research utilises a mixed-method approach that integrates qualitative inquiry, participatory action research, and quantitative assessment. This methodology guarantees contextual comprehension and quantifiable assessment of results throughout execution, adhering to established standards for case-based research and theory development in intricate organisational environments (Baxter & Jack, 2008; Ridder, 2017). The methodology was deliberately crafted to be lightweight, less intrusive, and appropriate for resource-limited SMEs.

The research strategy entailed establishing a baseline of operations and lifecycle activities, collaboratively devising ERM–CE interventions with SME owners and craftsmen, and subsequently applying these interventions throughout designated lifecycle stages. The impact of these interventions was assessed using a blend of observational, interview-based, and quantitative methodologies, facilitating triangulation and enhancing internal validity.

Context of Study

The pilot study was executed among eleven ceramic small and medium enterprises situated in West Bengal, India. These firms were chosen due to their significant material-resource requirements, considerable diversity in process norms, profound cultural and generational continuity, and constrained formal management frameworks. The cluster is marked by environmental and occupational health issues linked to burning techniques, dust exposure, and informal waste management, as recorded in similar ceramic contexts (Abo-el-Ata et al., 2023). These characteristics rendered the setting a suitable testing environment for examining ERM–CE integration under authentic restrictions.

Data Sources and Instruments

Various data collection methods were utilised to improve validity and triangulation. Semi-structured interviews with small and medium-sized enterprise owners, supervisors, and employees elicited perceptions of risk, production obstacles, cultural norms, and attitudes towards sustainability. Systematic observation and process mapping were employed to pinpoint waste, inefficiency, and risk exposure within the production and distribution chain. Weighted score sheets were created to record changes in performance throughout life-cycle stages, allowing respondents to indicate whether they observed an improvement, no change, or a decline in operational or environmental performance and to rate their confidence in these assessments. In certain instances, reflection diaries and informal discussions supplemented organised tools by documenting experiential learning and emerging concepts throughout the implementation phase.

All data were gathered with the voluntary involvement and informed agreement of small and medium-sized enterprise owners and employees. Anonymity was maintained in the reporting, and measures were implemented to guarantee that the findings could not be linked to specific organisations or individuals.

Variables and Measurement

Two principal performance dimensions were assessed: operational and financial enhancement, and environmental enhancement. Operational and financial enhancements encompassed modifications in efficiency, cost structure, yield, labour effort, and process stability. Environmental enhancement encompassed alterations in material utilisation, emissions, waste production, energy consumption, and perceived life-cycle impact. Respondents assessed performance at each life-cycle stage as better, unchanged, or deteriorated, and assigned a confidence level to each evaluation.

The integration of binary outcome categories and confidence weights was selected to mitigate random or speculative responses and to accurately represent the subjective certainty of respondents concerning observed changes. Weighted ratings were generated by multiplying categorical assessments by their corresponding confidence levels, resulting in a detailed representation of perceived influence across stages.

Analytical Process

The analysis was conducted in three parts. During the initial phase, binary responses from the five life-cycle stages—procurement, production, distribution, consumption, and reverse logistics—were consolidated to identify preliminary patterns of perceived enhancement. In the subsequent step, confidence-weight multipliers were utilised on these aggregates to enhance the signal and mitigate the impact of ambiguous responses. In the third phase, the weighted patterns were analysed through thematic cross-case comparison, connecting quantitative trends with qualitative observations and interview accounts.

This multi-phase analysis enabled the study to transition from descriptive reporting of improvement rates to interpretive elucidation of the reasons behind varying rates of progress, the influence of behavioural and institutional factors on outcomes, and the implications of these patterns for transferability to other sectors.

Validation

In accordance with reviewer feedback regarding methodological rigour, a validation matrix was created to compare three forms of evidence for each reported change: self-reported enhancements from scoring sheets, observed operational behaviour documented through process mapping and field notes, and corroborative or contradictory insights from cross-sector literature. When these sources converged, the findings were regarded as well validated; when they deviated, they were evaluated with caution and noted as context-dependent or for more investigation (Dey et al., 2019, 2022; Garza-Reyes, 2015).

Framework transferability was evaluated via sector mapping exercises, wherein the ERM–CE elements validated in pottery were conceptually applied to textiles, food processing, electronics assembly, and urban logistics, utilising existing literature on lean–green integration, circular strategies, and SME sustainability within these sectors (Dey et al., 2020; Gandhi et al., 2018; Paul et al., 2014).

Ethical Considerations

Participation in the study was voluntary, and respondents could quit at any moment without repercussions. All survey and interview data were anonymised, and no commercial or financial incentives were provided that could influence participation. The research honoured traditional craftsmanship and local community standards, aiming to establish ERM–CE integration as a beneficial addition rather than an external imposition.

Results

Table 1 shows the basic parts of the ERM–CE framework that were taken from the baseline and changed to be more useful to a wider audience. The idea is to connect ISO-aligned life-cycle stages to ERM routines and then match each junction with a CE lever that is possible given limited resources. The table is simple on purpose so that it can be used as a planning tool. The talk that follows will include examples from other sectors. By linking ERM routines with CE levers, we can get around known barriers to adoption. This is done by tying CE actions to leadership-driven risk processes, routines, and communication. This makes culture, context alignment, and iteration stronger, all of which the literature says are important for SMEs (Rizos et al., 2016; Ormazabal et al., 2018; Sharma et al., 2020; Nocco & Stulz, 2006; Sax & Torp, 2015; Yilmaz & Flouris, 2017; Furlan & Pacheco, 2021). The life-cycle anchoring follows the same reasoning as the ISO 14040/14001 standard utilised in the baseline.

Table 1. A transferable mapping of ERM routines to ISO-aligned CE life-cycle stages for resource-constrained firms

Life-cycle stage	ERM routine	Representative CE lever under constraints	Illustrative quick wins
Procurement	Risk identification and supplier risk analysis	Responsible sourcing, local material substitution, vendor diversification	Shift to regulated raw inputs; dual-source critical inputs to reduce disruption
Production	Risk analysis and treatment planning	Lean–green practices, energy efficiency, process yields, safe materials	Stove/kiln fuel upgrades, idle-time reduction, pigment substitution
Distribution	Performance monitoring and stakeholder communication	Load consolidation, route planning, modal shifts, packaging reduction	Neighborhood delivery pooling; low-emission vehicles where available
Consumption	Opportunity management and product–service design	Durability, reuse, refill, behavioral nudges	User instructions, design tweaks for longer use; micro-incentives
Reverse logistics	Risk evaluation and network development	Take-back design, disassembly, refill/reuse pathways	Partner with local collectors; scheduled returns synchronized with deliveries

Table 2 shows the pottery baseline's outcome structure, which shows the stage-by-stage profile of change that people in other fields may expect. The pattern—quick victories in consumption and distribution, slower progress in reverse logistics—fits with both the real-world evidence and the experience of many industries with small businesses on where coordination barriers are.

Table 2. Outcome profile from the pottery baseline that informs multi-industry adaptation

Stage	Share reporting enhanced operational/financial performance	Share reporting enhanced environmental performance	Noted friction point
Procurement	85%	–	Supplier practices and input substitution
Production	84%	–	Fuel and process changes
Distribution	73%	–	Routing and consolidation
Consumption	88%	100%	User behaviors and product tweaks
Reverse logistics	62%	61%	Take-back cost and coordination

Table 3 shows the mapping for four more industries and shows the most important risk categories and CE levers for first-wave adoption. These drawings aren't rules; they show how the same ERM–CE logic might be based on what really happens in a sector and what it can and can't do.

Table 3. Sectoral sketches for adapting the ERM–CE framework

Industry	Salient risks to map in ERM	High-leverage CE actions under constraints	Likely early indicators
Textiles & apparel	Chemical, water, labor, supplier reliability	Dye/pigment substitution, water reuse, cut-plan optimization, take-back pilots	Rework reduction, water/ power intensity, compliant supplier share
Food processing	Cold-chain, spoilage, packaging, regulatory	Cold-room setpoints, by-product valorization, lightweight packaging, local sourcing	Yield, waste-to-product ratio, material intensity, on-time delivery
Electronics assembly	Obsolescence, quality, hazardous substances	Design for repair/reuse, parts harvesting, ESD/quality controls, return loops	First-pass yield, RMA rate, recovery rate
Urban logistics	Routing, fuel costs, safety, client SLAs	Micro-hubs, route consolidation, EV two/three-wheelers, pack density	Drop density, km per drop, fuel cost per order, CO_2e per order

Figure 1 is a short bar chart that shows the stage-by-stage improvement percentages for operational and financial performance in the pottery baseline.

Figure 1. Stage-wise operational and financial performance enhancement in pottery SMEs

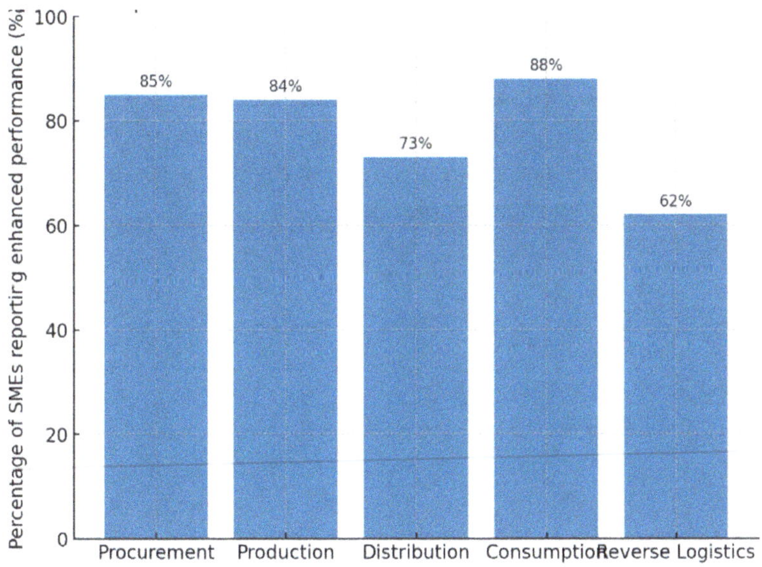

The way the bars are spread out shows why many small and medium-sized businesses (SMEs) start with the consumption and procurement stages. Reverse logistics, on the other hand, needs clear plans for dealing with risks and partners to get around cost and coordination problems.

Figure 2. Stage-wise environmental sustainability enhancement in pottery SMEs

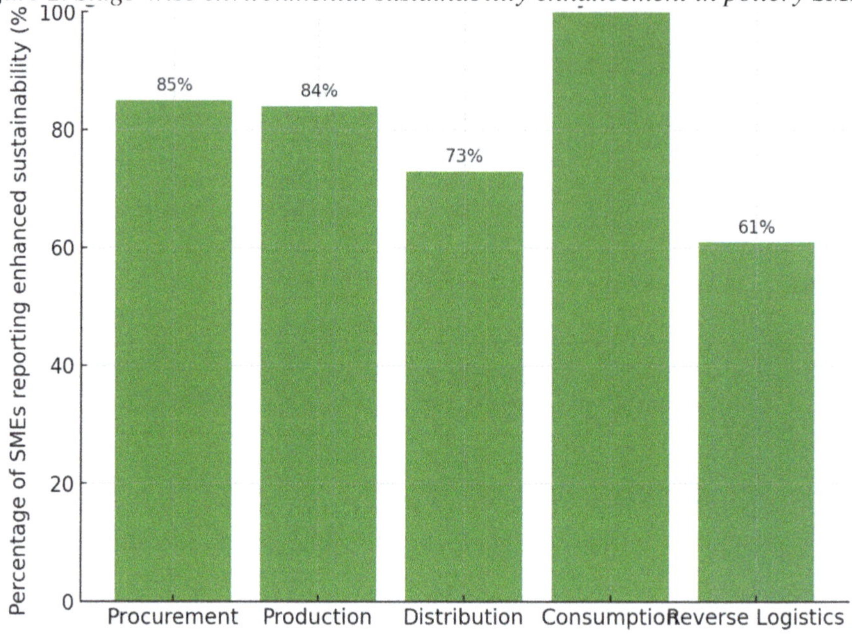

Figure 2 shows the percentage of pottery small and medium-sized businesses that said their environmental performance got better at each stage of the life cycle.

This again shows that cues during the use phase and changes to the product can happen quickly but take-back programs and recovery economics demand careful planning, phased trials, and sharing of risk.

Figure 3. Confidence-weighted impact scores for operational and financial performance

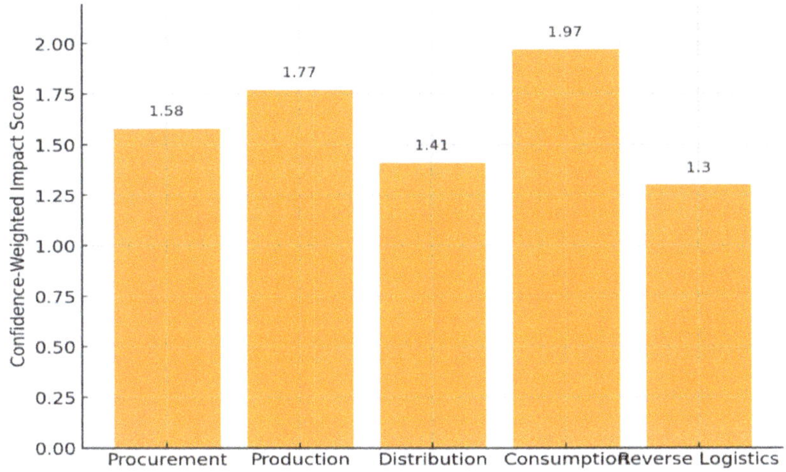

Figure 3 shows the confidence-weighted impact scores for improvements in operational and financial performance at different stages: The mean score is 1.61, with no-change and reduction weighted values average 0.30 and 0.19, respectively. The scores for procurement, production, distribution, consumption, and reverse logistics are 1.58, 1.77, 1.41, 1.97, and 1.30. This weighting improves signal quality by reducing outliers and bringing to light changes that owners were sure about after putting the measures in place.

Quantitative Outcomes

Implementation of the ERM–CE interventions led to measurable improvements across the five ISO-aligned life-cycle stages. In procurement, approximately 85 percent of respondents reported improved operational and financial performance, driven by more deliberate material selection, reduced wastage in purchasing, and better alignment between input quality and process requirements. In production, around 84 percent of respondents observed operational gains, reflecting more consistent process control and reduced rework. Distribution showed a lower but still substantial improvement rate of roughly 73 percent, as route consolidation and better coordination led to modest cost savings and reliability gains.

In the consumption stage, where product-use patterns and customer interactions were addressed, 88 percent of respondents reported operational benefits and all participants reported environmental improvements, particularly through design changes that reduced unnecessary material or improved durability. Reverse logistics exhibited the slowest progress, with about 62 percent of respondents perceiving operational improvements and 61 percent reporting environmental benefits. These figures reflect the complexity of organizing take-back practices and reuse systems in contexts where infrastructure and market mechanisms for recovered materials remain underdeveloped.

Across all stages, the average reported improvement was about 78 percent for operational and financial performance and about 76 percent for environmental performance. These results indicate that SMEs can achieve significant gains under resource constraints when CE practices are embedded within a structured ERM framework.

Confidence-Weighted Impact Scores

The use of confidence-weighted scores provided an additional layer of insight. In procurement, weighted improvement scores were high, with relatively low scores for no change or reduction, indicating not only frequent improvements but strong conviction among respondents that the changes were meaningful. Production followed a similar pattern, with the highest weighted improvement scores among all stages, suggesting that process-level interventions were both visible and well-understood. Distribution and reverse logistics showed more mixed but still positive patterns, with improvement scores exceeding no-change or reduction scores but with somewhat higher uncertainty. Consumption exhibited the strongest improvement profile, combining high improvement scores with low scores for reduction, reflecting a clear perception that product and communication changes had reduced environmental impacts and enhanced value for customers.

The mean improvement score across stages, after applying confidence weights, was about 1.61 on the chosen scale. This pattern confirms that reported improvements were not accidental or weakly held perceptions but reflected substantive and observable changes in practice.

Patterned Progression Across Stages

The results revealed a clear staged progression of change. Early gains emerged in procurement and consumption, where behavioural adjustments and relatively simple design choices produced immediate benefits without major infrastructure investment. Examples include more selective sourcing of clay and fuels, more careful storage and handling to reduce wastage, and modifications to product design that reduced material use while maintaining functionality.

Moderate gains emerged in production and distribution, where interventions required more technical adjustments or coordination. In production, changes such as better kiln loading practices, improved firing schedules, and systematic process checks led to more consistent output and reduced breakage. In distribution, improved route planning and coordination with buyers reduced transport inefficiencies and delivery failures. The slowest gains appeared in reverse logistics, where the absence of established take-back mechanisms, limited customer awareness, and weak markets for recovered materials constrained progress.

This sequencing aligns with global evidence indicating that reverse logistics and circular supply chains typically become viable only after upstream and midstream improvements stabilize and when supportive infrastructure and institutional arrangements emerge (Dües et al., 2013; Ormazabal et al., 2018; Rizos et al., 2016).

Sector Transferability Assessment

Cross-industry mapping suggested that the baseline ERM–CE pattern observed in the pottery cluster has broader relevance. In textiles, for example, procurement-stage gains can arise through more sustainable fibre choices and chemical substitution, mirroring the role of material shifts in pottery. In food processing, consumption-stage improvements can stem from packaging reduction, portion optimization, and shelf-life extension, reflecting similar dynamics to product design changes in pottery. Electronics assembly faces complex reverse logistics and end-of-life management issues that resemble the challenges pottery SMEs encounter when attempting to reclaim or repurpose discarded products. In urban logistics and last-mile delivery, distribution-stage optimizations such as route consolidation, modal shifts, and load optimization echo the transport-related improvements achieved in the pottery cluster (Dey et al., 2020; Gandhi et al., 2018; Tseng et al., 2018).

This pattern consistency strengthens the claim that the ERM–CE integration framework is transferable, although the specific levers and institutional conditions will necessarily vary across sectors.

Qualitative Observations

Qualitative data enriched the quantitative findings by illuminating behavioural and cultural dynamics. SME owners increasingly perceived circular economy activities as risk-reducing rather than cost-adding once these activities were framed within ERM logic and linked to concrete benefits such as reduced waste, more stable processes, and improved customer perception. Workers adapted rapidly to changes that simplified tasks or reduced avoidable rework but were more resistant to changes perceived as adding complexity without immediate payoff.

Cultural influences also played a significant role. Craft identity and pride in the quality and aesthetic of pottery products created a receptive context for improvements framed as protecting tradition while reducing waste and environmental harm. Informal networks of knowledge sharing, including family ties and peer-to-peer learning among artisans, accelerated the diffusion of successful practices. At the same time, the lack of documented procedures initially slowed implementation, as tacit knowledge had to be made explicit before it could be systematically improved. Over time, the normalization of ERM routines in the form of simple checklists, review meetings, and feedback loops helped institutionalize these changes, making them less dependent on individual champions.

DISCUSSION

Interpreting Results Through Dynamic Capabilities

The observed sequencing of improvements closely mirrors the sensing, seizing, and transforming mechanisms of Dynamic Capabilities Theory (Teece, 2007, 2018). In the early stages, SMEs sensed risks and opportunities by identifying waste hotspots, material inefficiencies, and regulatory vulnerabilities in procurement and consumption. As ERM processes structured these observations, firms were able to seize opportunities by implementing specific interventions in production and distribution, such as process adjustments and coordination improvements. Over time, attempts at reverse logistics and more advanced circular strategies signalled the beginning of transforming operations and stakeholder relationships, even though these efforts remained constrained by ecosystem readiness and infrastructure gaps.

This pattern suggests that circular economy initiatives are most successful when embedded within iterative learning cycles rather than imposed as one-time compliance exercises. ERM provides the scaffolding for these cycles by ensuring that sensing does not remain anecdotal, that seizing is prioritized and resourced, and that transforming is monitored and refined considering feedback.

Linking Outcomes to Institutional Pressures

Interview findings revealed growing awareness among SME owners of compliance risks related to emissions, waste burning, and occupational health, as well as market expectations from eco-conscious buyers and supply chain partners. These pressures correspond to the coercive, normative, and mimetic forces described by Institutional Theory (DiMaggio & Powell, 1983). Coercive pressures were evident in concerns about regulatory enforcement and potential sanctions. Normative pressures appeared in references to industry associations, NGO programmes, and academic guidelines that increasingly frame environmental performance as a criterion of responsible enterprise. Mimetic pressures emerged when SMEs compared themselves with peers or competitors that had adopted visibly greener practices or obtained certifications.

The introduction of ERM helped SMEs respond to these pressures systematically rather than reactively. Risk registers, mitigation plans, and review routines translated vague institutional expectations into concrete action items. This translation increased confidence among owners and workers in navigating policy and certification requirements, reducing the sense of being overwhelmed by external demands. The findings support the argument that institutional pressures become actionable and developmentally constructive when mediated through risk governance frameworks.

Why ERM Improved CE Adoption

The study confirms that ERM accelerated CE integration for several reasons. ERM provided a structured lens through which circular initiatives could be identified, evaluated, and sequenced. This structure helped SMEs avoid being overwhelmed by a proliferation of disconnected ideas and prevented misallocation of limited resources. By clarifying which risks and opportunities mattered most and at which life-cycle stages, ERM enabled prioritization of high-leverage, low-cost interventions that could deliver quick wins.

ERM also reduced uncertainty by framing circular actions as risk treatments rather than experiments detached from business objectives. This reframing increased willingness among owners to invest time and effort in CE initiatives, as they could see a clearer link between environmental and operational benefits. Documentation and communication routines associated with ERM strengthened alignment within the workforce and made it easier to maintain changes even when staff turnover or seasonal fluctuations occurred (Fraser & Simkins, 2016; Furlan & Pacheco, 2021).

Overall, the synergy between ERM and CE demonstrates that circular strategies are most effective when embedded within established decision-making frameworks rather than implemented as stand-alone projects.

Resource Constraint Advantage

While resource scarcity is often assumed to hinder sustainability transitions, the findings indicate that limited resources can, under appropriate governance, focus innovation. SMEs in the study adapted locally available materials instead of purchasing expensive new inputs, redesigned workflows rather than investing immediately in automation, and leveraged cultural and community knowledge as drivers of change. In this sense, the ERM–CE framework supported frugal and context-sensitive innovation, aligning with research showing that incremental, locally rooted solutions can be more sustainable and resilient than imported high-tech systems that do not fit local conditions (De et al., 2018; Ghazilla et al., 2015).

The case thus challenges deterministic assumptions that robust digital infrastructure and large budgets are prerequisites for meaningful circular transitions. Instead, it suggests that foundational risk governance and cultural alignment can unlock substantial gains even under tight constraints.

Behaviour and Culture as Hidden Variables

Qualitative evidence revealed that trust among craftspeople, shared identity within artisan clusters, intergenerational learning, and informal knowledge networks acted as hidden variables that significantly influenced the pace and depth of ERM–CE adoption. Where trust and shared purpose were strong, new practices spread more rapidly and met less resistance. Where relationships were fragmented or competition overshadowed collaboration, diffusion was slower and more uneven.

These socio-cultural dynamics functioned as silent catalysts for low-cost but high-impact transitions. They also imply that technical and financial interventions must be accompanied by relational and cultural strategies, such as peer-learning forums, recognition programmes, and co-created standards, if circular models are to be sustained over time.

Cross-Industry Translation Feasibility

The cross-sector comparisons suggest that the ERM–CE integration model is broadly adaptable, though the specific pathways and levers will vary. Industries with high material intensity, such as textiles and manufacturing, possess considerable scope for procurement and production-stage circular interventions, especially when combined with risk-based decision frameworks. Sectors with strong regulatory oversight, such as food processing and electronics, may experience powerful coercive and normative pressures that can be harnessed through ERM to accelerate CE adoption. Distributed logistics systems, including last-mile delivery, may benefit from ERM-informed circular strategies focused on route optimization, consolidation, and fuel-efficient operations. In sectors with weak ecosystem support, progress will likely be slower and more dependent on policy incentives, cluster-level infrastructure, and multi-stakeholder collaboration (Dey et al., 2020; Gandhi et al., 2018; Tseng et al., 2018).

IMPLICATIONS

Theoretical Implications

The study advances theory by explicitly linking ERM and CE in an SME context and demonstrating how this integration can be interpreted through the lens of dynamic capabilities and institutional theory. It shows that structured risk management processes can serve as drivers of circular capabilities, enabling firms to sense, seize, and transform in ways that align with external institutional expectations. The research also illustrates how coercive, normative, and mimetic pressures become operational when filtered through risk-based governance, offering a more granular understanding of institutional embeddedness in sustainability transitions.

Managerial Implications

For SME owners and operational leaders, the ERM–CE framework offers a practical roadmap that emphasizes low-cost, high-leverage steps. Managers can begin by focusing on procurement and consumption-stage interventions that yield quick, visible wins, such as material efficiency improvements and product design adjustments, and then use ERM cycles to progressively tackle more complex stages such as distribution and reverse logistics. Documenting and standardizing improvements through simple ERM tools builds repeatability and reduces dependence on individual champions. Engaging employees and stakeholders early, framing circular initiatives as risk-reducing and efficiency-enhancing, and establishing regular review mechanisms can reduce resistance and accelerate adoption.

Policy Implications

For policymakers and regulators, the findings suggest that ERM–CE integration can support CE transition efforts without imposing unsustainable burdens on SMEs. Incentive schemes can be designed to reward progression in risk governance maturity alongside compliance outcomes. Support for SME clusters can include shared reverse logistics infrastructure, resource centres for CE and ERM training, and accessible tools for ESG reporting. Expanding SEBI's BRSR guidelines in ways that accommodate SME realities, aligning NREP implementation with risk-based frameworks, and ensuring that EU-style circular requirements transferred through global value chains are accompanied by technical assistance can further enhance uptake (European Commission, 2024; Ministry of Environment, Forest and Climate Change, 2023; OECD, 2018; Securities and Exchange Board of India, 2024; United Nations, 2015).

Technology and Digital Strategy Implications

Although this study did not hinge on advanced digital infrastructure, the ERM–CE framework provides a platform for gradual digital augmentation. Once basic risk and circular routines are in place, SMEs can selectively adopt digital tools such as IoT sensors for material and energy monitoring, blockchain systems for supply chain traceability, AI forecasting for waste hotspots and demand prediction, and digital twins for lifecycle simulation and design optimization (Guo & Tsinopoulos, 2024; Tseng et al., 2018). These technologies should build upon existing routines rather than attempt to replace them wholesale. Aligning technology introduction with ERM cycles helps ensure that digital investments are targeted at the most material risks and opportunities, thereby increasing their likelihood of success.

Social and Cultural Implications

The study underscores the importance of social and cultural factors in shaping the success of ERM–CE integration. Trust, community learning, identity, and generational knowledge emerged as critical enablers in the pottery cluster, and similar dynamics are likely to exist in other traditional or networked SME sectors. Policy and managerial interventions that neglect these dynamics risk encountering resistance or superficial compliance. By contrast, peer-learning models, cluster-level demonstration projects, recognition-based motivation, and multi-stakeholder collaboration can align technical interventions with social realities, reinforcing both circular and risk management practices.

CONCLUSION

This study illustrates that the amalgamation of Enterprise Risk Management with Circular Economy concepts offers a pragmatic, scalable, and theoretically substantiated methodology for enhancing sustainability and operational resilience in resource-limited SMEs. The ERM–CE framework provides a systematic approach for shifting from linear production models to circular resource systems by integrating ISO 31000 and COSO-based risk governance with ISO 14040/14001 life-cycle thinking.

The field validation performed with eleven pottery SMEs in West Bengal demonstrated quantifiable enhancements in both operational and environmental performance. Substantial advancements were observed in initial interventions like procurement and consumption, however subsequent issues, especially reverse logistics, necessitated extensive ecosystem assistance and stakeholder collaboration. This adoption pattern closely resembles the sensing–seizing–transforming progression outlined by Dynamic Capabilities Theory and underscores the applicability of Institutional Theory in understanding CE implementation as a reaction to regulatory, market, and social pressures (DiMaggio & Powell, 1983; Teece, 2007, 2018).

In addition to the empirical findings, the research offers a versatile model applicable across several industries, such as textiles, food processing, electronics assembly, and logistics. The framework corresponds with international and national policy initiatives, including the EU Circular Economy Action Plan, India's National Resource Efficiency Policy, OECD due diligence guidance, SEBI's Business Responsibility and Sustainability Report framework, and the United Nations Sustainable Development Goals, thus facilitating both organisational and systemic sustainability results (European Commission, 2024; Ministry of Environment, Forest and Climate Change, 2023; OECD, 2018; Securities and Exchange Board of India, 2024; United Nations, 2015).

The research substantiates that sustainability and profitability do not have to be mutually exclusive objectives. Embedding circular economy activities inside established ERM processes enables SMEs to mitigate uncertainty, make educated trade-offs, and prioritise initiatives that provide both environmental and operational advantages. The paradigm facilitates gradual maturity via iterative implementation, learning, and institutional alignment, rather than necessitating substantial investment or extreme upheaval.

Limitations

The findings, while robust in pattern, are shaped by certain limitations. The study was conducted within one cultural and industry cluster, which may influence behaviour and adoption pathways in ways that differ from other settings. Measurement relied partly on self-reported confidence-weighted scoring, although this was mitigated through observation, process mapping, and triangulation with literature. Reverse logistics outcomes were strongly dependent on local infrastructure and ecosystem readiness, which may limit direct generalization to highly decentralized or fragmented supply chains. These limitations do not undermine the core findings but highlight contextual variables that should be considered in replication and scaling efforts.

Future Research Directions

Future research could build on this work in several directions. Longitudinal studies would help assess the durability of behavioural and operational improvements as well as the evolution of dynamic capabilities over time. Digital augmentation studies could explore how AI, IoT, blockchain, and digital twins interact with

ERM–CE frameworks to accelerate or reshape circular transitions, extending work by Tseng et al. (2018) and Guo and Tsinopoulos (2024). Financial modelling could quantify value creation, cost savings, and incentive alignment mechanisms arising from ERM–CE integration. Cluster and policy scaling studies could examine how government-backed circular ecosystems and risk-based support programmes accelerate adoption, while behavioural and cross-cultural comparative research could deepen understanding of how norms, identity, and governance structures influence adoption pathways across different regions and sectors.

Taken together, these lines of inquiry would contribute significantly to scaling circular transformation across industries, regions, and economic contexts. The integrated ERM–CE model is not just a framework; it is a transition roadmap. It shifts sustainability from a compliance-driven agenda to a strategic, operational, and cultural capability. As global policy, markets, and climate realities evolve, such frameworks will become increasingly essential tools for building resilient, circular, and future-ready enterprises.

REFERENCES

Abo-el-Ata, G. A., Abdellatif Abdelghany, F. H., Farghaly Ahmed, M., Abdelgwad, M., & Ramadan, M. A. (2023). Assessing neuropsychological disorders affecting pottery workers occupationally exposed to air pollutants. *Neurotoxicology*, *94*, 164–172. DOI: 10.1016/j.neuro.2023.01.011

Banjo, A., Aduloju, S. A., & Ajemunigbohun, S. S. (2022). Environmental risk and enterprise risk management in small and medium-sized enterprises: Evidence from Nigeria. *Annals of Spiru Haret University. Economic Series*, *22*(3), 217–240. DOI: 10.26458/22312

Baxter, P., & Jack, S. (2008). Qualitative case study methodology: Study design and implementation for novice researchers. *The Qualitative Report*, *13*(4), 544–559.

Brustbauer, J. (2014). Enterprise risk management in SMEs: A literature review and research agenda. *International Journal of Accounting & Information Management*, *22*(3), 297–316. DOI: 10.1108/IJAIM-02-2013-0008

Chakraborti, S., & Mishra, S. (2018). Environmental sustainability challenges faced by Indian SMEs: A review of policy and practice. *IUP Journal of Business Strategy*, *15*(1), 26–40.

Corvellec, H., Stowell, A., & Johansson, N. (2021). Critiques of the circular economy. *Journal of Industrial Ecology*, *25*(2), 359–368. DOI: 10.1111/jiec.13187

Dahlan, A., & Nurhayati, T. (2022). The effect of green innovation on organizational performance: The role of enterprise risk management as a mediation variable. *International Journal of Energy Economics and Policy*, *12*(4), 71–78. DOI: 10.32479/ijeep.13035

De, D., Chowdhury, S., Dey, P. K., & Ghosh, S. K. (2018). Impact of lean and sustainability-oriented innovation on sustainability performance of small and medium-sized enterprises: A data envelopment analysis-based framework. *International Journal of Production Economics*, *200*, 241–258.

Dey, P. K., Chowdhury, S., Rodríguez-Espíndola, O., Parkes, G., Tuyet, N. T. A., Long, D. D., & Ha, T. P. (2022). Circular economy and triple-bottom-line performance: Evidence from SMEs. *Journal of Business Research*, *149*, 278–293. DOI: 10.1016/j.jbusres.2022.05.036

Dey, P. K., Malesios, C., De, D., Chowdhury, S., & Abdelaziz, F. B. (2019). Could lean practices and process innovation enhance supply chain sustainability of small and medium-sized enterprises? *Business Strategy and the Environment*, *28*(4), 582–598. DOI: 10.1002/bse.2266

Dey, P. K., Malesios, C., De, D., Chowdhury, S., & Abdelaziz, F. B. (2020). Circular economy to enhance sustainability of small and medium-sized enterprises. *Business Strategy and the Environment*, *29*(6), 2145–2169. DOI: 10.1002/bse.2492

DiMaggio, P. J., & Powell, W. W. (1983). The iron cage revisited: Institutional isomorphism and collective rationality in organizational fields. *American Sociological Review*, *48*(2), 147–160. DOI: 10.2307/2095101

Draghici, A., & Iva cu, L. (2022). Pathways from green manufacturing to circular economy: A conceptual and empirical synthesis. *Sustainability*, *14*(1), 271. DOI: 10.3390/su14010271

Dües, C. M., Tan, K. H., & Lim, M. (2013). Green as the new lean: How to use lean practices as a catalyst to greening your supply chain. *Journal of Cleaner Production*, *40*, 93–100. DOI: 10.1016/j.jclepro.2011.12.023

European Commission. (2024). *Circular Economy Action Plan (2024 update)*. European Commission.

Fraser, J. R. S., & Simkins, B. J. (2016). The challenges of and solutions for implementing enterprise risk management. *Business Horizons*, *59*(6), 689–698. DOI: 10.1016/j.bushor.2016.06.007

Furlan, L. D., & Pacheco, A. S. V. (2021). Implementing risk management in organizations: Challenges and opportunities—A Brazilian case study. *Revista Ibero-Americana de Estratégia*, *20*(1), 1–20. DOI: 10.5585/riae.v20i1.18846

Galeazzo, A., Furlan, A., & Vinelli, A. (2014). Lean and green in action: Interdependencies and performance of pollution prevention projects. *Journal of Cleaner Production*, *85*, 191–200. DOI: 10.1016/j.jclepro.2013.10.015

Gandhi, N., Thanki, S., & Thakkar, J. (2018). Ranking the drivers for integrated lean–green manufacturing for Indian manufacturing SMEs. *Journal of Cleaner Production*, *171*, 675–689. DOI: 10.1016/j.jclepro.2017.10.041

Garza-Reyes, J. A. (2015). Lean and green—A systematic review of the state of the art literature. *Journal of Cleaner Production*, *102*, 18–29. DOI: 10.1016/j.jclepro.2015.04.064

Geissdoerfer, M., Savaget, P., Bocken, N. M. P., & Hultink, E. J. (2017). The circular economy—A new sustainability paradigm? *Journal of Cleaner Production, 143*, 757–768. DOI: 10.1016/j.jclepro.2016.12.048

Ghazilla, R. A. R., Sakundarini, N., Abdul-Rashid, S. H., Ayub, N. S., Olugu, E. U., & Musa, S. N. (2015). Drivers and barriers analysis for green manufacturing practices in Malaysian SMEs: A preliminary finding. *Procedia CIRP*, *26*, 658–663. DOI: 10.1016/j.procir.2015.02.085

Gonçalves, B. A., de Carvalho, F. M., & de Fiorini, P. C. (2022). Circular economy and financial aspects: A systematic literature review. *Sustainability (Basel)*, *14*(5), 3023. DOI: 10.3390/su14053023

Guo, H., & Tsinopoulos, C. (2024). Enabling a circular economy through green manufacturing in Chinese apparel manufacturers: Antecedents and outcomes. *IEEE Transactions on Engineering Management*, *71*, 9540–9554. DOI: 10.1109/TEM.2023.3296352

Henschel, T., & Lantzsch, A.-D. (2022). The relationship between ERM and performance revisited: Empirical evidence from SMEs. In Florio, C., Wieczorek-Kosmala, M., Linsley, P. M., & Shrives, P. (Eds.), *Risk management: Insights from different settings* (pp. 95–113). Springer., DOI: 10.1007/978-3-030-88374-4_5

Hoyt, R. E., & Liebenberg, A. P. (2015). Evidence of the value of enterprise risk management. *The Bank of America Journal of Applied Corporate Finance*, *27*(1), 41–47. DOI: 10.1111/jacf.12103

International Organization for Standardization. (2006). *ISO 14040: Environmental management—Life cycle assessment—Principles and framework*. ISO.

International Organization for Standardization. (2015). *ISO 14001: Environmental management systems—Requirements with guidance for use*. ISO.

International Organization for Standardization. (2018). *ISO 31000: Risk management—Guidelines*. ISO.

Lee, H., Kim, S., Park, J., & Yoon, B. (2023). ESG adoption constraints in developing-economy SMEs: A multi-country analysis. *Journal of Cleaner Production*, *412*, 137002. DOI: 10.1016/j.jclepro.2023.137002

Ministry of Environment. Forest and Climate Change. (2023). *National Resource Efficiency Policy (NREP): Policy framework 2023*. Government of India.

Nocco, B. W., & Stulz, R. M. (2006). Enterprise risk management: Theory and practice. *The Bank of America Journal of Applied Corporate Finance*, *18*(4), 8–20. DOI: 10.1111/j.1745-6622.2006.00106.x

OECD. (2018). *OECD due diligence guidance for responsible business conduct.* Organisation for Economic Co-operation and Development.

Ormazabal, M., Prieto-Sandoval, V., Jaca, C., & Santos, J. (2018). An overview of the circular economy among SMEs in the Basque country: A multiple case study. *Journal of Cleaner Production*, *220*, 333–343. DOI: 10.1016/j.jclepro.2019.02.045

Paul, R., Sahoo, A. K., & Routroy, S. (2014). Green manufacturing in small and medium enterprises: A case-study-based approach. *Procedia CIRP*, *15*, 325–330. DOI: 10.1016/j.procir.2014.06.021

Ridder, H.-G. (2017). The theory contribution of case study research designs. *Business Research*, *10*(2), 281–305. DOI: 10.1007/s40685-017-0045-z

Rizos, V., Behrens, A., van der Gaast, W., Hofman, E., Ioannou, A., Kafyeke, T., Flamos, A., Rinaldi, R., Papadelis, S., Hirschnitz-Garbers, M., & Topi, C. (2016). Implementation of circular economy business models by small and medium-sized enterprises (SMEs): Barriers and enablers. *Journal of Cleaner Production*, *132*, 70–82. DOI: 10.1016/j.jclepro.2015.11.093

Securities and Exchange Board of India. (2024). *Business Responsibility and Sustainability Reporting (BRSR): Updated reporting guidelines for listed entities*. SEBI.

Stahel, W. R. (2016). The circular economy. *Nature*, *531*(7595), 435–438. DOI: 10.1038/531435a

Teece, D. J. (2007). Explicating dynamic capabilities: The nature and microfoundations of (sustainable) enterprise performance. *Strategic Management Journal*, *28*(13), 1319–1350. DOI: 10.1002/smj.640

Teece, D. J. (2018). Business models and dynamic capabilities. *Long Range Planning*, *51*(1), 40–49. DOI: 10.1016/j.lrp.2017.06.007

Tseng, M.-L., Tan, R. R., Chiu, A. S. F., Chien, C.-F., & Kuo, T. C. (2018). Circular economy meets industry 4.0: Can big data drive industrial symbiosis? *Resources, Conservation and Recycling*, *131*, 146–147. DOI: 10.1016/j.resconrec.2017.12.028

United Nations. (2015). *Transforming our world: The 2030 Agenda for Sustainable Development*. United Nations.

Yong, R. (2007). The circular economy in China. *Journal of Material Cycles and Waste Management*, *9*(2), 121–129. DOI: 10.1007/s10163-007-0183-z

Zhu, Q., Sarkis, J., & Lai, K. H. (2007). Green supply chain management: Pressures, practices and performance within the Chinese automobile industry. *Journal of Cleaner Production*, *15*(11–12), 1041–1052. DOI: 10.1016/j.jclepro.2006.05.021

Chapter 4
Technology-Enabled Circular Economy

S. Haashni
https://orcid.org/0009-0004-2414-3839
Vimal Jyothi Convent Matriculation Higher Secondary School, India

T. C. Manjunath
https://orcid.org/0000-0003-2545-9160
Rajarajeswari College of Engineering, Bengaluru, India

K. L. Meera
Christ Academy Institute for Advanced Studies, India

S. Guru Basava Aradhya
https://orcid.org/0000-0002-3335-2681
Dr. D.Y. Patil B-School, Pune, India

P. K. Ranjitha
https://orcid.org/0009-0006-5550-4419
Jain University, India

M. R. Arun
https://orcid.org/0000-0002-8291-2956
Vel Tech Rangarajan Dr. Sagunthala R&D institute of Science and Technology, India

ABSTRACT

The concept of the circular economy has gained considerable global attention as a transformative framework for achieving sustainable development, reducing environmental impact, and enhancing economic resilience. Unlike the traditional linear model of "take-make-dispose," the circular economy is based on the regenerative principle of designing out waste and pollution, keeping products and materials in use, and regenerating natural systems. At its core, the circular economy aims to redefine growth by decoupling economic activity from the consumption of finite resources. It emphasizes a closed-loop system where resource input, waste, emissions, and energy leakage are minimized through intelligent design, reuse, remanufacturing, refurbishment, recycling, and upcycling. This approach not only extends the lifecycle of products but also facilitates the creation of a sustainable and restorative economic model that aligns economic, environmental, and social priorities.

DOI: 10.4018/979-8-3373-6435-3.ch004

UNDERSTANDING THE CIRCULAR ECONOMY: PRINCIPLES AND BENEFITS

The concept of the circular economy has gained considerable global attention as a transformative framework for achieving sustainable development, reducing environmental impact, and enhancing economic resilience. Unlike the traditional linear model of "take-make-dispose," the circular economy is based on the regenerative principle of designing out waste and pollution, keeping products and materials in use, and regenerating natural systems. At its core, the circular economy aims to redefine growth by decoupling economic activity from the consumption of finite resources. It emphasizes a closed-loop system where resource input, waste, emissions, and energy leakage are minimized through intelligent design, reuse, remanufacturing, refurbishment, recycling, and upcycling. This approach not only extends the lifecycle of products but also facilitates the creation of a sustainable and restorative economic model that aligns economic, environmental, and social priorities. One of the most compelling benefits of the circular economy lies in its potential to mitigate climate change and reduce ecological degradation. By extending the lifespan of products and materials and reducing the need for virgin resource extraction, the circular economy significantly lowers greenhouse gas emissions. Similarly, reducing food waste through improved storage, redistribution, and composting strategies not only alleviates pressure on agricultural systems but also minimizes methane emissions from landfills. The circular model also reduces environmental pollution, including plastic waste in oceans and toxic runoff from industrial processes, by promoting non-toxic, biodegradable, or recyclable materials and closed-loop production systems. Socially, the circular economy also presents significant advantages. It can contribute to greater equity and inclusion by enabling access to affordable products through sharing platforms and refurbished goods.

Community-based initiatives, such as repair cafés, tool libraries, and cooperative reuse schemes, empower individuals and foster collective responsibility for sustainable consumption. Additionally, by reducing dependency on imported raw materials and volatile commodity markets, the circular economy enhances national and regional self-sufficiency, which is particularly beneficial for developing countries or vulnerable economies. It calls for collaboration among diverse stakeholders—including governments, businesses, academia, and civil society—to co-create enabling conditions such as supportive infrastructure, financing mechanisms, education, and market access. Digital technologies such as the Internet of Things (IoT), blockchain, and artificial intelligence (AI) are increasingly playing a pivotal role in this transition, enabling better tracking of materials, optimizing reverse logistics, and enhancing circular value chains. In conclusion, the circular economy represents more than just a sustainable alternative to the linear model—it is a comprehensive, regenerative system that seeks to harmonize economic development with environmental preservation and social well-being. By prioritizing the responsible management of resources, extending product life cycles, and embedding restorative practices into industrial processes, the circular economy offers a viable pathway toward a low-carbon, inclusive, and prosperous future. While challenges remain, the opportunities for innovation, collaboration, and systemic transformation are vast, making the circular economy not only a practical necessity in the face of planetary limits but also an inspiring vision for a resilient and equitable global society.

THE ROLE OF TECHNOLOGY IN ENABLING THE CIRCULAR ECONOMY

The circular economy, with its core principles of eliminating waste, keeping materials in use, and regenerating natural systems, is deeply reliant on the intelligent deployment of technology. As the global community moves away from the traditional linear "take-make-dispose" model, technology has emerged as a vital enabler of this systemic shift. From smart product design and predictive maintenance to digital platforms facilitating reuse and remanufacturing, modern technologies play a pivotal role in unlocking the full potential of the circular economy. Innovations in digital tools, material science, artificial intelligence (AI), blockchain, and the Internet of Things (IoT) are accelerating the transition by increasing transparency, efficiency, and connectivity across circular value chains (Han Y., et al, 2023). These technological enablers not only facilitate better resource management but also create new business models, consumer behaviors, and governance mechanisms aligned with circular principles. Artificial Intelligence (AI) further amplifies the potential of the circular economy by enabling data-driven optimization and automation. AI algorithms can analyze vast amounts of data generated by IoT systems to forecast maintenance needs, predict product failures, and optimize resource utilization. In manufacturing, AI can help design products for disassembly and recyclability by analyzing material choices, lifecycle impacts, and end-of-life scenarios. Machine learning models are also used to improve sorting systems in recycling facilities, identifying and separating materials more accurately and efficiently than manual processes. Moreover, AI-driven platforms can support circular business models such as product-as-a-service or sharing economy solutions by managing dynamic pricing, user behavior analysis, and demand forecasting. Digital platforms are revolutionizing how circular practices are implemented at scale (Shojaei A., et al, 2021). Platforms that support product sharing, resale, rental, and repair empower consumers to access goods without ownership and reduce demand for new resources. For example, online marketplaces for refurbished electronics or peer-to-peer sharing apps for tools and vehicles make it easier to keep products in use longer. B2B platforms facilitate industrial symbiosis by allowing companies to trade waste materials, excess energy, or by-products that would otherwise be discarded. In the construction sector, digital material passports document the specifications and potential reuse of building components, enabling circular construction and deconstruction processes.

Material science and biotechnology are instrumental in developing new materials that are safer, more durable, and more compatible with circular systems. Innovations in biodegradable plastics, bio-based composites, and recyclable polymers allow for the creation of products that can return safely to the environment or be reused in multiple cycles without degradation. For instance, researchers are creating packaging materials from agricultural waste, fungal mycelium, or seaweed, offering sustainable alternatives to petroleum-based plastics. Enzyme-based recycling and bio-upcycling technologies are being developed to break down complex waste materials, such as textiles or multilayer plastics, into reusable components (Neri A., et al, 2023). These advances are critical in overcoming some of the technical barriers to circularity in industries heavily reliant on complex or hazardous materials. The integration of these technologies is what ultimately unlocks systemic circularity. When IoT sensors feed real-time data into AI systems, which then interact with blockchain-based supply chain ledgers and digital platforms, the result is a highly intelligent, transparent, and adaptive circular system. Businesses can automate reverse logistics, governments can enforce extended producer responsibility with precision, and consumers can participate in circular actions with ease and trust. This convergence of technologies—often referred to as "Industry 4.0" or "Digital Circular Economy"—is not just enhancing circular processes; it is enabling entirely new ecosystems of collaboration, innovation, and value creation. In conclusion, technology is not a peripheral enabler but a central pillar of the circular economy. It enhances the efficiency, transparency, and scalability of circular practices, bridging the gap between theory and implementation (Cagno E., et al, 2021). From sensor-based waste management systems to AI-driven remanufacturing, from blockchain-verified supply chains to bioengineered recyclable materials, technological innovations are creating new pathways for circularity in every sector. The strategic and ethical deployment of these technologies—supported by cross-sectoral collaboration and inclusive access—will determine how successfully societies transition to a truly circular and sustainable future. As we stand at the intersection of environmental necessity and technological possibility, embracing and innovating with technology is imperative for the realization of a regenerative economic model that serves both people and the planet.

DIGITAL PLATFORMS FOR RESOURCE SHARING AND COLLABORATION

Digital platforms have become foundational tools in advancing the circular economy by enabling large-scale resource sharing, collaborative consumption, and inter-organizational cooperation. At their core, these platforms leverage the power of digital connectivity, data analytics, and user-centric interfaces to facilitate the efficient use, redistribution, and regeneration of materials, products, services, and knowledge. By transforming how individuals, businesses, and communities interact with resources, digital platforms reduce the demand for virgin materials, minimize waste, and foster a culture of sharing over ownership (Abbas Y., et al, 2022). Their versatility allows them to operate across multiple sectors—ranging from transportation and housing to manufacturing and agriculture—effectively breaking down traditional silos and catalyzing collaborative ecosystems aligned with circular principles. One of the most visible impacts of digital platforms is in the sharing economy, where users access goods and services temporarily instead of owning them outright. Platforms such as Airbnb, Uber, and Zipcar exemplify how digital tools can unlock underutilized assets—homes, vehicles, tools, or even skills—by matching supply with demand in real time. These platforms reduce redundancy, increase asset utilization, and generate economic value from resources that would otherwise sit idle. In doing so, they decrease the environmental footprint associated with production and disposal while promoting more equitable access to goods and services. For example, car-sharing platforms reduce the number of vehicles needed per capita, thereby lowering emissions, traffic congestion, and resource consumption linked to car manufacturing (Reim W., et al, 2023). These platforms centralize data on material composition, repair history, performance metrics, and recovery options, enabling informed decision-making on reuse, remanufacturing, and recycling. For example, manufacturers can use PLM software to redesign products for easier disassembly, while service providers can use it to offer predictive maintenance and extend product life spans. By embedding traceability and accountability into product systems, PLM platforms foster circularity at scale and support compliance with regulations such as extended producer responsibility (EPR) and eco-design directives.

Digital repair and refurbishment marketplaces are also gaining popularity, facilitating circular consumption by extending the life cycle of products. Platforms like iFixit or Back Market connect users with repair guides, spare parts, or refurbished electronics, making it easier for consumers to repair instead of replace. These platforms empower individuals to take responsibility for product longevity while also supporting a growing ecosystem of technicians and small businesses specializing in repair services. In this context, digital platforms serve as educational and economic enablers, nurturing skills, reducing e-waste, and strengthening local economies. In agriculture, agri-tech platforms are fostering circularity by connecting farmers with suppliers, processors, distributors, and even consumers in digital ecosystems that minimize waste and optimize resource use (Hui W., 2025). Platforms that match supply with real-time demand reduce post-harvest losses, while apps that connect farmers with organic waste processors facilitate composting and soil regeneration. Additionally, precision farming platforms use data analytics to optimize water, nutrient, and pesticide application, reducing environmental impacts while maintaining productivity. These applications illustrate how digital collaboration can revitalize traditional sectors with circular principles. While the benefits of digital platforms for resource sharing and collaboration are extensive, their development and deployment must be approached with careful consideration. Issues such as data privacy, digital inclusion, platform monopolization, and technological lock-in can create barriers to equitable participation in circular systems. For example, if a platform excludes low-income users or small businesses due to subscription costs, lack of digital literacy, or limited internet access, its potential to promote circularity is constrained. Similarly, centralized control over user data or market access can create imbalances that stifle innovation and reduce user trust (Chowdhary S., et al, 2025). Therefore, digital circular platforms must be designed with principles of accessibility, interoperability, transparency, and community engagement. In conclusion, digital platforms are indispensable catalysts for resource sharing and collaboration in the circular economy. They enhance efficiency, transparency, and participation across diverse sectors by connecting people, products, materials, and data in innovative ways. From enabling peer-to-peer sharing and industrial symbiosis to powering lifecycle management and transparent supply chains, digital platforms bring circular economy principles to life on a practical and scalable level. By designing these platforms thoughtfully—with inclusivity, interoperability, and sustainability at their core—society can unlock new pathways to regenerative, low-carbon, and resilient futures. As digital technologies continue to evolve, their strategic integration into circular systems will remain a cornerstone of environmental and economic transformation.

ARTIFICIAL INTELLIGENCE FOR WASTE REDUCTION AND MANAGEMENT

Artificial Intelligence (AI) is emerging as a transformative force in the quest for sustainable development, particularly in the domain of waste reduction and management. As global populations grow and consumption patterns intensify, waste generation has reached unprecedented levels—posing severe environmental, health, and economic challenges. Traditional waste management systems, often reactive, labor-intensive, and inefficient, struggle to address the complexities of modern waste streams, including electronic waste, construction debris, plastic pollution, and organic matter. AI offers an intelligent, data-driven approach to optimize every stage of the waste lifecycle—from generation and collection to sorting, recycling, and disposal—ultimately enabling more circular, efficient, and environmentally sound systems (Coates J M., et al, 2025). At the waste generation stage, AI can be used to predict and reduce waste before it occurs. By analyzing data from consumer behavior, supply chain operations, and production processes, AI systems can identify inefficiencies and recommend adjustments to prevent waste. In manufacturing, for instance, machine learning algorithms help reduce material overuse and detect defects in real time, thereby minimizing production scrap and rework. In retail and food industries, AI can forecast demand with greater accuracy, enabling just-in-time procurement and inventory management that prevents overproduction and spoilage. These predictive capabilities empower businesses to shift from reactive to preventive waste strategies, aligning closely with circular economy principles that prioritize waste elimination at the design and planning stages. One of the most innovative applications of AI is in material flow analysis and circularity tracking. AI algorithms can process vast amounts of data from supply chains, production facilities, and waste management systems to map the flow of materials, identify points of leakage, and uncover opportunities for resource recovery (Pouri & Hilty, 2021). For example, AI can analyze construction and demolition waste to identify salvageable components such as steel beams, tiles, or wood panels, which can then be reclaimed and reused. In fashion and electronics, AI platforms help track product lifecycles and facilitate take-back schemes, resale, or remanufacturing. By making material flows visible and intelligible, AI enables decision-makers to close resource loops and design out waste more effectively.

E-waste management—an increasingly critical area—has also benefited from AI intervention. Electronic waste poses unique challenges due to its complex composition and hazardous components. AI systems are being used to automate the dismantling of electronic devices, identify valuable rare earth metals, and safely separate toxic materials. Machine learning models can forecast the obsolescence of electronic products, enabling proactive collection and recycling efforts. Moreover, AI helps design modular electronics that are easier to upgrade, repair, or recycle, reducing the accumulation of end-of-life devices (Fang B., et al, 2023). Companies using AI to manage e-waste are not only mitigating environmental harm but also capturing significant economic value from the recovery of high-value materials. In the broader scope of citizen engagement and behavioral change, AI-powered mobile applications and platforms are facilitating more sustainable choices. Apps that scan product barcodes and provide information on recyclability, carbon footprint, or disposal instructions empower consumers to act more responsibly. AI chatbots and voice assistants offer instant waste sorting guidance, reducing contamination and improving recycling rates. Gamified experiences and incentive programs driven by AI analytics encourage participation in community cleanups, circular exchanges, or zero-waste challenges. These interactive tools play a critical role in shifting mindsets and fostering a culture of environmental stewardship. In terms of global equity, developing countries often face barriers in accessing AI-based waste solutions due to limited digital infrastructure, funding, and technical expertise. International partnerships, capacity-building programs, and open-source AI tools can bridge this gap and enable low- and middle-income countries to leapfrog traditional waste systems (Huang & Koroteev, 2021). Pilot projects in smart waste management can serve as learning hubs, demonstrating scalable and context-specific applications of AI that align with local needs and resource constraints. In conclusion, Artificial Intelligence stands at the forefront of the evolution toward smarter, more sustainable, and circular waste management systems. Its applications span the entire waste lifecycle—preventing unnecessary waste generation, optimizing collection and sorting, enhancing material recovery, and informing strategic decisions. By enabling precision, efficiency, and scalability, AI not only reduces the environmental and economic costs of waste but also paves the way for regenerative practices that contribute to a healthier planet. To fully harness this potential, it is essential to foster inclusive innovation, ethical AI development, and multi-stakeholder collaboration. In doing so, AI can transform waste from a global challenge into a valuable resource and a cornerstone of circular prosperity.

INTERNET OF THINGS (IOT) FOR RESOURCE EFFICIENCY

The Internet of Things (IoT) is rapidly transforming the way resources are monitored, managed, and optimized across industries, making it a critical enabler of resource efficiency in a world grappling with climate change, resource depletion, and unsustainable consumption patterns. By embedding sensors, actuators, and connectivity into physical assets—ranging from household appliances and vehicles to industrial machinery and infrastructure—IoT allows for real-time data collection and intelligent decision-making. These capabilities enable the dynamic management of energy, water, materials, and waste, helping businesses, cities, and consumers reduce resource consumption, increase productivity, and minimize environmental impacts (Alsabt R., et al, 2024). As global economies transition toward more circular and sustainable models, IoT stands at the forefront of innovation, offering the digital infrastructure required to monitor and optimize resource flows continuously and automatically. In the energy sector, IoT has revolutionized how power is generated, distributed, and consumed. Smart grids—networks of connected devices and sensors across the electrical grid—enable real-time load balancing, remote fault detection, and dynamic pricing models that encourage off-peak energy use. By integrating renewable energy sources like solar and wind, smart grids stabilize supply and demand with minimal human intervention. IoT-enabled thermostats, lighting systems, and HVAC controls in buildings can adjust automatically based on occupancy, temperature, and energy price signals, significantly reducing unnecessary energy consumption. In industrial settings, predictive maintenance driven by IoT sensors ensures that machinery operates efficiently and avoids costly energy losses due to malfunctions or wear (Yu K H., et al, 2021). In manufacturing and industrial processes, IoT technologies enable precision monitoring and control over resource-intensive operations. Connected machinery can track raw material usage, detect bottlenecks, and adjust parameters in real time to minimize scrap and waste. IoT sensors embedded in production lines help reduce downtime through predictive maintenance, ensuring that equipment operates at optimal performance and reducing energy use. In the context of Industry 4.0, digital twins—virtual replicas of physical systems powered by real-time IoT data—allow engineers to simulate, test, and optimize processes digitally before implementation, saving both materials and energy. Additionally, IoT supports closed-loop production by tracking materials through multiple lifecycles and facilitating their reintegration into the manufacturing process.

In construction and building management, IoT technologies enable resource-efficient design, operation, and retrofitting. Smart building systems can automatically adjust lighting, ventilation, and temperature settings based on occupancy and environmental conditions, significantly lowering energy consumption. IoT-enabled sensors embedded in building materials can track structural health and material degradation, informing timely repairs and extending the lifespan of assets. During construction, IoT systems optimize material delivery and usage, preventing over-ordering and waste generation. Furthermore, digital material passports powered by IoT data provide insights into the composition, history, and reuse potential of building components, facilitating circular construction and deconstruction practices. Waste management and recycling are other key areas where IoT supports circularity and efficiency (Jagtap S., et al, 2021). Smart bins equipped with fill-level sensors notify waste collectors when they need to be emptied, reducing unnecessary collection trips and optimizing routing. IoT devices in recycling facilities monitor material composition and equipment performance, increasing the quality and efficiency of sorting and processing. RFID tags and GPS tracking help monitor the movement of waste materials, ensuring accountability and enabling more effective reuse or recycling. These systems contribute to a cleaner environment and more efficient resource recovery by reducing contamination and improving logistics. Despite its vast potential, the implementation of IoT for resource efficiency is not without challenges. Cybersecurity is a major concern, as the increased connectivity of critical systems opens new avenues for cyberattacks. Ensuring data protection and system resilience requires robust encryption, authentication protocols, and regular system audits. Interoperability is another issue, as many IoT devices and platforms use proprietary protocols that hinder seamless integration. Developing open standards and regulatory frameworks is essential to ensure compatibility and encourage widespread adoption. Furthermore, digital equity remains a barrier, especially in developing regions where infrastructure, connectivity, and technical expertise may be lacking. Addressing these challenges requires collaborative efforts among governments, industries, academia, and civil society to create inclusive, secure, and interoperable IoT ecosystems (Avalekar U A., et al, 2025). In conclusion, the Internet of Things is a powerful enabler of resource efficiency across sectors and scales. From optimizing energy use in smart buildings to enhancing water conservation in agriculture, from reducing waste in manufacturing to improving logistics in supply chains, IoT empowers stakeholders to make informed, real-time decisions that conserve resources, cut costs, and reduce environmental footprints. As the world moves toward more sustainable and circular models of development, IoT offers the digital intelligence necessary to turn ambitious environmental goals into actionable, measurable outcomes. By investing in secure, inclusive, and integrated IoT solutions, societies can harness their full potential to build a future defined by efficiency, resilience, and ecological balance.

CLOSED-LOOP PRODUCTION AND RECYCLING SYSTEMS

Closed-loop production and recycling systems represent a cornerstone of the circular economy, redefining the traditional model of industrial manufacturing and consumption. In essence, a closed-loop system ensures that products, materials, and resources are continually cycled back into the production and consumption chain, rather than being discarded as waste after use. This regenerative approach minimizes resource extraction, mitigates environmental harm, and fosters economic sustainability by maintaining the value of materials and components for as long as possible. Unlike the linear “take-make-dispose” model, closed-loop systems strive for zero waste, where outputs from one process become inputs for another, creating a self-sustaining loop of materials and energy. One of the key enablers of closed-loop production is eco-design or design for disassembly, which involves creating products that are easy to take apart, repair, upgrade, and recycle (Lokhande M., et al, 2025). This design philosophy considers the entire lifecycle of a product—from raw material sourcing and manufacturing to usage, recovery, and reintegration into new products. Modular construction, standardized components, and minimal material diversity are common strategies used in eco-design to facilitate efficient recycling and reuse. For instance, companies in the electronics and automotive sectors are increasingly adopting modular designs that allow damaged parts to be replaced without discarding the entire product, thus extending product life and enabling component-level recycling. Industrial symbiosis—a system where the waste or by-product of one industry becomes the input for another—is an exemplary model of closed-loop thinking. This collaborative approach is often seen in eco-industrial parks, where companies co-locate to share resources such as water, energy, and materials. For example, in Kalundborg, Denmark, a pioneering eco-industrial park has created a network where excess heat from a power plant is used by a fish farm and district heating system, while gypsum from flue gas desulfurization is used by a wallboard manufacturer (Tanveer M., et al, 2022). Such symbiotic relationships reduce environmental impact, cut operational costs, and increase systemic efficiency. In practice, many industries are successfully implementing closed-loop systems to improve sustainability and reduce costs. In the textile sector, brands like Patagonia, H&M, and Eileen Fisher have launched garment take-back and recycling programs, where old clothing is collected, sorted, and either resold, upcycled, or fiber-to-fiber recycled into new garments. In the automotive industry, companies like Renault and BMW operate remanufacturing plants that recover valuable components—such as engines, transmissions, and electronics—refurbish them to like-new condition and reintegrate them into new vehicles. The electronics industry is also moving toward closed loops through initiatives such as Apple's Daisy robot, which disassembles iPhones for component recovery.

Figure 1. Circular economy business models

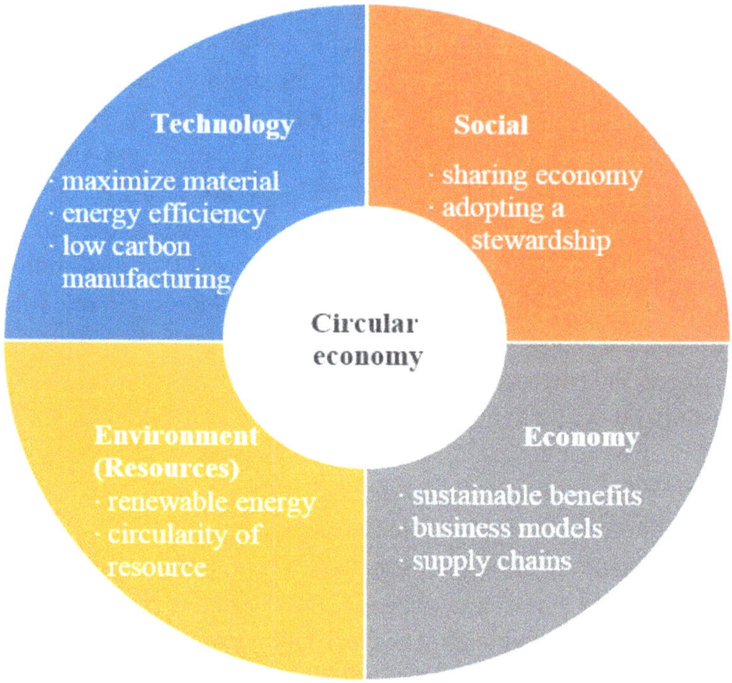

Despite its advantages, implementing closed-loop systems presents several challenges. First, the economic feasibility of recycling varies significantly across materials, depending on market prices, technological efficiency, and regulatory incentives. For some materials, especially mixed or contaminated waste streams, the cost of recycling exceeds the value of the recovered material, creating disincentives for collection and processing. Second, product complexity can hinder effective disassembly and material separation. Many modern products, such as smartphones or composite packaging, are made from multiple materials tightly bonded together, making them difficult and expensive to recycle. Governments and international bodies are increasingly recognizing the importance of closed-loop systems in achieving sustainability goals (Farhan L., et al, 2021). The European Union's Circular Economy Action Plan prioritizes product durability, repairability, and recyclability, while setting ambitious targets for packaging and electronic waste recycling. China's Circular Economy Promotion Law mandates resource recovery and recycling in industrial operations. In India, the E-Waste Management Rules compel producers to set up collection centers and ensure environmentally sound recycling of electronic waste. These regulatory frameworks are crucial in creating an enabling environment for closed-loop implementation across sectors. In conclusion, closed-loop production and recycling systems are a transformative component of the circular economy, offering a pathway to decouple economic growth from resource depletion and environmental degradation (Mao W., et al, 2021). By redesigning products for recovery, investing in advanced recycling technologies, leveraging digital tools, and fostering systemic collaboration, societies can build resilient, efficient, and low-waste economies. While technical, economic, and behavioral challenges remain, the momentum toward closed-loop thinking continues to grow across industries and regions. The transition to such systems is not only a moral imperative in the face of global environmental crises but also a strategic opportunity to unlock innovation, generate green jobs, and create a truly sustainable future.

ENVIRONMENTAL BENEFITS OF TECHNOLOGY-ENABLED CIRCULAR ECONOMY

The integration of advanced technologies into circular economy frameworks has significantly amplified their environmental benefits, transforming how societies consume, produce, and manage resources. By shifting from the traditional linear model of "take-make-dispose" to a circular model powered by digital innovation, economies are not only reducing waste and conserving natural resources but also tackling some of the most pressing environmental issues of our time—such as climate change, pollution, and biodiversity loss (Hagelüken & Goldmann, 2022). The technology-enabled circular economy leverages tools like artificial intelligence (AI), Internet of Things (IoT), blockchain, robotics, big data, and digital platforms to improve resource efficiency, minimize emissions, and drive sustainability across industrial sectors and supply chains. In addition to waste management, resource conservation is another major environmental gain facilitated by circular technologies. Smart material tracking systems, enabled by IoT and blockchain, provide real-time visibility into the flow of raw materials, components, and products. This enhanced traceability ensures that materials can be efficiently recovered and reused, reducing the need for virgin material extraction. For example, mining and logging—two resource-intensive activities—can be offset by the recovery of metals from electronic waste or wood from deconstructed buildings. Technologies like sensor-based sorting and automated disassembly systems further improve the precision of material recovery, ensuring that high-quality materials are preserved and reintroduced into production cycles with minimal degradation (Rosenberg S., et al, 2023). Another substantial benefit lies in pollution prevention. Circular systems, supported by technology, limit pollution at every stage of the product life cycle. IoT sensors in manufacturing plants can monitor emissions in real-time, enabling operators to make immediate adjustments to reduce pollutants like NOx, SOx, and particulate matter. Smart packaging technologies, such as biodegradable sensors or RFID tags, help track the condition and shelf life of products, minimizing food waste—a major source of methane emissions. Similarly, technologies in chemical recycling can decompose complex plastics into their original monomers, reducing plastic pollution in oceans and waterways. Robotics and automated systems are also being deployed for waste sorting, ensuring that hazardous substances are safely isolated and treated, thereby protecting soil and groundwater from contamination.

The preservation of biodiversity and ecosystems is another overarching environmental advantage of circular technologies. By reducing the demand for raw materials like timber, metals, and fossil fuels, circular practices help conserve natural habitats that are often degraded by extraction activities. For instance, the adoption of recycled construction materials reduces the need to quarry limestone or mine aggregates, activities that destroy local flora and fauna. Smart agriculture, which utilizes sensors and AI to optimize input usage, reduces pesticide runoff and soil degradation, thereby protecting ecosystems and pollinator populations (La Rosa A D., et al, 2021). Additionally, reduced waste and emissions in urban areas contribute to cleaner air and water, which benefits both human health and wildlife. Digital product passports (DPPs) are a promising innovation supporting environmental transparency and accountability. A DPP stores detailed information about a product's composition, environmental impact, repairability, and recycling pathways. This enables consumers and recyclers to make informed decisions, ensures compliance with environmental regulations, and facilitates high-quality material recovery. Such systems are crucial for transitioning to a product lifecycle model that is both circular and environmentally responsible. Importantly, the shift toward a technology-enabled circular economy also contributes to the resilience of natural systems. By decoupling growth from resource consumption, circular systems reduce the pressure on planetary boundaries—such as climate stability, freshwater availability, and land system change. This resilience is especially vital in the face of climate change, which exacerbates environmental degradation and resource scarcity (Nisha Angeline M., et al, 2025). Circular practices, supported by technology, allow societies to better adapt to changing environmental conditions, ensuring sustainable production and consumption even under stress. In conclusion, the environmental benefits of a technology-enabled circular economy are vast, transformative, and urgently needed. From minimizing waste and emissions to conserving water, protecting biodiversity, and enhancing resource efficiency, digital innovations are elevating circular economy strategies to unprecedented levels of impact. As climate change, pollution, and ecological degradation continue to threaten global stability, the convergence of technology and circularity offers a powerful, systems-level response that aligns environmental integrity with economic innovation. Embracing this paradigm shift is not merely a matter of sustainability—it is an imperative for the survival and flourishing of both people and the planet.

OVERCOMING BARRIERS TO TECHNOLOGY ADOPTION IN THE CIRCULAR ECONOMY

The adoption of technology in the circular economy holds immense promise in transforming how resources are utilized, waste is managed, and sustainability goals are achieved. However, despite the evident benefits, there are significant barriers to the widespread implementation of technology-driven solutions in circular systems. These barriers span economic, technical, institutional, cultural, and social dimensions. Understanding these obstacles is essential for devising strategic solutions that not only accelerate the adoption of enabling technologies but also ensure inclusive, resilient, and effective circular economy transitions. Economic and Financial Barriers: A primary obstacle in technology adoption for circular economy initiatives is limited financial resources. Many circular technologies—such as reverse logistics systems, advanced recycling infrastructure, AI-based sorting mechanisms, and IoT-enabled monitoring tools—require substantial upfront investment (Upadhyaya A N., et al, 2025). For small and medium-sized enterprises (SMEs), which form the backbone of many economies, the lack of access to affordable financing and perceived financial risk can deter adoption. Additionally, the traditional linear economic model often benefits from subsidies and established infrastructure, making it financially more viable in the short term. This creates an uneven playing field where circular alternatives appear costlier, even if they offer long-term savings and environmental benefits. Furthermore, financial decision-making processes in organizations often focus on short-term returns on investment, undermining longer-term sustainability-oriented innovation. Solutions: Public-private partnerships (PPPs) can de-risk investments in circular technology by pooling resources and spreading financial risk. Green finance instruments, such as sustainability-linked loans, green bonds, and impact investing, can provide targeted capital for circular initiatives. Government subsidies and incentives, including tax breaks, grants, and low-interest loans, can encourage businesses to invest in sustainable technologies. Promoting life-cycle cost analysis (LCCA) over traditional cost-benefit analysis can help shift focus from short-term costs to long-term value creation. Technical Challenges and Infrastructure Limitations: Many regions, particularly in developing and emerging economies, face infrastructure deficiencies that inhibit the deployment of circular technologies. Efficient waste collection systems, reverse logistics networks, smart grid capabilities, and digital connectivity are foundational to a tech-enabled circular economy, yet often underdeveloped or absent (Andersen A

L., et al, 2023). Additionally, interoperability and standardization issues pose significant problems. Circular technologies, especially those based on data (AI, IoT, blockchain), require interconnected platforms, shared data formats, and integrated systems. The lack of common standards makes it difficult for stakeholders across supply chains to coordinate effectively, share data, or scale up circular solutions. Moreover, technological maturity can be a barrier. Emerging technologies such as chemical recycling or AI-based waste recognition may still be in the pilot stage and not ready for broad-scale implementation, particularly in cost-sensitive sectors. Solutions: Invest in smart infrastructure development, such as waste collection and tracking systems, broadband connectivity, and renewable energy grids. Encourage open-source platforms and industry-wide protocols to improve interoperability and data exchange. Promote research and development (R&D) through public funding to accelerate the maturity and scalability of critical circular technologies. Develop technology roadmaps that outline phased, step-by-step adoption paths for organizations transitioning to circular models.

Regulatory and Policy-Related Barriers: In many jurisdictions, existing regulatory frameworks were built around the linear model and are ill-suited to facilitate or encourage circular technology adoption. Regulations may restrict innovative practices such as material reuse, remanufacturing, or secondary raw material trading due to outdated standards, liability concerns, or lack of definitions. Moreover, data privacy regulations can limit the use of digital technologies like IoT and blockchain that rely on continuous data sharing for transparency and traceability. In some cases, the lack of policy clarity or coherence across different levels of governance (local, national, international) adds to the uncertainty, discouraging investment and innovation. Solutions: Develop and implement comprehensive circular economy policies that include clear guidelines for technology integration. Establish regulatory sandboxes to allow pilot testing of circular innovations within flexible legal environments (Frank A G., et al, 2025). Harmonize data governance frameworks to balance privacy with operational transparency and accountability. Promote international cooperation to align standards and facilitate cross-border circular flows and digital interoperability. Fragmentation and Lack of Collaboration: Circular economy solutions often require multi-stakeholder coordination across sectors, supply chains, and geographies. Yet a major hurdle lies in the fragmented nature of industries, where stakeholders operate in silos and may be reluctant to share proprietary data, resources, or expertise. This inhibits the integration of circular strategies that depend on transparency, cooperation, and shared platforms. The lack of collaborative frameworks also limits systems thinking, which is vital for identifying circular opportunities that span beyond individual firms—such as industrial symbiosis, where the waste of one industry becomes the resource of another. Solutions: Develop collaborative platforms and consortia that bring together industry, government, academia, and civil society to co-design solutions (Agarwal S., et al, 2025). Encourage data-sharing agreements and blockchain solutions to build trust and transparency across stakeholders. Support cluster-based innovation ecosystems, such as circular hubs and living labs, that facilitate regional collaboration and experimentation. Foster policy dialogues and public-private forums to align objectives and strategies for scaling technology-enabled circular practices. Technology holds transformative potential for enabling the circular economy, from intelligent resource management to new business models and data-driven decision-making. However, the path to widespread adoption is fraught with economic, technical, regulatory, cultural, and institutional challenges. Overcoming these barriers requires a holistic, systemic approach—one that combines innovation with inclusivity, collaboration with regulation, and vision with practical action. By investing in infrastructure, skills, policy reform, collaboration, and cultural change, societies can unlock the full potential of technology to accelerate the circular transition. The rewards are multifaceted: resilient economies, inclusive societies, thriving ecosystems, and a regenerative relationship with the planet.

SCALING UP TECHNOLOGY-ENABLED CIRCULAR ECONOMY SOLUTIONS

The transition to a circular economy, driven by technology, presents an unprecedented opportunity to rethink and redesign the global economic system. While isolated pilots and initiatives have demonstrated the transformative potential of technologies like artificial intelligence (AI), the Internet of Things (IoT), blockchain, and digital platforms in facilitating circularity, the real challenge lies in scaling up these solutions across industries, regions, and value chains. Scaling is not merely about expanding reach but about creating systemic change—reshaping production and consumption patterns, enabling new business models, and embedding circularity into economic infrastructure and societal behavior (Trevisan A H., et al, 2023). Achieving this at scale requires strategic coordination, robust policy frameworks, capacity building, investment mobilization, and multi-stakeholder engagement. Understanding the Need for Scaling Up: Initial technology-enabled circular economy initiatives have successfully demonstrated significant environmental, economic, and social gains. For example, smart waste-sorting systems powered by AI can drastically improve material recovery rates, while IoT sensors can optimize energy and resource usage in industrial processes. However, these successes often remain localized or sector-specific, failing to reach the scale needed to make a significant impact on global sustainability goals. Scaling up is essential to maximize the return on innovation and infrastructure investment. Drive system-wide reductions in emissions, waste, and resource use. Normalize circular business practices across entire markets. Accelerate the pace toward achieving SDGs, Paris Agreement targets, and net-zero commitments. Key Challenges in Scaling Circular Technologies: Fragmentation and Lack of Integration: Many technology-driven circular solutions operate in silos—specific companies, municipalities, or industries—without integration across supply chains or ecosystems. Fragmentation limits data flow, collaboration, and scalability. Financial and Investment Gaps: While pilot projects often receive grant funding or temporary subsidies, long-term financial models for large-scale deployment are frequently underdeveloped (Mubarik M S., et al, 2024). Risk-averse investors may be hesitant to commit to unproven circular business models at scale. Infrastructure Limitations: Physical and digital infrastructure—ranging from broadband access and logistics networks to recycling plants and data centers—is often inadequate to support scalable circular operations. Skills and Capacity Constraints: A shortage of skilled labor and technical expertise in deploying and maintaining circular technologies can hinder scalability, particularly in developing countries. Policy Misalignment: Existing regulatory frameworks may not support, or may even hinder, the scaling of circular models. Misaligned incentives, unclear legal definitions, or lack of standards for circular products can all create friction.

Digital Technologies as Scaling Catalysts: Cloud Computing and Big Data Analytics: Scalable computing infrastructure enables real-time monitoring, analysis, and optimization of circular systems across regions and operations. Cloud-based platforms also support software-as-a-service models for circular supply chain management. Artificial Intelligence (AI): AI can scale solutions by automating processes like product sorting, maintenance prediction, energy efficiency monitoring, and market demand forecasting. As data volumes grow, AI becomes more accurate and valuable at scale. Blockchain: Blockchain enables traceability and trust in circular supply chains (Kandasamy J., et al, 2023). By creating transparent records of material provenance, usage, and recycling, it can support global product passports and certifications. Internet of Things (IoT): Scalable IoT networks allow for real-time tracking of materials, energy, water, and emissions. By embedding intelligence in products and systems, IoT facilitates usage-based pricing, predictive maintenance, and end-of-life resource recovery. Digital Twins: Digital twins can simulate circular processes at scale—allowing companies to model and optimize resource use, emissions, and supply chain performance across their entire operations before implementation. Enabling Social and Behavioral Shifts: Scaling circular solutions is not just technical—it requires consumer buy-in and cultural transformation. Public engagement strategies must: Build trust in new business models such as leasing, sharing, or refurbished products. Provide clear and compelling value propositions (e.g., cost savings, convenience, environmental impact). Engage communities through participatory design processes, storytelling, and local ownership of circular projects. Educational institutions can help scale adoption by embedding the circular economy and sustainability in school and university curricula (Mehra J., et al, 2025). Offering professional training and reskilling programs for workers in linear industries. Scaling up technology-enabled circular economy solutions is a critical imperative for addressing the dual crises of environmental degradation and resource scarcity. It offers a pathway to decouple growth from resource consumption, create resilient systems, and generate inclusive economic opportunities. However, achieving this transformation at scale demands far more than innovation alone. It requires coordinated action across stakeholders, long-term policy support, inclusive financing mechanisms, and a willingness to rethink traditional business paradigms. Through the intelligent application of digital technologies, the formation of enabling ecosystems, and a focus on equity and education, the circular economy can evolve from isolated projects to a global norm—transforming how we produce, consume, and thrive on a finite planet.

EMERGING TRENDS AND INNOVATIONS IN CIRCULAR ECONOMY

The circular economy is rapidly evolving from a conceptual sustainability model into a dynamic system of innovation and technological transformation. Driven by the urgency of environmental crises, volatile resource markets, and shifting consumer expectations, new trends and emerging innovations are redefining how resources are used, reused, and regenerated across industries and societies. These developments go beyond traditional recycling or waste management—embracing digitalization, decentralization, product redesign, platform thinking, and systemic transformation (Limbore N V., et al, 2025). This evolution is ushering in an era where the circular economy becomes not only an ecological necessity but also a catalyst for inclusive economic growth, industrial resilience, and global competitiveness. Digital Circular Economy: From Concept to Implementation: One of the most defining trends is the digitization of the circular economy. Advanced digital technologies such as Artificial Intelligence (AI), Internet of Things (IoT), blockchain, cloud computing, and digital twins are being integrated to create smart circular systems. This convergence enables real-time tracking of materials, predictive analytics for maintenance and lifecycle extension, and automated decision-making to optimize reuse, repair, and recycling. AI-driven material flow analysis helps organizations reduce inefficiencies and identify hidden value in waste streams. Digital product passports store essential data about product origin, composition, environmental impact, and end-of-life options. IoT-enabled resource monitoring is used in smart buildings, supply chains, and cities to manage circular resource use more precisely. (Thirumal S., et al, 2024)B These tools are transforming circular economy strategies from reactive and linear into predictive and intelligent systems capable of self-optimizing for sustainability and profitability. Biological and Regenerative Innovation: Beyond Recycling: Innovations rooted in biological cycles and regenerative practices are gaining momentum as part of the circular transition. These include: Biodegradable materials derived from agricultural or industrial waste (e.g., mushroom packaging, algae-based bioplastics). Compostable electronics and textiles are designed to break down into non-toxic organic compounds. Regenerative agriculture that rebuilds soil health, biodiversity, and water cycles. Instead of relying solely on mechanical recycling, these approaches restore ecosystems and close the loop in nature-inspired ways. Regenerative design principles are being applied to architecture, infrastructure, and even urban planning, signaling a shift toward systems that give back more than they take.

Circular Economy in Consumer Behavior and Retail Innovation: Consumer participation is increasingly central to the circular transition. Digital innovation is reshaping how people shop, use, and dispose of products. New trends include: Re-commerce platforms (resale, refurbishment, and rental) scaling rapidly in fashion, electronics, and furniture. Gamification and behavior nudges are embedded in apps and platforms to encourage sustainable actions. Smart labeling using QR codes and NFC tags to provide circular information at the point of sale. Brands are also creating circular loyalty programs to reward customers for returning items or choosing circular alternatives, embedding circularity into the customer experience (Hernandez et al., 2025). Global Circular Policy Innovation and Regulation: Governments around the world are introducing circular economy roadmaps and regulatory frameworks to guide systemic change. New trends in policy include: Extended Producer Responsibility (EPR) expansion to include electronics, textiles, and packaging. Right to repair laws empower consumers and ensure product longevity. Mandatory recycled content standards in construction, plastics, and other materials. Global trade policy is also shifting to accommodate secondary material flows, cross-border EPR, and circular certification schemes. These changes are driving innovation and accelerating industry alignment with circular practices. The circular economy is moving beyond pilot projects and theoretical models to become a mainstream driver of innovation, economic renewal, and climate action (Kio P N., et al, 2022). These emerging trends reflect a holistic transformation—powered by digital intelligence, regenerative thinking, and collaborative ecosystems. From product design and supply chains to cities, finance, and cultural shifts, the next generation of circular innovation is systemic, data-driven, and inclusive. As these trends accelerate, the circular economy will evolve into a resilient, adaptive, and self-optimizing system capable of meeting the challenges of the 21st century. Its future depends on continued experimentation, investment, cross-sector collaboration, and most importantly, a collective commitment to building economies that thrive within planetary boundaries. The emerging circular landscape promises not only to decouple growth from resource consumption but also to redefine progress as a regenerative and equitable force for people and the planet.

CONCLUSION

The intersection of advanced technology and sustainable development has given rise to one of the most transformative economic paradigms of the 21st century: the technology-enabled circular economy. In contrast to the traditional linear economic model of "take, make, dispose," the circular economy emphasizes resource efficiency, waste elimination, and product longevity through continuous loops of reuse, repair, remanufacturing, and recycling. While the concept of circularity has been present for decades, it is only through the recent advent of cutting-edge digital technologies that large-scale, systematic adoption of these principles has become feasible. Harnessing technology to build a more circular and sustainable economy is not only an ecological imperative but also a strategic pathway toward economic resilience, innovation, and equitable growth in a resource-constrained world. Technology as an Enabler of Circular Innovation: At its core, technology acts as an enabler, amplifier, and accelerator of circular strategies. Digital tools provide the intelligence, visibility, and efficiency required to track material flows, optimize product lifecycles, and manage complex reverse logistics networks. From the micro level of individual product tracking to the macro level of industrial ecosystems, technology equips stakeholders with the insights needed to design, implement, and scale circular practices. Artificial Intelligence (AI) facilitates predictive maintenance, smart sorting of waste, and advanced analytics for resource optimization. Internet of Things (IoT) sensors allow real-time monitoring of products, energy use, and material conditions, enabling timely interventions for reuse or repair. Blockchain provides secure, transparent, and immutable records of materials' provenance, ensuring traceability throughout supply chains. Cloud computing allows decentralized data access and real-time collaboration across geographies and industries. These technologies collectively form the backbone of intelligent circular systems—systems that are adaptive, transparent, and capable of minimizing waste and maximizing value creation. Digital Marketplaces and Sharing Platforms: Technology has also given rise to platform-based business models that facilitate resource sharing, peer-to-peer transactions, and secondary markets. These platforms foster collaborative consumption, reduce idle capacity, and extend the life of products by connecting users with underutilized resources. Sharing economy platforms (e.g., tool-sharing, carpooling, office space rental) help reduce the need for new resource extraction. Re-commerce platforms (for refurbished electronics, fashion, furniture, etc.) are booming globally, supported by AI-driven pricing, quality control, and customer experience tools. Digital product passports and resale certification systems ensure that second-hand goods retain quality assurance and trust. By leveraging cloud services, mobile apps, and AI algorithms, these digital platforms promote circular consumption habits at scale, making sustainability more convenient and accessible to the public.

Policy and Regulatory Alignment through Digitalization: Governments and regulators are leveraging technology to support circular policy implementation and compliance monitoring. Through digitized reporting systems, real-time audits, and automated enforcement mechanisms, technology ensures more effective governance of sustainability standards. Digital compliance platforms streamline reporting under Extended Producer Responsibility (EPR) and carbon disclosure regulations. Open-data initiatives support public access to information on material flows, corporate circularity metrics, and environmental impacts. Geospatial data and remote sensing technologies enable monitoring of illegal dumping, deforestation, and other harmful practices. Technological integration in policy ecosystems fosters trust, transparency, and accountability—key elements in driving collective circular progress. Future Outlook: Toward Intelligent Circular Ecosystems: The next frontier in technology-enabled circular economy is the emergence of intelligent, self-regulating ecosystems. These systems will combine digital twins, AI, IoT, and real-time data to manage urban resources, industrial processes, and consumption behaviors autonomously. Circular smart cities will dynamically adjust resource flows based on weather, population, and waste data. Autonomous circular factories will use real-time data to adapt production schedules and minimize material inputs. Consumer platforms will offer personalized sustainability insights, nudging users toward low-impact choices. The convergence of these technologies will lead to a more agile, responsive, and regenerative economy—capable of adapting to climate shocks, resource volatility, and global disruptions. Conclusion: Toward a Regenerative Digital Economy: Harnessing technology for a more circular and sustainable economy represents both a challenge and an unprecedented opportunity. The digital revolution offers tools not just for efficiency, but for transformation—for reinventing how we design, produce, use, and reuse everything. This shift demands bold leadership, inclusive innovation, cross-sector collaboration, and public engagement. As data becomes a resource, platforms become marketplaces, and waste becomes value, the future of sustainability lies not only in doing less harm but in doing better. A circular economy empowered by technology will not only reduce environmental degradation but also create resilient industries, vibrant communities, and a thriving planet. In the age of AI, IoT, and digital intelligence, the path to sustainability is circular—and the time to embrace it is now.

REFERENCES

Abbas, Y., Martinetti, A., Rajabalinejad, M., Schuberth, F., & van Dongen, L. A. M. (2022). Facilitating digital collaboration through knowledge management: A case study. *Knowledge Management Research and Practice*, *20*(4), 606–616. DOI: 10.1080/14778238.2022.2029597

Agarwal, S., Kweh, Q. L., Goh, K. W., & Wider, W. (2025). Redefining marketing strategies through sustainability: Influencing consumer behavior in the circular economy: A systematic review and future research roadmap. *Cleaner and Responsible Consumption*, *9*, 100143. DOI: 10.1016/j.clrc.2025.100298

Alsabt, R., Alkhaldi, W., Adenle, Y. A., & Alshuwaikhat, H. M. (2024). Optimizing waste management strategies through artificial intelligence and machine learning-An economic and environmental impact study. *Cleaner Waste Systems*, *8*, 100158. DOI: 10.1016/j.clwas.2024.100158

Andersen, A. L., Brunoe, T. D., Bockholt, M. T., & Andersen, S. (2023). Changeable closed-loop manufacturing systems: Challenges in product take-back and evaluation of reconfigurable solutions. *International Journal of Production Research*, *61*(11), 3727–3747. DOI: 10.1080/00207543.2021.2017504

Avalekar, U. A., Mehta, V., Mishra, B. R., Prasanna, L., Rajlakshmi, P. V., Manjunath, T. C., & Selvakumar, P. (2025). The current advances in AI-driven education. In Babulak, E. (Ed.), *Educational AI humanoid computing devices for cyber nomads* (pp. 173–196). IGI Global Scientific Publishing., DOI: 10.4018/979-8-3693-8985-0.ch008

Cagno, E., Neri, A., Negri, M., Bassani, C. A., & Lampertico, T. (2021). The role of digital technologies in operationalizing the circular economy transition: A systematic literature review. *Applied Sciences (Basel, Switzerland)*, *11*(18), 8330. DOI: 10.3390/app11083328

Chowdhary, S., Selvakumar, P., Sangeetha, B., Sarvaani, V. V., Anute, N., & Sharma, M. (2025). The role of social capital in entrepreneurial development. In Martínez-Falcó, J., Sánchez-García, E., & Marco-Lajara, B. (Eds.), *Support networks for entrepreneurs: Social capital and strategic collaborations* (pp. 311–338). IGI Global Scientific Publishing., DOI: 10.4018/979-8-3693-9836-4.ch013

Coates, J. M., Santhanalakshmi, K., Dinesh, N., Gandhimathi, S., & Chandel, P. S. Mohit, Selvakumar P, (2025). "Assessment and evaluation in transnational staff development." In Morris G, Ye J (Eds.), Training and development in transnational higher education (pp. 387-406). IGI Global Scientific Publishing. DOI: DOI: 10.4018/979-8-3373-1897-4.ch019

Fang, B., Yu, J., Chen, Z., Osman, A. I., & Farghali, M. (2023). Artificial intelligence for waste management in smart cities: A review. *Environmental Science and Pollution Research International*, *30*, 82435–82458. DOI: 10.1007/s11356-023-28188-5 PMID: 37362015

Farhan, L., Hameed, R. S., Ahmed, A. S., Fadel, A. H., & Gheth, W. (2021). Energy efficiency for green internet of things (IoT) networks: A survey. *Network (Bristol, England)*, *1*(3), 104–118. DOI: 10.3390/network1030007

Frank, A. G., Batista, L., Chaudhuri, A., Subramaniam, N., Subramanian, A. M., Dora, M., & Ritala, P. (2025). Circular Economy and Digital Technology-Enabled Innovation: Advances, Applications and Prospects. *R & D Management*, *55*(2), 215–233. DOI: 10.1111/radm.70003

Hagelüken, C., & Goldmann, D. (2022). Recycling and circular economy—Towards a closed loop for metals in emerging clean technologies. *Mineral Economics*, *35*(3-4), 105–119. DOI: 10.1007/s13563-022-00319-1

Han, Y., Shevchenko, T., Yannou, B., Ranjbari, M., Shams Esfandabadi, Z., Saidani, M., Bouillass, G., Bliumska-Danko, K., & Li, G. (2023). Exploring how digital technologies enable a circular economy of products. *Sustainability (Basel)*, *15*(13), 10248. DOI: 10.3390/su15032067

Hernandez, J. C. R., Villa-Enciso, E., Cardona-Acevedo, S., Valencia, J., & Velasquez Salas, S. (2025). Smart Innovation for a Circular Economy: A Systematic Review of Emerging Trends and the Future of AI in the Sustainable Economy. *Sustainability (Basel)*, *17*(7), 3670. DOI: 10.3390/su17135793

Huang, J., & Koroteev, D. D. (2021). Artificial intelligence for planning of energy and waste management. *Sustainable Energy Technologies and Assessments*, *47*, 101572. DOI: 10.1016/j.seta.2021.101426

Hui, W. (2025). Research on cross-organizational integration and sharing strategies of digital health resources in the context of cloud platforms. *Scientific Reports*, *15*(1), 5498. DOI: 10.1038/s41598-025-01520-2 PMID: 40389484

Jagtap, S., Garcia-Garcia, G., & Rahimifard, S. (2021). Optimisation of the resource efficiency of food manufacturing via the Internet of Things. *Computers in Industry, 129*, 103447. DOI: 10.1016/j.compind.2021.103397

Kandasamy, J., Venkat, V., & Mani, R. S. (2023). Barriers to the adoption of digital technologies in a functional circular economy network. *Operations Management Research, 16*, 387–400. DOI: 10.1007/s12063-023-00314-9

Kio, P. N., Anumba, C. J., & Ali, A. K. (2022). Circular economy trends–potential role of emerging technologies. *IOP Conference Series. Earth and Environmental Science, 1101*(6), 12030. DOI: 10.1088/1755-1315/1101/6/062005

La Rosa, A. D., Greco, S., Tosto, C., & Cicala, G. (2021). LCA and LCC of a chemical recycling process of waste CF-thermoset composites for the production of novel CF-thermoplastic composites. Open loop and closed loop perspectives. *Journal of Cleaner Production, 324*, 129285. DOI: 10.1016/j.jclepro.2021.127158

Limbore, N. V., Anute, N., Ahluwalia, G. K., Chordiya, P. B., Kalshetti, P., & Selvakumar, P. (2025). Ethical leadership and corporate responsibility. In Harriott, S. (Ed.), *Innovative approaches to managing conflict and change in diverse work environments* (pp. 193–218). IGI Global Scientific Publishing., DOI: 10.4018/979-8-3693-9556-1.ch009

Lokhande, M., John Deva Prasanna, D. S., Selvakumar, P., Ragupathi, T., Vaissnave, V., & Manjunath, T. C. (2025). "Novel ML algorithms for healthcare." In Ajibesin A, N G, Thangavel S (Eds.), Synergizing data envelopment analysis and machine learning for performance optimization in healthcare (pp. 113-134). IGI Global Scientific Publishing. DOI: DOI: 10.4018/979-8-3373-0081-8.ch005

Mao, W., Zhao, Z., Chang, Z., Min, G., & Gao, W. (2021). Energy-efficient industrial internet of things: Overview and open issues. *IEEE Transactions on Industrial Informatics, 17*(7), 4637–4647. DOI: 10.1109/TII.2021.3067026

Mehra, J., Madaan, S., & Balu, N. Mohit, Selvakumar P, Manjunath T C, (2025). "The ubiquitous internet and smart computation available 24/7 for anyone and anywhere." In Babulak E (Ed.), Educational AI humanoid computing devices for cyber nomads (pp. 247-270). IGI Global Scientific Publishing. DOI: DOI: 10.4018/979-8-3693-8985-0.ch010

Mubarik, M. S., Kontoleon, A., & Shahbaz, M. (2024). Beyond the hurdles: Exploring policy obstacles in the path to circular economy adoption. *Journal of Environmental Management, 344*, 119030. DOI: 10.1016/j.jenvman.2024.122667 PMID: 39369529

Neri, A., Negri, M., Cagno, E., Franzò, S., Kumar, V., Lampertico, T., & Bassani, C. A. (2023). The role of digital technologies in supporting the implementation of circular economy practices by industrial small and medium enterprises. *Business Strategy and the Environment*, *32*(1), 93–110. DOI: 10.1002/bse.3388

Nisha Angeline, M., Selvakumar, P., Gupta, M. K., Nazar, N., Elumalai, D., & Bhattacharya, S. (2025). Emerging materials for energy storage and environmental applications. In Muduli, K., Moharana, B., Ales, S., & Biswal, D. (Eds.), *Using computational intelligence for sustainable manufacturing of advanced materials* (pp. 53–72). IGI Global Scientific Publishing., DOI: 10.4018/979-8-3693-7974-5.ch003

Pouri, M. J., & Hilty, L. M. (2021). The digital sharing economy: A confluence of technical and social sharing. *Environmental Innovation and Societal Transitions*, *38*, 127–139. DOI: 10.1016/j.eist.2020.12.003

Reim, W., Andersson, E., & Eckerwall, K. (2023). Enabling collaboration on digital platforms: A study of digital twins. *International Journal of Technology Management*, *91*(3–4), 187–213. DOI: 10.1504/IJTM.2023.129150

Rosenberg, S., Kurz, L., Huster, S., Wehrstein, S., Nowak, D., Hesse, R., & Friedrich, B. (2023). Assessment to evaluate environmental benefits of recycling—A case study for direct and hydrometallurgical closed-loop recycling of electric vehicle battery systems. *Waste Management and Recycling*, *164*, 106321. DOI: 10.1016/j.wasman.2023.106321

Shojaei, A., Ketabi, R., Razkenari, M., Hakim, H., & Wang, J. (2021). Enabling a circular economy in the built environment sector through blockchain technology. *Journal of Cleaner Production*, *292*, 125021. DOI: 10.1016/j.jclepro.2021.126352

Tanveer, M., Alkhayyat, A., & Kumar, N.. (2022). REAP-IIoT: Resource-efficient authentication protocol for the industrial Internet of Things. *IEEE Internet of Things Journal*, *9*(1), 114–127. DOI: 10.1109/JIOT.2022.3188711

Thirumal, S., Udawatta, N., Karunasena, G., & Al-Ameri, R. (2024). Barriers to adopting digital technologies to implement circular economy practices in the construction industry: A systematic literature review. *Sustainability (Basel)*, *16*(4), 1577. DOI: 10.3390/su16083185

Trevisan, A. H., Lobo, A., & Guzzo, D. (2023). Barriers to employing digital technologies for a circular economy: A multi-level perspective. *Journal of Cleaner Production*, *408*, 137161. DOI: 10.1016/j.jclepro.2023.137161 PMID: 36801533

Upadhyaya, A. N., Debnath, P., Rani, P., Selvakumar, P., Manjunath, T. C., & Bhattacharya, S. (2025). AI-based disaster prediction and early warning systems. In Mobo, F. (Ed.), *AI and emerging technologies for emergency response and smart cities* (pp. 55–74). IGI Global Scientific Publishing., DOI: 10.4018/979-8-3693-9770-1.ch003

Yu, K. H., Zhang, Y., Li, D., Montenegro-Marin, C. E., & Kumar, P. M. (2021). Environmental planning based on reduce, reuse, recycle and recover using artificial intelligence. *Environmental Impact Assessment Review*, *90*, 106650. DOI: 10.1016/j.eiar.2020.106492

Chapter 5
Good Manufacturing Practices and Knowledge-Centric Approaches as Drivers of Circular Economy in Local Industrial Systems

Debashree Chakravarty
https://orcid.org/0009-0007-7677-8627
KIIT University, India

Ipseeta Satpathy
https://orcid.org/0000-0002-0155-5548
KIIT University, India

B. C. M. Patnaik
KIIT University, India

Abhiskek Kumar
KIIT University, India

ABSTRACT

Local industries remain crucial to regional economies but often face sustainability challenges such as inefficient resource use and poor waste management. Traditional Good Manufacturing Practices (GMP) focus primarily on compliance and product quality, offering limited scope for environmental responsibility. This chapter pro-

DOI: 10.4018/979-8-3373-6435-3.ch005

poses a Sustainability-Oriented GMP Framework that embeds circular economy principles, carbon reduction, and responsible resource use into local production systems, particularly for SMEs. Grounded in industrial ecology, knowledge-based theory, and stakeholder responsibility, the framework redefines GMP as a pathway to achieve the UN Sustainable Development Goals (notably SDG 12 and 13). By integrating energy efficiency, recycling, and knowledge-driven innovation with digital tools, industries can shift from compliance-based models to sustainable ecosystems. The chapter offers conceptual insights and practical directions for policymakers and managers, positioning GMP as a transformative tool for advancing sustainability in local industries.

INTRODUCTION

The industrial systems represent the essence of economic developments and environmental issues. Since the local industries play an important role in the way of employment provision, territorial development and creativity, they also consume excessive natural resources, produce waste, and increase carbon footprints (Rockström et al., 2009). This dual character requires the reconsideration of how industrial systems are employed so that they are not geared towards economic growth dimensionally at the cost of ecological deterioration. Amidst this need, the concept of sustainability has become a new paradigm of reshaping industrial process, supply chain and production practice to conform to the Sustainable Development Goals (SDGs), especially the aspect of responsible consumption and production (SDG 12), climate action (SDG 13) and partnerships to achieve the processes (SDG 17) (United Nations, 2015).

Clean technologies, circular economy principles and eco-efficient management strategies are to be integrated into the local industrial systems in a holistic manner in order to make it a sustainable process. Yet, even more so than the involvement of technology adoption and regulatory mechanisms, the skill transfer and organizational competence, as well as inter-organizational communications are also identified as critical enabling factors in creating sustainable transitions (Nonaka, 1994; De Marchi, et al., 2013). Knowledge-based practices would ensure that best practices, innovations and sustainability-driven standards can disseminate within and among industries and bring about collective environmental responsibility.

In this light, the environmentally friendly Good Manufacturing Practices (GMP) is an attractive conceptual approach to local industrial systems. Typically, GMP is used in conjunction with quality checks and safety in the pharmaceutical applications and food manufacturing. By incorporating environmental sustainability into this model, industries can develop a system that operates with due regard to the environmental impact of all manufacturing processes, such as the sourcing of raw materials, the production process itself, the management of waste products and their recycling. In such a way, GMP will be not only a compliance- and quality-related course, but also a means of improving climate neutrality, responsible production, and green supply chains (Aivazidou et al., 2018).

The paper, thus, tries to identify how eco-based GMP may be used as a sustainability focused model to local industrial eco-systems. In particular, it combines views of the circular economy, knowledge-based management, and technological innovation to suggest a Sustainability-Oriented GMP Framework. The framework focuses the attention on knowledge partnerships, green practices and digital technologies as enablers of responsible production and consumption at the local level. This model can be theorized to contribute to the understanding of the academic discourse, but also lead to practical avenues available to policymakers, managers, and researchers interested in the premises of environmental sustainability.

2. LITERATURE REVIEW AND THEORETICAL FOUNDATIONS

2.1 Sustainability Practices at Local Industrial Systems

In the recent literature it is noted that sustainability at a local level draws on SME capacities, local support, and networked learning. Systematic reviews indicate that Industry 4.0 tools (IoT, analytics, CPS) can help SMEs to minimise waste by monitoring energy/material flows, however adoption is limited due to skills, finance and ecosystem support (e.g. local incubators, university partners). A 2024 SLR reveals how SMEs are using I4.0 to embrace energy efficiency, pollution prevention and eco-design-but cautions that there are capability gaps that local policy must fill (Santos & Sant'Anna, 2024; Martins et al., 2022).

Meta-analytic evidence through 2023-2024 supports that green supply chain management (GSCM) enhances environmental, operational, and financial performance adding assured conviction that GSCM is a proven channel through which local firms can scale sustainability. A 2024 meta-analysis makes this distinction, based on >25,000 observations, and finds both green procurement and GSCM are positively correlated with performance; similar 2023 meta-analytic evidence also points in the same direction (Liu et al., 2024; Galdos-Urbizu et al., 2024; Holling, H., & Backhaus, L., 2023).

In circularity in SMES, the enablers (top management support, ecosystem partnerships), practices (remanufacturing, reverse logistics, product-service systems), and barriers (costs, regulatory ambiguity and data) were the subject of recent reviews in the field. A 2023 SLR framing SME transition in multi-level insights; a 2025 maturity model suggests stepped roadmaps to SME circular uptake and is helpful in local policy design (Ahmadov et al., 2025; Bandeira et al., 2025; Silva et al., 2022).

2.2 CE, IE in Local Systems

The CE reinvents value creation by uncoupling growth and resource throughput (preserving resources) through the likes of narrowing (efficiency), slowing (product life extension), closing (recycling), and regenerating flows (Lewandowski, M., 2016; Geissdoerfer et al., 2017). On the local level, CE exists in the form of industrial symbiosis- the deliberate trade of materials, energy, water, and byproducts between firms that are positioned next to one another to minimize virgin material and waste (Chertow, M. R.,2000). Industrial ecology seminal work (Frosch, R. A., & Gallopoulos, N. E.,1989) gives the systems perspective behind CE since it was seen that the systems could be mapped using ecological principles.

The CE terminology in the recent years has shifted towards the mechanics of implementation and not definitions- data visibility, business models, and coordination. The mechanism by which loops are closed repeatedly mentioned is through digitalization in the form of high-resolution tracking, predictive maintenance and material passports. An earlier CIRP article planted this seed; more current reviews since then demonstrate maturing toolchains to bring digital CE into production.

On online platforms in industrial symbiosis (IS), a 2022 study reports how the locality constraint is alleviated through online platforms mediating cross-site waste/energy/water flows, overcoming information asymmetry and trust-related barriers, to a large extent relevant in dispersed local networks. A 2025 synthesis lists 23 challenges (technical, economical, legal, social) and 20 opportunities to implement IS and gives a checklist of local cluster programs to follow (Sellitto et al., 2025; Silva et al., 2022).

2.3 Good Manufacturing Practice (GMP)

In response to product quality, safety, and consistency, standards were developed as Good Manufacturing Practices (GMP) codified in WHO guidelines, the EU's EudraLex, Volume 4, and the U.S. FDA 21 CFR Part 210/211. Traditional GMP aspects are documented procedures, validated processes, calibration of equipment, hygiene, training, deviation management and correction-prevention actions (World Health Organization, 2014; European Commission, 2022; U.S. Food and Drug Administration, 2022).

Historically, GMP is compliance-oriented, but its process discipline, meticulous documentation and relentless pursuit of quality improvement offer rich soil to plant environmental objectives-what we call eco-oriented GMP. Integrated quality management with environmental quality (i.e., ISO 9001 and ISO 14001) and energy quality (i.e., ISO 9001 and ISO 50001) have synergetic benefits of performance assimilation through shared procedures, audits, and learning loops (Zeng, Shi, and Lou, 2007; Asif, Fisscher, de Bruijn, and Pagell, 2010). The results of cleaner production studies and lean-green integration also discuss that waste reduction and energy efficiency initiatives become less difficult with standard work, visual control and root-cause analysis (King & Lenox, 2001).

Yet there is little explicit theoretical cross-reference between GMP and CE. New proposals include the addition of environmental critical control points (ECCPs) to GMP checklists, lifecycle thinking to change control, adding a green criterion to supplier qualification and material/energy KPIs to deviation/ CAPA capture systems (Garza-Reyes, J. A.,2015; Jabbour et al., 2018). It is this literature that inspires a conceptual reorganization: GMP as the functioning bedrock of sustainability as opposed to a side- by- side, supplemental initiative.

GMP still forms the backdrop of process discipline. In extant sources, operationalizations of GMP to the presumable environment of modern plants exist in the living GMP resources or the EudraLex Volume 4, with its Annex 1, on sterile manufacture, being fully applicable as of 25 August 2024 and directly extendable to ecologically relevant controls (e.g., ECCPs). An updated Part I guide was issued in August 2023 and retains international convergence. These governance updates demonstrate how GMP audibility, change management, deviation / CAPA and supplier qualification can imbed environmental requirements along with quality/ safety (World Health Organization, 2023; Pharmaceutical Inspection Co-operation Scheme, 2023).

The trend toward integrated management systems is also recent: in 2024, work on Integrated Management Systems (IMS) based on the ISO high-level structure, (Annex SL) contains firm suggestions to harmonize quality, environment, energy, and health/safety systems, the logical scaffold that would be needed to add environmental KPIs, green supplier audits, and lifecycle-aware change control to GMP.

2.4 Knowledge-Centric Approaches: Digital Knowledge, Traceability and Learning

The sustainable production systems based on the knowledge of how to perform given processes have increasingly become digital. To achieve traceability and trust, 2023-2025 updates note that blockchain helps CE by checking secondary materials authenticity, awards returns in token, and demonstrates recycling content- crucial to local reverse commerce and green buying. IoT can provide energy dashboards, anomaly detection, analytics/AI can recommend process adjustments and remove scrap; digital twins can design scenario analyses reuse/remanufacture all under conditions of SME (limited) when aligned with ecosystem support (e.g., shared testbeds, training hubs). There are also even more critical of technology-driven sustainability in SMEs seriatim: the collaboration with universities/platform providers and governmental schemes prove to be decisive in scaling pilots to everyday practice which the institution of a GMP-style governance shell can codify.

2.5 Climate Context, CSR and Stakeholder Pressure

Although they are coexist with the stakeholder theory and the traditional approaches to CSR (Freeman, 2010; Carroll, 1999), the current climate science suggests an increased sense of urgency. The IPCC has reported in its AR6 synthesis reports that there is a narrowing timeline to mitigation and call to action that must be facilitated at systemic levels swiftly- the IPCC reports also support the argument that there is need to incorporate sustainability as core to any operating system including GMP rather than as an afterthought. The warning and policy implications of AR6 are summarized in contemporary commentary (Intergovernmental Panel on Climate Change, 2023; Kolbert, E., 2023; The Guardian, 2023).

2.6 Gap and Direction

The evidence base to guide the current recommendation in 2021-2025 is categorical on what does and does not work (IS platforms, GSCM, digital CE enablers), and what are the show-stoppers (capability gaps, costs of coordination, and data/traceability). Recent updates to GMP/IMS provide a ready-made governance carcass-audits, change control, supplier qualification, deviation/CAPA, and KPI dashboards to support environmental critical control points (ECCPs), circular KPIs and digital traceability. What has still eluded is one that comprises an overall theoretical model. Redesigns GMP scopes to be expressly inclusive of circularity and environmental performance, the coupling between GMP routines with knowledge architectures (learning, absorptive capacity, inter-organizational transfer). Formalizes partnerships in technology (IoT, AI, blockchain, digital twins) with the potential to make local circularity possible.

3. THEORETICAL FRAMEWORK: SUSTAINABILITY-ORIENTATED GMP

The theoretical frames have taken a pivotal role in constructing a solid sustainability paradigm on local industrial systems. This chapter advents the Sustainability-Oriented Good Manufacturing Practices (SO-GMP) Framework, a theoretical paradigm that reconciles tried GMP aspects and ecological sustainability doors. GMP, historically aimed at promoting the safety of products, conformity, and quality, can be restructured in terms of environment- and socially conscious status such as maximizing eco-efficiency and waste reduction. However, by integrating the dimensions of sustainability in GMP, industries no longer have to limit themselves with compliant measures but take part in a systemic contribution to climate resilience and the circular economy.

The framework is based on four overlapping pillars that include (1) Environment friendly GMP practices, (2) Integration of circular economy, (3) Knowledge-centric approaches, and (4) Technology-enabled sustainability partnerships.

Environment Friendly GMP Practices

Traditionally, GMP framework focuses on hygiene, process control and standardization. A more prudent environmental responsibility must be incorporated into these trends. As an example, resource efficiency, renewable energy use, or reduction of pollution can be included as compliance criteria and be traded off against the quality of products. This theoretical modification places GMP as a duplicitous assurance system-not only as part of product integrity, but also as part of the ecological guardian.

Circular Economy Integration

Circular economy theories focus on the regeneration of resources, the expansion of the lifecycle of products, waste valorization (Geissdoerfer et al., 2017). Inserting such principles into GMP permits re-engineering of processes in which impediments are not only reduced but converted into input streams. This brings the industrial systems into line with the Sustainable Development Goals (SDGs), especially SDG 12: Responsible Consumption and Production, and SDG 13: Climate Action.

Knowledge-Centric Approaches

Theories based on organizational learning and knowledge transfer indicate that sustainability transitions need shared knowledge and effective communication lines between stakeholders (Nonaka, I.,2009; Saz-Carranza & Ospina, 2011). In this context, knowledge is considered as an enabler of sustainable GMP- where inter-firm communication encourages eco-innovation and best practices diffusion across sectors. This conceptual interconnection emphasizes the aspect of knowledge as a green capital.

Technology-Enabled Sustainability Partnerships

Based on the theories of digital innovation and socio-technical transitions (Geels, 2002; Frank et al., 2019) the framework suggests that sustainability-oriented GMP cannot exist outside of the technological partnerships. Circular GMP practices can be operationalized by employing digital platforms, blockchain to track traceability and AI to enable process optimisation. Technological and regulatory collaborations between firms, technology providers and regulators are necessary in scaling these innovations locally.

To conclude the, Sustainability-Oriented GMP Framework places environmental responsibility as an inevitable aspect of production excellence. By resolving the classical standards of GMP with circular economy thinking, knowledge-sharing theory and technology partnerships, the framework offers a road map that can help the local industries to work towards reconciliation of their production activities with planetary boundaries.

4. OBJECTIVES OF THE STUDY

- To understand and restructure Good Manufacturing Practice (GMP) as a model of sustainability, including environmental responsibility, the level of circular economy, and responsible production at the local industrial scale.
- To analyse a theoretical framework - the Sustainability-Oriented GMP Framework - that indicates the importance of knowledge management, inter-organizational communication, and a partnership with technology to enable industries to operate in consistent with the aims of global sustainability, e.g., the SDGs and carbon neutrality.

The first aim of the proposed research is to reconstruct Good Manufacturing Practice (GMP) as a sustainability-based framework which would combine environmental responsibility, principles of the circular economy and the notion of responsible production within the local industrial systems. GMP has traditionally been used as a regulatory approach of maintaining the safety, quality, and efficacy of products mostly within the industry pertaining to pharmaceuticals, food processing, etc. It is however traditionally concerned, to a large extent, with conformity with safety and consumer protection standards in regards to the products and less with the general ecological or social impact of the production processes (World Health Organization, 2020). Although this has been an effective way of upholding product integrity, it has restricted the capacity of GMP to spearhead sustainability shifts in industrial eco-systems.

Re-engineering GMP as a model that revolves around sustainability would recognize the pressing new environmental challenges of the 21 st century, namely CC, resource-depletion, and waste. The integration of environmental responsibility into GMP will empower industries to move beyond compliance toward the pro-active upholding of green operations through carbon-footprint reduction, minimization of hazardous waste, energy and water conservation and optimization. This change enables GMP to cease being a quality assurance methodology and become an innovative mechanism that can align industrial practices in the world with the sustainability agendas like the Sustainable Development Goals (SDGs) of the United Nations and in particular its Goal 12 (Responsible Consumption and Production) and Goal 13 (Climate Action).

It is also critical that the principles of a circular economy are incorporated into GMP. The most defining element in the circular economy is the design out waste, extend the life of materials, and regenerate natural systems (Macarthur & Heading, 2019). This redefinition of GMP promotes the reconsideration of product life cycles by industries, investments in recycles or bio-degradable component materials, and closed production. When it comes to local industrial systems that tend to be resource inefficient and linear with a take-make-dispose approach, a sustainability-driven GMP can be used as a systematic process of transitioning towards a circular system.

In addition, it is not only the impact of production on the environment that defines production responsibility, but also the ethical sourcing as well as fair and ethical labor practices, as well as open supply chains. Reconstituting GMP to fit this multidimensional duty makes local industries work in responsibility to their environment and the society. This not only makes production systems more sustainable but socially more acceptable and competitive in the global market where ethical and green products increasingly are demanded by the consumers (OECD, 2022).

On the whole, this aim reflects the general vision of redesigning GMP as a holistic, visionary program that reflects the needs of local industry and, at the same time, complies with the global sustainability agenda. It guarantees that industries cease being the passive adherers of regulatory principles but proactive movers of sustainable development that can convert the production systems into environment-friendly and socially equitable models.

4.1 Re-Engineering GMP Through Sustainability-Based Model

This study aims, in the first objective, to transform Good Manufacturing Practice (GMP) as a model of sustainability, which will merge environmental care, principles of a circular economy and responsibilities of production into local industrial systems.

To date, the concept of GMP is characterized mostly in terms of product quality, safety, and compliance with regulatory requirements (World Health Organization, 2014). Although these are essential, we need to take into account the heightening environmental quality- put in simple terms climate change, resource depletion, and industrial pollution- the critical issue that requires augmentation of GMP to turnover and operational construct of high sustainability.

GMP in its classical version never has been designed as a sustainability instrument. Rather, it was meant to ensure that the quality of products remained the same and that the consumers were kept safe in areas such as pharmaceutical industry, as well as the food processing, and manufacturing industries. Yet, the industrial ecosystem in the 21 st century has to bear pressures well beyond its control over quality: industries now form central nodes in the delivery of United Nations Sustainable Development Goals (SDGs), especially SDG 12 (Responsible Consumption and Production) and SDG 13 (Climate Action) (United Nations, 2015). Such a paradigm shift calls for a redesign of a concept of GMP that incorporates good corporate social responsibility and a circular economy in all levels of production and operations.

Why Redesign of Good Manufacturing Practices?

Environmental Imperative

Domestic industries tend to have resource-intensive production processes, which have a substantial effect on carbon emissions, waste production, and ecological deterioration (OECD, 2022). With a GMP that is sustainability-oriented, in terms of reducing the eco-footprint, industries will be inducted into such practices as close-loop production, energy efficiency and eco-friendly sourcing.

The Integration of Circular Economy

The circular economy (CE) focuses on waste minimization, resource re-utilization and recycling of materials in production systems. Operating through GMP, industries will be able to break the linear paradigm of take-make-dispose and transform into the circular paradigm of design-use-recover through the integration of CE principles (Geissdoerfer et al., 2017).

Responsible Production

Compliance is not going to be sufficient in the globalized world, and industries will have to learn to adopt the principle of social responsibility by ensuring labor fairness, community health, and responsible supply chains. A new GMP would include environmental, ethical, and social specificities of industrial standards.

Proposed Conceptual Model: Sustainability-Based GMP

Table 1. The redesigned GMP framework can be visualized as an integrated sustainability model, shown below

Table 1.

Dimension	*Traditional GMP Focus*	*Sustainability-Based GMP (Proposed)*	*Expected Impact on Local Industries*
Quality & Safety	Compliance with product safety standards	Maintains safety while reducing harmful chemicals and pollutants	Safer for consumers & environment
Resource Use	Efficiency in production processes	Circular economy: reduce, reuse, recycle, eco-design	Reduced waste & cost savings
Energy Management	Not a core concern	Renewable energy adoption, energy-efficient equipment	Lower carbon footprint
Waste Management	Disposal compliance	Zero-waste targets, industrial symbiosis, green packaging	Circular production systems
Knowledge Systems	Training for compliance	Knowledge sharing on sustainability, inter-organizational learning	Technology transfer & eco-innovation
Social Responsibility	Limited scope	Worker well-being, ethical supply chains, community engagement	Stronger social license to operate

This model shifts GMP from a compliance checklist to a strategic sustainability framework where quality and safety are preserved but expanded to include environmental integrity and social accountability.

Local industries play a crucial role in the economic growth of a region, but it experiences inefficiency towards the use of resources and pressure on the environment. Examples of such ground-breaking wastewater solutions include processes like Zero Liquid Discharge (ZLD)-based advanced filtration and resource recovery systems, which allow recycling of water and minimizing environmental risks to textile mills. In India, one constraint of moving towards circular economy practices is the fragmentation of supply chains, though research identifies digitization -better traceability and coordination- as a potential enabler of sustainable upgrading

Correspondingly, the European Farm to Fork Strategy as part of the European Green Deal presents a systematic transformation to decarbonize food system towards fair, healthy food systems. It is supportive of circular packaging, minimized waste, and energy efficiency throughout the agrifood supply chain (AGRINFO., 2023). The implementation of the Green Deal policy in general also helps transition to more circular models of industrial activity and greener product life-cycles.The examples described in our times confirm the fact that the redesign of GMP to sustainability is not such an abstract concept, as there has been a number of practical solutions noticed even to change local industrial systems.

Figure 1. The above sustainability-based GMP framework has redefined GMP not only as a compliance-based system but also as a change agent model to the local industries

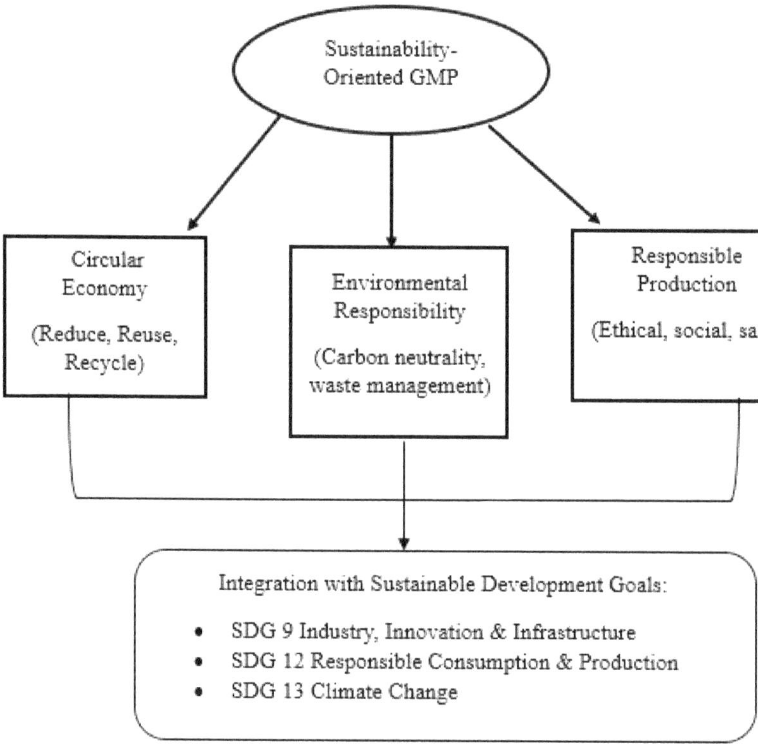

- **Environmental responsibility:** The industries incorporate minimal-carbon production, renewable sources, and eco-efficiency, and directly align with climate objectives and carbon neutrality directions.
- **Circular Economy:** Wastes are reduced by incorporating recycling, reuse, and optimization into the production cycle so that industries can run more closed-loop systems rather than the linear "take-make-dispose" cycle.
- **Responsible Production:** A Comprehensive approach goes beyond compliance by encouraging ethical labor practices, workplace safety and accountability with stakeholders in its supply chains.
- **About SDGs integration:** The framework is consistent with some Sustainable Development Goals, especially SDG 9 (Industry, Innovation, Infrastructure), SDG 12 (Responsible Consumption and Production), and SDG 13 (Climate Action).

Combined, this model guarantees that local industrial systems transform into environmentally sustainable, socially responsible and economically responsible ecosystems that can drive global sustainability transitions, but which will remain competitive.

4.2 Strengthening Local Industrial Sustainability Through Innovation and Technology

The second goal is to expand the traditional knowledge about the Good Manufacturing Practices (GMP) and to frame it in a wider sustainability-related theoretical context. Conventional GMP is mainly concerned with the products quality, safety and meeting the regulatory standards. Nonetheless, in a time of growing climate-related pressures and ambitious global environmental goals (United Nations SDGs and carbon neutrality commitments), there is a fundamental need to expand the scope of GMP beyond compliance. It has to be developed into a dynamic system, knowledge-based, and technologically-empowered system that will enable the local industries to reach its products as the leaders of sustainability.

This goal is justified with the help of three important pillars:

- **Knowledge Management (KM):** local industries are usually not exposed to high knowledge and long-term production strategy. By incorporating KM into GMP, companies will be able to document sustainability knowledge, disseminate and implement it systematically--including the practices of raw materials selection, increasing energy efficiency and minimizing waste. It is proven that KM is a key facilitator of sustainable innovation and adopting a circular economy in the industries (Centobelli et al., 2022).
- **Inter-Organizational Communication and Collaboration:** Sustainability can neither be attained in a vacuum. SMEs, as the bones of local industrial ecosystems, have to engage significantly with their supply chains partners, research entities and policymakers. Collaborative GMP promotes the transparency of information, collaborative problem-solving, and resource-sharing, that is needed to develop localized circular economies (Arsawan et al., 2024, Palombi et al., 2024).
- **Cooperation with technologies:** technology and green innovation like IoT-powered energy monitoring, AI-powered predictive maintenance, blockchain for supply chain visibility, and technologies that allow eco-friendly production can transform GMP. The besetment of technological systems within sustainability-based GMP gives the industries quantifiable and integratable tools that are capable of helping them attain environmental standards without undermining their competitiveness (Jamwal et al., 2021; Agrawal et al., 2023).

In this way, this goal supports the building of a Sustainability-Oriented GMP Framework that does not just make compliance mandatory but incorporates eco-care, vision-based interaction, and technological partnerships as new beliefs. This context places local industries as active participants in the global sustainability initiatives including the Paris Agreement, Net Zero by 2050, and the SDGs (namely SDG 9 – Industry, Innovation, and Infrastructure; SDG 12 -Responsible Consumption and Production; and SDG 13 -Climate Action).

DISCUSSION

The redesigning of Good Manufacturing Practices (GMP) in the context of sustainability has the potential to be a great change in the way the global industries and at the level of local and SME could approach climate responsibility as an operation standard. The conventional GMP models have in the past paid little attention to the wider environmental, social, and economic aspects of sustainability industrial practice that should be considered in any GMP model. It is through the proposed sustainable GMP Framework that this gap would be narrowed by incorporating the concept of environmental responsibility, circular economy, and local adaptability in their quality management system.

Table 2. Classic GMP vs. sustainability-oriented GMP

Dimension	*Classic GMP*	*Sustainability-Oriented GMP (Redesigned)*
Primary Focus	Product quality, safety, and compliance	Triple bottom line: Quality + Environmental responsibility + Social impact
Environmental Responsibility	Limited to waste avoidance in processes	Energy efficiency, water recycling, renewable sourcing, carbon footprint audits
Circular Economy	Not explicitly considered	Incorporates resource recovery, eco-design, waste valorization, green packaging
Local Adaptability	Uniform global standards, less context-specific	Customizable to local industries, SMEs, and sector-specific constraints
Regulatory Orientation	Compliance-driven (reactive)	Innovation-driven (proactive) with alignment to SDGs and climate commitments
Stakeholder Involvement	Primarily internal (quality teams)	Multi-stakeholder: regulators, communities, supply chains, customers
Technology Integration	Traditional monitoring systems	AI, IoT, and blockchain for traceability, energy monitoring, and LCA

1. The Transition of Classical GMP to Sustainable GMP

This classic GMP centered mostly on least safety, hygiene and quality compliance of products, which in particular industries like food, manufacturing and pharmaceutical industries and sectors, maintained products(s) on consistent and safe basis. Nevertheless, it was quite linear in its design, with little regard to both environmental footprints and the resource efficiency and effects on the surrounding communities.

In contrast, Sustainable GMP (S-GMP) has three more pillars:

- **Environmental Responsibility** - minimization of energy, water and waste afoot.
- **Circular Economy Integration** - using of by-products, recycling of waste streams, and designing closed-loop productions.
- **Social and Community Responsibility** - balancing the production with the local local development purposes and ethical labor.

2. Viability of SMEs Regarding Various Sectors

Whereas MNCs usually have the capabilities to invest in sustainability, SMEs are the engines of the local economies and can perhaps most effectively implement S-GMP in a context sensitive and scalable manner.

- **Textile Industry:** Textile small businesses also have the ability to introduce efficient dyeing solutions, wastewater management solutions and renewable energy to their production lines. A major challenge is the apparent high cost of such investments although in the long term it makes operational cost saving, penalties on pollution, and access to the market is better (because of eco-labeling).
- **Food Processing:** Food SMEs are able to transition to sustainable packaging, solar energy drying processes and traceability systems which are based on digital supply chains and result in compliance with food safety and sustainability objectives. This two-tier compliance creates consumer confidence as well as reduces carbon footprints
- **Pharmaceuticals:** Integration of greener solvents, energy-efficient HVAC and waste-to-energy technologies are all viable within the current GMP guidelines in pharma SMEs. Such innovations save money in the long-run, at the same time as enhancing reputation in international supply chains that have an increasing demand of green practices.

This suggests that SMEs not only can and should be potential implementers, but also excellent change agents to Sustainable GMP because they are more proximate to communities, human resources pool, and supply lines that, more and more, prioritize eco-responsibility.

3. Challenges

At that, the road to S-GMP adoption is not an easy one:

Cost Implications: The cost of initial investment on solar panels, water-recycling systems, or packaging materials that are biodegradable is rather high making it challenging to SMEs with a small margin.

- Awareness / Training: Most SMEs still perceive sustainability as a compliance burden instead of seeing it a source of value creation thus, leading to a cultural and managerial deficit.
- Regulatory gaps: Existing GMP norms lack integration with sustainability with the target of carbon neutrality or the circular economy or the UN SDGs.
- Cultural Resistance: In heavily linear industries there is quite often an issue with resisting cultural change, putting a high premium on the short term efficiency rather than igniting a change in culture to be more resilient.

4. Opportunities and Enablers

These barriers have not stopped the S-GMP adoption, even though there are a number of systemic enablers that quicken the adoption:

- **Policy Support:** This can include subsidies, tax incentives, and the development of public-private partnerships that lower initial SME costs, in addition to aligning GMP with the sustainability policies (such as the EU Farm-to-Fork Strategy and India Make in India + Green Growth).
- **Technological Innovation:** Access to low-cost tools, like IoT-powered energy monitors, AI-powered efficiency platforms and blockchain-based supply-chain visibility solutions allows sustainable GMP to be viable even to resource-starved SMEs.
- **Funding the SMEs and Large-Scale Partnerships:** MNCs can also absorb SMEs into green supply network by financing the sustainability upgrades of suppliers.

Informed Consumers The emergence of the environmentally conscious consumer allows SMEs with sustainability certification to develop a more favorable position and global competitiveness.

Sustainable scope for GMP rollout. We can imagine a three-layered framework - that of three concentric circles. GMP Fundamentals unchanged: Core (Product Quality and Safety). Middle Layer (Sustainability Pillars): Environmental responsibility (energy, waste, water) social responsibility (labour well-being, impact to the communities) and circular economy principles. Outer Layer (Enablers): Policy overall performance, technological acceptance and industry motivations that produce launching likely. This conceptual structure reveals that sustainability does not replace GMP but to add it to one complete future-oriented industrial standard.

CONCLUSION

As practiced today, the Sustainable Good Manufacturing Practices (S-GMP) can be pictured as a very relevant step in the evolution of the traditional Good Manufacturing Practices (GMP). Conventional GMP has always been concerned with the safety aspects of products, product hygiene and regulatory adherence but the inclusion of sustainability has added another dimension that is in line with the current global problem of climate change and depletion of various commodities. The proposed framework proves that S-GMP will not be a substitute to GMP but an expansion to make sure that the foundational quality standards will be present, and the ecological responsibility, social well-being, and circular economy will be part of the production ecosystem.

A policy opportunity is also available to integrate GMP standards with sustainability goals to achieve greater harmonization between industrial regulations and wider sustainability priorities of the United Nations Sustainable Development Goals (SDGs), the Paris Climate Agreement, and the European Union Green Deal priorities. The governments and regulatory authorities can play a key role in shifting their GMP guidelines to add game-changing sustainability benchmarks and offer subsidies, tax incentives and public-private partnerships to ease the shift for SMEs. Such conformity not only would alleviate compliance pressures but also would generate a level playing ground in which eco-responsibility is viewed as a source of competitiveness (rather than cost).

Among industry stakeholders, especially the SMEs, the possibility of sustainable GMP has been laid out covering sectors such as textiles and food processing to pharmaceuticals. Their closeness to communities and value chains gives them a unique position to act as agents of change in the green manufacturing practices. Nonetheless, some barriers that cannot be neglected to include the cost of purchasing, ignorance and legal inrerfence. The way forward to break such barriers will involve joint strategies in which SMEs, large firms, policy formulators, and consumers have a common sustainability agenda. The aforementioned competitiveness of SMEs that implement sustainable GMP is further boosted by the emerging global trends in the demand of environmentally-friendly products, which allows them to enter more high-end markets and make them more resilient.

REFERENCES

Agrawal, R., Majumdar, A., Kumar, A., & Luthra, S. (2023). Integration of artificial intelligence in sustainable manufacturing: Current status and future opportunities. *Operations Management Research : Advancing Practice Through Research, 16*(4), 1720–1741. DOI: 10.1007/s12063-023-00383-y

AGRINFO. (2023). *Overview: EU Farm to Fork and Green Deal Initiatives*. CO-LEAD. Retrieved [Insert retrieval date], from https://agrinfo.eu/book-of-reports/overview-eu-farm-to-fork-and-green-deal-initiatives/

Ahmadov, T., Durst, S., Gerstlberger, W., & Kraut, E. (2025). SMEs on the way to a circular economy: Insights from a multi-perspective review. *Management Review Quarterly, 75*(1), 289–322. DOI: 10.1007/s11301-023-00380-2

Aivazidou, E., Tsolakis, N., Vlachos, D., & Iakovou, E. (2018). A multi-objective optimization model for designing sustainable supply chain networks. *Journal of Cleaner Production, 174*, 473–486. DOI: 10.1016/j.jclepro.2017.10.318

Arsawan, I. W. E., Koval, V., Suhartanto, D., Hariyanti, N. K. D., Polishchuk, N., & Bondar, V. (2024). Circular economy practices in SMEs: Aligning model of green economic incentives and environmental commitment. *International Journal of Productivity and Performance Management, 73*(3), 775–793. DOI: 10.1108/IJPPM-03-2022-0144

Asif, M., Fisscher, O. A. M., de Bruijn, E. J., & Pagell, M. (2010). Integration of management systems: A methodology for operational excellence and strategic flexibility. *Operations Management Research : Advancing Practice Through Research, 3*(3–4), 146–160. DOI: 10.1007/s12063-010-0037-z

Bandeira, G. L., Ferasso, M., & Tortato, U. (2025). Circular economy maturity framework for SMEs. *Resources, Conservation & Recycling Advances, 27*, 200275. DOI: 10.1016/j.rcradv.2025.200275

Carroll, A. B. (1999). Corporate social responsibility: Evolution of a definitional construct. *Business & Society, 38*(3), 268–295. DOI: 10.1177/000765039903800303

Centobelli, P., Cerchione, R., Esposito, E., & Oropallo, E. (2022). Knowledge management in sustainable supply chains: A systematic literature review. *Journal of Cleaner Production, 331*, 129–153. DOI: 10.1016/j.jclepro.2021.129994

Chertow, M. R. (2000). Industrial symbiosis: Literature and taxonomy. *Annual Review of Energy and the Environment, 25*(1), 313–337. DOI: 10.1146/annurev.energy.25.1.313

De Marchi, V., Di Maria, E., & Micelli, S. (2013). Environmental strategies, upgrading and competitive advantage in global value chains. *Business Strategy and the Environment, 22*(1), 62–72. DOI: 10.1002/bse.1738

European Commission. (2022). EudraLex – Volume 4: Good manufacturing practice (GMP) guidelines. Retrieved from https://health.ec.europa.eu

European Green Deal. (2025). In *Wikipedia*. Retrieved from https://en.wikipedia.org/wiki/European_Green_Deal

Frank, A. G., Mendes, G. H., Ayala, N. F., & Ghezzi, A. (2019). Servitization and Industry 4.0 convergence in the digital transformation of product firms: A business model innovation perspective. *Technological Forecasting and Social Change, 141*, 341–351. DOI: 10.1016/j.techfore.2019.01.014

Freeman, R. E. (2010). *Strategic management: A stakeholder approach*. Cambridge university press. DOI: 10.1017/CBO9781139192675

Frosch, R. A., & Gallopoulos, N. E. (1989). Strategies for manufacturing. *Scientific American, 261*(3), 144–153. DOI: 10.1038/scientificamerican0989-144

Galdos-Urbizu, L., Zubeltzu-Jaka, E., Erauskin-Tolosa, A., & Etxeberria, I. A. (2024). The impact of green supply chain management on corporate environmental performance and the moderating roles: A meta-analysis. *Business Strategy & Development, 7*(1), e351. DOI: 10.1002/bsd2.351

Garza-Reyes, J. A. (2015). Lean and green–A systematic review of the state of the art literature. *Journal of Cleaner Production, 102*, 18–29. DOI: 10.1016/j.jclepro.2015.04.064

Geels, F. W. (2002). Technological transitions as evolutionary reconfiguration processes: A multi-level perspective and a case-study. *Research Policy, 31*(8-9), 1257–1274. DOI: 10.1016/S0048-7333(02)00062-8

Geissdoerfer, M., Savaget, P., Bocken, N. M., & Hultink, E. J. (2017). The Circular Economy–A new sustainability paradigm? *Journal of Cleaner Production, 143*, 757–768. DOI: 10.1016/j.jclepro.2016.12.048

Holling, H., & Backhaus, L. (2023). A meta-analysis of green supply chain management practices and firm performance. *Sustainability (Basel), 15*(6), 4730. DOI: 10.3390/su15064730

Intergovernmental Panel on Climate Change. (2023). Climate change 2023: Synthesis report of the IPCC Sixth Assessment Report (AR6). IPCC. https://www.ipcc.ch/report/ar6/syr

Jabbour, C. J. C., Jabbour, A. B. L. S., Foropon, C., & Godinho Filho, M. (2018). When titans meet—Can industry 4.0 revolutionize the environmentally-sustainable manufacturing wave? The role of critical success factors. *Technological Forecasting and Social Change, 132*, 18–25. DOI: 10.1016/j.techfore.2018.01.017

Jamwal, A., Agrawal, R., Sharma, M., & Giallanza, A. (2021). Industry 4.0 technologies for manufacturing sustainability: A systematic review and future research directions. *Applied Sciences (Basel, Switzerland), 11*(12), 5725. DOI: 10.3390/app11125725

King, A. A., & Lenox, M. J. (2001). Lean and green? An empirical examination of the relationship between lean production and environmental performance. *Production and Operations Management, 10*(3), 244–256. DOI: 10.1111/j.1937-5956.2001.tb00373.x

Kolbert, E. (2023, March 20). The IPCC's final warning on climate change. The New Yorker. https://www.newyorker.com

Lewandowski, M. (2016). Designing the business models for circular economy—Towards the conceptual framework. *Sustainability (Basel), 8*(1), 43. DOI: 10.3390/su8010043

Liu, W., Cao, Y., Hou, J., Cheng, Y., Chan, H. K., & Tang, O. (2024). Green procurement or green supply? A meta-analysis of their impacts on firm sustainability performance. *International Journal of Logistics Research and Applications*, 1-35.

Macarthur, E. L. L. E. N., & Heading, H. E. A. D. I. N. G. (2019). How the circular economy tackles climate change. *Ellen MacArthur Found, 1*, 1–71.

Martins, A., Branco, M. C., Melo, P. N., & Machado, C. (2022). Sustainability in small and medium-sized enterprises: A systematic literature review and future research agenda. *Sustainability (Basel), 14*(11), 6493. DOI: 10.3390/su14116493

Nonaka, I. (1994). A dynamic theory of organizational knowledge creation. *Organization Science, 5*(1), 14–37. DOI: 10.1287/orsc.5.1.14

Nonaka, I. (2009). The knowledge-creating company. In *The economic impact of knowledge* (pp. 175–187). Routledge.

Organisation for Economic Co-operation and Development (OECD). (2022). OECD guidelines for responsible business conduct. OECD Publishing. https://www.oecd.org/

Palombi, G., Schiaroli, V., Fraccascia, L., & Nonino, F. (2024). Identifying barriers and good practices for implementing circular economy principles in small and medium enterprises. *Business Strategy and the Environment*, *33*(7), 6773–6794. DOI: 10.1002/bse.3835

Pharmaceutical Inspection Co-operation Scheme. (2023). *PI 006-4: Good Manufacturing Practice (GMP) Guide, Part I* (Effective August 2023). Retrieved July 18, 2025, from https://picscheme.org/docview/6606

Rockström, J., Steffen, W., Noone, K., Persson, Å., Chapin, F. S.III, Lambin, E. F., Lenton, T. M., Scheffer, M., Folke, C., Schellnhuber, H. J., Nykvist, B., de Wit, C. A., Hughes, T., van der Leeuw, S., Rodhe, H., Sörlin, S., Snyder, P. K., Costanza, R., Svedin, U., & Foley, J. A. (2009). A safe operating space for humanity. *Nature*, *461*(7263), 472–475. DOI: 10.1038/461472a

Santos, A. D. M., & Sant'Anna, Â. M. O. (2024). Industry 4.0 technologies for sustainability within small and medium enterprises: A systematic literature review and future directions. *Journal of Cleaner Production*, *467*, 143023. DOI: 10.1016/j.jclepro.2024.143023

Saz-Carranza, A., & Ospina, S. M. (2011). The behavioral dimension of governing interorganizational goal-directed networks—Managing the unity-diversity tension. *Journal of Public Administration: Research and Theory*, *21*(2), 327–365. DOI: 10.1093/jopart/muq050

Sellitto, M. A., de Lima, M. S., Ackermann, A. E. F., Kadel, N.Jr, & Butturi, M. A. (2025). Exploring Industrial Symbiotic Networks: Challenges, Opportunities, and Lessons for Future Implementations. *Sustainability (Basel)*, *17*(4), 1509. DOI: 10.3390/su17041509

Silva, M. G., de Carvalho, T. S., Castagna, A. G., do Rocio Strauhs, F., & Piekarski, C. M. (2022). The role of online platforms to enable the process of industrial symbiosis: An analysis of tools available in the market. *Cleaner Production Letters*, *3*, 100021. DOI: 10.1016/j.clpl.2022.100021

The Guardian. (2023, March 20). IPCC climate report: The world must act now. The Guardian. https://www.theguardian.com

United Nations. (2015). Transforming our world: The 2030 Agenda for Sustainable Development. United Nations. https://sdgs.un.org/2030agenda

U.S. Food and Drug Administration. (2022). Current good manufacturing practice (CGMP) regulations. 21 CFR Parts 210 and 211. Retrieved from https://www.fda.gov

World Health Organization. (2014). WHO good manufacturing practices for pharmaceutical products: Main principles. (WHO Technical Report Series, No. 986, Annex 2). Geneva: World Health Organization. Retrieved from https://www.who.int

World Health Organization. (2020). Good manufacturing practices and inspection. World Health Organization. https://www.who.int/

World Health Organization. (2023). *Good manufacturing practices (GMP)*. Retrieved July 18, 2025, from https://www.who.int/teams/health-product-policy-and-standards/standards-and-specifications/gmp

Zeng, S. X., Shi, J. J., & Lou, G. X. (2007). A synergetic model for implementing an integrated management system: An empirical study in China. *Journal of Cleaner Production*, *15*(18), 1760–1767. DOI: 10.1016/j.jclepro.2006.03.007

Chapter 6
Enabling Environmental Sustainability in Local Industrial Systems Through Management Information System

A. Rahmath Nisha
https://orcid.org/0009-0001-5146-2411
Saveetha School of Management, Saveetha Institute of Medical and Technical Sciences, Chennai, India

A. K. Subramani
https://orcid.org/0000-0003-0783-7113
Saveetha School of Management, Saveetha Institute of Medical and Technical Sciences, Chennai, India

K. C. Ranjithaa
https://orcid.org/0000-0003-2445-0623
Saveetha School of Management, Saveetha Institute of Medical and Technical Sciences, Chennai, India

ABSTRACT

The paper lingers on the critical role that the Management and Information System (MIS) play in enhancing the activities of the local industrial systems in the environmental sustainability. It submits the significance of the MIS and other online platforms to the ecological accountability by incorporating environmental observation, compliance system and business sustainability. The research article will talk about the use of Environmental Management Systems (EMS) like, ISO 14001, Environmental

DOI: 10.4018/979-8-3373-6435-3.ch006

Management Information System (EMIS) and Sustainability Information System (SIS) in the attainment of resource efficiency, reduction of emissions and responsibility of an organization. The paper identifies the increased importance of the new technologies of cloud-based platforms, the Internet of Things (IoT), digital twins, predictive analytics, and the tools of sustainability planning. It also shows that these technologies will be useful in making real-time decisions, enhancing the ability to assess a risk, and making more informed and sustainable decision-making.

1.0. INTRODUCTION

The industrial systems in the world are progressively being put under strain to reduce the environmental externalities of the production processes that are resource intensive. With decarbonization, circularity and ecological responsibilities rapidly becoming the main focus of global policy regimes, organizations are moving to embracing sophisticated information infrastructures that can effectively monitor, manage and enhance their environmental impacts. The life-cycle assessment indicates that among the various activities that the industries are involved with, there are harmful effects on the environment at all the stages including the pre-production and manufacturing process, through to the usage and the final disposal of the products. These results highlight the urgency of the implementation of the overall and data-driven sustainability strategies (Bonamente et al., 2014).

In this scenario, the Management Information Systems (MIS) have been developed to be more than the traditional administration tools, but a socio-technical mechanism, which are able to facilitate the large-scale transformation of the environment. The information systems contribute significantly to the role of enhancing ecological responsibility through sensing, reporting, and making informed decisions. With the help of these functions, IS contribute to the minimization of emissions, resource optimization, and waste of operations. (Elliot, 2011; Melville, 2010; Watson et al., 2010).

Such systems enable the organizations to work with high volume of environmental data as well as incorporate sustainability concerns in core processes. Green IS also helping in changing organizational behavior, influencing environmental attitudes (Melville, 2010), enables virtual work arrangements which consume less energy (Bose and Luo, 2011), and also helps in real time monitoring of carbon through advanced analytics and business intelligence (Seidel et al., 2013). MIS now has an increased role in environmental risk assessment (ERA), whereby the ability to combine spatial, temporal, and ecological data is used to identify hazards, the exposure pathways, and mitigation strategies (Ghorbanzadeh et al., 2018; Amiri et al., 2024).

At the institutional level, Environmental Management System (EMS) exemplified by ISO 14001 offer systematic frameworks, which are useful in supporting governance, adherence to regulations, and ongoing performance enhancement. Such standards assist organizations in institutionalizing accountability, transparency, and continuous improvement and create internal and external gains such as improved productivity, cost savings, and improved stakeholder relationships (Machado et al., 2013; Gavronski et al., 2008; Arimura et al., 2016). However, the challenges that organizations are still experiencing during implementation are related to cost of infrastructure, training needs, and changes in regulations that change frequently (Oliveira and Serra, 2010).

2.0. OBJECTIVES OF THE CHAPTER

This chapter is based on the developments and therefore aims at achieving four goals:

1. To investigate how MIS, EMIS, and SIS improve the environmental performance, compliance, and sustainability results in the local industrial systems.
2. To study digital technologies as an IoT, cloud platforms, digital twins, and predictive analytics as the enhancers of real-time monitoring of the environment and making decisions.
3. To determine organizational, technical and governance issues that affect MIS adoption and scaling towards sustainability.
4. To offer policy-specific information and a map to incorporate MIS frameworks into sustainable industrial development.

This chapter demonstrates that MIS can create local industrial system as resilient, transparent, and eco-friendly systems by connecting technological infrastructures and governance systems.

3.0. METHODOLOGY

This chapter presents a structured narrative review of literature designed to integrate current understanding of how management information systems, environmental management information systems and emerging digital technologies support environmental sustainability in local industrial settings. Systematic search of databases, including Scopus, Web of Science, IEEE Xplore, ScienceDirect, and Google Scholar, with keywords, in particular, MIS, EMIS, sustainability information systems, ISO 14001, IoT, digital twins, and environmental monitoring, gave the literature review an added strength. Priority was given to peer-reviewed studies, reviews, and standard documents between 2018-2025, but it retained earlier foundational work for theorizing the review.

Primarily, screened sources were related to the chapter objectives; and synthesized thematically into certain key domains: MIS functions, EMS/ISO frameworks, digital sensing technologies, governance and data ethics; and organizational adoption challenges. It helped in creating a coherent analytical framework to incorporate some diverse empirical and conceptual insights. Thus, while systematic in search and selection, the review does not claim to be exhaustive; rather, it informs into a well-furnished, but conceptually dense synthesis, to buttress the arguments and recommendations made in this chapter.

4.0. LIMITATIONS AND ETHICAL CONSIDERATIONS

This chapter gives an integrative synthesis of the literature on MIS, EMIS, and EMS including new digital technologies, but there are quite a few limitations that should be accepted. First, the review is by no means comprehensive because of the broad and interdisciplinary nature of the topic. This is despite attempts made to follow systematic search procedures, but it is possible that valuable studies in specialized engineering or region-specific databases may not have surfaced. Second, the heterogeneity of the sources ranging from conceptual papers to policy documents and technical reports will limit the systematic comparison of findings. Third, the chapter focuses largely on digital sustainability within industrial scenarios; thus, its relevance regarding public sector or community-level environmental systems may be limited.

There are also ethical factors when digital technologies are incorporated in environmental management procedures. The application of IoT sensors, cloud systems, AI analytics, and digital twins brings up the issues of privacy, data ownership, transparency of algorithms, and cybersecurity. The environmental information can comprise sensitive operation information and organizations should make sure that the law of data protection in the country and international standards are met. Furthermore, using digital surveillance systems may introduce unequal power relationships between managers, employees, and regulators by introducing the necessity to provide transparent principles of governance of data access, consent, and responsible use. Lastly, ethical practice means that sustainability technologies must not have a disproportionately impact on smaller companies or marginalized groups and that any digital solution must be reachable, fair, and socially responsible.

5.0. LITERATURE REVIEW

Environmental sustainability is becoming a pressing issue as industrial systems are among the biggest contributors of carbon emissions, resource use and waste. Life-cycle analyses have revealed that industrial buildings have environmental effects at all stages, including raw material mining and final demolition pointing to the necessity of developed environmental control systems (Bonamente et al., 2014).

The research in the area of Information Systems (IS) tends to introduce technology as a radical option of enhancing ecological performance. The literature proposes that IS can assist organizations to minimize their footprint on the environment by facilitating sensing, reporting and decision support functions (Elliot, 2011; Melville, 2010; Watson et al., 2010). Modern research also underlines the idea that Green Information Systems can be adopted strategically to improve the performance of manufacturing, transparency and readiness to sustainability to the stakeholders (Bai et al., 2023). Green IS will help in organizational sustainability because it influences the behavior of employees, allowing them to work virtually and remotely and have real-time environmental analytics (Bose and Luo, 2011; Seidel and others, 2013).

Environmental management systems (EMS) especially the ISO 14001 supplements IS, and gives systematic governance structures of continuous improvement. It has been demonstrated that the ISO 14001 certification has been positively associated with improved environmental performance, operational effectiveness, and stakeholder trust, despite the challenges of implementation that can arise due to the lack of resources, changing laws, and the need to train employees (Machado et al., 2013; Oliveira and Serra, 2010; Arimura et al., 2016). Environmental Management Information System (EMIS) and Sustainability Information System (SIS) increase the capacity of the organization by automating the process of data collection, audit and sustainability report (Rikhardsson, 2001; El-Gayar and Fritz, 2006).

The latest studies emphasize that MIS in collaboration with AI-assisted analytical models contributes to the accuracy, reliability, and transparency of sustainability reporting to a significant degree, enhancing the accountability of organizations (Sharma and Frias-Aceituno, 2024). Advanced digital technologies like the cloud computing, IoT, and edge computing have changed environmental monitoring. The IoT sensor systems can now be used to track pollutants, emissions, energy use, and water quality in real-time, thus shedding light on the more widespread environmental management (Onukwulu et al., 2021; Ajayi et al., 2022; Zhao et al., 2019).

Digital twins complement the sustainability modeling even more by providing the virtual replicas of the industrial systems that make the predictive maintenance and scenario planning as well as optimization of the emissions possible (Negri et al., 2017; Wu et al., 2023). Environmental decision-making with the use of AI has become one of the primary tendencies, and systematic reviews have proven its usefulness in risk prediction, energy optimization, and production efficiency (Zhou et al., 2025).

Nonetheless, such developments present a governance issue, digital sustainability involves proper regulations about data ownership, provenance, compliance, accountability, and privacy, which are well-known problems in the literature about data governance (Lee et al., 2017). Ethical contradictions are observed when there is a collision between transparency and data security and privacy rights of individuals (Solomon et al., 2024; Uzougbo et al., 2024).

Digital literacy rate of people comprising the workforce, as well as the structural design of an organization, is a decisive factor in the successful implementation of MIS. The bureaucratic hierarchies can limit the knowledge circulation and reduce innovativeness, whereas inequalities in the access to digital devices and abilities can weaken the efficiency of systems (Martínez-Leon and Martinez-Garcia, 2011; Eshet, 2004; Hargittai, 2010). Besides, the complexity of procurement, ambiguity of IP, and the lack of stakeholder involvement are still significant obstacles (Penn et al., 2019).

Altogether, the literature indicates that technologies, including MIS, EMIS, SIS, IoT, digital twins, and AI have a major effect on better environmental sustainability results. The success of their implementation and longer-term effectiveness, though, is being reliant on the existence of good governance frameworks, policy consistency, organizational capacity, and labor readiness.

6.0. CONCEPTUAL FRAMEWORK

Figure 1. Conceptual model of digital information systems supporting environmental management

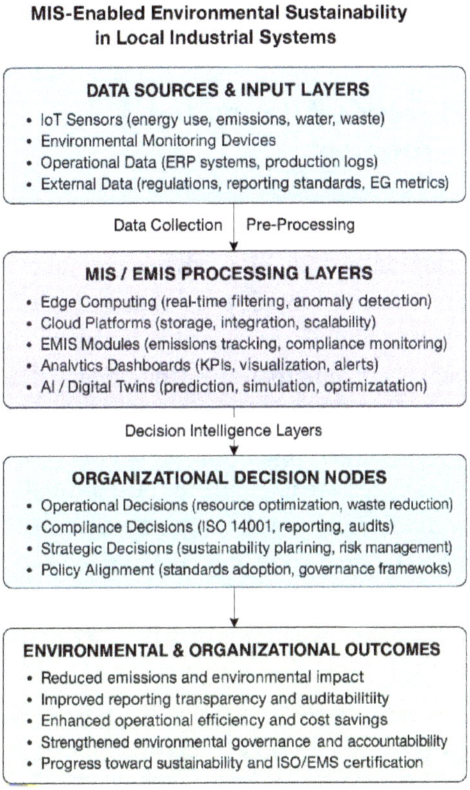

7.0. DISCUSSION

It reveals that MIS goes a long way in enhancing environmental sustainability through better monitoring, compliance, and decision-making.

7.1 MIS as a Facilitator of Evidence-Based Sustainability Decision-Making

MIS gives organizations the ability to convert unprocessed environmental information into operational knowledge. Combined with other EMS frameworks, including ISO 14001, MIS helps to comply, improve continuously, and verify the performance. However, there is still a significant issue of necessity to incorporate a variety of different data sources in many organizations, which leads to disjointed and even less coherent environmental data. This loophole reduces the capacity of MIS to aid strategic decision-making, especially in complicated industrial environments where environmental risks are multi-dimensional.

7.2 Technological Trends Altering the Governing of Environment

New technologies such as IoT, cloud analytics, and AI are transforming the way industries are monitoring and controlling environmental performance. Real-time emission and resource monitoring is made possible through IoT, whereas digital twins have the ability to predict modeling that is critical in ensuring that failures or risks to sustainability of systems are anticipated. AI-based analytics have the potential to detect inefficiencies and predict the environmental effects more accurately, yet also bring their own risks into existence, including algorithmic bias, lack of transparency, cybersecurity threats, and excessive dependence on data quality. These trends reflect opportunity, however, the necessity of a strong system of digital governance that helps to use environmental data responsibly and ethically.

7.3 Systemic Loopholes and Systemic Difficulties

Although the trend is expected to improve with digital advancement, companies still experience ongoing issues such as low levels of digital literacy, change resistance, complexity in procurement, and technology ownership issues. Bureaucracies can be counterproductive to sharing knowledge and implementations of the sophisticated environmental systems. Moreover, the issue of adoption is usually exaggerated within the local industrial systems when the resources are scarce and technological readiness uneven, which restricts the potential gains of MIS.

8.0 ENVIRONMENTAL MANAGEMENT SYSTEMS (EMS) IN LOCAL INDUSTRIAL SYSTEMS

8.1 Principles and Standards (e.g. ISO 14001)

In this context, the principles developed by (IBGC,2015) are:

Disclosure: Entails of the aspiration to disclose to the interested party's information, which is beneficial to the latter and not necessarily that imposed by the laws or regulations

Fairness: It is represented by the just and reasonable behavior of each of the partners and other stakeholders in consideration to their rights, needs, responsibilities, interests and expectations

Accountability: Governance agents should be liable to their actions in an explicit, succinct, understandable and timely manner, on pure assumption of the effects of their actions and failures and be absolutely careful and responsible on their part

Compliance: The Governance agents basically screen the economic and financial viability of the organizations, push down the negative impacts of the business and activity as well as escalate the positive ones, considering their business model the multiple resources (human, intellectual, manufactured, financial, social, environmental, reputational etc.) in the short-term, the medium term and the long-term.

The ideologies of corporate governance declared by the IBGC, in particular, the principles of transparency, including the active sharing of the information that would be required by the laws, and accountability, including the imperative that the organization is responsible to society in the way its agents and its structure operate, propagate long-term sustainability by adversarial social and environmental issues.

The idea of sustainability is based on the fact that every single entity such as a public and even a private organisation or a person has to have a responsibility to the society, and the more people become aware of the ecological effects of corporate and human activities, the more sustainable their deeds will be. Corporate sustainability is typically used to describe the position of a firm in sustaining itself in economic terms and having a seamless relationship with the society and environment (BM&F-BOVESPA, 2015).

Strategic investments in sustainability benefit businesses by generating competitive advantages and value in a business, thus improving environmental and social outcomes. Therefore, to develop the vision and strategy of the firm, it is crucial to generate value on the basis of social and environmental aspects, taking into consideration internal processes and culture and enhancing relations with the stakeholders (Martínez et al., 2016).

In that respect, (Morgan et al.,2009) highlight that one of the societal responsibilities of a firm is to create wealth by protecting and conserving the natural resources of a growing scarcity. According to (Baumgartner and Rauter, 2017), a sustainability-oriented company is defined by its readiness to focus on social and environmental effects by engaging in activities that respect the environment, adjusting its products, processes and organisational patterns to demands by the stakeholders in its operations.

The ISO 14000 series, created by the International Organization Standardisation (ISO) in 1947 out of 162 member countries, gives organisations of all categories viable resources that help them manage their environmental performance (ISO, 2014). In that context, the Environmental Management System (EMS), as per NBR 14001, is an effective management tool that organisations that aim at attaining their environmental objectives should adopt to benefit the society as well as environment conservation.

(Machado et al.,2013) observe that those companies that are certified under the NBR standard ISO 14001 are more committed to environmental management through programmes, initiatives, and management of resources; such organisations therefore distinguish themselves in the market by providing ecological benefits to the society as a whole. (Gavronski et al.,2012) found three sources of ISO 14001 certification, which include firms with internal orientation, firms with external orientation, and a cluster of companies with integrated orientation. Firms with core emphasis are characterised by the centrism of processes and internal resources; the outward orientation face organisational demands that dictate the social environment; the holistic approach taken by an organisation lays much focus on all motivational factors, whether internally or externally based.

According to (Oliveira and Serra, 2010), the main benefit of adopting the EMS regarding the ISO 14001 requirements is the promotion of environmental programs that minimize unexpected costs. The further benefits are reduced use of water, electricity, oil, etc., a positive impact of the EMS on organisational processes, reputation, as well as compliance with environmental laws and environment-related standards imposed by customers. In the same study conducted by (Oliveira & Serra, 2010) regarding the challenges of implementing the EMS in accordance with ISO 14001, it is stated that there was a rise in expenses for the companies involved, including the necessity to modify the organizational structure, provide training, implement awareness programs, upgrade equipment, and perform audits.

Lastly, as noted by ISO (Arimura et al., 2016), users of ISO 14001 have indicated that the standard assists in demonstrating compliance with legal and regulatory obligations; enhances the involvement and commitment of both leadership and employees; boosts the company's reputation and stakeholder confidence; achieves strategic business goals by integrating environmental considerations into corporate management; offers a competitive and financial edge through improved efficiency and cost savings; and promotes superior environmental performance among suppliers by incorporating them into the organization's operational frameworks.

9.0. ROLE OF INFORMATION SYSTEMS IN SUSTAINABILITY MONITORING AND CONTROL

9.1 Enterprise Resource Planning (ERP)

Enterprise resource planning systems are tailored information system packages that integrate information and information-based processes within and between different functional departments of an organization. The recent generation of the ERP systems also has reference models or process templates which claim to reflect the best current business practices. Already in 1969, Blumenthal suggested an integrated architecture and organizational information system structure (Blumenthal, 1969). The major organizational and technical complexities associated with their development, however, are involved. and implementation has rendered it difficult to achieve integrated enterprise-wide systems in. practice.

Considering the challenges faced by various organizations in coming up with in-house integrated systems, first generation enterprise resource planning (ERP) system packages came into view in the manufacturing industry. The ERP systems in the leading manufacturing companies like Boeing, Mercedes-Benz, BMW, and Ford gained their foundational bases on European companies, including SAP and Baan, which was known to be skilled in industrial engineering, manufacturing, and operational management.

9.2 Environmental Management Information Systems (EMIS)

The ISO 14001 standard defines environmental management system (EMS) as the part of the total management system, which includes the structure of the organization, planning activities, responsibilities, practices, procedures, processes, and resources to develop, implement, achieve, review, and maintain environmental policies (ISO 14001). Environmental information management is also a central element of the EMS because it involves the management of the environmental information of the organization. Thus, an Environmental Management Information System (EMIS) is aimed at automating and streamlining environmental management operations, auditing and compliance measures (Rikhardsson, 2001).

Such environmental management systems standards as ISO14001 (ISO,2002) and the European Eco-management and Audit System (EMAS) (EMAS,2006), are designed to provide an effective practical base to the management of the environment in organisations. These have been illustrated by the example of Siemens who was able to create an eco-efficiency assessment and trade-off matrix, or the case of SC Johnson and Company which used a scorecard-based approach to appraise the performance of its downstream partners in the supply chain. To achieve product stewardship the environmental issues must directly influence the research and development as well as the design decision-making. In recent times, sustainable supply chain (SSC) management has received enormous attention from academics and industries (Bhalaji et al.,2024).

Also, the example of BASF and Siemens integrating the concept of product life-cycle analysis into their design development is an example of the anticipatory aspect of this stage of development. The idea of the triple bottom line is anticipated and then executed at the ideal level of sustainable development. Although at this point, we cannot provide specific real-life examples, we believe that the strategies of such companies as Nike, Statoil, BP and Chiquita Brands International are groundbreaking and pioneering steps toward the resolution of the delicate challenge of the social aspect of the triple bottom line and corporate social responsibility (El-Gayar and Fritz, 2006). Goals for the sustainability have triggered the emphasis of Economic Production Quantity models, which was previously limited to financial considerations to include the environment aspects as well (Sugapriya et al., 2024).

9.3 Sustainability Information Systems (SIS)

The global climate change, scarcity of water, minerals, energy, and land, degradation of biodiversity, acid rain, poisoning with hazardous and toxic substances, and a variety of other environmental problems will influence our lives and survival in the short and in the long term. Such issues demand a reconsideration of how we produce, consume and use goods and services and technology. Sustainability requires that we put into consideration the requirements of the present generation as well as the future generation (Sarkis et al., 2013).

The concept of green growth, which can be described as the separation of economic growth and the environment, where the goal of the government is to achieve a win-win nexus between economic development and environmental improvement, has become the focus of the policy agenda (Vázquez-Brust & Sarkis, 2012). Within the context of business and organisational management, greening of supply chains, operations, strategies, human resources, marketing and technology are already being implemented to help industries to negotiate these issues of the environment and social responsibility.

Some of these areas have researched greening concerns decades ago, which has laid the relatively progressive research agenda. On the other hand, the Information Systems (IS) and Information Technology (IT) management has shown a mixed development and lack of research in regard to greening and sustainability. There has been a notably inadequate involvement of the IS and IT scholars in this discourse. It is in realisation of this lack that Green IS and IT have received increased consideration as a viable issue in recent years, with attendant research efforts (Jenkin et al., 2011).

Green IS about improving the information flow and management, but Green IT focuses on hardware and other supporting infrastructure that can be optimized and designed to consider the environmental factor (Sarkis and Zhu, 2008). The voices of socially responsible scholars have been raised with the formation of special interest groups (SIGs) devoted to Green IS and IT as this community grows larger and larger (Sarkis et al., 2013).

10.0. BUILDING SHARED ENVIRONMENTAL INFORMATION PLATFORMS

10.1 Case Study Examples

Emission monitoring in the industrial parks through cloud-based approaches and the implementation of Internet of Things (IoT) tools to monitor the effluents.

Monitoring of emissions in industrial parks through cloud-based platform and monitoring of effluent with Internet of Things (IoT) technologies. One of the most current technologies that offer the opportunity to monitor the carbon footprint in real-time is the Internet of Things that is highly sophisticated with the scope of tracking and improving numerous environmental sustainability efforts (Onukwulu et al., 2021; Elete et al., 2023).

By combining smart sensors, wireless communication technologies and advanced data analytics, the IoT can be used to monitor carbon emissions in a variety of industrial and infrastructural segments, including transportation, industrial production, and urban infrastructure. The paper discusses the main peculiarities of the IoT application in environmental monitoring, sensor and network functions, the role of wireless communication and edge computing, the difficulties of the implementation of large-scale IoT-based carbon tracking. Environmental monitoring IoT system could be defined as the implementation of connected gadgets that gather, disseminate, and process data about the environment to implement stewardship of the same (Oluokun, 2021). Figure 2 shows the use of IoT to measure the carbon-footprint.

Figure 2. The core components of IoT-based carbon footprint monitoring (Ojadi et al.,2023)

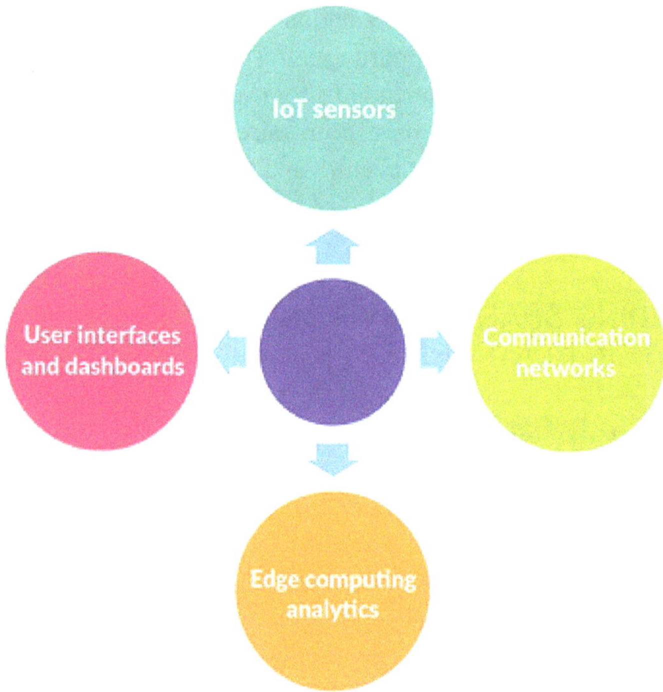

The IoT sensors are devices that will be utilized to monitor different environmental conditions such as CO2 emissions, air quality, temperature, and energy usage (Onukwulu et al., 2022). Infrastructures of communication are a wireless and wired structure that is utilized to relay accrued data either to the cloud-based or edge-computing system that is utilized to perform the analysis. Edge computing and cloud analytics can be defined as the type of computing models that receive and process real-time data streams, therefore, generating actionable data to minimize carbon emission (Ajayi et al., 2022).

User interfaces and dashboards are viewed as avenues where visualized data can be availed to the stakeholders such as policymakers, businesses, and researchers in order to help them in the process of decision making (Akhigbe et al., 2021). The IoT can help in sustainability efforts by ensuring the continuity of surveillance and a responsive reaction to the emerging trends of carbon emission hence the need to use a combination of these divergent aspects. The structure and implementation of the IoT-based detector networks have a capability to encompass the colourful environmental parameters such as the air quality, water quality and the biodiversity that can reduce the energy usage and environmental footprint (Banu et al.,2024).

One of the aspects of carbon footprint monitoring that can be enhanced with the introduction of IoT technologies is the implementation of advanced sensors and network infrastructure to gather environmental data in real-time (Ilager et al., 2020). Various sensor typologies are greatly applied in environmental surveillance and these include CO_2 sensors that measure the level of carbon dioxide found in the industrial environment, in cities, as well as in transit systems. Air-quality sensors are used to measure the amount of particulate matter (PM), nitrogen oxides (NOx), and volatile organic compounds (VOCs) to determine the level of pollution. Smart grids and energy meters are the systems that monitor energy consumption and emissions produced by the power plants, buildings, and electric vehicles (Diahovchenko et al., 2020).

These sensors are effectively placed in urban centres, industrial areas and supply chains in order to obtain high resolution environmental information. The information obtained is then relayed through IoT services like 5G, Wi-Fi, and Low Power Wide Area Network (LPWAN) and, therefore, facilitates smooth communication between the distributed sensor nodes. The IoT system of environmental monitoring cannot be supported by wireless communication (Tao, 2020).

In smart buildings and factory premises, localized monitoring using short range wireless technology Bluetooth and Zigbee are used. In addition to the good communication structures, edge computing has become more significant in the process of tracking the IoT. Edge computing is one of the processing types, which is applied at the sensor or gateway nodes instead of transmitting the information to the central cloud servers (Zhao et al., 2019).

This solution will improve real time decision making, minimize latency and bandwidth is also conserved. An example of this would be an artificial intelligence system that is edge based and analyses the data on the spot emissions and sends warnings or optimization procedure when it surpasses the threshold of the statute. Combined with the wireless communication and edge computing, the IoT systems become more responsive and performant to reduce the carbon footprints (Poongodi et al., 2020).

In data governance, the balance between privacy, security, and openness is a very tough challenge that requires keen consideration of the interconnections and possible viewing of the three variables into conflict. Ethical data management should include transparency, security, and privacy as its key components (Solomon et al., 2024b; Uzougbo et al., 2024c). To establish a healthy balance between privacy and security and transparency in data management, organisations can play the role of initiating a conception framework or a guideline, which ethics issues will be brought to the forefront and the path leading to logical data management behaviour will be interred. Another of such measures is the implementation of privacy-by-design which involves incorporation of both security and privacy protection in the software and processes in the early phase of data systems design and development. Organisations can reduce the risks of privacy breaches and security breaches by taking the privacy and security issues into consideration throughout the data life-cycle and by holding the accountable and the transparency (Uzougbo et al., 2024a; Yeung and Bygrave, 2022).

To identify weaknesses in how organisations, handle their data, organisations can also undertake privacy impact assessment and security risk assessment and devise mitigation strategies. Organisations can minimise the risk of data breaches and privacy violations and at the same time promote accountability and transparency in their governance procedures by actively responding to the threats and vulnerabilities (Uzougbo et al., 2024b; Wirtz et al., 2022).

The academic and business communities agree that a conceptual gap between management and governance must be accepted (Khatri and Brown, 2010; Lee et al., 2017; Weill& Ross, 2004, ISO).

Table 1. Data governance factors for platform ecosystem (Lee et al., 2017)

Factor	Description
Regulatory environment	Laws, statutes, or court rulings that could impact on ownership and use of data.
Data ownership and access	Identification of the parties that own the platform data, use it, and access it.
Data use case	The intended use of the data collected by PEs, which is a description of the use of the data.
Contribution measurement	Nanometres of measuring contribution in comparison to the value generated by provision of data.
Conformance	An audit for compliance based on strict processes and rules.
Monitoring	Systems of tracking a data supply chain and all activities related to data
Data provenance	Furnishes the means of tracing the history of derivation of the data in an open manner

11.0. ASSIMILATING MANAGEMENT AND INFORMATION SYSTEMS FOR A SYNCHRONIZED ACTION

11.1 Connecting Digital Platforms to Decision-Making and Structures of Governance

(Simon, 2013) also pointed out that the sheer presence of a volume of data cannot be deemed a sufficiently good reason to gather it in one, encompassing information system, which this paper loosens. Therefore, managers no longer need to pre-set the information that could be of value in decision-making in a digital age of cheap storage and machine intelligence. Therefore, data acquisition maximisation has become one of the guiding principles to various organisations.

11.2 Organizations as Coordinated Communication Systems

Organizations use their technological and human resources to gather unstructured data around them, organise the data to make it readable, turn it into information by giving it context and assigning significance, and finally convert the information to be valuable and utilisation (Sproull and Kiesler, 1991; Turnbull, 2017). This process necessitates synchronisation of both the artificial and human agents of the organisation, i.e., the sensors, algorithms and software on the artificial side and the members, workers and users on the human side as shown in Figure 3.

Figure 3. Data-to-knowledge transition: Organization as a coordinated communication system (TAPUS & MANOLACHE)

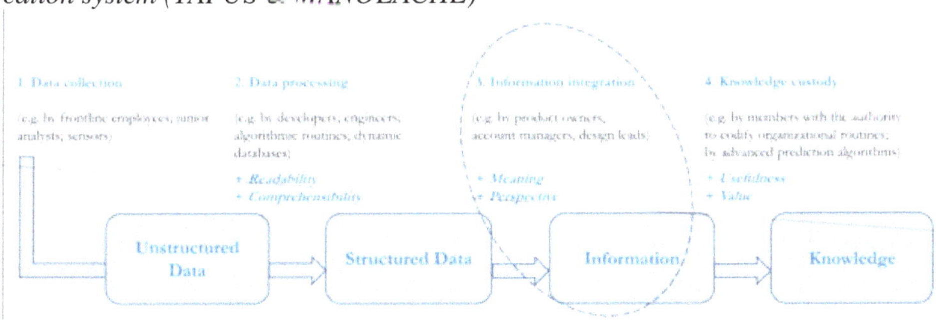

11.3 Decision-Making Systems Based on Businesses

Finally, unstructured data is scaffolded into decision-making by organisations, which gradually turns into structured data to be a part of the communication system. Organizations should therefore, allocate their resources strategically in an effort to attain their goals. These decisions are affected by the sometimes-conflicting motives of the organisational members that are involved in decision-making, which are repeated over time under confusing situations (Shapira, 2002). Whereas in small organisations, any decision can be made by the founder or after consultation with everyone, in large organisations, both approaches cannot be approached. It is not a single person within a group of hundreds of people who has time and mental capability to take into consideration all options, and a group-wide consultation can trigger decision paralysis (TAPUS & MANOLACHE).

11.4 Digitized Twin System and Situation Modelling of Sustainability Planning

The technology of digital twins, predictive analytics, and sustainable project-management approaches is transforming the supply-chain activities as it allows more visibility, decision-making, and a stronger effort towards sustainability. Digital twins create virtual models of physical supply chains (the description is presented in Figure 4) that deliver real-time performance data, as well as simulate possible disruptions (Negri et al., 2017).

Digital twin as an example can simulate factory processes to predict maintenance periods, which minimizes the expenses and downtime. The predictive analytics involves using data in large numbers so that it can predict future risks and optimise operations within the supply chain. The predictive models have the ability to prevent the occurrence of inefficiencies and make changes case by case using both past and real-time data (Waller and Fawcett, 2013). Therefore, the predictive analytics have been used in inventory management where companies have reduced the occurrence of over stocks and stock outs, increased customer satisfaction and saved on costs. (Ortega-Jimenez et al., 2020).

Figure 4. Integration of digital twin technology, predictive analytics, and sustainable project management in supply chain optimization (Wu et al., 2023)

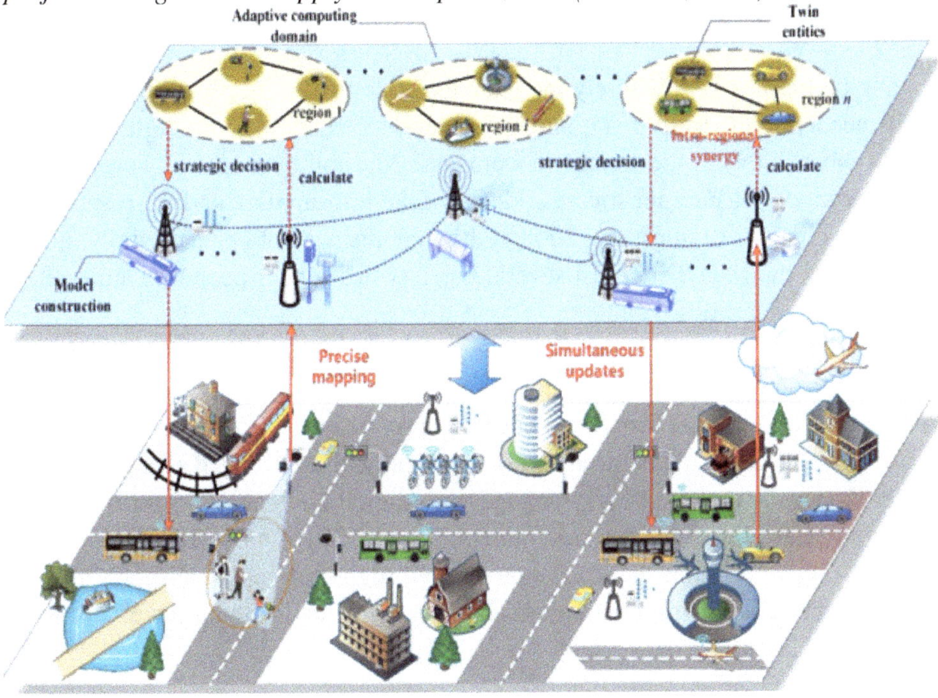

Figure 4 shows how the supply-chain operations have been transformed in collaboration with digital-twin technology, predictive analytics, and sustainable-project management. The idea of a digital-twin technology is developed as an active virtual layer which simulates a physical system including urban infrastructure and transport networks, thus allowing real-time updating and accurate mapping of space. It can be used to model likely disruptions and come up with the appropriate mitigation measures, e.g. bus optimization or factory optimization, which eventually result in cost savings and increased operational efficiency. (Balamurugan et al.,2024) stated that the sun is the source of solar energy which is clean, sustainable, and renewable source of energy which can be converted into electricity by using the solar panels, which are made up of photovoltaic cells

12.0 IMPLEMENTATION CHALLENGE AND ENABLERS

Digital Literacy

Digital literacy is not merely the mere technological proficiency rather it is the skills needed to critically read, judge, and produce content in the digital environment (Eshet, 2004). It additionally encompasses the ability to critically evaluate the trustworthiness of internet sources, identify fake information, and act responsibly about the content (Livingstone, 2008). Although the need of being digitally literate has been on the rise, there exist several barriers that hinder people to attain these life-sustaining abilities.

One of the main challenges is the unequal access to education and digital tools which are also often related to socioeconomic status, geographic location and age (Hargittai, 2010). These disparities have a disproportionate impact on marginalised groups such as older adults, rural and economically disadvantaged groups, and thus affect their ability to effectively engage with digital content. Most individuals in the groups have limited access to high-speed internet, the modern digital devices, or educational programs that can improve their digital literacy (Helsper, 2012). In addition, the deficit in digital literacy is directly associated with low levels of formal education (Hargittai, 2010).

Transparency

(Alsaedi et al., 2022): There are two main types of risks, including social transparency asymmetric and transparency symmetric. Asymmetric social transparency takes place when one party gives more information than the other, which leaves one side having an unequal perception and understanding of the other. Symmetric social transparency is defined by equal disclosure whereby both the parties will be transparent in their information and that there is enough mutual perception. Table 2 provides examples of the risks and risk factors.

Table 2. Examples of risks and risk factors related to transparency sharing practice

Risk Factors	Risks
Symmetric Transparency	Conditional reciprocity between organizational members can be based on the symmetric transparency. Social transparency is demonstrated by employees when their co-workers also demonstrate the same. On the other hand, when the colleagues fail to be open on a regular basis, they get a bad reputation, and other staff members avoid openness with them. Such dynamic puts pressure on workers to maintain both-way transparency and to overcome negative feeling towards others when they believe the transparency has not been achieved.
Asymmetric Transparency	The unequal application of social transparency signifies unequal information dissemination between the team members. In the event some of the members share the information, it can lead to power imbalances where other employees would use the information to enhance their position or to enrich themselves. On the other hand, denial of information to other people is another type of asymmetric social transparency. These unfair disclosure activities may shake up the group cohesion in terms of information asymmetry amongst workers.

Structure of Organization:

The organizational structure outlines the way task distribution and coordination and supervision are ordered in achieving the organizational goals (Mintzberg, 1984). It is described by (Chen and Huang, 2007) as the system of relationships, communication channels, and decision-making processes that are formalized and which affect the allocation of resources as well as the coordination of social interaction among the personnel. (Martínez-León & Martínez-García,2011) also note that the organization structure may have either positive or negative impacts on knowledge creation and information exchange (Nicolini and Meznar, 1995).

The case of specific interest is the example of the bureaucratic model, among the numerous structural variations, which is characterized by centralized decision-making. A bureaucratic system is characterized by a very formalized, hierarchical structure where the power is concentrated at the top thus requiring that all key decisions be implemented by managers at the top. Such institutions have a high degree of specialization that interferes with the appearance of new competencies (Jiménez-Jiménez and Sanz-Valle, 2011) and is counterproductive to the creation of new knowledge, which limits the personal development of staff (Martínez-León & Martínez-García,2011)

Enablers: Capacity building, policy incentives, vendor partnerships, and open-source tools (Penn et al., 2019) are the key drivers in the current information system development.

Development and Updating of the MIS

The management information system (MIS) development and upgrading involve consideration of procurement such as cost complexity and ownership and specificity of the system be it bespoke or generic and the level of stakeholder involvement.

Procurement of the MIS, Including Cost Complexity and Ownership

MIS Procurement, cost complexities evaluation, ownership, etc. are under within the procurement rules under the government of the procurement of the MIS. The issue of cost of MIS development often comes out very strong as a prevailing theme and it shows the balance between idealistic vision and reality. A managerial participant foregrounded ownership issues whereas intellectual property is not clear, especially as MIS development advances with time.

Stakeholder Involvement in the Development of the MIS

The aspect of stakeholder participation during MIS development by the respondents is also universal as they state that the participation of different stakeholders must start at the commencement of the project. Such involvement, however, is inhibited by practical limits of delivery schedules, even though its desirability is recognised. The literature also explains the need to focus on the multi-stage approach of development stating that the strategic role of MIS is not frequently considered.

Open-Source Tools

Vendors have a long-term history of active participation in open-source software (OSS) projects and communities, and have implemented different types of engagement strategies. One of them is using or sending staff to work on targeted OSS projects. As an illustration, a statement released in 2021 stated that Intel has over 100 OSS projects that are worked on by their software engineers such as the Linux kernel, Chromium, and TensorFlow (Intel Corporation, 2021; Business Wire, 2022).

The other approach is the open-source of proprietary software and the development of an adjacent community. In 2015, Microsoft released Visual Studio Code as open-source under a permissive MIT license, keeping the ownership but encouraging community participation (Microsoft, n.d. b; Microsoft, n.d.a). The 2023 studies of Stack Overflow have put VS Code the utmost popular development tool amongst developers (Stack Overflow, 2023). Another approach is the cooperation between several vendors such as Kubernetes project, which is an initiative of Cloud Native Computing Foundation (CNCF) hosting Google, IBM, Microsoft, and VMware

(Cloud Native Computing Foundation, 2015). The figure 5 explains about the different types of OS foundation where an organization can effectively implement.

Figure 5. Classification of open-source foundations (Yenişen Yavuz et al., 2025)

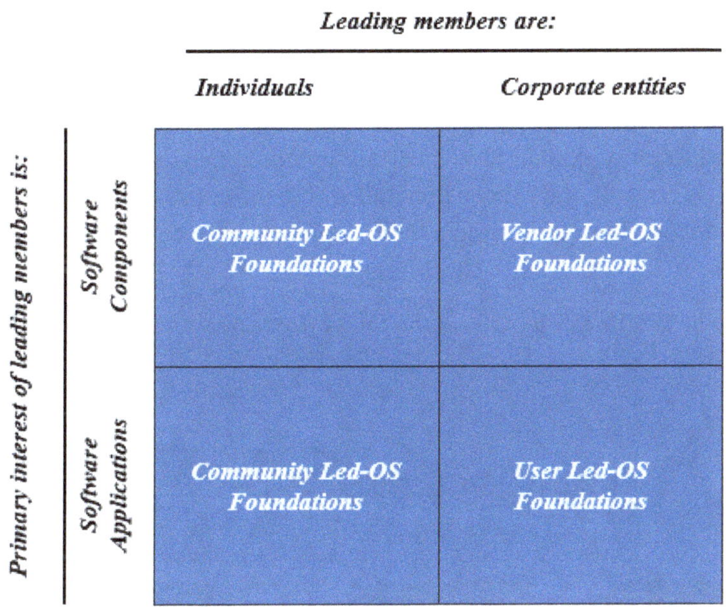

Classification of Open-Source Foundations Explanation (Figure 4)

The Community-Led OS Foundation is made up of individual members who can come up with software to the developers and users. Vendor-Led and User-Led OS Foundations, in comparison, are mostly made up of corporate organizations. Vendor-Led OS Foundations release software ingredients that are meant to be utilized by the developers, whilst User-Led OS Foundations release software applications meant to be consumed by the end users.

13.0 ROADMAP AND TOOLKIT FOR INTEGRATION

The foundation to the analysis of the design process of the information systems, in the contemporary practice, is shifting further forward to the control of the work, the workflow creation tools development, and their management. It is necessary to mention that the notions of business process management and workflow management are different. A business process defines what has to be done, whether it is in the form of creating ratios of inputs and outputs. A business process structure can have a manual aspect, which uses any form of resource (Bass, 2012; Kulvatunyou & Wysk, 2000). Workstreams explain how a certain outcome may be attained (O'Brien et al., 2008). As a result, it has been concluded that the correlation between business process and workflow is too simple to be depicted graphically in Figure 6.

Figure 6. The relationship between business processes and workflows (Shendryk & Boiko, 2015)

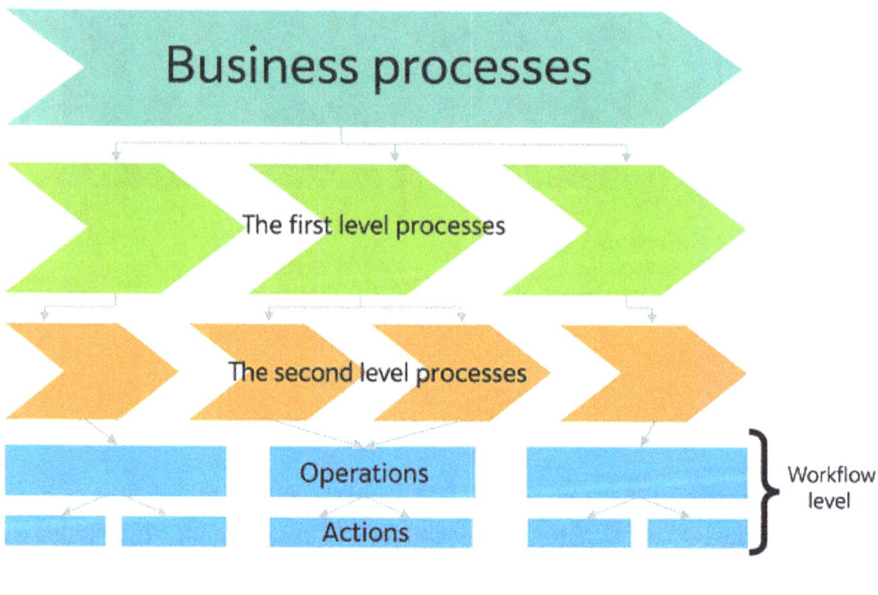

Information systems (IS) developed under the category of manufacturing execution system (MES), however, do not fulfill the present trend of network information systems founded on the virtualization. Virtualization also enables optimization of resource utilization of servers by reduction of equipment down time thereby reducing costs on IT hence centralized management and enhancing security of IT infrastructures (Shendryk & Boiko, 2015).

13.1 Policy Implications

The composite study of MIS, EMS, EMIS, and digital sustainability technologies in the local industrial systems can have a clear policy implication. To begin with, the governments and regulatory agencies ought to concentrate on standardized environmental data protocols to achieve uniformity in reporting across any industry. The Monitoring using MIS depends on the harmonized measures, data format interoperability and the adherence to the national and international reporting standards such as ISO 14001, GRI and the ESG indicators.

Second, the policymakers need to develop digital environmental governance schemes that relate to data ownership, privacy, cybersecurity, and accountability of the system. With the IoT sensors, digital twins, and cloud-based architectures introduced to industries, the regulations should specify roles in the field of data provenance, accessibility, ethical use especially when environmental information affects the health and the community at large.

Third, incentive measures such as fiscal benefits, grants or tax credits should be included in the public policy to industries that implement MIS based systems of sustainability. These subsidies have the potential to hasten the shift in industries with low digital capacity and encourage their adoption by small and medium enterprises.

Fourth, capacity-building programs should be incorporated by regulatory bodies to overcome digital illiteracy and reinforce technical skills of industrial employees. MIS and related technologies have the potential to be utilized to their full potential without proper competencies.

Finally, the policy frameworks should facilitate the collaborative data ecosystems, enabling the inter industry and inter-agency exchange of environmental data. This type of cooperation promotes shared situational awareness, better planning of regional sustainability, and ecosystem resilience in local industry.

14.0. FUTURE RESEARCH DIRECTIONS

Based on the implications of MIS, EMS, EMIS, IoT, and digital governance in the local industrial systems, there are multiple future research opportunities. To begin with, researchers ought to study comparative research on MIS facilitated sustainability results in various areas of the industrial sector, including manufacturing, energy, waste management, and construction to find out where digital systems can create the largest environmental value.

Second, it (and next generation) research can look at integrated MIS-AI systems, which support predictive environmental risk management, carbon-footprint forecasting and automated sustainability decision making. Although, at the moment, there is some evidence of the transformative nature of AI and digital twins, more empirical analysis is required to comprehend the issues of implementation, ethical risks, and long-term environmental gains.

Third, there is need to conduct research about the challenges faced by small and medium enterprises (SMEs) in adopting MIS. Most of the SMEs are faced with the challenge of limited resources, infrastructural barriers, and lack of digital literacy. The knowledge of these barriers will assist in creating more inclusive digital sustainability structures that can support the different organizational capacities.

Fourth, more exploration of environmental data governance models is required, especially on the data provenance, ownership, privacy, and interoperability in the multi-stakeholder industrial systems. The higher the level of granularity in environmental data, the higher the regulatory frameworks need to be altered to provide ethical and transparent practices.

Lastly, longitudinal studies determining the actual effects of the digital environmental systems will be essential. Follow-ups on the performance of the sustainability initiatives by MIS with time will be valuable insights into resilience in the systems, industrial transformation, and the efficacy of the digital policies.

15.0. CONCLUSION

Consolidation of MIS, EMS, EMIS, and innovating digital platforms is a revolution route towards promoting environmental sustainability in the local industrial framework. The increased reviews in this chapter indicate that MIS is not merely an informational technology but an enabling business strategy that can transform the way industries learn, observe and handle their environmental obligations. MIS assists organizations to move beyond the reactive compliance mode to the proactive culture of sustainability stewardship by promoting real-time data acquisition, predictive analytics, transparent reporting and structured decision-making.

The introduction of policy implications supports the necessity to have consistent regulatory frameworks, standard data protocols, and capacity building programs that would help the MIS-based sustainability systems to scale effectively. The increasing multifacetedness of digital environmental regulation that includes the ownership of data, privacy, cybersecurity, and interoperability requires a concerted policy response. Coupled with reinforced areas, industries will be in a position to maximize the opportunities of MIS and other technologies like IoT and digital twins in ethical, transparent and accountable governance systems.

The future research potential is seen in the further development of MIS that will be used to sustainability systems. Sectoral comparisons, studies on SME-adoption, MIS models that are AI-integrated, and further understanding of environmental data governance will also be very informative to the enhancement of the technological and policy interventions. These research directions emphasize the dynamism of sustainability issues and the necessity of constant academic and industrial partnership.

Altogether, the results and improvements mentioned in this chapter confirm that MIS can be an important tool in helping local industrial systems to find their way to more resilient, intelligent, and sustainable futures. To achieve this potential technology innovation is insufficient but policy alignment, organizational preparedness, and long-term research participation are needed. All this being the case, MIS can be seen as an effective catalyst towards bringing about a significant environmental change.

REFERENCES

Ajayi, A. J., Akhigbe, E. E., Egbuhuzor, N. S., & Agbede, O. O. (2022). Economic analysis of transitioning from fossil fuels to renewable energy using econometrics. *International Journal of Social Science Exceptional Research, 1*(1), 96–110. DOI: 10.54660/IJSSER.2022.1.1.96-110

Akhigbe, E. E., Egbuhuzor, N. S., Ajayi, A. J., & Agbede, O. O. (2021). Financial valuation of green bonds for sustainability-focused energy investment portfolios and projects. *Magna Scientia Advanced Research and Reviews, 2*(1), 109–128. DOI: 10.30574/msarr.2021.2.1.0033

Alsaedi, T., Sherief, N., Phalp, K., & Ali, R. (2022). Online social transparency in enterprise information systems: a risk assessment method. Information Technology and Management, 23(2), 95-124. Digital divide: Impact of access. *The international encyclopedia of media effects,* 1, 1-11. DOI: 10.1007/s10799-021-00347-3

Amiri, A., Soltani, K., Ebtehaj, I., & Bonakdari, H. (2024). A novel machine learning tool for current and future flood susceptibility mapping by integrating remote sensing and geographic information systems. *Journal of Hydrology (Amsterdam), 632*, 130936. DOI: 10.1016/j.jhydrol.2024.130936

Arimura, T. H., Darnall, N., Ganguli, R., & Katayama, H. (2016). The effect of ISO 14001 on environmental performance: Resolving equivocal findings. *Journal of Environmental Management, 166*, 556–566. DOI: 10.1016/j.jenvman.2015.10.032

Bai, C., Quayson, M., & Sarkis, J. (2023). Evaluating green information systems adoption in manufacturing: A review and future research agenda. *Journal of Cleaner Production, 416*, 137926. DOI: 10.1016/j.jclepro.2023.137926

Balamurugan, K., Alexander, T. J., & Karthikeyan, V. (2024). Application of Solar Energy in Sustainable Mobility. In *Energy Efficient Vehicles* (pp. 250–264). CRC Press., DOI: 10.1201/9781003464556-13

Banu, E. A., Venkateswaran, N., & Mageshwari, P. S. L. and 2 more (...) (2024). Sustainable and Green IoT Solutions for Environmental Monitoring. *Proceedings of 9th International Conference on Science, Technology, Engineering and Mathematics: The Role of Emerging Technologies in Digital Transformation, ICONSTEM 2024*. DOI: 10.1109/ICONSTEM60960.2024.10568826

Bass, L. (2012). *Software architecture in practice*. Pearson Education India.

Baumgartner, R. J., & Rauter, R. (2017). Strategic perspectives of corporate sustainability management to develop a sustainable organization. *Journal of Cleaner Production, 140*, 81–92. DOI: 10.1016/j.jclepro.2016.04.146

Bhalaji, R. K. A., Bathrinath, S., Ali, S. M., & Koppiahraj, K. (2024). Risk assessment in sustainable supply chain: Theoretical and managerial implications for circular economy in emerging economies. *International Journal of System Assurance Engineering and Management, 15*(10), 4966–4981. DOI: 10.1007/s13198-024-02507-4

Blumenthal, S. C. (1969). Management information systems: A framework for planning and development.

BM&FBOVESPA. (2015). *Novo valor: sustentabilidade nas empresas como começar, quem envolver e o que priorizar*. BM&FBOVESPA.

Bonamente, E., Merico, M. C., Rinaldi, S., Pignatta, G., Pisello, A. L., Cotana, F., & Nicolini, A. (2014). Environmental impact of industrial prefabricated buildings: Carbon and energy footprint analysis based on an LCA approach. *Energy Procedia, 61*, 2841–2844. DOI: 10.1016/j.egypro.2014.12.319

Bose, R., & Luo, X. (2011). Integrative framework for assessing firms' potential to undertake Green IT initiatives via virtualization–A theoretical perspective. *The Journal of Strategic Information Systems, 20*(1), 38–54. DOI: 10.1016/j.jsis.2011.01.003

Business Wire. (2022) Intel accelerates developer innovation with open software-first approach. Business Wire. Retrieved June 20, 2024, from https://www.businesswire.com/news/home/20220928005326/en/ Intel-Accelerates-Developer-Innovation-with-Open-Software-First-Approach

Chen, C. J., & Huang, J. W. (2007). How organizational climate and structure affect knowledge management –The social interaction perspective. *International Journal of Information Management, 27*(2), 104–118. DOI: 10.1016/j.ijinfomgt.2006.11.001

Cloud Native Computing Foundation. (2015) New Cloud Native Computing Foundation to drive alignment among container technologies. Retrieved June 20, 2024, from https://www.cncf.io/announcements/ 2015/06/21/new-cloud-native-computing-foundation-to-drive-alignment-among-container-technologies/

Diahovchenko, I., Kolcun, M., Čonka, Z., Savkiv, V., & Mykhailyshyn, R. (2020). Progress and challenges in smart grids: Distributed generation, smart metering, energy storage and smart loads. *Iranian Journal of Science and Technology. Transaction of Electrical Engineering, 44*(4), 1319–1333. DOI: 10.1007/s40998-020-00322-8

El-Gayar, O., & Fritz, B. D. (2006). Environmental management information systems (EMIS) for sustainable development: A conceptual overview. *Communications of the Association for Information Systems*, *17*(1), 34. DOI: 10.17705/1CAIS.01734

Elete, T. Y., Nwulu, E. O., Omomo, K. O., Esiri, A. E., & Aderamo, A. T.Tari Yvonne EleteEmmanuella Onyinye NwuluKingsley Onyedikachi OmomoAndrew Emuobosa EsiriAdeoye Taofik Aderamo. (2023). Alarm rationalization in engineering projects: Analyzing cost-saving measures and efficiency gains. *International Journal of Frontiers in Engineering and Technology Research*, *4*(2), 22–35. DOI: 10.53294/ijfetr.2023.4.2.0022

Elliot, S. (2011). Transdisciplinary Perspectives on Environmental Sustainability: A Resource Base and Framework for IT-Enabled Business Transformation1. *Management Information Systems Quarterly*, *35*(1), 197–236. DOI: 10.2307/23043495

EMAS. (2006) "Introducing EMAS," http://www.emas.org.uk/aboutemas/ mainframe.htm (current Feb. 16, 2006).

Eshet, Y. (2004). Digital literacy: A conceptual framework for survival skills in the digital era. *Journal of Educational Multimedia and Hypermedia*, *13*(1), 93–106.

Gavronski, I., Paiva, E. L., Teixeira, R., & Andrade, M. C. F. (2012). ISO 14001 certified plants in Brazil: Taxonomy and practices. *Journal of Cleaner Production*, *39*, 32–41. DOI: 10.1016/j.jclepro.2012.08.025

Ghorbanzadeh, O., Feizizadeh, B., & Blaschke, T. (2018). Multi-criteria risk evaluation by integrating an analytical network process approach into GIS-based sensitivity and uncertainty analyses. *Geomatics, Natural Hazards & Risk*, *9*(1), 127–151. DOI: 10.1080/19475705.2017.1413012

Hargittai, E. (2010). Digital na (t) ives? Variation in internet skills and uses among members of the "net generation". Sociological inquiry, 80(1), 92-113.). Internet (non-) use types and motivational access: Implications for digital inequalities research. *New Media & Society*, *19*(8), 1157–1176. DOI: 10.1111/j.1475-682X.2009.00317.x

Helsper, E. J. (2012). A corresponding fields model for the links between social and digital exclusion. *Communication theory, 22(4),* 403-426.https://doi.org/ . 01416. xDOI: 10.1111/j.1468-2885.2012

Ilager, S., Muralidhar, R., & Buyya, R. (2020, October). Artificial intelligence (ai)-centric management of resources in modern distributed computing systems. In 2020 IEEE Cloud Summit (pp. 1-10). IEEE. DOI: 10.1109/IEEECloudSummit48914.2020.00007

Instituto Brasileiro de Governança Corporativa – IBGC. (2015). Código das melhores práticas de governança corporativa (5th ed.). Retrieved in 2016, April 5, from https://www.ibgc.org.br/inter.php?id=18180

Intel Corporation. (2021) Open source software: fact sheet. Intel Corporation. Retrieved June 20, 2024, from https://download.intel.com/newsroom/2021/client-computing/OSS-Fact-sheet.pdf

International Organization for Standardization – ISO. (2014). The ISO survey of management system standard certifications: executive summary. Retrieved in 2016, April 5, from http://www.iso.org/iso/iso_survey_executive-summary.pdf?v2014

ISO. (2002) "The ISO 14000 Family of International Standards," https://www.iso.org/iso/en/prodsservices/otherpubs/iso14000/index.html (current Feb. 16, 2006).

ISO. (available at https://www.iso.org/standard/56639. html)

Jenkin, T. A., Webster, J., & McShane, L. (2011). An agenda for 'Green' information technology and systems research. *Information and Organization*, *21*(1), 17–40. DOI: 10.1016/j.infoandorg.2010.09.003

Jiménez-Jiménez, D., & Sanz-Valle, R. (2011). Innovation, organizational learning, and performance. *Journal of Business Research*, *64*(4), 408–417. DOI: 10.1016/j.jbusres.2010.09.010

Khatri, V., & Brown, C. V. (2010). Designing data governance. *Communications of the ACM*, *53*(1), 148–152. DOI: 10.1145/1629175.1629210

Kulvatunyou, B., & Wysk, R. A. (2000). A functional approach to enterprise-based engineering integration. *Journal of Manufacturing Systems*, *19*(3), 156–171. DOI: 10.1016/S0278-6125(00)80009-6

Lee, S. U., Zhu, L., & Jeffery, R. (2017). Data governance for platform ecosystems: Critical factors and the state of practice. *arXiv preprint arXiv:1705.03509.*

Livingstone, S. (2008). Engaging with media—A matter of literacy? *Communication, Culture & Critique*, *1*(1), 51–62. DOI: 10.1111/j.1753-9137.2007.00006.x

Machado, C.Jr, Mazzali, L., Souza, M. T. S., Furlaneto, C. J., & Prearo, L. C. (2013). A gestão dos recursos naturais nas organizações certificadas pela norma NBR ISO 14001. Produção, 23(1), 41-51. http://dx.doi.org/ » http://dx.doi.org/DOI: 10.1590/S0103-65132012005000005

Martínez, J. B., Fernández, M. L., & Fernández, P. M. R. (2016). Corporate social responsibility: Evolution through institutional and stakeholder perspectives. *European journal of management and business economics, 25(1),* 8-14. DOI: 10.1016/j.redee.2015.11.002

Martínez-León, I. M., & Martínez-García, J. A. (2011). The influence of organizational structure on organizational learning. *International Journal of Manpower, 32*(5/6), 537–566. DOI: 10.1108/01437721111158198

Melville, N. P. (2010). Information systems innovation for environmental sustainability. *Management Information Systems Quarterly, 34*(1), 1–21. DOI: 10.2307/20721412

Microsoft. (n.d.a) Visual Studio Code license and privacy statement. Retrieved June 20, 2024, from https:// code.visualstudio.com/license

Microsoft. (n.d.b) Visual Studio Code [Repository]. GitHub. Retrieved June 20, 2024, from https://github. com/microsoft/vscode

Mintzberg, H. (1984). Power and organization life cycles. *Academy of Management Review, 9*(2), 207–224. DOI: 10.2307/258435

Morgan, G., Ryu, K., & Mirvis, P. (2009). Leading corporate citizenship: Governance, structure, systems. *Corporate Governance (Bradford), 9*(1), 39–49. DOI: 10.1108/14720700910936047

Negri, E., Fumagalli, L., & Macchi, M. (2017). A review of the roles of digital twin in connected systems. *Procedia Manufacturing, 11*, 939–948. DOI: 10.1016/j.promfg.2017.07.198

Nicolini, D., & Meznar, M. B. (1995). The social construction of organizational learning: Conceptual and practical issues in the field. *Human Relations, 48*(7), 727–746. DOI: 10.1177/001872679504800701

O'Brien, W. J., Hammer, J., Siddiqui, M., & Topsakal, O. (2008). Challenges, approaches and architecture for distributed process integration in heterogeneous environments. *Advanced Engineering Informatics, 22*(1), 28–44. DOI: 10.1016/j.aei.2007.08.008

Ojadi, J. O., Onukwulu, E., Odionu, C., & Owulade, O. (2023). Leveraging IoT and deep learning for real-time carbon footprint monitoring and optimization in smart cities and industrial zones. *Iconic Research and Engineering Journals, 6*(11), 946–964.

Oliveira, O. J., & Serra, J. R. (2010). Benefícios e dificuldades da gestão ambiental com base na ISO 14001 em empresas industriais de São Paul. Produção, 20(3), 429-438. http://dx.doi.org/DOI: 10.1590/S0103-65132010005000013

Oluokun, O. A. (2021). Design of a power system with significant mass and volume reductions, increased efficiency, and capability for space station operations using optimization approaches (Doctoral dissertation, McNeese State University).

Onukwulu, E. C., Agho, M. O., & Eyo-Udo, N. L. (2022). Circular economy models for sustainable resource management in energy supply chains. *World Journal of Advanced Science and Technology, 2(2),* 034-057.

Onukwulu, E. C., Dienagha, I. N., Digitemie, W. N., & Egbumokei, P. I. (2021). Predictive analytics for mitigating supply chain disruptions in energy operations. *Iconic Research and Engineering Journals*, *5*(3).

Ortega-Jimenez, C. H., Garrido-Vega, P., & Torres, C. A. C. (2020). Achieving plant responsiveness from reconfigurable technology: Intervening role of SCM. *International Journal of Production Economics*, *219*, 195–203. DOI: 10.1016/j.ijpe.2019.06.001

Penn, L., Goffe, L., Haste, A., & Moffatt, S. (2019). Management information systems for community-based interventions to improve health: Qualitative study of stakeholder perspectives. *BMC Public Health*, *19*(1), 105. DOI: 10.1186/s12889-018-6363-z

Poongodi, T., Ramya, S. R., Suresh, P., & Balusamy, B. (2020). Application of IoT in green computing. In *Advances in Greener Energy Technologies* (pp. 295–323). Springer Singapore., DOI: 10.1007/978-981-15-4246-6_19

Rikhardsson, P. M. (2001). Corporate environmental management and information technology. Eco-Management and Auditing. *Eco-Management and Auditing*, *8*(2), 90–99. DOI: 10.1002/ema.149

Sarkis, J., Koo, C., & Watson, R. T. (2013). Green information systems & technologies–this generation and beyond: Introduction to the special issue. *Information Systems Frontiers*, *15*(5), 695–704. DOI: 10.1007/s10796-013-9454-5

Sarkis, J., & Zhu, H. (2008). Information technology and systems in China's circular economy: Implications for sustainability. *Journal of Systems and Information Technology*, *10*(3), 202–217. DOI: 10.1108/13287260810916916

Seidel, S., Recker, J., & Vom Brocke, J. (2013). Sensemaking and sustainable practicing: Functional affordances of information systems in green transformations. *Management Information Systems Quarterly*, *37*(4), 1275–1299. DOI: 10.25300/MISQ/2013/37.4.13

Shapira, Z. (Ed.). (2002). *Organizational decision making*. Cambridge University Press.

Sharma, A., & Frias-Aceituno, J. V. (2024). Digital sustainability reporting: Integrating AI and MIS for enhanced transparency. *Sustainable Technology and Entrepreneurship*, *7*, 100142.

Shendryk, V., & Boiko, A. (2015). Stages of information system development in the process approach. *Procedia Computer Science*, *77*, 98–103. DOI: 10.1016/j.procs.2015.12.365

Simon, H. A. (2013). *Administrative behavior*. Simon and Schuster.

Solomon, N. O., Simpa, P., Adenekan, O. A., & Obasi, S. C. (2024). Circular Economy Principles and Their Integration into Global Supply Chain Strategies. *Finance & Accounting Research Journal*, *6*(5), 747–762. DOI: 10.51594/farj.v6i5.1133

Sproull, L., & Kiesler, S. (1991). *Connections: New ways of working in the networked organization*. MIT press.

Stack Overflow. (2023) Stack overflow developer survey 2023: most popular technologies - Integrated development environment. Retrieved June 20, 2024, from https://survey.stackoverflow.co/2023/#sectionmost-popular-technologies-integrated-development-environment

Sugapriya, C., Fariya Azleena, A., Nagarajan, D., & Kavikumar, J. (2024). A sustainable fuzzy economic production quantity model (SFEPQM) with sporadic machine failures, inspection and prepayments under carbon tax and cap policy. *International Journal of Systems Science: Operations & Logistics*, *11*(1), 2419398.

Tao, Z. (2020). Advanced Wavelet Sampling Algorithm for IoT based environmental monitoring and management. *Computer Communications*, *150*, 547–555. DOI: 10.1016/j.comcom.2019.12.006

Tapus, N., & Manolache, M. A. Multilayered integrated decision-making platform architecture.

Turnbull, S. (2017). Grounding a theory of firms in the natural sciences. *Humanistic Management Journal*, *1*(2), 159–186. DOI: 10.1007/s41463-016-0017-z

Uzougbo, N. S., Ikegwu, C. G., & Adewusi, A. O. (2024a). Enhancing consumer protection in cryptocurrency transactions: Legal strategies and policy recommendations. .DOI: 10.30574/ijsra.2024.12.1.0801

Uzougbo, N. S., Ikegwu, C. G., & Adewusi, A. O. (2024b). Regulatory Frameworks for Decentralized Finance (DeFi): Challenges and opportunities. *GSC Advanced Research and Reviews*, *19*(2), 116–129. DOI: 10.30574/gscarr.2024.19.2.0170

Vazquez-Brust, D. A., & Sarkis, J. (2012). Green growth: managing the transition to sustainable economies. In *Green Growth: Managing the Transition to a Sustainable Economy: Learning by Doing in East Asia and Europe* (pp. 1–25). Springer Netherlands., DOI: 10.1007/978-94-007-4417-2_1

Waller, M. A., & Fawcett, S. E. (2013). Data science, predictive analytics, and big data: A revolution that will transform supply chain design and management. *Journal of Business Logistics*, *34*(2), 77–84. DOI: 10.1111/jbl.12010

Watson, R. T., Boudreau, M. C., & Chen, A. J. (2010). Information systems and environmentally sustainable development: Energy informatics and new directions for the IS community. *Management Information Systems Quarterly*, *34*(1), 23–38. DOI: 10.2307/20721413

Weill, P., & Ross, J. W. (2004). IT governance on one page. Available at *SSRN* 664612.DOI: 10.2139/ssrn.664612

Wirtz, B. W., Weyerer, J. C., & Kehl, I. (2022). Governance of artificial intelligence: A risk and guideline-based integrative framework. *Government Information Quarterly*, *39*(4), 101685. DOI: 10.1016/j.giq.2022.101685

Wu, H., Ji, P., Ma, H., & Xing, L. (2023). A comprehensive review of digital twin from the perspective of total process: Data, models, networks and applications. *Sensors (Basel)*, *23*(19), 8306. DOI: 10.3390/s23198306

Yenişen Yavuz, E., Riehle, D., & Mehrotra, A. (2025). Why do companies create and how do they succeed with a vendor-led open-source foundation. *Empirical Software Engineering*, *30*(1), 40. DOI: 10.1007/s10664-024-10588-9

Yeung, K., & Bygrave, L. A. (2022). Demystifying the modernized European data protection regime: Cross- disciplinary insights from legal and regulatory governance scholarship. *Regulation & Governance*, *16*(1), 137–155. DOI: 10.1111/rego.12401

Zhao, Y., Wang, W., Li, Y., Meixner, C. C., Tornatore, M., & Zhang, J. (2019). Edge computing and networking: A survey on infrastructures and applications. *IEEE Access : Practical Innovations, Open Solutions*, *7*, 101213–101230. DOI: 10.1109/ACCESS.2019.2927538

Zhou, Y., Wang, L., & Koo, C. (2025). Artificial intelligence-enabled environmental decision-making: A systematic review. *Information Systems Frontiers*. Advance online publication. DOI: 10.1007/s10796-025-10456-2

KEYWORDS AND DEFINITIONS

Environmental Management Systems (EMS): An organized system of policies, processes and practices that allow organizations to regulate, measure and enhance their environmental performance by ensuring that they are within the regulations as well as contribute to sustainable development objectives.

Environmental Sustainability: The sustainable utilization and management of natural resources in order to satisfy the current needs without affecting the capacity of future generations to satisfy own needs in order to ensure ecological stability, conservation and health of the environment in the long term.

ISO 14001: This is an internationally accepted standard that offers organizations a guideline on how they may develop and execute an Environmental Management System (EMS) with the aim of enhancing their environmental performance, legal adherence and encouraging continuous sustainability enhancement.

Management Information Systems (MIS): This is a complex combination of people, processes and technology which is created to gather, process, store, and disseminate information to aid decision-making, coordination, and control in an organization.

Sustainability Information Systems (SIS): Information systems that are used to gather, analyze and report data concerning social, environmental and economic aspects of sustainability and are used to assist an organization in performance monitoring, decision making and in meeting sustainable development objectives.

Technology: Technology is the practical application of the scientific knowledge, tools, and techniques in order to create the solutions to solve problems and improve the human life.

Chapter 7
Environmentally Sustainable Event Management for Industry Innovations and Growth

Senthilmurugan Paramasivam
https://orcid.org/0000-0003-3249-8088
Christ University, India

Shikha Bhagat
https://orcid.org/0000-0001-8745-5558
Christ University, India

Shilpa Sarvani Ravi
https://orcid.org/0000-0003-0950-1760
Christ University, India

N. Mohan
Vasavi Vidya Trust Group of Institutions, India

Abdud Daiyaan Khan
Christ University, India

ABSTRACT

This chapter investigates the convergence of event management operations and sustainable practices to guide their function of stimulating innovation and development within the event management sector. With increasing necessity, sustainability becomes central and unavoidable to the world economy, and event professionals are increasingly embracing green practices in order to minimize environmental

DOI: 10.4018/979-8-3373-6435-3.ch007

footprints while enhancing overall experiences for attendees. This chapter gives an extensive overview of important sustainability concepts, energy efficiency, sustainable procurement and encompassing waste minimization and outlines their day-to-day applicability in the planning, staging and post-event assessment of events It also centers the role of stakeholder engagement, ranging from suppliers through to guests, in influencing sustainable action. The chapter ends with suggesting future research directions for embedding sustainability into event management, focusing on the imperative of innovation, cooperation, and constant industry-wide dedication to protecting the planet.

1. INTRODUCTION

Events play very significant symbolic role in people's life across the world. It delivers a stage for people to get together and share values, attitudes, customs, ideas, celebrations and memories. Nevertheless, events also have a substantial effect on the society, economy and environment from resource consumption to waste generation and emissions. Sustainability is not a phenomenon that one can artificially carrying out into actions of individuals and the means of society functions, but it is understood in today's world, sustainability is a contemporary issue and the lack of sustainability in the event management industry is closely associated with social, economic and legal problems. The lack of sustainability produces undesirable consequences in the fast-growing event industry directly. Events with absence of sustainable practices, careless decision making and modes of operations that consume aplenty of resources that are posing a risk to the society and its future. (Mooney et al., 2011). The World Commission on the environment and development defines sustainability as meeting the needs of the present without compromising the ability of future generations to meet their needs. A sustainable event is one that is designed, organized and implemented in a way that minimizes negative impacts and leaves a beneficial legacy for the host community and all involved.

Sustainability and event management are two relative concepts when combined, that's how the concept of sustainable event management comes along. (Jones, 2023) Sustainability in the event is achieved through integration of sustainable development suitable to reduce the negative impact that beget by event. Sustainable development isn't a new conception in achieving sustainability. But in order to attain the best results, both sustainability and event management have to be managed responsibly by all stakeholders of the event management industry, which includes the organizers, vendors, suppliers, and even the audience. The British standards institution described the sustainable event industry as a means of coping with social progress, the economic activity and responsibility towards the climate. The sustainability of the event management industry is the merger of sustainability with the whole process, the project planning of event management. It creates a better more sustainable future for the economy, the community and even the rest of the community. Event organizers constantly seeking to improve and make more responsible, sustainable choices and create more sustainable events is also very popular with the media and the public (Laing & Frost, 2010).

Applying the notions of sustainability in the event sector is a necessity for a responsible events profession operating within the 21st century (Musgrave & Raj, 2009). Hediger (2000) proposes that a component of sustainable principles is to induce cultural change within society, and thus within the events sector. In connection with this thought, Foley et al. (2009) state that events must be used to engage and empower community groups to contribute to community building long after the event itself has moved on. Looking at the wider picture, practitioners and supporters of sustainable practice are aware of the need to pay attention to the triple bottom line of economy, society, and environment, but in practice, this is a difficult balancing act.

The event management professionals are profoundly focusing on sustainability as a core value in their planning and execution to mitigate adverse impacts on society, economy and environment. Sustainability in event management endeavors to produce an event that produces positive impacts on the society and planet without compromising to deliver memorable experiences to the spectators and participants. The process of sustainable event management begins with planning the event and concludes with post-evaluation of the event. This chapter enumerates practices and strategies to develop and execute sustainable event management.

In recent times the sustainability endeavours in event management have stretched outside conventional environmental issues to integrate health and safety, digitalisation, social inclusion, and legacy building (Mair et al., 2024; Park et al., 2023). Research exhibits that post-pandemic event organisers have begun to gradually be embracing virtual and hybrid event formats to lower emissions and resource usage while providing engagement (García-Melón et al., 2023; Ramos-Soler et al., 2024). The triple bottom line approach remains to control sustainability evaluations, with growing focus to carbon neutrality and community impact (Walters et al., 2023). Moreover, scholars highlight how circular economy principles — such as renewable energy integration, material reuse, and waste minimization, — are changing the event setting (De Brito & Luna, 2024). These developments prove that sustainable event management now targets not just to reduce harm but to create recreating value for the economy, society, and environment (Jones, 2023).

Figure 1. The three pillar impacts of events (Musgrave & Raj, 2009, p. 5)

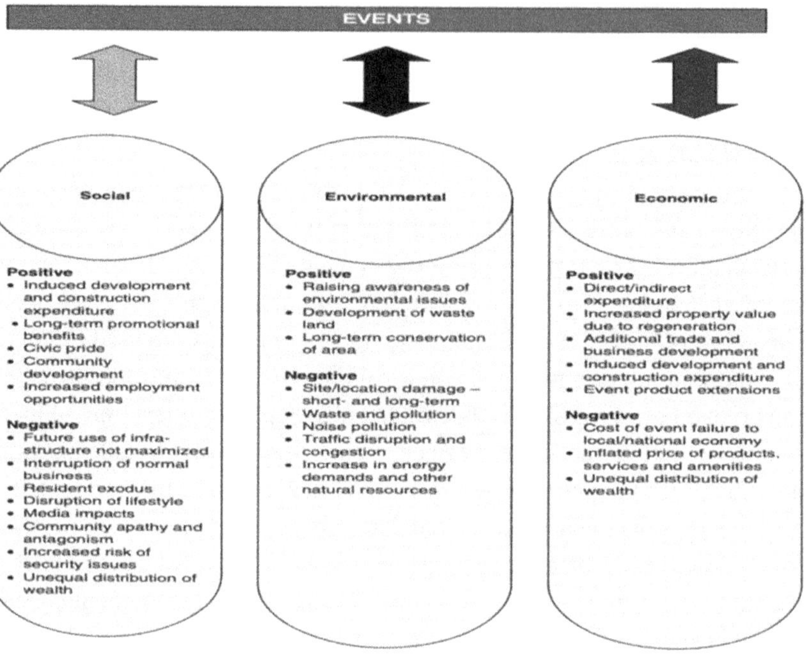

2. DEVELOPING SUSTAINABILITY IN EVENT MANAGEMENT

Planning: Planning and formulating goals are vital to achieve sustainability in event management. An effective way to realize sustainability in the event management is to setting clear goals that are compliance with the United Nations' Sustainable Development Goals (SDGs). SDGs offer an outline for global sustainability. An event can positively contribute to the world, when events goals are aligned with SDGs goals. Sustainable event management goals need to be measurable, specific, relevant, achievable, and time-bound (SMART). These goals will ensure to reducing waste generation, promoting responsible sourcing, reducing greenhouse gas emissions, and enhancing social inclusivity.

Design: Event professionals need to implement sustainable practices and technologies in the design and construction of events. The design and construction of sustainable event involves deploying sustainable materials, such as recyclable products, biodegradable and the adaptation of sustainable event design principles, such as minimizing water, materials and energy usage. Embracing the emerging technologies facilitate to reduce the environmental impact of events, such as by utilizing virtual and hybrid events. The virtual events and related technologies help to minimize the physical footprint of an event, travel requirements, energy and material resources by which it reduces the carbon footprint of events.

Sourcing and Waste Reduction: The environmental impact can be significantly reduced when an event is implemented with sustainable strategies. Event professionals can outsource the local service providers, encourage public transportations, carpooling and ride sharing can significantly reduce the pollutions and number of traffic flow in the roads for attending event by the attendees. In addition to that usage of electric and hybrid vehicles can also facilitate to decrease emissions and showcase sustainable technology. The event venues can be equipped with solar energy lightings and other electrical devices. The event organizer also uses non plastic materials. Sustainable sourcing includes procuring products and services from suppliers who adopt sustainable criteria, such as those who prioritize environmental and social responsibility. This contains sourcing seasonal food, recycled and biodegradable products that reduces the carbon footprint of the event by reducing transportation emissions. Waste reduction can be achieved by executing strategies such as recycling and composting programs, minimizing packaging and utilizing reusable materials. Sustainable food practices such as donating surplus food to local charities and food banks can reduce food wastages.

Social inclusivity and Diversity: Encouraging social inclusivity and diversity is indispensable to expanding sustainability in event management. Event professionals can promote inclusivity and diversity by confirming that all participants will have equal access to the event and its facilities. This comprises accessible amenities for people with differently-abled, gender-neutral restrooms, and other accommodations that endorse inclusivity. Event organizers can also arrange hiring and working with vendors who promote diversity and inclusivity in their processes.

Post-event Evaluation: Eventually, post-event evaluation is important to measuring the success of sustainability in event management. Appraising the sustainability performance of an event can facilitate to identify the areas of enhancement and direct future planning efforts. This comprises tracking and gauging sustainability metrics, such as greenhouse gas emissions, waste generation.

Figure 2. Sustainable event management wheel (Musgrave & Raj, 2009, p. 26)

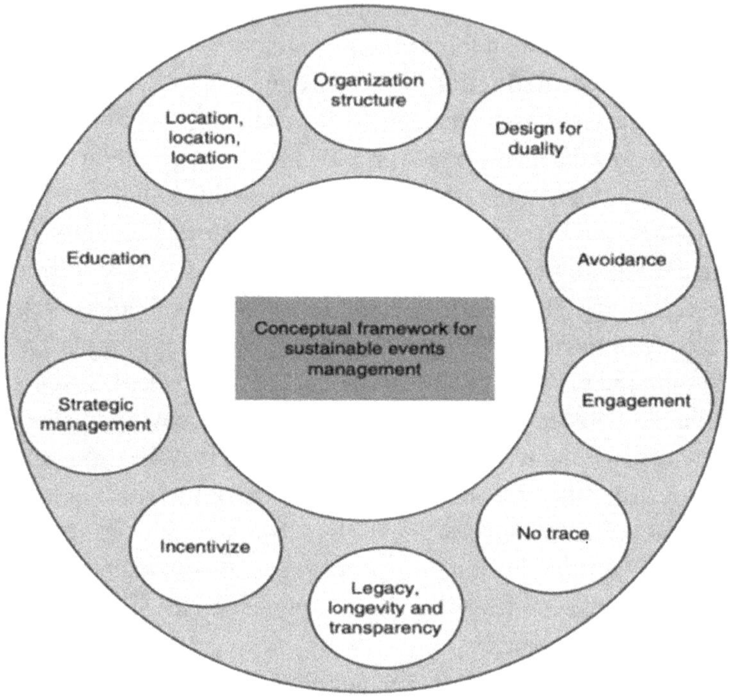

Musgrave & Raj (2009) developed a conceptual framework for sustainable event management which integrates economic, environmental, and social considerations into the planning, organization, and implementation of the events.

They suggested some practices that can be followed for sustainable event management:

i. Selection of a sustainable venue: Choosing a venue that is committed to sustainability practices and adopting environmental policies.
ii. Reduction of waste: Replacing single-use plastics and paper products by using reusable or compostable materials. Placing compost and recycle bins throughout the event space.
iii. Usage of sustainable transportation: Efforts can be made to encourage attendees to utilize public transportation or carpooling options to reduce carbon emissions.
iv. Serving sustainable food and beverages: Serving organic and locally sourced food and beverages. Discouraging usage of disposable utensils and dishes.
v. Usage of sustainable energy sources: Using self-generated renewable energy sources such as solar or wind power along with energy-efficient lighting and equipment.

vi. Digital communication: Digital communication can be used instead of papers by sending invitations, event schedules, and other information through email or mobile apps.
vii. Sustainable accommodations: Create awareness and encourage attendees to stay in hotels which abide sustainable accommodations and environmental policies.
viii. Engagement with the local community: Extending support to local businesses and organisations that are committed to sustainability.
ix. Measuring and reporting on sustainability: Evaluating the event's sustainability performance using metrics such as carbon footprint, waste diversion rate, and water usage.

These practices will ensure the organizers to conceptualize and execute sustainable practices in the event that significantly reduce environmental impacts.

3. SUSTAINABLE EVENT MANAGEMENT: LITERATURE REVIEW

Sustainable event management has developed as a serious research sphere, shifting from isolated "green" practices toward holistic agendas incorporating environmental, social, and governance (ESG) dimensions. The following review covers current literature, directing on social legacy, measurement,, strategic governance, practicality and future research tracks.

Environmental practices and measurement

The modern-day scholars stress that environmental sustainability in the event industry remains disintegrated in measurement and inconsistent in practice. For example, Toscani, Vendraminelli & Vinelli (2024) showed a systematic review and recognised that while interventions such as transport reduction, waste management, and energy efficiencies are present, their measurement absences longitudinal tracking and standardisation. They contend for life-cycle standard indicator and assessment sets to enable comparation across event types. The gap remains in consistent practice of metrics, benchmarking and transparency.

Social sustainability and legacy

Apart from environmental measurements, recent works highlight the significance of social sustainability: cultural preservation, community inclusion, sense of place, and long-term legacy. For example, Dolasinski, Roberts & Young (2025) recommend a conceptual model (SBSC-ET) that comprises a "community perspective" highlighting cultural preservation and social equity alongside environmental and economic factors. The research recommends that sustainable event tourism must embed stakeholder roles and responsibilities, rather than treating legacy as an after-thought. In addition, Li, Choe, Kang & Lee (2025) contextualise events under external shocks (COVID-19) and highlight how behavioural intentions (of attendees) influence long-term social sustainability and resilience. The social measurement, while progressively recognised, remains under-measured, especially longitudinally.

Governance, strategic integration and frameworks

A solid subject in recent literature related to sustainability of event is the drive toward integrating sustainability into governance, strategy and performance measurement within events. Dolasinski et al. (2025) clearly convey sustainability into the balanced scorecard approach for event tourism, thus providing a holistic framework (SBSC-ET) that affiliates stakeholder perspectives (community, providers, host location, tourists) with sustainability metrics and strategic roles. This approach indications a change from isolated "green event" projects to enterprise-level strategic management of events with sustainability devised into planning, execution and evaluation. The literature gives this as a key enabler of scalable and repeatable sustainable event management practice.

Barriers, drivers and practical implementation

The research outcomes indicates that in-spite of increasing interest, practical obstacles continue: high cost of sustainable technologies, stakeholder alignment challenges, supplier readiness, and measurement difficulties. Toscani et al. (2024) pinpoint these as major inhibitions to the commitment of sustainable practices in events. Meanwhile, research on hybrid event formats (Santos et al., 2024) discloses that while hybrid modes reduce environmental impact (by lowering travel/footprint), they also increase questions about attendee decision-making, the changed nature of participation and inclusivity. Additionally, resilience against external shocks (e.g., climate risk, pandemic,) has entered sustainability discussion: Li et al. (2025) discovered that event attendance intentions are appreciably influenced by perceived risk and prior behaviour, underlining the significance of risk management as part of sustainable event frameworks

Overall, the leading descriptions of recent literatures emphasise moving beyond token "green" practices to incorporated frameworks that comprise economic, environmental, and social dimensions. There is strong thrust toward strategic governance models (e.g., balanced scorecards), but practical implementation remains hampered by measurement inconsistency, cost and risk, stakeholder complexity. Future research should prioritise longitudinal and comparative designs, standard measurement tools, effective strategies and resilience integration for scaling sustainable practices in event contexts

4 DEVELOPING SUSTAINABILITY IN EVENT MANAGEMENT

4.1 Economically Sustainable Practices

Economically profitable sustainability involves effective utilization of accessible sources and ensures that all plutocrat- related commitments can be met on time. Social and economic considerations must also be taken into account with the responsibility of a possible recovery. A common theme of latest event reconstruction literature is that metropolises will try to draw on established profitable capital. The small businesses and vulnerable people are ensured that they can be yield the profit and benefit from futures contracts that are generally established by large events. Small businesses are to be protected from the exploitation by some large events by eroding their profits. Even though the Barcelona Olympics are extensively celebrated, they had outraged hundreds of small businesses from places that have traditionally provided low rents. London Olympics organizers also had to bowed the relocation of 284 companies in medication for the 2012 Games. (Higham, 2005).

The recent researches endorse that such mega-events often emphasise spatial and social disparities rather than mitigate them. For example, a 2023 study on large-scale urban regeneration projects discovered that deprived communities are frequently displaced before and after event-driven regeneration, while employment and income benefits tend to accumulate disproportionately to middle- and higher-income groups (Alberts & Guaralda, 2023). Similarly, research in a major European city exposed that mega-events contribute to revamping and tourism-led redevelopment, often favouring high-value firms and real estate investors over small local vendors that conventionally serve local populations (Cargnello et al., 2021).

Moreover, rent-seeking behaviour fixed within event-host economies signifies that mega-events can act as vehicles for elite capital accumulation, emphasising existing inequalities in access and benefit distribution (Osei-Assibey & Osei, 2025). Post-COVID-19 research also focuses that inclusivity and resilience should be central to sustainable event planning to avoid creating "legacy deserts" that weakness small enterprises and local communities. The evolving trend promotes "scaling-down" or "spacing-out" event strategies to ensure better integration and participation of local stakeholders (Capolongo et al., 2022). The research indication from recent Olympic host cities further illustrate that large-scale infrastructural investments do not essentially lead to increased profitability for smaller business actors, signifying that the supposed trickle-down benefits of mega-events remain limited (Rewilak, 2023).

By assisting the development of profitable practices through association and connection, equal profitable impact, transparenc, original support, and responsible governance, sustainable events involve that businesses will align to become economically viable and that it will take decades to change previously artificial economic and operational styles. Similarly, event itineraries must integrate long-term perspectives in sustainable event operation to ensure profitable sustainability.

4.2 Socially Sustainable Practices

Profitable and social equivalency are balanced in social sustainability practices. To combat poverty, this entails respecting community rights and granting each member of the community equal rights. People must make assumptions about their needs and desires, raise the megacity's standard of living, manage and cherish their lives, and recognize and appreciate its various societies. Mortal rights, the impact on the environment, labor practices, artistic respect, safety, and well-being are examples of social factors. Meeting the needs of all those involved in planning or directing an event in reverse is how social sustainability is achieved. contends that socially sustainable associations will guarantee the positive impact of operational procedures on available goods.

The business will take into account the internal effects on employees and direct stakeholders in addition to the external effects on host communities, original metropolises, and laterally impacted populations. Normal social chops can include equal employment, regular work, good working hours, living stipends, health and education, and low-paid leave. Particularly in the areas of governance, social and artistic life, social development and public services, civic real estate, transportation and convenience, profitable reform, and public structure development, external social sustainability conditioning will significantly contribute to the social sustainability of the association's host megacity (Ngan, et al., 2022).

Organising the large-scale events must progress beyond infrastructure expansion and concentrate on creating long-term social legacies through participatory governance and inclusive planning. Empirical research findings from the 2002 Winter Olympics indicated that community satisfaction can decline once event activities conclude, disclosing that event-driven initiatives alone may not guarantee sustained social unity (June et al., 2024). Similarly, studies of the FIFA World Cup Qatar 2022 highlight how accessibility and inclusion for people with disabilities improved significantly as part of the planning process, further the need is to establishment of stronger inclusive policies and continuous supervision to ensure that these advantages are consistent (Al Thani et al., 2025).

Social sustainability has also transformed to embrace technology and innovation for inclusion and community engagement. The recent days initiatives including immersive VR-based training and digital participation platforms have been recognised as influential instruments to increase diversity, equality, and participation in event spaces (Byers et al., 2021). Meanwhile, human-rights issues continue like labour exploitation, forced evictions,, and the marginalisation of vulnerable groups—signifying that governance reforms, transparency, and participatory decision-making remain necessary to realising true social sustainability (Suresh, 2024). Overall, socially sustainable event management necessities transparent governance, community participation, inclusive employment and education policies, accessible infrastructure, and long-term legacy evaluation rather than treating social sustainability as an addition to economic or environmental targets.

4.3 Environmentally Sustainable Practices

Environmental sustainability requires decision-making and action to stop planetary degradation. Products used, produced, displayed, and acquired by the event's organizers, participants, and other stakeholders are the main objects of environmental concern. Some of these factors are more the suppliers' responsibility, but others may directly affect the event planner. It's important to remember that organizing a "green" event doesn't necessarily mean spending more money.

Maintaining the ecosystem requires making decisions and implementing actions that reduce environmental degradation. Sustainability factors, which are mostly related to products, are used by the organizer, attendees, and other stakeholders when organizing, carrying out, and participating in events. It is important to note that planning a "green" event does not necessarily mean that it will be more expensive. In actuality, many environmental plans can increase efficiency and reduce costs if properly implemented. In particular, it makes references to significant events like conferences and athletic competitions.

An event that is organized, planned, and executed in a way that minimizes its negative impacts on the environment and enhances the host community is considered green. There are many green practices, like recycling programs, trash reduction strategies, and water conservation initiatives, but there are also more creative ways that events can lead to greener activities. Frost and Laing (2010).

Green practices include things like recycling and waste reduction, energy management, pollution control, zero emissions, water conservation, reduced material use, material life cycle assessment, and toxicity reduction. Basic environmental activities include things like recycling resources, managing waste, capturing and storing carbon, managing supply chains, sourcing ethically, and protecting biodiversity. In 2007, Braungart et al.

The large-scale events can play as catalysts for comprehensive "green urban transformation" when properly designed with urban policy and infrastructure. A Chinese study in 2024 identified that hosting mega sports events significantly augmented green total factor productivity in host cities, through mechanisms such as digital economy development, increased public transport uptake, and industrial upgrading (Tongji University et al., 2024). Furthermore, research of global mega-events reveals that visitor exposure to sustainable technologies and design—especially when built into the event infrastructure—can increase environmental awareness among attendees, with positive behavioural effects post-event (Frontiers, 2023). Technological developments have enabled event organisers who increasingly utilize blockchain-verified carbon offset systems, real-time carbon-footprint tracking, AI/IoT-driven venue energy and resource optimisation and thus augmenting transparency and effectiveness of 'green' event practices (Sportsyncr, 2025). However, this study also cautions that such practices do not automatically guarantee positive results: the legacy impact of events heavily relies on state structures, stakeholder commitment, and post-event management (Frontiers, 2025). In brief, environmentally sustainable event practice is progressing from isolated 'eco-friendly' actions to being deeply embedded in event strategy, participant engagement, infrastructure legacy, and city-wide systems.

4.4 Marketing Tools

Marketing has been defined as 'The management's process of being responsible for identifying, anticipating and satisfying a customer's requirement in a way that is profitable to the customer'. In simpler terms marketing deals with products, experiences and ideas that satisfy the needs and the wants of their customers. It is very important to understand that marketing is not only something that initiates after all the details have been confirmed, it is way more than advertising. Marketing plays a very vital role before the event, during the event and even after the event. (Kailash,.al, 2012).

Venue operators frequently employ a variety of marketing techniques, including word-of-mouth, traditional media, trade exhibitions, digital media, television ads, the internet, public relations, and social media. These resources stimulate consumer interest, spread knowledge, and motivate action. Word-of-mouth is the most crucial factor that conference organizers should take into account when selecting a location. Venue operators consider each meeting organizer and attendee as a potential source of marketing influence. Meeting planners notify other planners about the locations, amenities, fees, and various meeting services. Sales leads and brand recognition are impacted by this recommendation.

Conventional media, including magazines, radio and television commercials, brochures, and newspaper ads, are widely used by venue operators. Using conventional channels, local marketing can be effective. According to Chiu and Ananzeh's (2012) study, local respondents placed a higher value on TV/radio and newspaper ads than did foreign respondents, who placed a higher value on the internet, public relations, and word-of-mouth. However, location managers today have used this traditional marketing approach less frequently. Conventional meeting outlets, such as brochures and newspaper ads, have the least effect on meeting planners, according to an empirical study of 200 meeting professionals. Additionally, collateral materials, direct mail, and printed materials are the least effective media types that influence an event planner's decision regarding the venue for a meeting or conference.

For venue owners, trade shows remain a popular marketing tactic. While venue operators engage with meeting planners and present new products, venues, and amenities, meeting planners attend trade shows to generate fresh concepts for suggested events and search for new locations, venues, and vendors.

Digital media includes both mobile devices and the internet. Venue operators are increasingly using internet marketing and information technology. based on Fesenmaier and Wang (1998). The site selection process benefits from having sufficient information about the facilities, online registration options, and visual content. The same study found that among the most helpful components of a conference venue website were details about accommodations and activities/attractions. 58% of marketers, Solaris reports. Prior to, during, and following their events, use social media for marketing. Real-time communication between meeting planners and venue operators is made possible by social media platforms such as Facebook, Instagram, LinkedIn, Twitter, and smartphone applications. This allows venue operators to instantly share venue information with thousands of meeting planners.

4.5 Renewable Energy Sources

Renewable energy sources are refers to energy sources are means of energy that is unlimited, and is more sustainable as these energy sources are natural and can be produced again my natural phenomenon's. A lot of the energy sources being used to produce electricity are non-renewable like nuclear power, coal, natural gases, hydropower, wood fucls and peat. The wind power is still quite small compared to other sources but has a lot of cultivation potential.

Hydropower is the binding of naturally flowing water to produce hydroelectricity. This hydroelectricity is generated for useful purposes like powering mechanical devices. This hydropower is created by using naturally flowing water like rivers or even manmade installations, the water goes through a turbine with a generator and thus energy is produced.

The use of hydropower has both advantages and disadvantages, some of the advantages being that it is easily storable in large tanks and can be harnessed from the large storage tanks when consumption is at its peak but a disadvantage is that it is pretty dependent on weather conditions. Irrespective of these factors, hydropower is a clean solution for the environment as it causes only a minimum impact on the environment and affects only those water areas that are situated close to the turbines. It does not negatively impact the environment by producing any concrete waste and there are absolutely no emissions to air water or the soil. It also prevents floods and the damages that could be caused by floods.

Solar Power is solar energy created by using the sun's heat and the light radiate by the sun using different technologies. The radiations caused by the sun hold an enormous amount of energy even though only a small part of it can be exploited for energy production. This solar energy can be used to provide electricity, heat, lighting and even mechanical power. The solar energy nowadays is being stored even in the form of batteries so it can used at times with no sunlight such as night time and during rainfall. The disadvantage of this solar energy is it is high in costs and extremely dependent on weather conditions and the limited sources that can benefit from it, while the benefits of solar power is the decrease of fossil fuels as well as energy costs.

Wind power uses windmills also wind turbines to capture kinetic energy and then convert it into electric energy. The most common place where these wind plants can be constructed is on hills and mountains and places on higher altitudes where wind is plenty and found more easily in comparison to places lying on lower altitudes. A lot of wind power plants abroad can be found on the coast near the sea as well. The production of wind powered energy depends majorly on the amount of wind found in that area which varies hour to hour and day to day. It is a power which is renewable, clean, abundant and does not produce greenhouse gases or any harmful emissions to land, air and water. It functions with the least amount of negative impact on the environment. The only disadvantages of producing wind power is the damage to scenery caused by windmills and its high in noise pollution. As of 2014, wind power was not financially lucrative but had quite the promising future.

Bio Energy and Geothermal energy are two other forms of man-made energy which can be produced with the help of materials found naturally in the environment. Bio energy is a form of energy that is formed from biomasses like wood, manure and other by products of agricultural processes. This form of energy releases carbon di oxide in to the air. However, it recycles this carbon di oxide from the atmosphere and reduces its own negative impact. Some benefits of using this wind energy is that the carbon di oxide from the atmosphere can be reduced and this also helps to slow down climate change. This technology uses combustion processes for producing bioenergy and therefore can produce electricity at any time unlike wind energy and solar energy that depend majorly on weather conditions.

Geothermal energy is got from heat that is stored below the surface of the earth. The heat from the earths molten core which is formed by radioactive decaying of minerals and then transformed into electric energy. This heat circulates within the rocks and then transferred into water reservoirs that spread deep underground into the earth's crust. This is cheap form of energy which can be produced close to the surface where the heat can be located. One disadvantage to produce geothermal energy is that the capitol costs of constructing these geothermal plants are expensive even though the operation and maintenance costs are not and there is no cost for fuels.

The comprehensive recent reviews underline that renewable technologies—hydropower, solar, wind, bioenergy and geothermal—are converging towards economically viable, supported by advances in storage, grid integration, scalable deployment, policy frameworks and financing mechanisms (Sharma et al., 2025). For instance, research indicates that renewable energy systems are vital for attaining global decarbonization targets, suggesting that renewable energy technologies could reduce up to 90% of electricity-sector carbon emissions by 2050 if incorporated effectively with storage and demand response strategies (Marouani, 2024). Development in smart grid and artificial intelligence (AI) usages are boosting the deployment of solar and wind assets by enhancing forecasting, control optimisation and asset management—which helps to reduce the intermittency challenge and enhance reliability (Razak et al., 2025). In the domain of bioenergy, emerging feedstocks such as agricultural residues, algae, and industrial by-products are creating new pathways for bioenergy generation with lower lifecycle emissions and greater circular-economy integration (Kumar et al., 2024; "Advances in bioenergy," 2025). Meanwhile, geothermal energy deployment is gaining momentum, facilitated by innovations in drilling and heat-exchange technologies, though capital costs remain a barrier in many regions (Chen, 2024). Overall, the empirical research highlights that while each renewable source has leverages and limitations, the collective deployment of multiple technologies—along with storage, grid flexibility, digital optimisation and supportive policy and finance frameworks—is key to opening the full potential of renewable energy in the 21st century.

5 SUSTAINABLE MEASURES IN THE INDUSTRY

5.1 Sustainable Heating and Ventilation

Heating and ventilation contribute to a major part of any facility's utility costs. Modern technology enables more products that are energy efficient and need lesser maintenance which in turn reduces costs. In climates that are moderate, the natural outside air can be used for ventilation. Outdoor events are most sustainable but in case of an indoor event, it is more sustainable to choose a venue with maximum number of windows to maximize ventilation and not suffocate the audience.

5.2 Insulation

Checking that the insulation of the building is up to standards is another way to save up a significant amount of energy which helps reduce the negative impact of that event on the environment. The venue owner could consider upgrading the single- glazed windows to double or even triple so as to prevent the heat from escaping. There are multiple methods to maximize the efficiency of the windows and doors without replacement by making sure the frames of the windows are sealed properly. (Okokpujie, 2022).

5.3 Lighting

Energy efficient lighting consists of replacing all lighting bulbs with energy saving and more efficient LED's or CFL's. these lights last almost ten times longer than the incandescent light bulbs and use approximately 75-80% lesser energy in comparison to normal light bulbs. LEDs last from 100,000 hours to 1,000,000 hours compared to a fluorescent bulb's maximal life of 30,000 hours. (Sloan et. al 2013) CFLs are the best choice for general lighting and places that require a lot of light. For areas that require color options or various effects, LEDs are the best choice. (Pérez Gosende, Pablo. (2019).

The lighting fixtures should be chosen according the lighting levels needed. When they are useful, using motion sensors, timer switches, and dimmers for the illumination is a cheap and efficient way to save money and energy. as well as the effects on the ecosystem. Candles and natural lighting should always be used, whenever feasible and weather permitting. No illumination could be used 24 hours a day! The management might consider organizing a switch-off campaign at the event, and turning off the lights whenever feasible is a good and simple way to contribute to sustainability. (Pérez Gosende, Pablo. (2019).

5.4 Laws and Legislations

A rule prohibiting manufacturers from producing or importing incandescent lamps was enacted in 2009. It should be noted that although this law forbids the sale of bulbs, their use is still allowed. The only issue with CFL lights is that they contain mercury, which makes them hazardous waste. The recycling of LEDs and energy-efficient light bulbs was made simpler for the consumer in 2013. because these bulbs can be returned to the shop where they were purchased for free and without making a new commitment. Due to their poor power factor, CFLs are not one of the most efficient methods to light up an outdoor event that uses mobile power generators. The savings produced by using CFLs are long-term and can be seen as a decrease in the electricity cost. Consequently, they are not the ideal option for short- term events. Due to their minimal loads, LED bulbs are ideal for pairing with alternative energy sources like wind, solar, water, etc. (Tracey & Ala-Juusela, 2014).

5.5 Waste Management

Events will create waste. In the worst-case situation, waste could be a source of air, water, or ground pollution that poses various health risks because of noxious materials or living microorganisms. It can also interfere with the environment and harm and the scenery. The event planner should concentrate on how to control and minimize waste. The various methods of handling garbage, including reuse, recycling, composting, and landfill deposit, should be maximized. In order to decrease waste production, the event organizer should prioritize ethical buying and the use of sustainable and recycled materials. The event manager should be aware that waste management is governed by laws and rules. The Waste Act in Finland seeks to prevent the risk and harm that waste and waste management pose to human health and the environment, reduce the amount and harmfulness of waste, encourage the sustainable use of natural resources, ensure effective waste management, and prevent littering. (Lodh, Debasree. (2024).

Ways of Handling Waste

Event trash can be divided into litter, standard, recyclable, and biodegradable groups. Waste bins must be provided and placed in the event location by the event manager. Going through the event step-by-step can help you identify. what rubbish is created and where. The waste planning should account for the number of participants, the different waste kinds, the size and layout of the event, as well as the entrance and exit locations when calculating the volume capacity. The event organiser will benefit from these calculations in terms of waste bin volume, placement, and quantity. The trash cans need to be properly labelled, placed, and kept. Local law enforcement can be called to inquire about pre-existing color-coding or labelling methods for various recyclables. Composting biodegradable waste directly on the land where the event is held could aid improving plant growth and add organic matter to soil. Waste from the audience and the production should be recycled, and all extra materials should be reused and saved to the greatest degree possible. Requesting a list of items that can be recycled from the materials recovery center is advised. The event manager will benefit from this when making purchasing and waste management choices. A waste management firm can assist in the planning of waste operations for bigger events. (Lodh, Debasree. (2024).

5.6 Sustainable Cuisine

According to Sustainable Development Commission of UK Government, sustainable food as safe, healthy, and nutrient-rich food that supports rural economies, offers a feasible means of subsistence, and complies with biophysical and environmental constraints. The gathering manager has the option to choose sustainable food and drinks served at the event by choosing greener ingredients. Foods that are sustainable may be naturally occurring, organically grown, locally obtained, or seasonal.

Organic, Local, Natural, and Seasonal Cuisine

Organic food is food that was grown without the use of herbicides or synthetic fertilisers. and without using genetic modifications. Foods that can be gathered according to the season are referred to as seasonal foods. The use of seasonal meals is advantageous because they are more readily available. when the item is at its best. On the other hand, natural food is made in a natural and minimally processed manner. The event organizer is ensuring nature conservation and the preservation of biodiversity as well as lowering water usage by choosing to use organic food products. Local and seasonal food products should be preferred because they have a lower environmental effect during procurement. Local food is frequently fresher and the ingredients can be verified as being produced locally. Utilizing local food items not only reduces the carbon footprint of food miles or the distance that food travels, but it also saves money. it also advocates small local businesses. Using local, natural and seasonal Food producers can be advantageous to both sides. because the local producer has an opportunity to become visible and attract repeat customers. (Seyfang, Gill, 2007).

Ethically Produced Cuisine

The goal of the fair trade system is to give producers a minimum amount that covers the costs of sustainable production while also ensuring that they have decent working and living conditions. Additionally, one of the terms of fair trade is investing in initiatives that will improve workers' quality of life and appreciate! relationships that are mutually advantageous over time, as well as economic, environmental, and social responsibility Event managers are participating in morally and socially accountable behavior. purchasing when using the fair trade system. Regarding animal rights the issue gets more complicated. It is possible to buy meat, fowl, and eggs that have been raised outdoors, but the producer must be able to show that this is the case. Animals weren't confined in cages if the meat was labelled as "cage- free," which means. Although they may be more expensive, cage-free and free-range goods are a more ethical option.

The growing global effort on incorporating eco-friendly cuisine practices as both an ethical and strategic imperative. Research shows that sustainable catering now forms a key performance benchmark for event planners, with over 70% of event industry planners prioritizing waste minimization, eco-conscious sourcing, and fair-trade compliance (Hannon, 2025). Sustainable cuisine is also being formed by consumer demand for transparency, where attendees expect event menus to replicate values such as local economic support, biodiversity preservation, and humane animal welfare (EventFlare Journal, 2024). Digital innovations are inducing these practices — for instance, immersive and augmented-reality applications like *DinAR* promote environmentally responsible food preferences among guests (Johns et al., 2024). Furthermore, initiatives such as Sustainable Gastronomy Week (2024) and Sustainable Gastronomy Day (FAO, 2025) have emphasized the significance of local, organic, and ethical food systems in reducing environmental footprints and supporting community resilience. Research also highlights that sustainable cuisine must extend beyond ethical sourcing to contain full supply-chain transparency, waste-reduction strategies, vendor education, and menu design optimization to achieve genuine sustainability effects (Kk Event Management Services, 2024; MDPI, 2020).

5.7 Presentation and Packaging

In order to comply with sustainable practices, packaging for food items should be kept to a minimum. Whenever possible as an illustration, buying condiments, buying bulk goods should be preferred over buying numerous tiny packages. for example, drinks, salt and pepper, or ketchup. It is best to use recycled materials, biodegradables, and bioplastics. Veg ware products provision of biodegradable flatware to compostable tableware made of cassava starch and plant fibres, which ranges from products made of potatoes and maize. as well as hot and chilly cups Innovative options should also be taken into consideration, such as edible tableware.

The sustainable packaging in event planning has become a crucial element of environmental responsibility, with a transformation toward closed-loop packaging systems and circular economy models (Chen et al., 2024). Events are embracing compostable biomaterials like sugarcane bagasse, cornstarch, and mycelium-based packaging, which significantly reduce landfill waste and carbon emissions (Li & Zhao, 2023). The application of smart labelling technologies—including QR-coded packaging—has also evolved, enabling guests to trace the sustainability credentials of packaging materials (Green & Taylor, 2024). The increasing number of hybrid events now generate lowest presentation strategies, such as refillable stations and reusable service ware, aligning with the United Nations' Sustainable Development Goal 12 on responsible consumption (UNEP, 2024). Research from the *Journal of Sustainable Event Design* further highlights that incorporating sustainable packaging design not only increases environmental outcomes but also reinforces the event's brand equity among eco-conscious participants (Rahman & Prasad, 2025).

5.8 Sustainable Transportation Methods

Moving people, products, and equipment to the event is crucial and required for it to happen. The largest carbon dioxide contributor to activities is usually transportation. The event manager must concentrate on reducing transportation and finding new methods to reduce air emissions, as well as creating strategies that will affect the participants' travel choices.

The mobility-related emissions reason for up to 70% of total event-related carbon output (Silva & Diniz, 2024). The incorporation of green mobility systems—such as electric shuttle buses, shared bicycles, and carpooling networks—has confirmed successful in reducing the carbon footprint of large-scale events (Kumar & Ahmad, 2023). Hybrid event models combining in-person and virtual participation have also considerably reduced long-distance travel and associated emissions (Bauer et al., 2024). The event planners are gradually adopting mobility-as-a-service (MaaS) applications that permit participants to prefer low-emission routes, track carbon savings, and access incentives for sustainable travel (Nguyen & Lee, 2025). Cities conducting mega-events are introducing transit-oriented designs that connect event venues with public transportation hubs, promoting the use of metro, electric buses, and walking routes (Rahim et al., 2023). These strategies associate with global sustainability frameworks like the ISO 20121 Event Sustainability Management System, which prioritizes eco-friendly logistics and participant mobility planning (ISO, 2024).

Freight

Getting goods and equipment to and from the venue site in a sustainable manner is the greatest challenge. Planning carefully will help minimize any negative effects on the ecosystem. The areas of emphasis are the lengths and types of transportation. The procurement strategy should include local sourcing. All goods should be purchased from suppliers that are situated as near to the event as is practical, as should the hiring of contractors. When feasible, it is preferable to select sustainable shipping firms that utilize environmentally friendly vehicles or other green modes of transportation. If possible, in addition to having a sustainable transportation strategy, suppliers and contractors should also think about combining freight shipments. (Pfoser, Sarah. (2022).

Participant and Staff Transport

The location of the event should be considered from the logistics side. Ideally, the location should be approachable. the location somewhere handy location that is central to mass transport, cycling or walking or by car. To be encouraged is mass transit. Moreover, the event manager has the option to include public transportation advice and transport schedules on the invitation to the event. Cycling and walking should be encouraged. It is important to encourage these health advantages. Furthermore, the event organizer may provide participants with discounts on ticket prices, for instance. Choosing this mode of transport. The benefits of carpooling should also be taken into account. To cars that are full, ticket rates may be provided. Allocated parking spaces and parking fees may be taken into account in order to reduce this method of transportation. To ensure that every participant, the event organizer could, for instance, arrange for shuttle service for those arriving by train or plane.

6. CONCLUSION

The events industry confronts both an opportunity and a challenge for innovation and expansion of environmentally sustainable event management. Event planners are expected to create a balance between the need to make memorable experiences and the accountability to reduce ecological footprints since environmental awareness grows worldwide. Further, to preserve the environment, event planners can accelerate favourable social and economic results by implementing sustainable practices, such as reducing the waste, using limited energy, and sourcing materials ethically.

Adapting sustainability into event management process is not just a trend; rather, it is a calculated attempt that enhances stakeholder loyalty, builds brand recognition, and creates new business prospects. Sustainability is witnessed as a vital distinctive factor that provides competitive leverages in a developed and dynamic market, as exhibited by a number of case studies. Furthermore, as innovations and technologies advance, so will the methods and resources available to produce events that are more impactful, efficient, and sustainable.

In the future, the events sector needs to adopt a comprehensive, cooperative strategy for sustainability that includes all parties involved, from vendors to guests. Continuous innovation, the exchange of best practices, and an unwavering dedication to bringing about constructive, long-lasting change are key components of the future of environmentally sustainable event management. In the end, sustainable events will be crucial in creating a more resilient, sustainable, and responsible global economy in addition to fostering industry growth.

7. LIMITATIONS AND SCOPE FOR FUTURE RESEARCH

In-spite of increasing efforts on conducting sustainable event management, there are several constraints on its integration and assessment. The recent research has not explored much on this area, for instance, they are focusing on environmental dimensions while underexploring the long-term social and economic impacts. There is also a lack of standardized sustainability measurement tools and benchmarking frameworks that could facilitate to measure the performance across diverse event verticals and scales. Furthermore, many studies rely on qualitative or case-based approaches, limiting generalizability across regions and event contexts.

Future research needs to concentrate on developing standard metrics and cross-cultural comparative frameworks to assess the sustainability performance comprehensively. More empirical research studies are needed to assess the long-term social legacies of sustainable events, particularly in post-pandemic and hybrid formats. Investigating the role of emerging technologies—such as digital twins, and blockchain, artificial intelligence,—in improving transparency and efficiency could also advance sustainable event practices. Moreover, future studies could explore behavioural aspects, including how stakeholders' attitudes, motivations, and cultural contexts influence the embracing of sustainable event management practices. Addressing these research shortcomings will improve the theoretical and practical foundation for creating inclusive, resilient, and sustainable event ecosystems.

REFERENCES

Al Thani, S., Al Kuwari, H., & Elmaghraby, H. (2025). Social sustainability, inclusive cities, and the legacy of FIFA in Qatar: The case of Education City. *Social Sciences (Basel, Switzerland), 14*(3), 171. DOI: 10.3390/socsci14030171

Alberts, M., & Guaralda, M. (2023). *Mega-events and urban inequality: Rethinking regeneration after COVID-19*. EWA Proceedings of Contemporary Hospitality Research.

Bauer, K., Jensen, L., & Ortiz, P. (2024). Hybrid event design and its role in reducing travel emissions: A sustainability assessment. *Event Management, 29*(2), 89–104. DOI: 10.3727/152599524X

Byers, T., Hayday, M., Mason, D. S., Lunga, S., & Headley, D. (2021). Innovation for positive sustainable legacy from mega sports events: Virtual reality as a tool for social inclusion legacy for Paris 2024 Paralympic Games. *Frontiers in Sports and Active Living, 3*, 625677. DOI: 10.3389/fspor.2021.625677 PMID: 33969293

Capolongo, S., Buffoli, M., Oppio, A., & Rebecchi, A. (2022). Rethinking event planning for inclusive urban resilience in post-pandemic cities. City. *Territory and Architecture, 9*(4), 1–12. DOI: 10.1186/s40410-022-00175-0

Cargnello, M., Pizzi, G., & Visentin, M. (2021). Mega-events, tourism, and gentrification: Understanding the economic reshaping of urban space. *Sustainability, 13*(22), 12827. DOI: 10.3390/su132212827

Chen, L., Wu, Y., & Singh, M. (2024). Circular packaging systems for sustainable event catering: A life-cycle perspective. *Journal of Cleaner Production, 456*, 144322. DOI: 10.1016/j.jclepro.2024.144322

Chen, W.-H. (2024). Progress in Green Energy and Fuel for Sustainability. *Green Energy and Fuel Research, 1*(1), 13–22. DOI: 10.53941/gefr.2024.100003

De Brito, R., & Luna, C. (2024). Circular economy strategies for sustainable event management: Integrating environmental innovation and stakeholder collaboration. *Sustainability, 16*(7), 3412. DOI: 10.3390/su16073412

Dolasinski, M. J., Roberts, C., & Young, L. (2025). Sustainability, the Balanced Scorecard, and Event Tourism: The SBSC-ET Model. *Sustainability (Basel), 17*(5), 2174. DOI: 10.3390/su17052174

Event, Kk. Management Services. (2024, October 1). *Sustainable catering: Eco-friendly options for your event.* https://kkeventsindia.com/2024/10/01/sustainable-catering-eco-friendly-options-for-your-event/

EventFlare Journal. (2024). *Sustainable catering practices for corporate events: Eco-friendly approaches to modern dining.* https://eventflare.io/journal/sustainable-catering-practices-for-corporate-events

FAO. (2025). *Sustainable Gastronomy Day – 18 June.* Food and Agriculture Organization of the United Nations. https://www.fao.org/sustainable-gastronomy-day/en

Foley, M., McGillivray, D., & McPherson, G. (2009). Policy, politics and sustainable events. In *Event management and sustainability* (pp. 13–21). CABI. DOI: 10.1079/9781845935245.0013

Frontiers. (2023). *Sustainable technologies, mega-events, and environmental awareness – Lessons from Expo 2020 Dubai.* Frontiers in Sustainable Cities. DOI: 10.3389/frsc.2023.1192295

Frontiers. (2025). *Green and resilient hotel operations through mega-event legacies.* Frontiers in Sports and Active Living, 1604131. DOI: 10.3389/fspor.2025.1604131

García-Melón, M., Lozano, R., & Molina-Moreno, V. (2023). Measuring sustainability performance of hybrid events: A multi-criteria decision-making approach. *Journal of Cleaner Production, 416*, 137892. DOI: 10.1016/j.jclepro.2023.137892

Green, T., & Taylor, J. (2024). Smart labeling for sustainable packaging in the event industry. *Sustainable Materials and Technologies, 39*, 120875. DOI: 10.1016/j.susmat.2024.120875

Hannon, P. (2025, June 19). *Sustainability in event catering: How eco-friendly practices are reshaping the industry. Catersource.* https://www.catersource.com/business-operations/sustainability-in-event-catering-how-eco-friendly-practices-are-reshaping-the-industry

Hediger, W. (2000). Sustainable development and social welfare. *Ecological Economics, 32*(3), 481–492. DOI: 10.1016/S0921-8009(99)00117-2

Higham, J. (2005). *Sport tourism destinations: Issues, opportunities and analysis.* Elsevier Butterworth-Heinemann.

ISO. (2024). *ISO 20121: Event sustainability management systems – Guidelines for sustainable event operations.* International Organization for Standardization. https://www.iso.org/standard/

Johns, M. J., Choi, E. S., & Baskaran, D. (2024). DinAR: Augmenting reality for sustainable dining. *arXiv*. https://arxiv.org/abs/2404.13272

Jones, M. (2023). *Sustainable event management: A practical guide* (5th ed.). Routledge.

June, H. M., Kernan, A. R., Sumsion, R. M., Cope, M. R., Sanders, S. R., & Ward, C. (2024). Mega events, sustainable social legacies, and community experience: A longitudinal analysis following the 2002 Winter Olympics. *Sustainability (Basel)*, *16*(20), 8993. DOI: 10.3390/su16208993

Kumar, R., & Ahmad, S. (2023). Green mobility and low-carbon logistics in sustainable event management. *Transportation Research Part D, Transport and Environment*, *128*, 104353. DOI: 10.1016/j.trd.2023.104353

Laing, J., & Frost, W. (2010). How green was my festival: Exploring challenges and opportunities associated with staging green events. *International Journal of Hospitality Management*, *29*(2), 261–267. DOI: 10.1016/j.ijhm.2009.10.009

Li, B., Choe, Y., Kang, S., & Lee, J. (2025). Sustainable Event Tourism: Risk Perception and Preventive Measures in On-Site Attendance. *Sustainability (Basel)*, *17*(8), 3455. DOI: 10.3390/su17083455

Li, X., & Zhao, Y. (2023). Biodegradable biomaterials and innovations in event packaging design. *International Journal of Hospitality Management*, *113*, 103623. DOI: 10.1016/j.ijhm.2023.103623

Mair, J., Trendafilova, S., & Werner, K. (2024). Environmental Sustainability in Event Management: A critical commentary. *Event Management*, *28*(3), 477–484. Advance online publication. DOI: 10.3727/152599524X17046754077334

Marouani, I. (2024). Contribution of renewable energy technologies in combating phenomenon of global warming and minimizing GHG emissions. *Clean Energy Science and Technology*, *2*(2), 164. Advance online publication. DOI: 10.18686/cest.v2i2.164

Mooney, S. D., Harrison, S. P., Bartlein, P. J., Daniau, A. L., Stevenson, J., Brownlie, K. C., & Williams, N. (2011). Late Quaternary fire regimes of Australasia. *Quaternary Science Reviews*, *30*(1-2), 28–46.

Musgrave, J., & Raj, R. (2009). Introduction to a conceptual framework for sustainable events. In *Event management and sustainability* (pp. 1–12). Cabi. DOI: 10.1079/9781845935245.0001

Ngan, H. M., Ong, T. S., & Wong, K. S. (2022). Social sustainability in event management: Balancing equity and inclusion in community engagement. *Journal of Sustainable Tourism and Events*, *10*(3), 215–230.

Nguyen, H., & Lee, S. (2025). Mobility-as-a-service applications for sustainable event transportation: A behavioral study. *Journal of Sustainable Tourism*, *33*(4), 577–593.

Osei-Assibey, E., & Osei, C. (2025). Mega-events, rent-seeking and urban inequality in emerging economies. *Journal of International Relations and Development*. Advance online publication. DOI: 10.1057/s41268-025-00342-z

Park, S., Park, E., & Kim, Y. (2023). Advancing sustainable practices in mega-events: Stakeholder collaboration and policy implications. *Sustainability*, *15*(9), 7144. DOI: 10.3390/su15097144

Rahim, A., Choudhury, M., & Patel, D. (2023). Public transit connectivity and carbon mitigation in urban event planning. *Sustainable Cities and Society*, *102*, 104378. DOI: 10.1016/j.scs.2023.104378

Rahman, F., & Prasad, D. (2025). Eco-design and branding through sustainable event packaging. *Journal of Sustainable Event Design*, *2*(1), 18–29.

Ramos-Soler, I., Martínez-Sala, A. M., & López-Cantos, F. (2024). Hybrid events and digital sustainability: Towards a low-carbon future in event management. *Sustainability*, *16*(14), 6013. DOI: 10.3390/su16146013

Razak, T. R., Ismail, M. H., Darus, M. Y., Jarimi, H., & Su, Y. (2025). *Artificial Intelligence in Renewable Energy: A systematic review of trends in solar, wind, and smart grid applications*. Research and Reviews in Sustainability., DOI: 10.5334/rss.6

Rewilak, J. (2023). *Do mega-events boost local business investment? Evidence from the Olympics*. Nottingham Trent University Research Repository. https://irep.ntu.ac.uk/id/eprint/46595

Santos, J. A. C., Fernández-Gámez, M. A., Puig-Cabrera, M., & Santos, M. C. (2024). Sustainability in Business Events: How Hybrid Formats Shape Attendee Decision-Making. *Journal of Tourism and Services*, *15*(29), 320–348. DOI: 10.29036/jots.v15i29.1006

Silva, F., & Diniz, T. (2024). Decarbonizing event transport: Strategies for low-emission mobility transitions. *Sustainability*, *16*(5), 2292. DOI: 10.3390/su16052292

Sportsyncr. (2025). *Technology is increasingly becoming the backbone of sustainability in mega sporting events*. Sportsyncr Features. (https://www.sportsyncr.com/features-5.html)

Suresh, S. (2024). Human rights and mega sporting events: Impacts and the possibility of redress. *International Journal of Social Science and Economic Research, 9*(11), 5428–5442. DOI: 10.46609/IJSSER.2024.v09i11.030

Toscani, A. C., Vendraminelli, L., & Vinelli, A. (2024). Environmental sustainability in the event industry: A systematic review and a research agenda. *Journal of Sustainable Tourism, 32*(12), 1–35. DOI: 10.1080/09669582.2024.2309544

UNEP. (2024). *Responsible consumption and production (SDG 12): Packaging and waste in global events.* United Nations Environment Programme. https://www.unep.org/resources

Walters, G., Mair, J., & Deery, M. (2023). The triple bottom line revisited: Measuring sustainability outcomes of community festivals. *Event Management, 27*(4), 563–579. DOI: 10.3727/152599523X16748370839457

Chapter 8

Ecoleviant Governance: A Model for Environmental Sustainability in Local Industrial Systems

Filiz Resuloğlu

https://orcid.org/0000-0001-6457-1572

Kocaeli University, Turkey

ABSTRACT

This chapter introduces the Ecoleviant Governance Model, a novel conceptual framework developed through doctoral research, designed to address the pressing need for environmentally sustainable governance in terms of crisis governance. The model can be applied to local industrial systems as well. Based on interdisciplinary inquiry the model emphasizes participatory, adaptive, and multi-scalar mechanisms of environmental decision-making. It responds to the complex ecological interdependencies and socio-economic challenges faced by industrial zones, especially in emerging economies. The chapter will first outline the model's key components. It then discusses its empirical grounding in case studies from China and Italy. The chapter concludes with practical recommendations for policymakers and industry leaders seeking to localize sustainability efforts through integrated governance strategies.

INTRODUCTION

The Covid-19 pandemic, which first emerged in China in 2019 and quickly turned into a global crisis, deeply affected not only countries' healthcare systems but also their economic structures, social relations, and governance models. As a result, the world was not merely drawn into a temporary crisis but entered a profound transfor-

DOI: 10.4018/979-8-3373-6435-3.ch008

mation process. This transformation, like many pivotal moments in human history, signals a crossroads. The discourse that "nothing will ever be the same again" has become increasingly prevalent.

The experience of the pandemic made it clear that states cannot build a sustainable future by dealing only with domestic issues. Humanity has come to a greater realization that we are all in the same boat. The global climate crisis, the loss of biodiversity, the breakdown of supply chains, environmental injustices, and social inequalities became more visible during the pandemic, revealing that these problems transcend national borders and must be addressed through a global approach.

In this new post-pandemic era, the concept of governance itself requires rethinking. The traditional understanding of "government" has given way to a more inclusive, participatory, and multi-layered structure known as "governance." Today's problems are multi-actor, multi-scale, and interdisciplinary—therefore, the solutions must also reflect these characteristics. Within this framework, two dominant phenomena have gained increasing importance at the core of effective governance: digital surveillance and sustainability.

Digitalization, accelerated by the pandemic, has created multifaceted effects—not only in communication and data management but also in public health monitoring, influencing consumer behavior, automating production processes, and enabling continuous monitoring of individuals. While digitalization facilitates faster and more efficient decision-making, it also brings issues like privacy, ethics, and inequality. Meanwhile, sustainability has evolved into a holistic approach that goes beyond eco-friendly practices to include economic, social, and ecological dimensions. Taken together, these two phenomena—digital surveillance and sustainability—form the indispensable dynamics of contemporary governance. Therefore, actors who lack knowledge and awareness of these issues cannot develop effective governance models that ensure environmental sustainability, particularly in local industrial systems.

To make sense of the post-pandemic era, not only technical or sociological analyses but also literary and cultural texts serve as powerful resources. E.M. Forster's 1909 short story *The Machine Stops* offers a striking example in this context. The story presents a dystopia in which people live underground, all their needs are met by a "machine," face-to-face contact has disappeared, and individuals live disconnected from one another. Although this machine-based life creates a comfort zone, it suppresses human will and critical thinking, turning into an unsustainable order. Forster's foresight strongly parallels today's themes, such as social isolation due to the pandemic, dependency on digitalization, and total submission to centralized authorities' decision-making processes.

Similarly, dystopias like Orwell's *1984* and Huxley's *Brave New World* depict how digital surveillance, manipulation of consumer habits, and limitations on individual freedoms can unfold. These dystopian narratives have gained more relevance than ever in the post-pandemic period. The balance between technology-driven governance models and the fight against environmental crises has become crucial for humanity's future.

Current environmental management approaches are typically defined either by state-centered structures or market mechanisms. In these models, nature is often seen as an "external element" to be managed; decision-making processes are reduced to technical expertise, data, and cost calculations. However, complex issues such as the climate crisis, biodiversity loss, social inequalities, and resource depletion clearly reveal the inadequacies of these approaches. Most current environmental management systems rely either on state-controlled structures or market-based mechanisms. These frameworks often treat nature as something external that needs to be managed, reducing decision making to matters of technical expertise, data-analysis, and cost-benefit calculations. However, the growing challenges of climate change, biodiversity loss, social inequality, and resource depletion make it clear that such approaches are no longer sufficient.

The pandemic made these shortcomings even more visible—exposing the fragility of societies, the reactivity of nature, and the centralization of governance systems. This fragility has underscored the need for a new management paradigm. Additionally, most current models treat sustainability not as a normative goal but as an instrumental one—in other words, as a tool for the system's continuity. This perspective overlooks the ethical dimension and the depth of our relationship with nature. For these reasons, a new model was necessary because:

- Existing management models do not respect ecological limits.
- Decision-making processes are mostly top-down and lack participation.
- Ethical, cultural, and ontological dimensions are excluded; nature is viewed only as a resource.
- Sustainability is represented symbolically rather than structurally.

The *Ecoleviant Governance Model* aims to address these shortcomings by offering both a philosophical reconsideration and a conceptual framework.

The Ecoleviant Governance Model emerged as the intellectual product of an effort to interrogate new forms of interaction between sustainability and governance in the era of global crises accelerated by the pandemic. The model is inspired not only by conceptual foundations but also by concrete contemporary governance practices. Two specific case studies, representing opposing poles, were particularly influential in shaping the model: China's *Social Credit System* and Italy's *Capraia Smart Island Project.*

Developed by the People's Republic of China, the Social Credit System is a management model based on the continuous monitoring of citizens' behaviors via digital data and rewarding or penalizing these behaviors accordingly. In this pilot-stage system, various actions—including environmental behaviors—are to be evaluated through "credit" scores, potentially enabling direct interventions into individuals' social lives.

This model demonstrates how digital surveillance can become a tool for norm production, how public policies can be internalized via technological infrastructure, and how citizenship rights might be redefined. For instance, harming the environment could be penalized, while recycling might be rewarded. However, this system is criticized as a product of disciplinary and centralized sustainability, rather than voluntary engagement.

In contrast, the *Smart Island Project* carried out on Capraia Island off the coast of Tuscany, Italy, envisions sustainability not as a means of control through technology, but as a participatory and cooperative lifestyle. In this project, residents' and visitors' energy consumption, waste production, and interactions with nature are monitored using sensors and digital tools. However, this monitoring process functions as a practice of collective learning and awareness, rather than surveillance.

The Capraia model turns technology into a tool that supports individuals in establishing more conscious relationships with nature. Data serves community-based feedback and adaptive planning processes rather than centralized control. In this regard, it aligns closely with Ecoleviant's vision of sustainability as an ethical consensus and partnership.

These two case studies highlight why a unique approach like the Ecoleviant model is needed. China's model shows how surveillance can function as a disciplinary tool, while Capraia's model reveals how it can be transformative and participatory. The Ecoleviant model seeks a balance between these extremes—exploring how digital technology can be ethically, voluntarily, and pluralistically harnessed in favor of sustainability.

Thus, the Ecoleviant model delineates the ethical boundaries of digital surveillance tools while examining how these tools can support nature-centered decision-making processes. It reframes digitalization not as an end, but as a means to build sustainable, just, and learning-oriented systems.

In this book chapter, in line with the above framework, the conceptual Ecoleviant Governance Model will be introduced. This multi-dimensional model aims to offer a perspective on how management mechanisms can be redesigned to ensure environmental sustainability.

BACKGROUND

Governance

At various scales—local, regional, national, and international—different social structures and corresponding management practices are evident within society. In what is now referred to as the Anthropocene, or the human-dominated era, these administrative practices are undergoing significant transformations in response to contemporary shifts and developments. Many traditional forms of management appear increasingly inadequate in addressing the complex demands of the present age and are thus entering processes of reform or transformation.

Global phenomena such as the climate crisis, rapid technological advancement, globalization, and democratization have blurred the previously distinct boundaries between administrative units. These conditions call for a more interactive and collaborative governance paradigm. Notably, the COVID-19 pandemic served as a critical juncture, testing the administrative capacities of governments worldwide. The heavy losses and systemic failures observed during and after the crisis have fueled widespread scrutiny of existing management models and generated a push toward more effective and resilient administrative systems.

The pandemic, recognized as a global crisis, has underscored the urgent need for anticipatory and well-prepared policy frameworks, particularly in the field of crisis management. As evidenced by ongoing research and academic inquiry, stakeholders have increasingly been drawn into a multifaceted process of evaluation and redesign aimed at enhancing the adequacy of current management models. Governments have also sought to improve internal institutional performance through reform initiatives, striving to discover more efficient modes of management (Kapucu, 2010: p.6).

In a global context characterized by the retreat of hegemonies, the erosion of borders (and the dismantling of the walls that enforced them), the emergence of civic uprisings in urban centers, and the declining relevance of military alliances, questions surrounding the future of global order and governance have gained unprecedented importance (Rosenau, 1992, p. 7). Contemporary transformations reflect a broader societal shift from an industrial society to an information society, from Fordist modes of production to more flexible manufacturing systems, from a world of nation-states to one shaped by globalization, and from modernist ideologies to postmodern perspectives (Tekeli, 1996: p.49).

In the 1930s, public administration was marked by the dominant presence of centralized institutions. At that time, information technologies were still in a very rudimentary state, communication between geographically distant regions was slow, and the overall education level of public officials was relatively low. Therefore, it was necessary to physically bring them together in order to collect data effectively and to deliver instructions efficiently. On the other hand, the system allowed ample time for information to flow from the top down and for decisions to move from the bottom up within the administrative hierarchy.

However, today's conditions differ significantly. Information sources have become virtually limitless, communication between distant regions is now instantaneous, public employees are generally well-educated, and working environments are characterized by rapid and unpredictable change. There is no longer sufficient time to wait for information to be transmitted from the top down or for decisions to ascend step-by-step through the administrative ranks. In the current context, the effective delivery of public services depends on empowering people working at the local level—such as those in schools, public housing, parks, and educational programs—to make decisions and actively participate in management processes (Osborne, 1992, p. 3).

Traditional approaches to addressing problems, formulating policies, and delivering public services—which typically treat complex issues as matters of intellectual design and place central importance on research and expert knowledge—are no longer sufficient. In today's complex societal landscape, the intricate challenges that confront governments, private enterprises, and civil society demand a fundamentally different approach. Rather than relying on hierarchical, top-down problem-solving models, there is a growing need for more horizontal, collaborative methods. This shift is often described as a transition from government to governance, reflecting a move toward shared responsibility and multi-actor engagement in public decision-making (Klijn & Koppenjan, 2015, p. 5).

Contemporary administrative systems, which increasingly struggle to meet growing and complex societal demands, are undergoing a significant transformation. Unlike earlier models that were centralized, hierarchical, and based on instrumental rationality—where a single authority formulated goals, made decisions, produced services, and monopolized both resources and power—there is now a shift toward a governance approach grounded in human rights and performance-based standards. This emerging model emphasizes multi-actor participation, localized and network-based relationships, and a communicative form of rationality. Rather than positioning itself at the center, this new understanding of governance seeks to empower societal actors, facilitate coordination, and enable the effective allocation of resources (Tekeli, 1996, pp. 50–51).

Following this period of rapid transformation, the governance paradigm that began to take hold in the 1990s has gained particular relevance in the context of globalization. It plays a key role in striking a balance between the opportunities and challenges that have emerged on a global scale (Özer, 2006, p. 59).

The broad and multifaceted nature of governance as a concept has prevented its scholarly literature from being confined within narrow boundaries. Various academic disciplines have approached governance from diverse perspectives, including development studies, economics, geography, international relations, planning, political science, public administration, and sociology. Consequently, it is evident that each academic community contributes uniquely to the body of governance literature. While social theorists often employ the term governance to conceptualize social coordination and organization for abstract analyses, other social scientists focus on examining shifts in governance patterns among corporations, public institutions, and global affairs. Ideally, governance theory is expected to encompass both abstract analyses that address organizational forms such as hierarchy, markets, and networks, as well as more empirical investigations into the evolving nature of social and political life (Bevir, 2013, p. 1).

Although the concept of governance was not originally coined by the World Bank, the institution played a significant role in bringing it to prominence and increasing its visibility. The World Bank first identified a crisis situation in Africa in 1989, characterizing it as a "governance crisis," which led to the conceptual evolution and subsequent adoption of governance by other financial and international organizations (World Bank, 1989, p. 2). The association of the term governance with values such as transparency, accountability, the rule of law, and participation has fostered its acceptance as a necessary policy framework for democratic functioning in national and international decision-making processes (Zabcı, 2015, p. 151).

Within this context, governance is linked to neoliberal public sector reforms advocated by the World Bank—namely marketization and New Public Management (an approach to public administration that applies business and market principles to the public sector) —which are believed to enhance efficiency. In contrast, an examination of the British Economic and Social Research Council's Local Government and Whitehall Programmes reveals that the term governance is employed to describe a new form of relationship between the state and civil society. Here, governance is associated with networks, defined in contrast to hierarchies and markets (Stoker, 1999, p. 15).

The concept of governance, which is employed across numerous disciplines and fundamentally means "managing together" (Toksöz, 2008, p. 17), is broader and more complex than the traditional notion of management. Due to its multidimensional nature, there is no universally accepted, comprehensive definition of governance. However, political scientist Fukuyama, after reviewing the political science literature on the subject, argues that there are at least three distinct definitions of governance:

1. The first pertains to international cooperation among bodies controlled by supranational or external political authorities, independent of the state system. This perspective, emerging alongside globalization, suggests that national sovereignty

is evolving toward horizontal collaborations with supranational entities such as the European Union and other informal institutions.

2. The second definition treats governance as synonymous with public administration, emphasizing the effective implementation of state policies. This interpretation arises from the recognition that global poverty stems from corruption and weak state capacity.
3. The third definition views governance as the regulation of social behavior through networks and non-hierarchical mechanisms (Fukuyama, 2016, pp. 1–6).

Similarly, Rhodes proposes that governance has at least six different definitions, categorizing them as minimal state, institutional governance, New Public Management (NPM), good governance, socio-cybernetic systems, and self-regulating networks (Rhodes, 1996, p. 652).

According to the United Nations, governance can be understood as the economic, political, and administrative authority exercised at all levels to manage a country's affairs. It encompasses the mechanisms, processes, and institutions through which citizens and groups articulate their interests, exercise their legal rights, fulfill their obligations, and mediate their differences. While governance includes the state, it also extends beyond to incorporate the private sector and civil society, all of which are essential for sustaining human development. The state provides an appropriate political and legal framework; the private sector generates employment and income; and civil society facilitates political and social interaction, enabling participation in economic, social, and political activities. Given that each sector possesses distinct strengths and weaknesses, the United Nations argues that its support for good governance aims to promote constructive interaction among all three (UNDP, 1997, pp. 2–3).

Governance is not synonymous with government; rather, it represents a more inclusive concept. While it encompasses government institutions, governance also involves informal, non-governmental mechanisms that enable individuals and organizations within its scope to advance, meet their needs, and fulfill their demands (Rosenau, 1992, p. 4).

Governance refers to the complex mechanisms, processes, relationships, and institutions through which citizens and groups articulate their interests, fulfill their rights and obligations, and deliberate over their differences (Rondinelli & Blunt, 1997, p. 9). The joint efforts of all relevant actors within a given social-political system are central to the implementation of governance. In this context, governance can be understood as coordinating conflicting actors and diverse networked structures. It emerges as a management process in which various societal actors are recognized as stakeholders, and governance is shaped through their collective efforts and interactions.

Sustainability

In September 1991, a group of eight scientists—four women and four men—were sealed inside a 13,000-square-meter glass dome known as Biosphere II, constructed in Oracle, Arizona. The goal of the experiment was to create a miniature model of Earth's primary ecosystems. However, by the end of the two-year mission, the artificial environment was on the brink of collapse, and the severely weakened researchers owed their survival to externally supplied clean air. Despite a budget of 200 million dollars, Biosphere II failed to produce adequate breathable air, potable water, and sufficient food—even for just eight individuals. In contrast, Biosphere I—the Earth itself—effortlessly provides these vital services to billions daily. It is thus particularly alarming that this very biosphere is now under threat. The true cost of degrading the ecosystem only becomes clear once the services it provides begin to falter (Lovins et al., 1999, pp. 145–146).

In recent years, the increasing frequency of climate-related disasters has prompted humanity to confront the reality of climate change and global warming, spurring urgent concern over how much time remains before irreversible consequences set in. At the 2007 Intergovernmental Panel on Climate Change (IPCC), carbon dioxide emissions were identified as the primary driver of climate change, with human activity pinpointed as the dominant source of this increase. This acknowledgment has led various actors—governments, societies, individuals—to engage more seriously with the concept of sustainability, and in turn, to develop aligned policies and strategies (Yavuz, 2010, p. 64).

As awareness of global warming and its perceived impacts has grown, academic research on sustainability has expanded significantly. Numerous disciplines have attempted to explore and define the concept of sustainability, which varies in scope depending on context and includes multiple dimensions. A few widely referenced definitions include:

1. *Brundtland Commission Definition (1987):* "Sustainable development is development that meets the needs of the present without compromising the ability of future generations to meet their own needs." This definition emphasizes the balance between economic, social, and environmental considerations for long-term well-being.
2. *Environmental Perspective:* Sustainability refers to practices and processes that minimize negative environmental impacts, conserve natural resources, and protect biodiversity to benefit both current and future generations (UNEP).
3. *Social Perspective:* Sustainability involves ensuring that basic human needs are met, communities are empowered, and cultural diversity is respected. It emphasizes the promotion of social justice, equity, and inclusiveness (UN DESA).

4. *Economic Perspective:* Sustainability relates to fostering economic resilience and long-term value creation by minimizing harmful impacts and promoting responsible resource management, innovation, and sustainable business practices (Elkington, 1998, p. 75).

It is well understood that change is a fundamental characteristic of natural systems; indeed, it is widely accepted that change itself is the only constant. From this viewpoint, development can be seen as both a form and a result of change. Accordingly, sustainability may be interpreted as the capacity to preserve the potential for transformation and evolution within both objects and processes (Çağlar, 1998, p. 61).

These definitions underscore that sustainability is a multidimensional concept—spanning environmental, social, economic, and systemic dimensions—aimed at ensuring a long-term harmonious balance between human activity and the natural world. Furthermore, the logic behind sustainability and sustainable development brings intergenerational responsibility to the forefront, with a particular emphasis on environmental stewardship. Ultimately, sustainability necessitates a long-term, systematic, and integrated balance among economic, ecological, and cultural resources (Fisunoğlu, 1997, p. 13).

The Systems Approach

The systems approach has recently gained considerable attention in the literature on management and systems theory. Despite its increasing popularity, the term is defined in various ways, largely due to the fact that the concept of a "system" carries different meanings across disciplines. For instance, while an engineer might interpret a system as a tangible physical structure, an economist may view it as an operational model such as the free market or inventory systems. In contrast, a computer scientist might conceive it as hardware architecture, software programs, or information systems (Chen, 1975, p. 32).

Despite these disciplinary differences, there is a shared understanding of some core principles when referring to the systems approach. One of the most prominent among these is the emphasis on *holism*—the idea that the subject or object in question should be considered as an integrated whole. What sets the systems approach apart from other methodologies is its simultaneous examination of the system's components and the interactions between them.

Viewed from a systems perspective, the analysis of any phenomenon generally revolves around three fundamental questions:

1. What are the key components of the system?

2. What are the main processes that connect these components and ensure their coherence?
3. What are the overarching goals the system aims to achieve? (Tecim, 2016, p. 80).

Charles West Churchman defines a system as "a whole composed of interrelated components coordinated to achieve a set of objectives" (Churchman, 1968, p. 29). This concept can also be reformulated as "structures consisting of organized elements" (Churchman, 1971, as cited in Berente, 2008, p. 1). According to Churchman, two essential characteristics define a system: every system has a specific purpose, and it consists of interacting elements that function as a unified whole.

The Austrian biologist Ludwig von Bertalanffy is widely recognized as the founder of the systems approach. In the 1920s, he introduced the General Systems Theory, which posits that it is possible to develop a comprehensive theory to explain all phenomena in the world—seen themselves as hierarchically organized systems—by examining them holistically, uncovering the relationships among them, and integrating the various scientific disciplines that study these phenomena separately into a meaningful framework of interconnected knowledge (Perker & Akıncıtürk, 2012, p. 198). In Bertalanffy's view, General Systems Theory is a logically and mathematically structured discipline that, while formally abstract, can be applied across all fields concerned with systems (Bertalanffy, 1950).

Three distinct modes of thinking have significantly contributed to the development of systems thinking: holistic thinking, teleological thinking, and synthetic thinking. Holistic thinking emphasizes the importance of understanding the whole rather than just its parts, based on the belief that a narrow focus on components may obscure essential aspects of reality. It critiques reductionist, mechanistic, and atomistic approaches in favor of perspectives that are holistic, ecological, and organic. While teleological thinking centers on grasping the common purpose or final cause, synthetic thinking highlights the dynamic interrelations among parts and maintains a focus on the system as a whole (Tecim, 2016, p. 80). For instance, in environmental policymaking, a teleological perspective would emphasize the ultimate goal—such as achieving sustainability or ensuring the well-being of future generations—and shape all actions toward that purpose. In contrast, synthetic thinking would focus on understanding how various factors, such as energy production, transportation, economic policies, and social equity, interact within the broader environmental system. Together, these two modes of thinking provide a more comprehensive understanding of complex problems, balancing clarity of purpose with awareness of interconnections.

DEVELOPMENT OF THE ECOLEVIANT GOVERNANCE MODEL

The Ecoleviant Governance Model is fundamentally a conceptual crisis governance framework, premised on the recognition that contemporary society inhabits an age defined by persistent and multifaceted crises—where crises have become the "new normal." This model emerges as a critical response to the inadequacies of traditional governance structures in managing the intertwined ecological, technological, and social challenges that characterize local industrial systems today.

In order to gain a deeper understanding of the term "leviant" in the model's name, it is useful to refer to Thomas Hobbes's well-known work Leviathan. In this work, Hobbes emphasizes the concept of the "state of nature," a condition in which everyone lives in fear and in conflict with one another in a struggle for survival. Hobbes described life as solitary, poor, nasty, brutish, and short. This view was shaped by the selfish human drive to assert dominance for survival. In Leviathan, Hobbes rationalizes a form of governance wherein people voluntarily submit to a greater power for their own protection. Having lived through the English Civil War, which exposed him to terror, Hobbes centered his political thought on the "fear of violence." Consequently, Leviathan advocates for individuals surrendering to a strong authority to safeguard their lives. Hobbes's image of the state is that of a colossal artificial man—a symbiotic construct where individuals collectively cede authority to preserve order. He argued that a strong authority is essential to maintain and uphold social order.

In contrast, the Ecoleviant Governance Model stresses the need for a well-defined governance structure coupled with a holistic approach to effectively manage crises. The "leviathan" in Ecoleviant is not a single authority but a coalition of stakeholders united as a collective force. Today, the prevailing sentiment in Ecoleviant has shifted from the "state of nature" and survival fear to what Giorgio Agamben calls the "state of exception" and a culture of fear (Agamben, 2020). The world is thought to be on the brink of numerous natural, financial, and socio-economic crises. Especially after the Covid-19 pandemic, people in the 21st century face significant biological, socio-economic, and mental challenges. These three dimensions are interdependent, and deficiencies in one can severely affect the others. Therefore, the Ecoleviant Governance Model advocates for a robust governance framework grounded in ethical practices and sustainability, enabling people to live harmoniously and navigate crises with minimal harm. Instead of Hobbes's call for submission to a singular great power, it promotes strong, collaborative governance practices to ensure order and stability. The "Eco" part of the model's name reflects the belief that ecological balance, cooperative working in harmony, and sustainability should be the "leviathan"—the power to which one surrenders.

Progress is often seen as a phenomenon scaled primarily through technology. In his book Cannibals with Forks, John Elkington aims to enlighten managers about their companies' blind spots (Elkington, 1996). In doing so, he tests best management practices against the imperative of sustainability. The book opens with a quote from Polish poet Stanislaw Lee: "Is it progress if a cannibal uses a fork?" This question challenges many to think beyond merely adopting and using new technological tools. True progress is not about the technologies themselves, but about how they are used. Ecoleviant Governance, which addresses crisis management within the framework of digitalization and sustainability, is driven by the belief that in today's era of technology, digitalization, and crises, technological advances must be combined with ethical practices, continuous education, and a commitment to the greater good of humanity. The issue is not just having the right tools but using them in a way that reflects the highest human values and principles.

E.F. Schumacher's Small Is Beautiful serves as a key reference for the Ecoleviant Governance model. Schumacher advocates for economic models aligned with nature and centered on human needs, opposing growth-driven capitalism. The model supports these ideas through a focus on sustainability and ethical governance. It prioritizes small-scale, local solutions over large, impersonal systems. Ecoleviant Governance emphasizes managing technological progress based on moral values and human well-being, aiming to empower local communities.

The conceptual foundation of the Ecoleviant Governance Model is informed by empirical insights drawn from diverse case studies, notably China's Social Credit System and the Capraia Smart Island project. The Social Credit System exemplifies the complexities and ethical challenges inherent in deploying advanced digital governance mechanisms within socio-political contexts marked by surveillance and control, highlighting the risks of exclusion and authoritarian overreach. In contrast, the Capraia Smart Island initiative illustrates a community-driven, technology-enabled approach to sustainability governance, emphasizing participatory decision-making, environmental monitoring, and local empowerment. By synthesizing lessons from these contrasting yet complementary cases, the Ecoleviant model articulates a governance framework that aspires to harness technological innovation for inclusive, ethical, and sustainable crisis management, while consciously mitigating the dangers of exploitation, opacity, and marginalization.

At its core, the model foregrounds a systems-oriented thinking that reframes industrial activity as part of a broader socio-ecological network rather than as isolated economic operations. This holistic vantage point facilitates integrated decision-making, whereby environmental impacts—ranging from resource extraction to waste disposal—are assessed not as discrete challenges but as interrelated dimensions of systemic transformation.

Moreover, the model's emphasis on the ethical deployment of digital technologies introduces new mechanisms for ecological accountability. Tools such as AI-assisted diagnostics, sensor-based monitoring, and predictive analytics are not merely valorized for efficiency gains, but rather instrumentalized to support transparency, traceability, and adaptive management within industrial operations. In this respect, technology functions not as an end in itself, but as a medium for reinforcing environmental stewardship and operational resilience.

Significantly, Ecoleviant Governance also challenges the technocratic insularity of conventional governance by privileging inclusive, participatory modalities. Through the intentional incorporation of local actors—including marginalized communities—into governance processes, the model fosters a deliberative democratic ethos that resists top-down impositions and instead grounds sustainability efforts in place-based knowledges and lived experiences.

In aligning itself with post-extractive logics such as circularity and regenerative design, the model offers a substantive alternative to linear, growth-oriented industrial paradigms. It calls for a structural reorientation toward practices that prioritize ecological regeneration, material recirculation, and long-term viability over short-term output maximization.

Taken together, the Ecoleviant Governance Model constitutes a strategic framework for reconfiguring local industrial systems in accordance with environmental imperatives and justice-based principles. Its integrative and reflexive character renders it particularly well-suited to navigating the entangled crises of our time, from climate change to socio-technical disruption, thereby positioning it as a necessary innovation in the discourse and praxis of sustainable governance.

CASE STUDIES

In an era where technological advancements contribute positively to society, this study introduces a new perspective on crisis management through a conceptual model called Ecoleviant Governance. This model envisions a future framework in which governance is guided by ethical digital practices and the goals of sustainable development. It aims to address the challenges of modern crises—whether environmental, economic, or social—through innovative, transparent, and human-centered approaches, while also seeking insulation from the harms of digital surveillance and exploitation. Ecoleviant Governance envisions a future where digital technologies are used not only for monitoring and control, but also to enable positive transformation, ensure sustainability, and uphold ethical standards in governance. Overall, the model proposes a responsible use of digital advancement—where technology serves the greater good by enhancing crisis management while maintaining ethical boundaries and a commitment to sustainability. In this context, case studies such as China's Social Credit System and the Capraia Smart Island Project offer valuable insights into both the potential benefits and risks of integrating digital technologies into governance and crisis response.

Social Credit System of China

The Social Credit System, selected as a case study within the model, contains elements that serve both as inspiration and caution in the development of the Ecoleviant Governance model. Although it is currently being implemented only in pilot regions and has not yet been extended nationwide, the system has drawn significant criticism—particularly from the West. These criticisms are reflected in academic literature and media coverage, where the system is frequently compared to Orwellian-style surveillance. On the other hand, there are also studies that argue the system, if implemented in line with its stated goal of building a "prosperous, ethically grounded, and harmonious society," could serve as a model for other countries.

The Social Credit System is considered to provide insight into the potential of digital technologies to monitor and influence individual behavior. Examining this system is important for understanding ethical concerns such as surveillance and privacy violations, as well as the risks of misuse. Its extensive data collection and monitoring practices offer valuable perspectives on the possible negative impacts such measures may have on personal freedoms and social trust. In this context, it becomes evident that intrusive surveillance should be avoided and the ethical use of data should be prioritized.

Furthermore, analyzing how the system uses rewards and penalties to influence behavior may offer useful lessons on how to implement positive reinforcement strategies that promote ethical conduct and respect for individual rights—without resorting to coercion.

The Social Credit System provides valuable insights into how data can be used for governance and behavioral regulation, which offers a useful reference point for the Ecoleviant Governance model, especially in promoting accountability and creating feedback mechanisms. However, elements such as dependency on surveillance, potential for misuse, and lack of transparency represent major flaws that the Ecoleviant model intentionally seeks to avoid. Instead, Ecoleviant Governance prioritizes ethical digital practices, focusing on the protection of individual freedoms and building governance systems rooted in sustainability and the well-being of humanity.

On the other hand, one of the key motivations behind the Social Credit System is Confucian philosophy, which emphasizes moral conduct, social harmony, and the importance of virtue in society. Its strong focus on moral responsibility, community cohesion, and the role of education offers meaningful insights for the Ecoleviant model. Integrating the positive aspects of this philosophy could enhance ethical leadership, collective responsibility, and continuous learning within the model. However, traditional Confucianism also promotes hierarchical structures, which contrasts with the collaborative and participatory nature central to Ecoleviant Governance. For this reason, while the ideas of virtue and responsibility can be adapted into the model, decision-making processes should favor more horizontal and inclusive structures.

Furthermore, Confucian philosophy's strong emphasis on social order and harmony may lead to tensions regarding individual liberties. While collective responsibility remains important in the Ecoleviant Governance model, it must not come at the expense of personal autonomy or freedom. Striking a balance between individual rights and social harmony is arguably the most distinguishing feature of Ecoleviant Governance when compared to the more restrictive tendencies observed in the Social Credit System, both in its general structure and in its Confucian influence.

Capraia Smart Island Project

American scientist Jared Diamond, known for his theories about the collapse of the civilization that once lived on Easter Island and the environmental destruction caused by its inhabitants, used the extinction of the island's native Polynesian people, the Rapa Nui, as a cautionary tale about the dangers of environmental degradation. In an article he wrote for *Discover* magazine in 1995, Diamond explained how the people of Easter Island destroyed their forests within just a few centuries, exhausted their plant and animal species, and how their once complex society descended into chaos and eventually cannibalism—until they disappeared from the pages of history. Following Diamond's account of Rapa Nui, he posed the question, "Are we following in their footsteps?", igniting new debates on environmental destruction and sustainability. According to him, the people who lived on Easter Island had no books or knowledge of other collapsed societies, and people of our time should learn from the fate of Rapa Nui (Diamond, 1995; Diamond, 2011: p. 79).

It is also recognized that small islands, as isolated systems, can serve as ideal laboratories for implementing innovative models in energy, water cycles, and waste management to address the world's most urgent and significant environmental problems. Consequently, many projects focusing on these areas have been carried out on such islands. Moreover, smaller islands are often the subject of studies conducted by the European Commission and major national and international research centers to test sustainability models. In fact, due to their clear geographical and social boundaries and their suitability for studying regional contexts more easily, islands are increasingly turning into true open-air laboratories (Chimica Verde Assoc., n.d.).

It can be said that Capraia Island, located in the western part of Italy's Tuscany region, is a place where significant sustainability practices have been implemented, and that it continues to take steps toward becoming a model for other regions and authorities in this field (Smart Island EU, 2023). Described as a small island striving to achieve great things, the volcanic island of Capraia has launched a series of sustainability initiatives aimed at creating a model that can be transferred to other small islands in the Mediterranean and adopted on a global scale. Although the Capraia Smart Island Project (CSIP) is an independent initiative aimed at transforming Italy's Capraia Island into a smart and sustainable community, it also offers valuable insights and practical implementations that can contribute to the Ecoleviant Governance model. In this context, the aspects of the project that could positively inform and support the Ecoleviant model can be explored under several key themes:

- Focus on Sustainability
- Community Engagement and Shared Interest
- Use of Technology
- Developing Resilience
- Potential for Scaling Up

The Capraia Smart Island Project embodies principles of good governance through its commitment to transparency, community engagement, sustainability, resilience, digital innovation, and collaborative partnerships. Therefore, it serves as a valuable example for other governance models. If it achieves full success, the project has the potential to significantly enrich and strengthen the Ecoleviant Governance model by providing practical examples in areas such as sustainability, community participation, digital integration, resilience, collaborative governance, and effective monitoring. It already offers a framework addressing these issues and contributes accordingly. In summary, the experiences and outcomes of the Capraia project can serve as exemplary cases to advance the principles and practices of Ecoleviant Governance, enabling a more cyclical and responsive approach to governance.

THE ECOLEVIANT GOVERNANCE MODEL

"No man is an island, Entire or itself, Every man is a piece of the continent, A part of the main'' John Donne

John Donne's famous lines, "No man is an island… Every man is a piece of the continent, A part of the main," emphasize the interconnectedness of human beings. The quote suggests that no one exists in complete isolation; our lives, actions, and experiences are deeply linked to others. It highlights the importance of community, cooperation, and empathy, reminding us that individual well-being is tied to the well-being of society as a whole. In a broader sense, this idea can apply to social, environmental, or global issues, where collaboration and shared responsibility are essential for addressing challenges that affect everyone.

Referring to Antonio Gramsci's redefined concept of *interregnum*—a term used to describe extraordinary periods in which the existing legal and social order has lost its effectiveness and functionality, while a new order, adapted to emerging conditions, is still in the process of being shaped and has yet to fully replace the old—Zygmunt Bauman argues that the current organizational and administrative regime is undergoing a profound legitimacy crisis. According to Bauman, the *interregnum* is not only a sign of the collapse of the old order but also marks a moment when new possibilities begin to emerge. In such periods, ideas previously deemed marginal or unacceptable in the old system can gain serious consideration within the emerging order (Bauman et al., 2024, p. 17).

In the context of contemporary crisis management, much like Gramsci and Bauman's assertions, governments, organizations, and international institutions are now confronted with highly complex global challenges that severely test their capacity to respond effectively. These challenges are not isolated; on the contrary, they are often interconnected, transcending borders and affecting multiple sectors simultaneously. Drawing from recent global crises, some of the most pressing challenges requiring immediate action within the framework of crisis management can be identified as follows:

- Climate Change and Environmental Crises
- Pandemics and Public Health Emergencies
- Cybersecurity Threats and Digital Disruptions
- Global and Regional Political Instability and Conflict
- Economic Volatility and Inequality
- Resource Scarcity and Food Security Challenges
- Gaps in Transnational Cooperation and Global Governance
- Population Growth and Urbanization Pressures
- Social Inequality and the Marginalization of Vulnerable Groups
- Technological and Scientific Challenges

The Ecoleviant Governance Model has emerged from the growing recognition that more coordinated, ethical, and sustainable approaches are urgently needed in the field of crisis management. Integrating key concepts such as technology, sustainability, and human well-being within a cohesive conceptual framework, models like Ecoleviant Governance are seen as vital for addressing today's complex and interconnected global crises. It is also increasingly acknowledged that effective solutions must prioritize cross-border cooperation, the inclusion of marginalized communities, and a dual focus on both immediate threats and long-term challenges.

Core Concepts of the Model

At its core, the Ecoleviant Governance Model aspires to establish a governance framework that leverages digital technologies while centering the well-being of humanity and the planet. By explicitly rejecting exploitative practices such as surveillance, environmental degradation, and social injustice, this model advocates for a more ethical, transparent, and sustainable approach to crisis response and management. Core concepts of the model are as follows:

Systemic and Holistic Approach: Rooted in systems theory, the Ecoleviant Governance Model conceptualizes crises as complex, interconnected challenges that require comprehensive strategies addressing root causes and broader impacts rather than short-term fixes. This perspective, if adopted across levels—from individuals to supranational institutions—enables more informed and coordinated responses to the multifaceted crises and challenges brought about by globalization.

Sustainability, Ethical Values, and Continuity of Relationships: A fundamental principle of the Ecoleviant model is its sustainability orientation. It is premised on the belief that advancing sustainability will minimize the long-term social and environmental damages caused by crises. Thus, environmental literacy should be cultivated from early ages and promoted among those lacking awareness. Leaders and citizens alike must be ecologically and digitally literate, understanding environmental dynamics and digital systems to make informed decisions. Furthermore, ethical and humanitarian values must underpin decisions toward sustainability. Recognizing the rise of transient, virtual, and short-term interactions as a "new normal" that potentially undermines sustainability practices, the model stresses the importance of durable, trust-based relationships among governments, institutions, and citizens as essential to effective crisis governance.

Digitalization and Human Welfare: The model advocates responsible utilization of digital technologies to foster human and societal development, grounded in transparency and accountability. Media and digital literacy should be widely promoted and institutionalized through dedicated resources and training programs. Especially crucial is the digital competence of decision-makers and leaders, enabling them to make conscious, foresighted decisions informed by technological realities.

Meritocracy and Effective Feedback Mechanisms: Ecoleviant supports the principle that competence should be central to governance roles. Tasks must be assigned to qualified and capable individuals or institutions. Systems that leverage real-time data and stakeholder feedback are essential for maximizing benefits, minimizing errors from communication breakdowns, and developing early-warning mechanisms.

Innovative Crisis Management: The model encourages deploying cutting-edge digital technologies for efficient crisis response. Data-driven forecasting aims to predict and resolve potential problems proactively. Simulation tools and artificial intelligence are vital for anticipating crisis scenarios, alongside training programs to ensure their responsible use. While the adoption of digital innovation is supported, it is imperative that technology is integrated into daily life without compromising societal and ethical values—human rights and social principles must not be subordinated to technological progress.

Supranational Authorities and Global Interconnections: Acknowledging the interdependence of nations in a globalized world, Ecoleviant highlights the critical role of international cooperation and organizations during crises. Given that national authorities often fall short in addressing global crises, and considering the near impossibility of isolation in an interconnected world, the model emphasizes the importance of communication with external actors. It advocates for strengthening international networks, frameworks, and alliances to facilitate decision-making processes oriented toward the collective benefit of humanity.

The Potential Actors of the Model

Potential actors of the model are as follows:

State Institutions: In the context of Ecoleviant Governance, state institutions play a role in policymaking, regulation, and supervision of digital applications and sustainability practices. At the same time, various state institutions can play critical roles in ensuring the effective implementation of the model's principles. Some possible institutions and their functions can be listed as follows:

1. Environmental Agency: Responsible for supervising and enforcing regulations related to environmental protection and sustainability. It promotes eco-friendly practices and technologies. It carries out advertising, educational, and advocacy activities to foster the understanding that humans are not masters of nature but a part of it.
2. Digital Governance Agency: Responsible for managing digital infrastructure and ensuring the ethical and transparent use of digital tools. Additionally, it may play a role in developing policies for data privacy, cybersecurity, and responsible digital governance. It can develop control and supervision mechanisms to combat misinformation. It can impose deterrent sanctions on harmful digital platforms and on individuals, institutions, and organizations operating on those platforms that encourage negative behavior. However, it is important to emphasize that personnel working in this agency must act impartially and ethically. Acting within an ethical framework by administrative staff and those in positions of authority in such agencies is a primary condition of the Ecoleviant Governance Model.

3. Crisis Management Agency: Responsible for coordinating responses to all types of crises such as natural disasters, pandemics, and economic crises, and for developing and implementing policies to improve preparedness for crises. This agency can also conduct activities aimed at increasing awareness about crises and crisis management among the public, organizing training and drills. Especially in times of crisis, it can coordinate with necessary individuals and institutions to plan training on search and rescue and first aid.
4. Public Health Agency: Responsible for promoting preventive medicine practices, spreading health literacy to broader audiences, and encouraging responsible and conscious use of technology in health services. Particularly during pandemics or health crises, it focuses on protecting the health of individuals in its scope. It can communicate and cooperate with relevant persons and organizations to plan joint training on first aid and preventive medicine. It can conduct educational and informative activities to promote healthy nutrition, clean food, and healthy living.
5. International Cooperation Agency: In the age of crises referred to as the Anthropocene, it may play a role in encouraging collaboration with other countries and international authorities via synchronized communication networks to combat global challenges. It can actively participate in coordinating cross-border efforts for sustainable development and effective crisis management. Additionally, it can be responsible for keeping information updated regarding global agendas, world events, and communication between countries.
6. Research and Development (R&D) Agency: Responsible for research and technological development in the field of sustainable technologies and practices. It can play an active role in training on the correct use of these technologies. It may conduct activities to raise awareness about ecological sustainability among broader audiences.
7. Literacy and Education Agency: Responsible for efforts aimed at achieving literacy outcomes within the framework of UN sustainable development goals and coordinating other literacy education effectively. The main types of literacy encouraged within Ecoleviant Governance include health literacy, environmental literacy, digital literacy, media literacy, critical literacy, and financial literacy.
8. Social Services and Assistance Agency: During crises, it is important to support vulnerable groups and ensure their access to resources. This agency can be responsible for such efforts. Additionally, it can play an active role in promoting social equity and participation in decision-making processes.
9. Infrastructure, Transportation, and Logistics Agency: Responsible for planning and implementing sustainable infrastructure projects prioritizing environmental concerns; integrating digital solutions for efficient and eco-friendly transportation systems. It can play an active role in developing projects to minimize

carbon emissions from individual to national levels. For example, it can conduct awareness-raising activities on emissions released through individual product shipping journeys or using private vehicles instead of public transportation when alternatives exist. It can encourage collective online shopping or increased use of public transport.

10. Local Governments Agency: Responsible for the implementation of Ecoleviant Governance principles at the community level, ensuring local participation and feedback. It can play an active role in structuring policies and initiatives to meet the specific needs of the local population.

The agencies listed above can cooperate with each other within the framework of Ecoleviant Governance to provide a holistic and effective approach in promoting crisis management and sustainability.

The Potential Tools of the Model

The tools of the Ecoleviant Governance Model are designed to align with fundamental principles such as sustainability, the ethical use of digital technologies, and transparent crisis management. These tools are expected to facilitate decision-making, coordination, monitoring, and training processes while addressing environmental and social needs during crises. Potential tools that could support the model can be categorized as follows:

Governance and Transparency-Focused Digital Tools: Decentralized decision-making platforms can be created to ensure transparent and participatory governance. Platforms are being developed to allow stakeholders—governments, private sector, and citizens—to make decisions collaboratively based on their inputs. Blockchain technology is at the forefront of these platforms and formations.

Real-time data monitoring and analytical tools can be used to provide immediate insights into environmental, social, and economic crises. The monitoring and analysis of environmental and social data can be enabled through Internet of Things (IoT) sensors, satellite data, and artificial intelligence analyses. A dashboard that collects and displays data such as carbon emissions, natural disaster alerts, or resource allocation can serve as an example of these tools.

The importance of stakeholder feedback for governance is well recognized. Therefore, digital feedback mechanisms can also be considered important tools.

Sustainability-Focused Tools: To ensure energy resilience and sustainability during crises, green energy crisis management tools should be developed. These tools should include systems such as solar-powered communication systems or renewable energy sources like wind power that can supply essential infrastructure during crisis times. The development of portable energy kits for natural disasters is also encouraged.

Circular economy platforms can be developed to promote responsible consumption and production. Especially during crises, digital platforms that facilitate resource sharing, waste reduction, and recycling can play a crucial role. In fact, platforms that coordinate the distribution and reuse of resources such as food, water, and medical supplies during crises are of great importance. When these resources are not properly coordinated, those in need continue to suffer shortages, and resources are wasted.

Considering that crises have become a part of our daily lives, incorporating sustainability-focused tools into everyday life practices can be seen as a proactive approach to crisis management. Instead of only thinking about moments of crisis, developing and spreading tools that serve or have the potential to serve sustainability will contribute to resilience in fighting crises. A simple example of integrating sustainability into daily life practices is encouraging the use of bicycles for short distances as a result of raising awareness about carbon emissions among individuals.

Education and Literacy-Focused Tools: To increase public awareness about digital and environmental systems, ecological and digital literacy platforms can be developed; e-learning platforms covering topics such as environmental sustainability, digital governance, and crisis management; and online educational platforms offering courses on climate change, digital ethics, and sustainable crisis response strategies can be created.

To educate and mobilize communities about crisis response, gamified platforms simulating crisis scenarios like natural disasters and pandemics can be developed, which help improve crisis management skills through sustainable practices.

Starting from primary education, individuals — the smallest units constituting society — can be given awareness and first aid training about disasters. Curriculum can include courses on sustainability, ethics, and preparedness for crises. Forest schools can be expanded; through collaboration with forest schools, children can spend more time in nature. This contributes to early environmental consciousness development.

Collaboration, Coordination, and Communication Tools: Crisis management agencies can be established to facilitate collaboration among governments, NGOs, and citizens; crisis coordination applications such as mobile and web apps can be developed to ensure effective responses among volunteers and local communities.

A public communication platform and public crisis communication application can be developed to provide verified information, crisis updates, and community engagement opportunities. Access to these tools can be facilitated for individuals in socio-economically disadvantaged groups. Education can be provided to these individuals on the correct use of the relevant tools.

Accountability and Ethical Oversight Tools: Blockchain-based public systems can be developed to ensure transparency and accountability in crisis management. From a critical perspective, developing public systems also carries risks of misuse. Therefore, pilot implementations can test whether blockchain-based public systems positively contribute to transparency and accountability. While not applicable to every case, developing blockchain-based public systems for coordination during disaster and crisis times can help better manage the process.

The credit system within the Social Credit System can also contribute to transparency and accountability if used within an ethical framework; however, since misuse of the system could lead to unjust exclusion and punishments, the reliability of social credit systems remains debatable.

An AI system that analyzes resource distribution during crises and ensures fair allocation among communities can be developed to ensure that AI systems used in crisis management operate in accordance with ethical principles.

Resilience, Preparedness, and Crisis Simulation Tools: Digital twin models of cities or ecosystems can be created to simulate crisis scenarios and increase preparedness potential; the governance model's responses can be tested through these simulations.

AI and sensor-based early warning and near-miss systems can be developed to predict and prevent hazards; these systems could detect early signs of crises ranging from climate-related threats to public health emergencies and send timely alerts to relevant authorities.

In summary, the Ecoleviant Governance Model envisages using a combination of digital technologies, collaborative platforms, sustainability-focused tools, and education-based mechanisms. These tools are thought not only to facilitate ethical crisis management but also to ensure that the governance process is inclusive, transparent, and aligned with sustainability goals. Through these tools, it is believed that crises can be addressed with an approach that supports both social well-being and long-term ecological resilience.

Policy & Practical Implications For the Industries

The coordination of implementation phases by domain experts is considered highly significant. The model's practical application can be conceptualized in the following stages:

Stage I: Pilot Projects

Pilot projects can be initiated at local or regional levels to test and refine the principles and applications of the model. Selecting pilot sites, defining objectives, and evaluating outcomes are key tasks in this phase. For instance, the case studies examined in this research—the Social Credit System in China and the Capraia Smart Island Project—are still at the pilot stage. Close monitoring of other pilot initiatives may provide valuable insights.

Stage II: Expansion and Adaptation

Successful pilot projects can be scaled to broader contexts, and the model can be adapted to different environments and scales.

Stage III: Policy Integration

The model's principles can be integrated into existing policies and frameworks at various governance levels. This may involve reviewing and updating current policies to ensure alignment with the model's sustainability and ethical guidelines.

Stage IV: Monitoring and Evaluation

Continuous monitoring and assessment of the model's implementation and impacts should be conducted. Establishing clear evaluation criteria and making necessary adjustments based on feedback are critical to maintaining effectiveness.

The Ecoleviant Governance model offers particular promise in transforming local and regional industrial systems by embedding sustainability and ethical digital governance at their core. Industrial sectors, as significant contributors to environmental degradation and resource consumption, require governance models that go beyond traditional regulatory approaches to foster systemic resilience and adaptive capacity. By adopting a systems-oriented and holistic perspective, Ecoleviant Governance encourages industrial stakeholders—including manufacturers, supply chains, labor groups, and local authorities—to collaboratively address ecological impacts while maintaining economic viability.

In industrial contexts, the model's emphasis on multi-actor coordination, real-time data integration, and ethical digital tools supports transparent and accountable decision-making processes that are vital for managing environmental risks and social equity concerns. Furthermore, the promotion of sustainability literacy among industrial workers and leaders nurtures an organizational culture aligned with long-term environmental stewardship and social responsibility.

Moreover, as industrial activities are increasingly integrated within globalized value chains, the model's advocacy for supra-national collaboration and policy coherence ensures that local industrial governance is not isolated but embedded within wider networks of sustainable practice. This is especially crucial for confronting cross-border environmental challenges such as pollution, resource depletion, and climate change.

In manufacturing industries, Ecoleviant Governance facilitates the integration of real-time environmental monitoring systems powered by digital technologies, enabling the continuous assessment of emissions, waste management, and resource consumption. This allows for more proactive and transparent management of ecological footprints. By fostering collaboration between factory management, local governments, workers, and community stakeholders, the model ensures that production processes align with sustainability goals without compromising economic performance. For example, pilot projects can focus on circular economy practices, where waste is minimized and by-products are reused within the industrial ecosystem.

In regional energy systems, particularly those reliant on fossil fuels or transitioning to renewable sources, Ecoleviant Governance promotes inclusive decision-making processes that incorporate local communities and environmental advocates alongside energy producers and regulators. Digital tools enhance grid management efficiency and support the integration of renewable energy sources, facilitating resilience in energy supply while reducing environmental risks. Ethical governance ensures that data privacy and equitable access to energy services are prioritized.

Agricultural and agro-industrial systems benefit from the model by promoting sustainable land-use practices, biodiversity conservation, and water resource management. Ecoleviant Governance encourages participatory platforms where farmers, local authorities, scientists, and consumers co-create solutions to challenges such as soil degradation and food security. Digital literacy programs support the adoption of precision agriculture technologies that optimize inputs and reduce environmental impact while safeguarding farmer rights and data security.

The model also supports the transformation of logistics and supply chains through enhanced transparency and traceability enabled by digital tools such as blockchain and IoT sensors. This increases accountability regarding carbon footprints, labor conditions, and sourcing ethics. Collaborative networks fostered by the model encourage cross-sector dialogue and joint initiatives to minimize environmental harm while maintaining operational efficiency.

By promoting sustainability literacy among industrial workers and leadership, the Ecoleviant Governance model cultivates an organizational culture that values environmental responsibility and social equity. This is crucial for achieving systemic change and long-term resilience.

Finally, recognizing the interconnectedness of local industries within global supply chains, Ecoleviant Governance emphasizes the need for alignment with international sustainability standards and cooperation among regional, national, and supranational institutions. Such multi-level governance enhances the capacity of local industrial systems to respond to transboundary environmental challenges such as climate change, pollution, and resource scarcity.

In essence, Ecoleviant Governance provides a dynamic, ethical, and scalable governance framework that equips local and regional industrial systems to effectively manage complex environmental and social challenges. By encouraging systemic collaboration, leveraging digital innovation responsibly, and prioritizing sustainability, the model contributes significantly to the urgent transformation toward environmentally sustainable industrial development in the Anthropocene era.

CONCLUSION

"Humanity lost its neutrality within the web of life a long time ago; for quite some time now, it has sided only with its own species — but it doesn't have to be that way. Once upon a time, it was not so." Şiirsel Taş

The quote above reflects on humanity's position within the natural world, arguing that humans have long abandoned a neutral role in the web of life and have prioritized only their own species. It suggests that this self-centered approach is not inevitable; in the past, humans coexisted more harmoniously with other forms of life. The statement calls attention to the possibility of change, reminding us that humans can choose to restore a more balanced and respectful relationship with nature.

This chapter begins with the premise that two dominant phenomena mark the 21st century. No form of governance can be considered independent from practices of surveillance and monitoring. Although surveillance has become more indirect and sometimes imperceptible over time, its presence continues to grow proportionally with the increasing human population on Earth. The data obtained through surveillance plays a crucial role in shaping the lives and governance of human societies.

The second dominant phenomenon is "sustainability," which has gained prominence particularly in response to the rise of environmental and climatic disasters and the breakdown of ecological balance—often referred to as the Anthropocene, or the "age of human dominance." In his book *Last Child in the Woods*, Richard Louv notes that one seldom-discussed cost of technological progress is the weakening of human senses (Louv, 2017: p. 71). Modern humans, living removed from the true nature of the world and immersed in technology, are increasingly alienated from nature and thus unprepared and uninformed about natural disasters. This disconnection from nature also conflicts with humans' own nature, leading to "nature-deficit disorder" and a weakening of resilience. According to the Anthropocene concept, humans are not masters of nature but part of it. The primary goal of ecological sustainability is to rekindle this awareness in people and take the necessary steps to live in harmony with nature.

The Ecoleviant Governance model represents a comprehensive and holistic management approach that integrates digital innovation, sustainability, and ethical principles. Built upon lessons drawn from existing systems and projects, the model aims to create governance that is responsive, transparent, accountable, and sensitive to the needs of humanity. It actively excludes harmful practices such as surveillance and exploitation. The model not only seeks to manage crises effectively but also promotes a sustainable future aligned with principles of equity and environmental responsibility.

In the Anthropocene era, where human actions shape the planet's future, being prepared for crises and practicing sustainable crisis management is no longer a luxury but a necessity. The world is shaken daily by various crises and is prone to new ones; thus, crises have become the "new normal" of our lives. Governance models must therefore combine resilience and the power to overcome challenges with environmental protection to strengthen their capacity to respond to crises while ensuring a just and healthy world for future generations. The technological and digital tools used in this process must be embedded within an ethical framework.

The Ecoleviant Governance model should be seen as a small but crucial step toward increasing the capacity to sustain a just and healthy world for future generations. As biologist Robert Michael Pyle eloquently puts it:

What is the extinction of a condor to a child who has never seen a wren?

REFERENCES

Agamben, G. (2020). *İstisna Hali (Çev. K. Atakay)*. Ayrıntı Yayınları.

Bauman, Z. Bauman, I., Kociatkiewicz, J. (2024). Akışkan Modern Dünyada Yönetim. İstanbul: Ayrıntı.

Berente, N. (2008). C. West Churchman: Champion of the Systems Approach. *Available at SSRN* 2411176. DOI: 10.2139/ssrn.2411176

Çağlar, Y. (1998). Sürdürülebilirlik ve Türkiye Ormancılığı. In *Şu kitapta: Sürdürülebilir Kalkınmanın Uygulanması*. Türkiye Çevre Vakfı Yayını.

Chen, G. K. C. (1975). What is The Systems Approach? *Interfaces, 6*(1), 32–37. https://www.jstor.org/stable/25059250. DOI: 10.1287/inte.6.1.32

Churchman, C. W. (1968). *The Systems Approach*. Dell Publishing Co.

Diamond, J. (1995). Easter's End. Discover Magazine. https://www.discovermagazine.com/planet-earth/easters-end / 9 Şubat 2025

Diamond, J. (2011). *Collapse: how societies choose to fail or succeed* (revised edition). Penguin.

Elkington, J. (1998). *Cannibals with Forks: The Triple Bottom Line of 21st Century Business*. Capstone Publishing Ltd.

Fisunoğlu, M. (1997). Sürdürülebilir Kalkınma ve Ekonomi. In *Sürdürülebilir Kalkınmanın Uygulanması, Tartışma Toplantısı 11-12 Aralık 1997, TÇV Yayınları*. Önder Matbaası.

Forster, E. M. (1909). *The Machine Stops*. Oxford and Cambridge Review.

Fukuyama, F. (2016). Governance: What Do We Know, and How Do We Know It? *Annual Review of Political Science, 19*(1), 89–105. DOI: 10.1146/annurev-polisci-042214-044240

Hobbes, T. (1992). *Leviathan (Çev. S. Lim)*. YKY.

Kapucu, N. (2010). *Governance Reforms Comparative Perspectives. Ankra*. Usak Publications.

Klijn, E. H. & Koppenjan, J. (2015). Governance Networks in the Public Sector. London: Routledge.

Louv, R. (2017). *Doğadaki Son Çocuk. Çocuklarımızda Doğa Yoksunluğu ve Doğanın Sağaltıcı Gücü (Çev. C. Temürcü)*. Tübitak Yayınları.

Lovins, A. B., & Hunter Lovins, L. & Hawken, P. (1999). "A Road Map for Natural Capitalism." Harvard Business Review, 77(3), 145-158. https://hbr.org/2007/07/a-road-map-for-natural-capitalism

Osborne, D. T. (1992). Reinventing Government: How the Entrepreneurial Spirit is Transforming the Public Sector. Boston: Addison Wesley Publishing Company.

Özer, M. A. (2006). Yönetişim Üzerine Notlar. *Sayıştay Dergisi*, (63), 59–89.

Perker, Z. S., Nilüfer Akıncıtürk. (2012). Sistem Yaklaşımı Bağlamında Geleneksel Anadolu Konutunun Güne Uyarlanmasını Etkileyen Üst Sistemler ve Sistemler Arası Etkileşime İlişkin Bir İrdeleme. *Elektronik Sosyal Bilimler Dergisi*, *11*(41), 195–213.

Rhodes, R. A. W. (1996). The New Governance: Governing Without Government. *Political Studies*, *44*(4), 652–667. DOI: 10.1111/j.1467-9248.1996.tb01747.x

Rondinelli, D. A. ve Blunt, P. (1997). Reconceptualising Governance: Discussion Paper 2. Department of Public Affairs. New York: UNDP.

Rosenau, J. N., & Czempiel, E.-O. (Eds.). (1992). *Governance Without Government: Order and Change in World Politics*. Cambridge University Press., DOI: 10.1017/CBO9780511521775

Schumacher, E. F. (1989). *Küçük Güzeldir (Çev. O Deniztekin)*. Cep Kitapları.

Stoker, G. (1999). *The New Management of British Local Governance*. Macmillan. DOI: 10.1007/978-1-349-27295-2

Taş, Ş. (2019). *Sekoyana'nın Kapıları. Bir Orman Öyküsü*. Doğan Egmont.

Tecim, V. (2016). Sistem Yaklaşımı ve Soft Sistem Düşüncesi. *Dokuz Eylül Üniversitesi İktisadi İdari Bilimler Fakültesi Dergisi*, *19*(2), 75–100.

Tekeli, İ. (1996). Yönetim Kavramı Yanısıra Yönetişim Kavramının Gelişmesinin Nedenleri Üzerine. *Sosyal Demokrat Değişim.*, *3*, 45–54.

The World Bank. (1989). *Sub-Saharan Africa: From Crisis to Sustainable Growth. A Long term Perspective Study*. Library of Congress Cataloging -in- Publication Data.

Toksöz, F. (2008). *İyi Yönetişim El Kitabı*. TESEV.

United Nations Development Programme. (1997). *Governance for Sustainable Human Development*. UNDP.

vonBertalanffy, L. (1950). An Outline of General System Theory. *The British Journal for the Philosophy of Science*, *1*(2), 134–165. DOI: 10.1093/bjps/I.2.134

Yavuz, V. A. (2014). Sürdürülebilirlik Kavramı ve İşletmeler Açısından Sürdürülebilir Üretim Stratejileri. *Mustafa Kemal Üniversitesi Sosyal Bilimler Enstitüsü Dergisi*, *7*(14), 63–86.

Zabcı, F. C. (2015). "Dünya Bankasının Küresel Pazar için Yeni Stratejisi: Yönetişim."Ankara Ünv. *SBD*, *57*(3), 151–179. DOI: 10.1501/SBFder_0000001758

Chapter 9
Greening the Workplace:
Evaluating the Impact of Environmental HR Practices on Employee Engagement and Organization

Revati Ramrao Rautrao
https://orcid.org/0000-0001-7803-8699
Dr. D.Y. Patil B-School, India

A. V. Senthil Kumar
https://orcid.org/0000-0002-8587-7017
Nehru Institute of Information Technology and Management, India

Sheetal Darekar
Dr. D.Y. Patil B-School, India

Sanjayan Thottapattunjalil Suseelan
GVM's Dr. Dada Vaidya College of Education, India

G. Vanishree
https://orcid.org/0009-0000-1335-2919
ICFAI Business School, India

Ankita Chaturvedi
https://orcid.org/0000-0002-0739-5792
IIS University, India

ABSTRACT

In response to increasing environmental challenges, organisations are quickly integrating stability into internal systems. The study examines Green Human Resource Management (GHRM)'s strategic role in increasing staff commitment and promoting organisational stability. Based on social exchange theory and ecological behavioural models, it evaluates the effect of green HR practices, such as environmentally focused training, green evaluation and participation initiatives, on the initiative approach and behaviour. By using a semi-practical design in two Indian IT companies, findings suggest that GHRM promotes an opinion of environmental awareness, commitment and environmental responsibility. This research advocates establishing HR as a catalyst for sustainable culture and ecological changes.

DOI: 10.4018/979-8-3373-6435-3.ch009

INTRODUCTION

Background on Environmental Sustainability in the Workplace

Greenness of the workplace has become a strategic necessity for organisations with an aim to streamline operational activities in tandem with the global ecological objectives. With the increasing environmental concerns, companies are taking affirmative actions to reduce their ecological footprint and achieve long-term sustainability. There must be a greater efficiency in energy usage by the implementation of schemes like smart lighting, HVAC optimisation, and the establishment of an atmosphere in which employees act in an energy-aware manner, among others. Waste minimisation (through the use of robust recycling systems) that includes the use of individual plastic equipment and a paperless transition towards paperless processes is also essential. They need to be more concentrated on eco-labelled products when dealing with businesses, life-cycle impacts, and developing moral supply alliances.

The application of green building principles through the means of integration - renewable materials ensures that environmental objectives are met by means of natural lighting enhancement and other water-saving fixtures. Sustainable commuting via public transport, cycling or interconnection encouragement will not only decrease emissions, but also enhance the best interests of the employee. Still, employee engagement forms a cornerstone, with awareness and involvement management being the pillars of the success of green initiatives. The sustainability of the employees will be the most significant stakeholder in the outcome by organising training and establishing environmental committees. The so-called overlooked water conservation is gaining momentum using low-current plumbing, rainwater harvesting and implementing smart irrigation systems. The method of reducing carbon emissions is by establishing average targets and establishing renewable energy solutions or programs for offsetting. Similarly, the parallelogram, permanent technology alterations, the energy-efficient server and cloud-based system, as a whole, fulfil the environmentally friendly digital infrastructure need. Lastly, the transparency is enlightened by reporting on environmental performance not only to the stakeholders but also to make the corporate administration structure sustainably built. Collectively, these combined practices make the workplace adaptable and environmentally friendly, which makes a meaningful contribution to the global sustainability objectives. Through this practice, organisations can, in many cases, benefit the environment by enhancing cost savings, enhancing employee satisfaction and building a better image to consumers and stakeholders.

The Importance of Employee Involvement in Organisational Sustainability

Influence of employees also is significant in the process of organisational sustainability, which is a Linchpin which links strategic environmental goals with behavioural action at the workplace. When the commitment of the employees is considered in the framework of green human resource management, the conventional productivity matrix is replaced by the transformational strength that fosters the environment management and organisational engagement. The attached employees will tend to be more internal in values of sustainability in the environment, are more meaningful in the environmental arts and also within and out of the workplace become promoters of sustainable practices. This adjustment between individual values and organisational vision improves the shared feeling of inspiration, job satisfaction and purpose. Through the organisation's cultural material attitude in the cultural material training programs, green teams, decision-making and recognition systems for partnerships, the HR functions catalyse among the employees. In addition, environmentally engaged employees contribute to new ideas, identify the limitations of the processes and support the continuous improvement in sustainability benefits. Such a commitment is not just a reactive response to the instructions from the top down, but a bottom-up approach that maintains long-term environmental changes. Empirical studies indicate a positive correlation between green HR practice - such as green recruitment, performance evaluation and levels of learning and employee involvement, which emphasises the strategic importance of human capital in sustainability infections. In addition, organisations that prioritise employees' involvement in sustainability often employers, are improving branding, reduce sales and improve stakeholder confidence, leading to competitive advantages. Thus, promoting employees' involvement is necessary not only to meet environmental purposes, but also to strengthen a flexible, value-oriented organisational identity. In this dynamic interaction, employees are no longer inactive implementing sustainability policy, but a green, more sustainable future co -co-construction for the organisation and society.

OVERVIEW OF ENVIRONMENTAL HUMAN RESOURCES

Green Human Resource Management (Green HRM) is a strategic management approach that incorporates the idea of environmental sustainability into HR policies and practices in order to promote pro-environmental employee behaviour and promote environmental ecology in organisations. The strategy covers the whole lifecycle of employees, starting with green recruitment that focuses on employees with environmental values. It goes as far as green training and development that instil sustainability competency and performance management systems where environmental objectives are factored in appraisal criteria. Also, eco-friendly compensation, e.g. rewards for eco-friendly measures, strengthens environmental responsibility. Green teams and engaging employees in the decision-making process contribute to improved transparency and encourage the ongoing environmental improvement.

With the paradigms of organisations moving to sustainable models, the HR role should move beyond the traditional functions of the role to incorporate environmental responsibility within its operations. Renwick et al. (2013) underline that the practices of Green HRM support the creation of a culture in which employees are enabled to become the champions of sustainability, matching personal values with the corporate environmental missions. On the same note, Jabbour and Santos (2008) assert that HRM plays a pivotal role in the creation of sustainable organisations that influence the attitudes and behaviours of employees towards environmental responsibility. This is becoming a strategic integration requirement due to the pressures to meet the regulations and the expectations of the stakeholders.

According to recent studies, the efficacy of the Green HRM is justified. It has been demonstrated that GHRM can have a positive influence on the engagement of employees and the sustainability of organisations through becoming environmentally conscious and integrating sustainability into business strategy. This gets a two-way culture that builds a commitment to sustainable practices and improves the overall environmental performance (Palupiningtyas & Wahono, 2023). The practices also eventually lead to a green work environment and improved corporate image (Abhisha & Melvin, 2024). Thus, Green HRM not only help to promote the ecological agenda, but also enhances organisational resilience, promotes innovation and improves the employer brand, which generates a loyal workforce that drives the organisation to its environmental goals.

Green HRM (Green Human Resources Management) is the composite of environmental HR practice that significantly influences on sustainability of the company and the participation of the employees. GHRM methods promote employees' commitment to the company and help them match environmental goals, which improves results and sustainability. Although GHRM has many benefits, obstacles can prevent its adoption, including strong corporate engagement and the need for potential employees' protest. These problems must be solved to adapt to the beneficial effects of GHRM on employee involvement and organisational sustainability. Green HRM practices that are sustainable affect organisational sustainability in a positive way by increasing the level of employee involvement, improving the performance of the organisation in environmental matters, and corporate social responsibility. However, Top management obligations to remove organisational obstacles and a strategic strategy are necessary for successful implementation. According to (Raajarajeseswari and Dhandayudam, 2024), research has to bring into green HRM strategies a favourable effect on employee involvement, which in turn improves organisational benefits. This emphasises the need for strategies to promote a permanent workplace and increase the sustainability goals with employees' commitment to promote and increase the overall organisational results. (2024 Palpinating). This paper explains the way Green Human Resource Management (GHRM) can enhance environmental and employee engagement. It also contributes to the sustainability of the organisation by minimising pollution and waste and establishing a safe working environment. Finally, GHRM matches HR practices with environmental considerations to prevent greenwashing. (Abdelhakim, 2024).

Impact on Employee Engagement

Green Human Resource Management (GHRM) practice acts as a catalyst for the cultivation of deep employee involvement by coordinating the individual purposes of organisational sustainability goals. GHRM promotes the culture of environmental responsibility and actively encourages employees to participate in the initiative of sustainability (Dhandayudam & Raajarajeswari, 2024). Using training programs in permanent practice increases employees' knowledge and commitment, leading to a more engaged workforce (Palupiningtyas, 2024). When employees are actively involved in the environmental initiative, they experience agency and a sense of shared responsibility, which improves their emotional and cognitive conditions in the workplace. (Palupiningtyas, 2024). When employees are actively involved in environmental initiatives, they experience agency and a sense of shared responsibility, which improves their emotional and cognitive conditions in the workplace. GHRM efforts like Green on Boarding, Eco-Challenge and Sustainability Training are a source of experience that is not only more awareness generating but also inspires long-term action. The more the internal environmental values of the employees in the organisation, the greater the likelihood of them desiring to engage in green programs and the culture of being ecologically minded. This interaction, in effect, shifts the positions of the transaction, and employees become Active agents of organisational change. Reducing Waste and Pollution Contributes to A Healthy Workplace, which can be added to gain morale and engagement levels. (Abdelhakim, 2024).

Leadership, especially transformational leadership and ethical leadership, is critical towards inculcating the values of greenness in organisational culture. Such leaders provide a vision of sustainable behaviour by expressing a vision of sustainability, modelling a sustainable practice, and creating a feeling of collective meaning. They generate trust by the mandate of having open lines of communication and involving employees in the process of making decisions that are not just required but also motivated to pursue green efforts outside the mandate. In the manufacturing industry, executives have led the way in cutting down on waste by incorporating lean-green procedures, as well as incentivising the cause of environmental friendliness. Ethical leadership in the IT sector has facilitated the physical sustainability of the digital environment by utilising energy-saving server technology and virtual teamwork software. The educational institutions demonstrate leadership by integrating environmental stewardship in the curriculum design and operations within the campus. In all industries, leadership has the common denominator of being able to translate the abstract sustainability goals into concrete, behavioural results. Through value congruence, policy and practice, transformational and ethical leaders will be agents of sustainable ecological change at the workplace. They are both strategic and highly human-centred.

Contribution to Organisational Sustainability

Green Human Resource Management (GHRM) Serves AS A Strategic Conduit for Embedding Sustainability with the Operational Fabric of an Organization. The city is integrating eco-conscious practices into recruitment, training, and performance management, and GHRM fosters a workforce aligned with environmental values. This adjustment translated into average reforms in environmental performance, energy efficiency and resource adjustment. GHRM practices are associated with better environmental performance, which is important for achieving stable goals (Nugraha et al., 2024). Organisations using GHRM adjust their human resource strategies according to environmental preferences, leading to better normal performance (Ramjani et al., 2024). Continuous practice not only benefits the environment but also increases corporate social responsibility, which can improve an organisation's reputation and ownership. (Dhandayudam & Raajarajeswari, 2024). In addition, strategic consistency improves HR and sustainability goals. Organisational agility in responding to ecological and regulatory challenges. Over time, such practice contributes to corporate social responsibility by strengthening moral behaviour, openness and stakeholders. Thus, GHRM transfers environmental compliance, sustainability for innovation, competitive advantage and long-term organisational flexibility in a rapid environmental economy.

In order to generate permanent effects, organisations cannot continue developing sporadic green activities but rather implement policy-related strategies that internalise sustainability. This is also associated with the integration of environmental KPIs into the HR performance systems, alignment of the GHRM to the long-term strategic planning as well as establishing a culture of continuous improvement. The implementation of sustainability as a part of the HR governance structure can help organisations to stay on course, address current ecological changes, and bring a strong, well-educated workforce dedicated to environmental protection.

PROBLEM STATEMENT

With the growing environmental degradation and mounting pressure exerted by the stakeholders, organisations have had to incorporate sustainability in their operations. Although there has been considerable improvement in the domain of green supply chains, energy-efficient technologies, the contribution of the human resource management (HRM) towards environmental sustainability development has not been sufficiently researched. To be more specific, the extent to which the Environmental Human Resource Practices (EHRP) can influence the degree of employee engagement and the sustainability goals on a large scale has not been researched in detail using empirical analysis. Research in this field has tended to consider strategic or policy perspectives in green HRM but without considering the processes of behaviour or psychology that could have EHRs implication on employee attitudes and behaviour. Although it is agreeable that employee engagement forms a fundamental pillar in organisational success, the connection between the notion of employee engagement and environmental sustainability and in particular, the context of HR instigated models, requires additional investigation. The literature gap is very big because no specific models can be found to relate EHRPs, employee engagement and perceived organisational support with sustainability. It is with this kind of understanding that organisations might still engage in the implementation of fragmented or symbolic environmental initiatives that do not achieve a long-term change in the behaviours or quantifiable impact.

The current study fills this research gap by establishing the efficacy of a planned GHRM intervention in improving environmental awareness, eco-behavioural intent, and perceived organisational support in employees. The research aims at producing evidence-based information that could help in filling the gap between environmental HR practice and sustainable organisational results by exploring the mediating and moderating roles of these constructs. In doing so, it aims to contribute both theoretically and practically to the evolving discourse on greening the workplace through strategic HR interventions.

Although short-term interventions can produce instant behavioural changes in the framework of Green HRM, their sustainability can be achieved only through the integration of the principles of sustainability into the organisational policies. GHRM initiatives without strategic incorporation may result in reduced complexity of either a symbolic or episodic nature. Hence, the current research highlights the importance of a long-term policy alignment, institutional reinforcement, and on-going reinforcement mechanisms, which would ensure that the environmental HR practices do lead to long-lasting change in the cultural and operational.

Research Gap

Even though the interest in Green Human Resource Management (GHRM) remains to have an ever-increasing rate, a research gap is eager on the psychological and behavioural impact the same has on employees. Whereas strategic and policy frameworks are talked about, there are limited empirical studies on the impacts of Environmental HR Practices on employee engagement, which is a key context in creating sustainable change. The research area is characterised by cross-sectional research, restricting causal assumptions, and experimental research has not proven to validate GHRM interventions. Also, the mediating factors, which are eco-behavioural intention and perceived organisational support of sustainability, are under-researched. This failure is further enhanced by the absence of robust theoretical alignment with well-known theories such as the Resource-Based View or Social Exchange Theory, and practitioners have no specific models that can be used to connect HR strategies with tangible results. The present study fills these gaps by suggesting a mixed-method approach to research the multi-level impact of a structured EHRP intervention.

Research Objectives and Significance

The key aim of the research, therefore, is to determine the impacts of human resource practice (green HRM) in the environment on employee engagement and sustainability of the organisation. This is an attempt to determine the influence of HR strategies on involvement in the approach, behaviour and sustainability initiatives of the environment and improve organisational ecological performance.

To determine the most significant environmental HR practices that are adopted in contemporary workplaces.

To study the connection between green HRM and the level of engagement of the employees.

To determine the input of engaged employees towards meeting sustainability objectives, and

To come up with a conceptual model that connects environmental HR practices, employee.

Behaviour and sustainable organisational results.

The study is important since it combines HRM, environmental management and organisational behaviour to fill a critical gap. It reveals the role of HR policies and employee engagement in bringing sustainable change as a theoretical and practical contribution to leaders and policy makers. The research highlights that the sustainable organisations are not only dependent on technology, but essentially on the active and responsible human capital.

LITERATURE REVIEW

Green HRM covers guidelines that encourage the permanent use of resources in business organisations. Research has identified the following major drivers: Top competition assistance, Green Management and employee Empowerment (Yaslaija et al., 2020). GHRM has shown positive effects on organisational citizens 'behaviour, employees' morale and green innovation (Kim et al., 2019).

Environmental Training

Green HRM is based on training employees on environmental issues, especially on how to make environmentally friendly decisions. It increases environmental sensitivity and literacy and creates a culture of sustainability in the organisation (Tang et al., 2018). HR is an essential facilitator who converts these green initiatives into actual sustainability results (Saeed et al., 2019). This training is not a compliance issue, as it enables employees to recognise sustainability issues and play a significant role in green solutions in their day-to-day activities. This employee development is based on the Social Exchange Theory; as an organisation is shown to have a real interest in being sustainable, employees will also respond by further engaging, working harder in a more discretionary and proactive manner to ensure the overall environmental performance of the organisation.

Environmental Sustainability in Organisations

Environmental responsibility has become a strategic concern as opposed to being a reactive compliance issue, and it has been incorporated into all the functions of the organisation. Actually, sustainable organisations balance economic objectives with environmental and social principles in order to minimise their activities on the planet by involving overall operational and cultural modifications (Dyllick and Hockerts, 2002). Competitive, legitimacy and responsibility pressures are some of the forces that facilitate this integration (Bansal and Roth, 2000). Most importantly, the importance of human capital is now appreciated as the key driver of these ecological initiatives. Sustainability has ceased to be a technical challenge but a well-entrenched organisational principle, and in this regard, the involvement of employees and effective human resource systems is the key to developing resilient and future-oriented organisations that manage to operationalise their environmental objectives.

Employee Engagement: Definitions and Importance

Employees' connection has proven to be an important construct in organisational behaviour, representing the emotional, cognitive and behavioural involvement of employees with their work and organisational goals. Kahn (1990), a leader in the principle of engagement, defines it for his roles for their own roles, where individuals express physically, cognitively and emotionally during performance. This multidimensional concept goes beyond job satisfaction or inspiration, which incorporates intensive psychological association that influences the adjustment of discretionary endeavours, progression and corporate vision. Saks (2006) differentiates further job engagement with organisational involvement, which emphasises the active behaviour of organisational involvement, knowledge sharing and organisational citizenship. Dedicated employees in working environments are regarded as an important resource that not only offers work productivity, performance, but also cultural flexibility and strategic changeover. In fact, their engagement in the initiative of sustainability has been associated with well-developed organisational cultures and enhanced organisational adaptability. Engaged employees will be more supportive of green practices, initiate environmentally friendly work and act as promoters of corporate values of sustainability. Moreover, the level of high engagement is low sales, better morale and longer employees, which is associated with branding work that is essential in the storage and organisational reputation of the genius. In the context of environmental HR practice, employee involvement becomes a mediated strength that translates sustainability policy into meaningful action. Thus, understanding and nourishing employees' commitment is not only for organisational efficiency, but also central to promoting organisational goals. Literature confirms that sustainable changes are most effective when employees are fully invested - emotionally and practically - in the organisation's assignments, especially when that assignment involves environmental management.

Environmental HR Practices

Green Human Resource Management (Green HRM) is a strategic approach that incorporates the environmental sustainability concept in the central HR activities. It involves green recruitment as the means of hiring talents with a high level of ecology-consciousness and environmental training as the means of improving performance and employee engagement (Jabbour, 2011; Renwick et al., 2013). This changes the organisational culture to value-based sustainability, which enhances long-term resilience. The contemporary studies emphasise the intensification of these initiatives by means of digital tools. Nasim et al. (2024) emphasise that Green HRM promotes eco-friendly behaviour by means of virtual teamwork and energy-saving systems. Moreover, AI-based HR is becoming a strategic workforce planning centre of sustainability (KPMG, 2025). According to Gartner (2024), responsible AI in HR will be widely adopted in 2025, and Forbes (2025) views this as the transformative aspect of facilitating predictive, environmentally friendly workforce analytics to combine the digital technology with its objectives.

Organisational Sustainability: Economic, Social, and Environmental Aspects

The concept of organisational sustainability is based on the triple bottom line (TBL), which combines the economic, social, and environmental aspects towards success through the long term (Elkington, 1998). The economic pillar entails efficiency of resources and profitability, and competitive advantage through innovation. The social aspect focuses on ethical labour practice, equity and community interaction, and, as a result of that, investor trust. The pillar of environment is concerned with the minimisation of ecological footprints through emission control and the minimisation of energy consumption. Sustainability demands a shift towards compliance, into responsible practices that form the core of business value creation (Dyllick and Muff, 2015). Such an organisational holistic strategy improves stakeholder relationships and organisational resilience. Importantly, such integration is realised in the form of internal processes, such as Green Human Resource Management (Green HRM), which reconciles the human capital and the sustainability objectives, turning out to be a strategic necessity in the contemporary business environment.

THEORETICAL FRAMEWORK

In this study, the theoretical basis of the research is a combination of the Ability-Motivation-Opportunity (AMO) model and the Stakeholder Theory, which illustrates how Environmental Human Resource Practices (EHRPs) contribute towards organisational sustainability. The AMO model assumes that the performance of employees is a variable based on the following factors: Ability (e.g., through green training), Motivation (e.g., through green rewards), and Opportunity (e.g., to work on environmental programmes) to help the organisation to achieve its goals and objectives (Appelbaum et al., 2000). Green environment EHRPs built on AMO enable employees to become change agents of the environment. To this end, the argument presented by Stakeholder Theory (Freeman, 1984) states that organisations should consider the needs of all stakeholders, defined as employees, communities and the environment. EHRP is an essential tool in attempting to align internal HR operations with these larger external demands and inculcate sustainability in organisational culture and governance. Collectively, the theories can be used as a multi-dimensional prism, where behavioural (AMO) and institutional (Stakeholder) arguments of embedding sustainability strategically into HRM would deliver long-term sustainability effects.

Social Exchange Theory

This paper analyses the idea of Environmental Human Resource Practices (EH-RPs) in the context of employee engagement due to relational reciprocity based on the concepts of the Social Exchange Theory (SET). According to SET, social relationships are ruled by rules of reciprocal good and debt (Blau, 1964). When an organisation allocates resources towards EHRPs like green training, environmental stewardship culture, or reward systems that are green, the employees feel that they have some support and dedication from the organisation. They, in turn, retaliate and, due to the norm of reciprocity, are driven to ensure that they reciprocate more earnestly, loyalty, and discretionary efforts that are geared towards the sustainability objectives of the organisation (Cropanzano and Mitchell, 2005). Within the workplace greening setting, such an exchange forms a psychological agreement in which workers perceive themselves as participants in an important green agenda. This promotes pro-environmental behaviours that are not necessarily in line with formal job responsibilities. Moreover, SET emphasises the importance of authenticity and consistency of these green HR efforts. In case the practices are real, then employees will see sustainability as a fundamental organisational principle, which will reinforce the motivational loop and increase engagement towards long-term environmental performance.

Stakeholder Theory

The Stakeholder Theory offers an effective model of comprehending the way organisations incorporate environmental responsibility by using Environmental Human Resource Practices (EHRPs). It was postulated by Freeman (1984) that organisations have to distribute the interests of a wide web of stakeholders-employees, communities and the natural environment in addition to shareholders. EHRPs prove their concern with these stakeholders by incorporating the values of the environment in the fundamental HR processes, such as recruitment, training, and performance management. This model makes employees crucial drivers towards the process of aligning internal processes with the external sustainability demands. When organisations have a true commitment to environmental and stakeholder well-being through such practices, they create a culture of trust between them and reciprocity. When stakeholder interests are part of governance, it increases the organisational legitimacy and effectiveness, as was argued by Donaldson and Preston (1995). In turn, there is a greater likelihood of employee response in the absence of diminished organisational commitment due to increased engagement, trust and organisational citizenship behaviours, which form a sense of mutual commitment to a common vision of sustainability and enhance the overall credibility of the organisation.

Resource-Based View of the Firm

The company's resource-based view (RBV) provides a strategic framework to understand how environmental human resource practice (EHRPS) contributes to competitive advantage and organisational sustainability. According to Barney (1991), organisations receive constant competitive advantages when using these resources and using resources that are valuable, rare, inseparable and non-substitutable (VRIN) effectively. Within this paradigm, human capital - especially when combined with sustainability goals - appears as a strategic trait. EHRPs such as green recruitment, environmental training, green performance management and sustainability prizes play an important role in developing a working group that is not only environmentally conscious, but also linked, innovative and committed to organisational goals. This practice helps to cultivate unique employees' abilities and environmental skills that competitors cannot easily repeat. Based on the Resource-Based View (RBV), Wright et al. (2001) hypothesise that strategic human resources develop distinctive resources. Green HRM produces a workforce that can lead to environmental responsibility and eco-innovation, which can be applied in a strategic way. This turns sustainability into a strategic investment instead of a moral imperative, with the attributes of engagement and long-term competitive advantages and organisational resilience.

Hypotheses

H1a: Employee awareness of environmental objectives will increase in the GHRM condition.
H1b: The increase in environmental awareness will be significantly greater in the GHRM condition than in the control group.
H2a: GHRM practices will increase employee eco-behavioural intentions.
H2b: The increase in eco-behavioural intentions will be higher in the GHRM condition.
H3a: Employees in the GHRM group will show higher perceived organisational support for sustainability.

METHODOLOGY

Method – Background: The growing imperative for environmental responsibility has been referred to as the strategic priorities of modern human resource management (HRM). Green Human Resource Management (GHRM) has proven to be a transformative approach that integrates sustainability principles into CORE HR functions such as training, results evaluation, recruitment and employee involvement. This study is based on the principle of ecological systems and social exchange structure, which recognises the mutual relationships between environmentally conscious organisational practice and active employee behaviour. To evaluate the impact of GHRM, a semi-process design was included in Pune, two medium-sized IT organisations located in India. The organisation has embraced a six-period GHRM intervention which comprises Green on Boarding, gamified eco-behaviour Tracking and Peer-led sustainability workshops. The organisation maintained their conventional practices on the part of the hr, which acted as a control group. Participating in the study was voluntary, and 120 employees (60 in each organisation, aged between 25 and 45 years) were used. The data were gathered using structured interviews and pre- and post-intervention surveys. The comparative design will allow evaluating the cognitive, behavioural and conceptual changes related to the green HR practice and offer an empirical understanding of their effectiveness in facilitating the sustainability of the workplace.

Participants: This was analysed on 120 full-time employees of two similar medium-sized IT organisations based in Pune, India. The targeted tests were used to choose the participants to make them relevant to the sustainability-oriented purposes of the study. Each organisation donated 60 participants who were balanced based on job roles, employment and functional department. The age limit was between 25 and 45 years, with gender and experience in the profession in a balanced representation. The inclusion criteria will entail that the participants must have been in their respective organisations for at least one year and possess fundamental knowledge of the HR guidelines in the company. Every subject willingly participated in the study and promised confidentiality and withdrawal at any time. There was no previous risk of formal green HR intervention between subjects. This demographic sustainability in groups ensured internal validity in evaluating the effect of GHRM interventions.

INSTRUMENTS

In order to determine the impact of Green Human Resource Management (GHRM) interventions, a structured questionnaire was drawn up prior to post-intervention posting and after the intervention, comprising four key items. The survey questionnaire was sent in the form of an online survey and completed at two stages, namely, prior to intervention and upon completion of the intervention. The sample units were chosen regarding theoretical adaptability to the designs of the study and pilot samples with sub-names of 10 employees to plateau capacity and the pertinent fitness. The response made a slight adjustment to the language and expression to improve clarity.

1. Environmental Awareness Scale

Adapted from Norton et al. (2015), this 6-Utam-Under-scale measures cognitive understanding of the organisation's environmental goals and practice. Sample articles include the following:

"I am aware of the environmental goals of my organisation."

"I think my role helps reduce the company's ecological footprint."

The reactions were scored on a 5-point Likert scale (1 = firmly disagree, 5 = strongly agree).

This subscale received an alpha of 0.84, Cronbach's.

2. Eco-Behavioural Intention Scale

A custom 5-item measure was developed to assess the possibility of employees engaging in environmentally friendly behaviour in the workplace. This included practices such as reducing the use of paper, energy conservation and supportive sustainability initiatives. Example of items:

"I intend to reduce my use of resources at work."

"I would also be happy to be involved in green programs that are planned in my organisation."

The scale achieved a Cronbach's alpha of 0.81.

3. Perceived Organisational Support for Sustainability (POSS)

This 8-item scale was adapted from Eisenberger et al.'s perceived organisational support framework, contextualised to emphasise environmental dimensions. It measured employees' perception of how much the organisation values sustainability and supports green practices. Example items included:

"My organisation encourages environmentally responsible behaviour."

"Management actively supports sustainability-related suggestions from employees."

The POSS demonstrated high internal consistency with a Cronbach's alpha of 0.89.

4. Demographic and Control Variables

The questionnaire also included demographic items such as age, gender, tenure, department, and previous exposure to environmental training. These variables were controlled for in later analyses to ensure an unbiased interpretation of results. A short qualitative layer was introduced to balance the quantitative focus of the chapter with the help of the stories of the employees and their behavioural observations. Participants have indicated that they felt personally responsible for doing environmentally friendly things, including using less paper and promoting green behaviours to others. It was observed that spontaneous participation in sustainability-related talks and voluntary engagement in green efforts were observed. These reflections form the emotional and cognitive aspects of GHRM influence that contribute to the statistical data with personal experience and the supporting role of this chapter in the workplace culture and long-term behavioural change. All items retained the original 5-point Likert format for uniformity across constructs. Scales were presented in a mixed order to reduce response bias. Reverse-coded items were included where appropriate.

Procedure: Following ethical clearance and managerial consent from both participating organisations, the intervention phase was initiated over six weeks. Beliefs in both organisations were briefed through email before the implementation of the purpose of the study, its voluntary nature thereof and the guarantee of confidentiality. All the participants signed informed consent electronically.

In Organisation A (intervention group), the GHRM model was enacted with the help of a multi-layered framework that is based on constructivist-based learning and participation. The initial two weeks were dedicated to the Green Orientation Module, where employees were introduced to the environmental vision at the company, and thereafter, they were exposed to interactive e-learning on sustainable work practices. In weeks three and four, the employees were engaged in gamified eco-behaviour competitions via a mobile application, which provided them with rewards like less printing, energy conservation and sustainable commuting. Week five and six included sustainability circles that were conducted by peers, in which employees discussed the issue of resource optimisation and carbon footprint reduction together. HR managers were able to assist in periodic check-ins to provide participation and overcome technical problems.

Organisation B (control group) maintained the same routine HR functions throughout the same timeframe and did not obtain any training or interventions in the form of sustainability-specific interventions. The two groups, however, took part in pre- and post-interventions, which were carried out online via a secure survey tool. One session lasted approximately 20 minutes. Standardisation of survey instructions was done across groups in order to provide comparability. To prevent contamination, the participants were requested not to mention the details of interventions outside their teams, nor to contaminate them. The information was gathered anonymously with the use of unique participant IDs, which allowed a perfect match of the pre- and post-responses.

Data Analysis

The descriptive and inferential statistical methods were used to analyse the data collected to determine the efficacy of Green Human Resource Management (GHRM) intervention. Then the SPSS (version 27) and R software were used to analyse with advanced statistical modelling. Data cleaning consisted of the search for missing values, duplication of responses, and straight-lining patterns (i.e., the same response under the Likert items). Cases where more than 20 per cent of the values were missing or where pre- and post-matches were not consistent were omitted in the final data set, leaving 114 successful paired responses (58 in the intervention group, 56 in the control group). To test internal consistency in all scales, reliability analysis was conducted with the help of Cronbach's alpha. The constructs all met the criteria of more than 0.80, which means that the construct was highly reliable. The hypotheses were tested by a mixed-design analysis of variance. It allowed assessing the within-subject (pre- vs post-intervention) and between-group (intervention vs control) differences. The most important dependent variables included: Environmental Awareness, Eco-Behavioural Intention and Perceived Organisational Support of Sustainability (POSS). To establish the homogeneity assumption, the Levene Test of Equality of Variances has been performed. When assumptions were violated, Welch's ANOVA and non-parametric alternatives such as the Wilcoxon Signed-Rank Test were used. The effect sizes (Cohen's d and η^2) were reported to complement the p-values, providing insight into the practical significance of the findings. Correlation matrices were generated to explore the relationships between dependent variables. All statistical tests were conducted at the 95% confidence level ($\alpha = 0.05$).

RESULTS

Descriptive Statistics

Three main variables were made in the three main variables to check the pre-control and subsequent intervention points for control groups: environmental awareness, intentions from environmentally friendly behaviour and experienced organisational support for sustainability (POSS). Before the hypothesis test, before and subsequent intervention points were briefly described in both groups. In the intervention group, the average environmental awareness score increased from M = 3.12 (SD = 0.58) to M = 4.21 (SD = 0.45). Eco-behavioral intention rose from M = 3.18 (SD = 0.67) to M = 4.03 (SD = 0.51). A significant increase from alleged organisational support (possesses) for sustainability showed M = 3.06 (SD = 0.60) to M = 4.12 (SD = 0.49). Conversely, the control group exhibited negligible change across these variables, indicating the absence of external influencing factors and supporting the robustness of the intervention effects.

Correlation Analysis

Pearson correlation analysis revealed strong positive interrelationships among the post-intervention variables in the intervention group. Environmental Awareness correlated significantly with Eco-Behavioural Intention ($r = 0.71$, $p < .001$) and POSS ($r = 0.66$, $p < .001$). On the same note, POSS also had a significant relationship with Eco-Behavioural Intention ($r = 0.69$, $p < .001$). These correlations indicate that both awareness and organisational support can have a combined effect on the intentions of employees to operate in ways that are eco-friendly.

Hypothesis Testing: There was a mixed-design ANOVA that was done to assess the intervention in the short term. There were significant interaction effects between time (pre/ post) and group (intervention/ control) in all variables:

Environmental Awareness: $F\,(1{,}112) = 26.87$, $p < .001$, $\eta^2 = 0.19$

Eco-Behavioral Intention: $F\,(1{,}112) = 21.14$, $p < .001$, $\eta^2 = 0.16$

POSS: $F\,(1{,}112) = 33.45$, $p < .001$, $\eta^2 = 0.23$

These findings substantiate the postulations that the intervention is significant in the enhancement of the environmental awareness of the employees, their intention, and perception of organisational support on sustainability.

Mediation and Moderation Analysis: Eco-Behavioural Intention Mediation analysis with the PROCESS macro (Model 4) revealed that Eco-Behavioural Intention mediates the correlation between Environmental Awareness and POSS (indirect effect = 0.29, 95% CI [0.18, 0.41]), meaning that the increased awareness strengthens the intention, hence improving the perceived support. Further, moderation analysis (Model 1) showed that the relationship between Environmental Awareness and Eco-Behavioural Intention was greater in situations where POSS was high (interaction effect = 0.24, $p < .05$). This implies that pro-environmental intention is an enhancer of organisational support as a context.

Inferential Statistics

The ANOVA analysis was a mixed-design study that involved an interaction effect on three dependent variables, namely, Environmental Awareness, Eco-Behavioural Intention, and Perceived Organisational Support for Sustainability (POSS). The results of the analysis showed statistically significant interaction effects of all three variables, which implied that the GHRM intervention resulted in different changes in the two groups over time.

Environmental Awareness: F (1,112) = 26.87, $p < .001$, $\eta^2 = 0.19$

- **Eco-Behavioral Intention**: F (1,112) = 21.14, $p < .001$, $\eta^2 = 0.16$
- **POSS**: F (1,112) = 33.45, $p < .001$, $\eta^2 = 0.23$

These effect sizes (e2) reflect huge intervention effects, especially in the perception of the employees that the organisation supports sustainability as per the recommendations of Cohen's (1988) yardstick.

Table 1. Descriptive and inferential statistics for pre- and post-intervention scores

Variable	Group	Pre (M ± SD)	Post (M ± SD)	F (1,112)	p-value	η^2
Environmental Awareness	Intervention	3.12 ± 0.58	4.21 ± 0.45	26.87	< .001	0.19
	Control	3.10 ± 0.57	3.15 ± 0.54			
Eco-Behavioral Intention	Intervention	3.18 ± 0.67	4.03 ± 0.51	21.14	< .001	0.16
	Control	3.22 ± 0.64	3.25 ± 0.60			
Perceived Org. Support (POSS)	Intervention	3.06 ± 0.60	4.12 ± 0.49	33.45	< .001	0.23
	Control	3.08 ± 0.59	3.10 ± 0.55			

In order to detect these differences, Bonferroni-Right comparisons were held after the hog. The findings showed that there was a statistically significant improvement in all three areas of the intervention in the intervention group. (p <001). Conversely, no significant changes were observed in the control group (P > 0.05), indicating the effect of the green HRM intervention. These findings confirm the claim that targeted environmental HR practices can lead to meaningful changes in alleged organisational support for employee awareness, intentions, and sustainability. They also confirm the legitimacy to enter the GHRM initiatives into extensive strategies to strengthen employee involvement and pursue organisational sustainability goals.

Visualisation of Interaction Effects

In order to continue the statistical results, a visual representation was created to portray the internal effects of the intervention of green human resource management (GHRM) on three main structures: environmental awareness, intentions on environmental behaviour and experienced organisational support for sustainability. The plot captures temporary changes in intervention and control groups, comparing pre- and post-intervention points. Especially after the implementation of the EHRP graph, all three variables show a clear upward trend in the group, while the control group performed a minimum without any significant change. This visual distinction confirms the effect of GHRM intervention to achieve positive behaviour and conceptual results among employees.

In the intention of environmental awareness and environmental behaviour, the participants' sensitivity to the sustainability issues of the participants in the entrance following the intervention of the sensitivity of the participants and their increased preparedness to participate in the environment in organisational contexts. Similarly, reforms highlighted in the Poss role of human resource initiatives that are structured to increase the perception of employees are actually committed to permanent practice. In contrast, the flat trend lines of the control group in all variables confirm that the effects were intervention-specific, which supports the internal validity of the study. The visualisation serves as an intuitive and empirical confirmation of the mixed-design ANOVA findings, offering stakeholders an accessible means to comprehend the practical impact of GHRM. Thus, the graphical evidence strengthens the argument for adopting systematic environmental HR interventions as an effective strategy for driving employee engagement and organisational sustainability performance.

Figure 1. Interaction Effects of the Green HRM (GHRM) Intervention on Key Variables

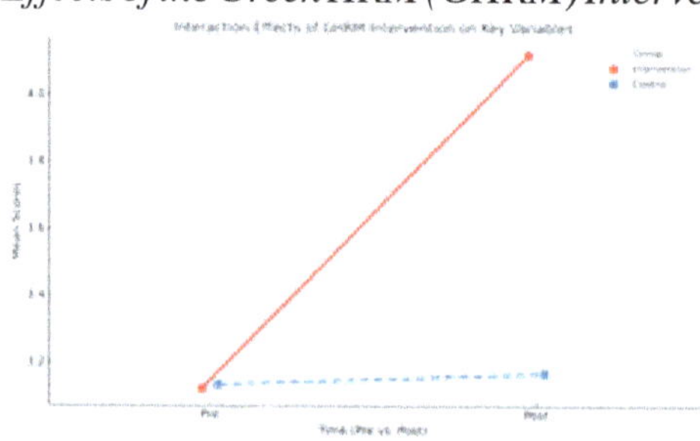

Discussion: The current study examined the effect of Green Human Resource Management (GHRM) practice on alleged organisational support for employees' environmental awareness, intentions of the environment and sustainability. The conclusions strongly support the conclusion that structured GHRM interventions can significantly increase sustainability-oriented results in the workplace. According to social exchange theory, data indicate that when organisations show a real commitment to environmental values through training, it is more likely that communication and reinforcement mechanisms are more likely to have reciprocity with active, environmentally friendly behaviour. Improvement observed in alleged organisational support for sustainability (POSS) has been observed in previous research (e.g., Saeed et al., 2019), strengthening the idea that human resource departments not only act as administrative units, but also as strategic drivers for cultural changes. A significant increase in environmental awareness among the participants in the intervention group validates the importance of participation and old forms of learning. These methods created cognitive engagement and reduced abstract consistency concepts with the realities of the workplace. In addition, an increase in intentions of eco-behaviour suggests that GHRM not only informs, but also inspires permanent tasks. Interestingly, the control group did not show any significant changes in this variable, which confirmed the conclusion that traditional HR practice, while stable, naturally does not stimulate environmental awareness. Poss. The environmentally friendly intentions relationship also highlights a crucial point: Employees are more likely to employ permanent behaviour when encouraged by their institutions. It is in this context that one grows a green culture to transcend politics, and it should become part of the lives of employees.

Interpretation of Findings

The results of the present research are strong arguments that Environmental Human Resource Practices (EHRP) are imperative in stimulating employee engagement and promoting organisational sustainability. The substantial positive change in the level of environmental awareness, environmental behavioural intention, and perceived organisational support of sustainability by the intervention group highlights the usefulness of the designed green HR interventions. These findings confirm theoretical propositions of both Social Exchange Theory and the Resource-Based View, which implies that in situations where organisations engage in sustainability-oriented HR practices, employees would respond with increased engagement and pro-environmental behaviours (Blau, 1964; Barney, 1991).

The results of the mixed-design ANOVA show that the effect sizes are very large, especially in terms of perceived organisational support in the area of sustainability, which is why it is possible to conclude that not only does the value of green initiatives get internalised by the employees, but it is also perceived as something that the organisation is seriously committed to. These impressions seem to be triggers of behavioural change. The mediation analysis proves that eco-behavioural intention is a mediator where the environmental awareness influences the organisational sustainability perceptions in line with current findings that highlight the significance of intentionality in the conversion of awareness to action (Pham et al., 2019). Moreover, moderation analysis indicates that the organisational support enhances the connection between the awareness and the intention, and that sustainable engagement is context-specific. Such results are added to the existing literature on empirical studies that argue that sustainability should be institutionalised using HR systems. Organisational commitment can be enhanced by integrating the concepts of environmental values into the recruitment, training and performance systems, as well as offering a strategic edge in the markets driven by sustainability demands (Jabbour and Santos, 2008). In this way, the article once again confirms the strategic and behavioural applicability of EHRPs to create meaningful and measurable sustainability outcomes.

This paper has a lot to say in the theoretical development of Environmental Human Resource Management (EHRM) as it supports and expands the use of the previously existing models, including the Social Exchange Theory (SET) and the Resource-Based View (RBV). The analysis, through the prism of SET (Blau, 1964), shows that, when organisations are interested in environmental sustainability, as expressed in HR-related activities, i.e., green training, green rewards and sustainability-related performance appraisals, employees react by becoming increasingly engaged and environmentally positive. This reciprocity process proves the behavioural essence of EHRPs and their ability to reinforce psychological contracts between employees and the organisation.

Furthermore, the empirical results that prove the effectiveness of the Resource-Based View (Barney, 1991) include the claim that EHRPs represent a strategic tool to generate rare and inimitable resources in the shape of environmentally conscious and engaged human capital. Not merely is the study located as a result, but it is also a dynamic capacity that enables organisations to turn sustainability strategies into operating success. These theoretical perspectives introduce the discourse in permanent HRM by amalgamating the HR practice on the micro-level with the organisational sustainability outcomes on the macro-level. This study brings up the conceptual distinction between environmental management and strategic HRM by introducing a single model, in which EHRP is a proactive advantage by virtue of the permanent behaviour of the employees. The study offers a good insight into how environmental values can be institutionalised within organisational systems by means of human resource structure, by integrating the psychological (SET) and strategic (RBV) approaches.

The research offers several action-oriented lessons to the HR executives and the organisational leaders who attempt to incorporate sustainability in the primary operations of human resources. Empirical studies underline that environmental human resource practice (EHRPS) is not merely symbolic, but it is a strategic force of involvement of employees and long-term sustainability outcomes. Human Resource Managers are encouraged to embrace green recruitment approaches to acquire sustainability-conscious talents, integrate sustainability modules in training and development initiatives, as well as incorporate eccentric objectives in performance assessment and rewarding systems. This is not only a practice of enhancing personal advocates of environmental behaviour but also developing the culture of collective responsibility of environmental management (Jabbour and Santos, 2008). Significantly, this study highlights the critical role that experienced organisational support of sustainability contributes to the reinforcement of the relationship between environmental awareness and environmental behaviour and reinforces the importance of open communication and involvement in sustainability government.

The organisations must also understand that employee engagement is also a mediating mechanism that occurs sustainably. In this regard, encouraging effective employee engagement in green projects, through inexperienced teams, comments, or sustainability issues in the workplace, could be a highly effective way to motivate and identify with an organisation (Renwick et al., 2013). Incorporating the EHRPs into the HRM architectures in a strategic manner can help the firms to fine-tune the values of the employees to the corporate sustainability missions, thus achieving ecological integrity and desirable performance. In addition, the integration of environmental objectives into HR coverage helps to meet the requirements of the global sustainability standards, including ISO 14001 and UN Sustainable Development Goals, to enhance the reputation of the firm and its position in the market. This is because such a look provides a strong reason why HR professionals should embrace sustainability as an essential gauge of strategic human resource management.

Limitations: While the findings from this study provide valuable insight into the role of Green Human Resource Management (GHRM) in promoting permanent workplace behaviour, many restrictions should be accepted. First, the relatively small sample of the study may limit the generality and strength of the short-term results of the intervention. A large, more miscellaneous sample can increase statistical power and achieve wider praise. Second, research was limited to a single field - the economy - and was conducted in a geographical field in India. Since organisational culture and environmental commitment can vary widely in industries and locations, these conclusions should be carefully about extracting extras without further recognition in other contexts. Third, the study lacked long-term behaviour tracking. While immediate reactions after the intervention were promising, the absence of longitudinal data prohibits conclusions on the sustainability and durability of observed behavioural changes. Over an hour, future studies will benefit from follow-up assessments to evaluate the firmness of green behaviour. These limitations provide a basis for further research aimed at the purpose of continuing in detail on the understanding of GHRM's influence in wider organisational ecosystems.

While this study provides valuable insights into the influence of environmental human resource practice (EHRPS) on employee involvement and organisational sustainability, many restrictions should be accepted. First, the research has used a semi-professional design with a limited sample size and organisational scope, which may interfere with the generality of findings in industries and cultural contexts. The intervention was performed in a specific organisational setting, which was probably affected by the already existing environmental values or management support, with partial participation. Secondly, dependence on self-reported measures for large constructions such as environmental awareness, environmental behaviour and alleged organisational support, may be socially desirable, exposed to prejudice, despite the insurance of oblivion, oblivion. In addition, the short duration between assessment before and after the intervention cannot fully capture the long-term behaviour and attitudinal changes of EHRPs. Sustainability-related changes often require continuous organisational involvement and extended employee involvement, which cannot reflect a short-term study design. Furthermore, the study was not responsible for external environmental pressure - for example, regulator mandates or expectations of ownership interest - which may affect the implementation of EHRP and employees' perceptions. Finally, while mediation and moderation analyses provided strong insight into the underlying mechanisms, they limited the use of quantitative methods alone in the depth of relevant understanding. Integrating qualitative approaches, such as an interview or focus group, can enrich interpretations and provide a nice approach to employees' experiences and motivations. The purpose of future research should address these restrictions by incorporating longitudinal designs, multimethod approaches and comparing them to the transverse sector to elaborate on the theoretical understanding of increasing the practical purpose of green HRM frameworks in various organisational ecosystems.

CONCLUSION

The paper is a contribution to the expanding body of knowledge on the permanent organisational practice that shows that Green Human Resource Management (GHRM) can be a potent driver of stimulating environmental awareness and behavioural change within the workforce. The conclusions indicate that GHRM interventions, which are used to improve wise design and strategic sustainability of employees, can significantly increase awareness of the goals if they can strengthen their desire to use environmentally friendly practices, and strengthen the perception of organisational support for environmental responsibility. Through a structured intervention associated with orientation, gamified learning and peer trapping, the research core emphasises the practical value of integrating sustainability into HR functions. It is important to confirm the role of HR not only as an administrative unit, but also as a transformational force that is capable of adjusting employees' values for broad ecological purposes. Although the extent of the study was limited, the results highlight the ability to form an organisational culture for GHRM and promote permanent growth from within. Since environmental ideas are reinforced, it is not just a moral imperative; This is a strategic requirement for long-term organisational flexibility.

Policy Recommendations

Depending on the empirical insight generated through this study, many political recommendations to organisations aimed at institutionalising sustainability through human resource management appear. First, organisations should develop a formal green HRM (GHRM) policy structure that integrates sustainability goals at all stages of employees' life cycle - revival, training, performance evaluation and compensation. These guidelines should prefer environmental literacy and behavioural skills in job descriptions and selection criteria, thus ensuring adaptation between organisational values and individual abilities. Secondly, policies should make the ongoing environmental workouts and capacity programs compulsory, and not only emphasise compliance, but also internal motivation for environmentally supportive behaviour. These programs must be in line with department requirements and supported by average learning outcomes.

Third, it is mandatory to institutionalise green results management systems that include sustainability indicators as important results in results evaluation. Connecting environmental performance with prizes and recognition mechanisms can help pursue environmentally conscious behaviour. In addition, organisations should install green teams or sustainability committees, including cross-functional staff, to work on the beginning and monitoring of green practice at the ground level. These teams should be strengthened through politics that promote innovation and response loops below. Political frameworks should also emphasise the importance of openness and communication, and ensure that sustainability goals, performance and challenges are regularly communicated to employees through internal reports, dashboards or sustainability portals. Finally, policies must be by extensive environmental rules and global standards such as ISO 14001 and the UN's goals for sustainable development (SDG), which allow organisational efforts within the global sustainability agenda. Such combined political strategies are not merely required to comply with the legislation and assure the stakeholders, but also to develop environmental management and culture to make the organisation flexible in the long term.

Implications and Future Scope: The study findings can be very important to the field of educational research as well as organisational behaviour. The governance perspective dictates that the outcomes of strategic regeneration of human resource management (HRM) should be established on its foundations. Organisations aimed at promoting a culture of organic responsibilities should include green training modules, eco-demonstration indicators and participation platforms that allow employees to participate in permanent behaviour. These interventions not only improve environmental outcomes, but also use engagement and organisational identity. For political decision makers and sustainability advisers, this study provides a replica framework to perform green HR practices in different fields. The conclusions confirm that environmental goals are achieved most effectively when learning, inspiring and social exchange of human-focused strategies. Future research can build on this foundation by increasing the extent of many industries and fields, and incorporating longitudinal tracking to evaluate the durability of behavioural changes. In addition, discovering the role of leadership styles, digital green tools, and generational differences in the design of environmental behaviour can enrich the understanding of GHRM efficiency. An approach that is not interested in involving employees, management and external sustainability auditors can be continued, and the impact of green HR interventions can be expanded in running long-term organisational changes.

REFERENCES

Abdelhakim, M. (2024). The Role of Green Human Resource Management in Enhancing Employee Engagement and Environmental Performance. *International Journal of Green Management and Business Studies*, *4*(1). Advance online publication. DOI: 10.56830/IJGMBS06202404

Abhisha, N., & Melvin, C. L. J. (2024). *Green human resource management practices for sustainability: a comprehensive review.*, DOI: 10.58532/V3BHMA24CH1

Appelbaum, E., Bailey, T., Berg, P., & Kalleberg, A. L. (2000). *Manufacturing Advantage: Why High-Performance Work Systems Pay Off.* Cornell University Press.

Bansal, P., & Roth, K. (2000). Why companies go green: A model of ecological responsiveness. *Academy of Management Journal*, *43*(4), 717–736. DOI: 10.2307/1556363

Barney, J. (1991). Firm Resources and Sustained Competitive Advantage. *Journal of Management*, *17*(1), 99–120. DOI: 10.1177/014920639101700108

Blau, P. M. (1964). *Exchange and Power in Social Life*. Wiley.

Cropanzano, R., & Mitchell, M. S. (2005). Social Exchange Theory: An Interdisciplinary Review. *Journal of Management*, *31*(6), 874–900. DOI: 10.1177/0149206305279602

Dhandayudam, P., & Raajarajeswari, M. A. (2024). *Exploring Sustainable Green HRM Practices: An Investigation of Employee Perspectives and Organisational Challenges*. Springer International Publishing., DOI: 10.1007/978-981-99-7798-7_25

Dyllick, T., & Hockerts, K. (2002). Beyond the business case for corporate sustainability. *Business Strategy and the Environment*, *11*(2), 130–141. DOI: 10.1002/bse.323

Dyllick, T., & Muff, K. (2015). Clarifying the Meaning of Sustainable Business: Introducing a Typology from Business-as-Usual to True Business Sustainability. Organisation & Environment, 29(2), 156-174. (Original work published 2016)DOI: 10.1177/1086026615575176

Elkington, J. (1998). *Cannibals with Forks: The Triple Bottom Line of 21st Century Business*. New Society Publishers.

Freeman, R. E. (1984). *Strategic Management: A Stakeholder Approach*. Pitman.

Gartner. (2024). Hype Cycle for HR Technology 2024: Responsible AI and employee experience. Gartner Press Release. https://www.gartner.com/en/newsroom/press-releases/2024-11-25-gartner-hype-cycle-for-hr-technology-highlights-innovations-that-deliver-greater-flexibility

Jabbour, C. J. C., & Santos, F. C. A. (2008). The significant role of human resource management in the search for sustainable organisations. *International Journal of Human Resource Management*, *19*(12), 2133–2154. DOI: 10.1080/09585190802479389

Jabbour, C. J. C., & Santos, F. C. A. (2008). Relationships between human resource dimensions and environmental management in companies: Proposal of a model. *Journal of Cleaner Production*, *16*(1), 51–58. DOI: 10.1016/j.jclepro.2006.07.025

Kahn, W. A. (1990). Psychological conditions of personal engagement and disengagement at work. *Academy of Management Journal*, *33*(4), 692–724. DOI: 10.2307/256287

Kim, A., Kim, Y., Han, K., Jackson, S. E., & Ployhart, R. E. (2019). Multilevel influences on voluntary workplace green behaviour: Individual differences, leader behaviour, and coworker advocacy. *Journal of Management*, *45*(5), 1519–1544. DOI: 10.1177/0149206317744508

KPMG. (2025). AI for the Chief Sustainability Officer: Understanding the intersection of AI and sustainability. KPMG Insights Report. https://www.accountingtoday.com/news/kpmg-report-encourages-ai-for-sustainability

Nasim, S., Verma, R., & Choudhury, A. (2024). Green HRM practices and digital sustainability: A post-pandemic perspective. *Cogent Business & Management*, *11*(1), 2418418.

Palupiningtyas, D. (2024). *Green HRM: Strategies for Sustainable Business Practices and Employee Engagement*. Productivity., DOI: 10.62207/dhfpj238

Palupiningtyas, D., & Wahono, S. M. (2023). *Green Human Resource Management: A Comprehensive Analysis of Practices*. Impacts, and Future Directions., DOI: 10.56910/ictmt.v1i1.6

Pham, N. T., Tučková, Z., & Jabbour, C. J. C. (2019). Greening the hospitality industry: How do green human resource management practices influence organisational citizenship behaviour in hotels? A mixed-methods study. *Tourism Management*, *72*, 386–399. DOI: 10.1016/j.tourman.2018.12.008

Renwick, D. W. S., Redman, T., & Maguire, S. (2013). Green Human Resource Management: A Review and Research Agenda. *International Journal of Management Reviews*, *15*(1), 1–14. DOI: 10.1111/j.1468-2370.2011.00328.x

Saeed, B. B., Afsar, B., Hafeez, S., Khan, I., Tahir, M., & Afridi, M. A. (2019). Promoting employees' pro-environmental behaviour through green human resource management practices. *Corporate Social Responsibility and Environmental Management, 26*(2), 424–438. DOI: 10.1002/csr.1694

Saks, A. M. (2006). Antecedents and consequences of employee engagement. *Journal of Managerial Psychology, 21*(7), 600–619. DOI: 10.1108/02683940610690169

Tang, G., Chen, Y., Jiang, Y., Paillé, P., & Jia, J. (2018). Green human resource management practices: Scale development and validity. *Asia Pacific Journal of Human Resources, 56*(1), 31–55. DOI: 10.1111/1744-7941.12147

Wright, P. M., Dunford, B. B., & Snell, S. A. (2001). Human Resources and the Resource-Based View of the Firm. *Journal of Management, 27*(6), 701–721. DOI: 10.1177/014920630102700607

Yusliza, M. Y., Osibanjo, O. A., Rampal, A., Mohamad, Z., Noor Faezah, J., & Shamsudin, M. F. (2020). Addressing the sustainable development goals through green human resource management practices: Do green psychological climate and employee green behavior matter? *Journal of Cleaner Production, 258*, 120968. DOI: 10.1016/j.jclepro.2020.120968

Chapter 10
Green Technologies

S. C. Vetrivel
https://orcid.org/0000-0003-3050-8211
Kongu Engineering College, India

T. P. Saravanan
https://orcid.org/0000-0002-6881-0929
Nandha Engineering College, India

V. Sabareeshwari
Amrita Vishwa Vidyapeetham, India

V. P. Arun
https://orcid.org/0009-0009-1211-2150
JKKN College of Engineering and Technology, India

ABSTRACT

The transition toward sustainable industrial practices is increasingly driven by the adoption of green technologies that minimize environmental impacts while enhancing resource efficiency. This chapter explores the critical role of green technologies in reshaping local industrial systems to align with global sustainability agendas. It examines how innovations such as renewable energy integration, energy-efficient manufacturing, waste-to-energy solutions, water recycling systems, and cleaner production techniques contribute to reducing carbon footprints and achieving circular economy goals. The discussion highlights the relevance of green innovations in addressing pressing challenges such as industrial pollution, excessive resource consumption, and ecological degradation. Case studies from both developed and developing economies illustrate how localized adoption of green technologies fosters competitive advantages, lowers operational costs, and promotes long-term environmental responsibility.

DOI: 10.4018/979-8-3373-6435-3.ch010

1. INTRODUCTION

1.1 Concept and Scope of Green Technologies

Green technologies represent a comprehensive set of innovations aimed at minimizing ecological damage while enhancing industrial efficiency and economic productivity. They are rooted in the principles of sustainability, circular economy, and clean production, and extend across energy systems, material use, and waste management strategies. Wu and Strezov (2024) emphasize that green technologies not only involve pollution control but also integrate sustainable practices within the entire value chain of industrial systems, thereby redefining competitiveness in the global market. In the Indian context, Sharma and Padhi (2025) show that industries have increasingly adopted measures to improve green total factor productivity, using advanced efficiency metrics that align industrial growth with reduced environmental impact. Their findings reveal that eco-efficiency improvements in manufacturing processes have been instrumental in balancing resource consumption and economic gains. Similarly, a comparative analysis across emerging economies highlights that green technological progress contributes significantly to long-term sustainable growth, though disparities remain in emissions reduction and innovation diffusion (Zhang et al., 2023). Together, these perspectives illustrate that the scope of green technologies is both conceptual—framing sustainability as a guiding principle—and practical, driving industrial transformation toward cleaner, resilient, and more sustainable systems.

1.2 Evolution of Green Innovations in Industrial Systems

The evolution of green innovation in industrial systems reflects a dynamic shift from isolated technological upgrades toward integrated, sustainability-driven redesigns across value chains. Recent research indicates that green innovation has become a foundational strategy for enhancing the competitiveness and resilience of Indian small and medium enterprises (SMEs), with the concept of a "green premium" enabling these firms to achieve sustainable transformation while commanding higher market value (Sharma et al., 2025). At the macro level, urban policy interventions—such as pilot programs for industrial clusters in India—have demonstrated measurable impacts by reducing energy intensity, spurring green technology innovation, restructuring industrial frameworks, and strengthening digital infrastructure, with notable spillover benefits across neighboring regions (Li et al., 2025). Meanwhile, a global study spanning emerging economies found that green innovation and institutional quality jointly influence green energy adoption and environmental outcomes, with methodological robustness provided by FMOLS, DOLS, and OLS regressions, highlighting that robust governance structures reinforce the trajectory of industrial greening (Gómez-Rubio et al., 2025). Together, these contributions trace the evolution of green innovation from enterprise-level strategies to policy-led systemic change, underpinned by institutional quality and enabling environments that foster sustainable industrial progression.

1.3 Role of Green Technologies in Environmental Sustainability

Green technologies are increasingly recognized as pivotal instruments in advancing environmental sustainability by mitigating ecological damage, bolstering resource efficiency, and enabling resilient industrial practices. Sharif and colleagues (2024) demonstrate that green technology deployment, coupled with renewable energy and globalization, significantly improves environmental outcomes in highly impacted national contexts, emphasizing reduction in carbon footprints and enhancement of ecological balance. In India, the development of sustainable materials has seen notable progress, particularly with initiatives such as IIT-Indore's creation of geopolymer concrete using industrial byproducts like fly ash and GGBS, which reduces CO_2 emissions by up to 80% while also curbing construction costs and water usage (Times of India, 2025) — these innovations underscore how material-level green technologies can transform infrastructure sustainability. Further, an extensive literature review by Bin Amin and Islam (2024) highlights that across renewable energy, green buildings, waste management, and sustainable transport domains, green technology innovations have delivered measurable environmental benefits, while also identifying persistent barriers such as high upfront costs and technological limitations. Together, these findings illuminate how green technologies act as essential levers—advancing cleaner production, lowering emissions, conserving resources, and reinforcing sustainable development pathways within industrial systems.

2. FOUNDATIONS OF GREEN TECHNOLOGIES

2.1 Principles of Cleaner Production and Resource Efficiency

Cleaner production (CP) and resource efficiency are recognized as cornerstones of sustainable industrial transformation, shifting the focus from reactive environmental controls to proactive prevention strategies. The United Nations Environment Programme defines CP as the continuous application of an integrated preventive strategy to processes, products, and services to increase efficiency and reduce risks to humans and the environment. Recent scholarship shows that CP principles operate on three levels: improving input efficiency, reducing waste and emissions at the source, and redesigning product life cycles for sustainability (Yacooub, 2024). This requires industries to monitor material and energy flows and adopt technologies that lower pollution loads without compromising output.

Empirical evidence demonstrates how these principles directly contribute to resource savings. In a study of textile and dyeing industries in South Asia, the application of CP led to a reduction of 1.3 million cubic meters of water and 1,441,500 kWh of energy annually, preventing thousands of tons of effluent discharge and saving nearly USD 0.79 million in costs (Wang & Hynes, 2023). Beyond cost savings, such interventions build industrial resilience to global resource scarcity and climate-related risks. In India, CP initiatives supported by the National Productivity Council (NPC) have been piloted in MSMEs, showing that the integration of resource-efficient technologies not only reduces production costs but also enhances compliance with environmental norms, thereby strengthening export competitiveness (Singh & Bansal, 2023).

Resource-efficient and cleaner production (RECP) approaches further highlight how small and medium enterprises can contribute to sustainability. A recent application in Jordan's MSME sector revealed that simplified implementation frameworks and capacity-building workshops enabled firms to cut waste and emissions by nearly 30 percent, showing that CP principles are scalable across industrial contexts (Fresner et al., 2025). Similar approaches are being increasingly adopted in Indian industrial clusters—such as leather, chemicals, and textiles—where resource-efficient innovations are being linked with circular economy practices to maximize material recovery and reduce landfill waste. These cases underline that CP and resource efficiency are not abstract concepts but practical, measurable strategies for industries striving toward low-carbon growth and sustainable competitiveness.

2.2 Life Cycle Assessment and Circular Economy Integration

Life cycle assessment (LCA) serves as a foundational methodology for quantifying environmental impacts across the entire lifespan of products, from raw material extraction to end-of-life disposal, making it indispensable for evaluating circular economy strategies. Recent advances in LCA methodology underscore the importance of extending product lifetimes through reuse, remanufacturing, and repair to better align with circular economy goals, where longer product lifetimes significantly lower environmental burdens per functional unit (Jerome & Ljunggren, 2025). In the high-impact domain of lithium-ion battery remanufacturing, applying consequential LCA reveals that second-life battery models achieve just 16 percent of the global warming potential of manufacturing new batteries—demonstrating the substantial emissions reductions achievable when reuse is optimized (Authors of the 2023 LI-ion study, 2023). From an Indian perspective, life cycle and circular economy integration has been examined in the renewable energy sector: Prabhu, Shrivastava, and Mukhopadhyay (2021) found that implementing robust end-of-life recovery for solar photovoltaic systems in India not only recuperates aluminum and glass—saving 70.3 GWh of energy—but also creates potentially USD 11.7 billion in recovered material value, reinforcing the economic and environmental advantages of circular recycling frameworks. Collectively, these studies illustrate that LCA is not just a diagnostic tool but a strategic enabler—quantitatively linking product longevity, waste reduction, and resource valorization—pivotal for steering industrial systems toward a more circular and sustainable trajectory.

2.3 Energy Efficiency and Carbon Footprint Reduction

Energy efficiency has become one of the most cost-effective pathways for industries to reduce greenhouse gas (GHG) emissions and improve competitiveness. The scope extends from process optimization and adoption of high-efficiency equipment to digital monitoring systems that minimize energy wastage. Globally, industries account for nearly 30% of final energy consumption, and improving efficiency can potentially reduce CO_2 emissions by 15–20% by 2030 (Schulze et al., 2025). The concept has shifted from being a technological intervention to a holistic practice that integrates energy management systems, life-cycle emission analysis, and employee behavioral changes within industrial ecosystems.

In India, initiatives like the Perform, Achieve, and Trade (PAT) mechanism under the National Mission on Enhanced Energy Efficiency have incentivized industries to lower their specific energy consumption while allowing trading of efficiency credits. The PAT scheme has been particularly effective in energy-intensive sectors like cement, steel, and textiles, leading to avoided emissions of nearly 100 million tonnes of CO_2 equivalents over successive cycles (NMEEE, n.d.). Case studies from Indian steel plants show that waste-heat recovery, variable frequency drives, and digital energy dashboards contribute to measurable reductions in both energy use and carbon footprints.

On the technological frontier, AI-driven frameworks are enabling predictive energy optimization. For instance, the DeepGreen-Opt model uses long short-term memory (LSTM) neural networks for real-time energy demand forecasting, coupled with adaptive particle swarm optimization for energy allocation, significantly reducing peak load demand in manufacturing environments (Bai et al., 2025). Moreover, integrating renewable energy within factories—such as rooftop solar PV and biomass cogeneration—further complements efficiency gains by displacing fossil fuel dependence. Together, these strategies highlight that reducing industrial carbon footprints requires a layered approach that merges policy incentives, digital technologies, renewable integration, and organizational commitment.

3. GREEN ENERGY SOLUTIONS IN INDUSTRY

3.1 Renewable Energy Integration in Industrial Processes

The integration of renewable energy within industrial processes is increasingly recognized as a transformative strategy, aiming not only to reduce carbon footprints but also to enhance operational resilience and efficiency. Hechelmann and colleagues (2020) document a compelling case in the food processing sector where renewable energy sources—such as solar thermal and biomass—are deployed effectively to supply steam, illustrating how green energy can seamlessly substitute conventional fossil-based heating systems in industry. At the systems level, the adoption of combined heat and power (CHP) configurations that incorporate renewable inputs significantly elevates overall energy recovery—enabling efficiencies as high as 90 percent—thereby proving that cogeneration remains a crucial pathway toward sustainable industrial energy systems (review by Energy Conversion and Management, 2020). In India, hydropower continues to play a critical role in system flexibility and reliability, with technological enhancements in turbine efficiency, storage solutions, and grid integration improving the adaptability of power systems; Singhania (2024) underscores how hydropower integration advances not only cost competitiveness but also enables ancillary service revenue streams that benefit industrial consumers. Together, these studies affirm that renewable energy integration in industry is not a peripheral intervention but a strategic reconfiguration of industrial energy ecology—positioning renewables as central components of heat, power, and operational planning that support decarbonization and long-term economic sustainability.

3.2 Solar, Wind, and Bioenergy for Sustainable Production

The amalgamation of solar, wind, and bioenergy technologies into industrial production systems is emerging as a cornerstone of sustainable energy strategies, offering a multi-faceted pathway for decarbonization, resilience, and rural empowerment. Globally, hybrid systems that blend photovoltaic (PV), wind turbines, and hydrogen fuel cell storage showcase enhanced continuity in energy supply and cost-effectiveness, especially under variable climate conditions (Al-Badwawi et al., 2025). Forecasting research reinforces the need for hybridization: solar and wind forecasting techniques, particularly those combining physical and machine learning models, are critical for smoothing renewable intermittency and ensuring grid stability in industrial settings (State-of-the-Art Review, 2025). From an Indian perspective, a techno-economic assessment of a hybrid renewable energy system in rural Bihar—comprising solar PV arrays, wind turbines, biogas generators, diesel backup, and battery storage—demonstrates not only feasibility but also substantial reductions in generation cost and CO_2 emissions compared to conventional diesel-based power (Patil et al., 2024). Furthermore, India's bioenergy potential is significant, with agro-residues such as bagasse, rice husk, and groundnut shells capable of delivering up to 16,000 MW of biomass-generated power and 3,500 MW through bagasse cogeneration, offering energy security and material circularity in industrial and agricultural domains (IREDA, 2024). These converging global and domestic studies underscore how combining solar, wind, and bioenergy fosters resilient, eco-efficient production systems—particularly vital in areas of low grid reliability and dispersed industrial activity.

3.3 Waste-to-Energy and Cogeneration Systems

Waste-to-energy (WtE) technologies represent an essential part of sustainable industrial energy solutions by enabling the conversion of municipal and industrial waste into usable energy forms such as electricity, heat, or fuels. Globally, WtE processes reduce landfill dependency and contribute to decarbonization strategies by displacing fossil-based energy sources (Allevi et al., 2023). The most commonly applied technologies include incineration, pyrolysis, gasification, and anaerobic digestion, each with distinct efficiencies and emission profiles. Studies highlight that integrating these WtE technologies into cogeneration systems—where both heat and power are produced—improves thermal efficiency by up to 70%, thus making them highly suitable for energy-intensive industrial ecosystems (Soni et al., 2025).

In the Indian context, WtE systems have gained increasing policy and industrial attention due to rapid urbanization and rising municipal solid waste generation. For instance, assessments of Delhi's Ghazipur landfill emphasize that WtE integration could recover significant energy while mitigating severe public health hazards linked to uncontrolled dumping (Kumar et al., 2024). Moreover, national initiatives under the Swachh Bharat Mission and National Electric Mobility Mission stress WtE adoption as part of India's broader green energy transition. Indian industries are also exploring biomass-based cogeneration systems, particularly in the sugar and paper sectors, to harness agricultural residues and reduce dependence on coal-based energy sources (Gupta & Bansal, 2023). These models not only address waste challenges but also create localized energy loops, supporting industrial resilience.

From an environmental standpoint, WtE combined with cogeneration reduces greenhouse gas emissions, diversifies industrial energy portfolios, and contributes to circular economy goals. Recent European case studies show that when coupled with district heating networks, cogeneration-based WtE plants provide stable grid support, enhance renewable integration, and improve financial viability under carbon pricing policies (Allevi et al., 2023). Parallel Indian research demonstrates that life cycle assessments of bio-based cogeneration reduce both direct emissions and the carbon footprint of industrial supply chains (Gupta & Bansal, 2023). These findings suggest that WtE and cogeneration technologies should be considered not merely as waste disposal alternatives, but as integral components of industrial energy systems aligned with sustainability transitions.

4. SUSTAINABLE MANUFACTURING PRACTICES

4.1 Eco-Design and Green Product Development

Eco-design, also referred to as design for environment (DfE), is a proactive approach in which environmental considerations are embedded at the earliest stages of product development. Unlike conventional design processes that focus primarily on functionality, cost, and aesthetics, eco-design evaluates the entire product life cycle—from raw material extraction to end-of-life disposal—to minimize ecological impact (Deng et al., 2023). This life cycle perspective ensures that products consume fewer resources, use recyclable or biodegradable materials, and generate less waste, ultimately aligning industrial production with sustainability principles. Research shows that adopting eco-design can reduce resource use by 20–30% in manufacturing sectors, making it a key strategy for balancing industrial competitiveness with environmental responsibility (Wang et al., 2023).

Green product development builds upon eco-design by integrating innovation, consumer preferences, and regulatory requirements into sustainable production processes. In the global context, industries are increasingly moving toward designing products with low carbon footprints, energy-efficient components, and circular economy compatibility. For example, electronics manufacturers are adopting modular designs to extend product lifespans and facilitate repairability, thereby reducing electronic waste (Huang et al., 2024). Similarly, in the automotive sector, green product design includes lightweight materials and renewable energy-based manufacturing processes to improve fuel efficiency and lower emissions. These innovations contribute not only to regulatory compliance but also to consumer trust and brand differentiation.

In India, eco-design is gaining traction, particularly in sectors such as textiles, packaging, and consumer goods. Indian companies are adopting biodegradable packaging materials, plant-based textiles, and eco-label certifications to meet both domestic and global sustainability demands. Studies indicate that India's eco-design adoption is strongly linked to export competitiveness, as international buyers increasingly demand compliance with green product standards (Kumar & Raj, 2023). Furthermore, government initiatives such as the Extended Producer Responsibility (EPR) framework encourage industries to rethink product design in order to minimize waste at the end-of-life stage. Green product development in the Indian industrial context is thus emerging as a driver of both environmental sustainability and economic growth.

Eco-design and green product development are not merely environmental strategies; they also represent pathways to innovation-driven competitiveness. By aligning industrial systems with principles of circularity, companies reduce production costs, improve resource security, and enhance resilience against environmental regulations and market shifts. Integrating digital technologies such as artificial intelligence (AI) and life cycle assessment tools further enhances eco-design decision-making, allowing industries to simulate environmental impacts and optimize design choices in real-time (Deng et al., 2023). Together, these advancements position eco-design as a cornerstone of green industrial transformations in both developed and emerging economies.

4.2 Energy-Efficient Manufacturing Technologies

Energy-efficient manufacturing technologies have emerged as a cornerstone of industrial sustainability, enabling industries to reduce their environmental footprint while simultaneously improving productivity and cost-effectiveness. These technologies encompass advanced production processes, automation, and clean energy integration that collectively minimize energy intensity per unit of output. Globally, studies indicate that energy-efficient innovations in manufacturing can reduce industrial energy use by 25–30%, significantly lowering greenhouse gas emissions (Zhang et al., 2023). Approaches such as process intensification, lean manufacturing, and adoption of energy recovery systems play critical roles in optimizing energy utilization across industries.

One of the most transformative aspects of energy efficiency in manufacturing is the integration of smart and digital technologies. The application of Industry 4.0 solutions, such as the Internet of Things (IoT), big data analytics, and artificial intelligence, allows real-time monitoring of energy consumption and predictive maintenance of equipment (Li et al., 2023). These technologies help industries detect inefficiencies, predict machine failures, and optimize production schedules, reducing unnecessary energy expenditure. Furthermore, innovations in machine design, such as energy-saving motors and variable frequency drives, enhance the efficiency of equipment in sectors ranging from textiles to automotive and heavy engineering.

In India, the promotion of energy-efficient manufacturing is closely linked to policies like the Perform, Achieve, and Trade (PAT) scheme under the Bureau of Energy Efficiency (BEE). Indian industries, particularly in steel, cement, and textile sectors, have achieved substantial energy savings by adopting waste heat recovery systems, efficient boilers, and renewable energy-based power supply. For example, studies reveal that the Indian cement industry has reduced its energy consumption by nearly 20% through the deployment of energy-efficient kilns and co-generation technologies (Gupta & Sharma, 2023). These practices not only support India's commitment to achieving net-zero emissions by 2070 but also strengthen industrial competitiveness in global markets where energy efficiency is increasingly a requirement.

On the global stage, manufacturing hubs such as Germany, Japan, and South Korea have pioneered the adoption of advanced energy-efficient technologies. For instance, German industries are leaders in applying energy management systems (EnMS) based on ISO 50001 standards, which provide a structured framework for continuous energy performance improvement. Similarly, Japan's emphasis on Kaizen and resource optimization has fostered highly efficient production systems that serve as global benchmarks. These international practices provide valuable models for emerging economies to replicate, adapting them to local industrial contexts.

Ultimately, energy-efficient manufacturing technologies are not merely about reducing costs or complying with environmental regulations; they represent a systemic shift toward sustainable industrial ecosystems. By integrating renewable energy, digital innovation, and circular economy principles into production processes, industries can significantly lower their carbon footprints while fostering long-term resilience. As industrial energy demand is projected to rise globally, the adoption of energy-efficient technologies will become central to balancing industrial growth with environmental sustainability.

4.3 Green Supply Chain Management

Green Supply Chain Management (GSCM) has become a critical strategy for industries aiming to align business operations with environmental sustainability goals. Unlike traditional supply chain models, which prioritize efficiency and cost, GSCM incorporates environmental considerations across procurement, production, distribution, and product end-of-life management. This approach integrates eco-friendly practices such as sustainable sourcing, energy-efficient logistics, waste reduction, and recycling initiatives. By embedding these practices, industries not only reduce their environmental footprint but also enhance long-term resilience and competitiveness (Mavi et al., 2023).

In the Indian context, industries have begun adopting GSCM under the combined influence of policy frameworks, consumer demand, and global supply chain pressures. Sectors like automotive, textiles, and pharmaceuticals are integrating eco-design principles, sustainable packaging, and renewable energy in their supply networks. For example, Indian automobile manufacturers are increasingly sourcing from suppliers who meet ISO 14001 environmental standards, ensuring compliance across the value chain. Studies also highlight that Indian firms adopting GSCM strategies improve operational efficiency while reducing emissions, especially in energy-intensive sectors such as steel and cement (Kumar & Anbanandam, 2022). These developments reflect India's broader climate commitments, including its pledge to achieve net-zero emissions by 2070.

Globally, multinational corporations are leveraging GSCM not only for compliance but also for brand differentiation and stakeholder engagement. Companies like Toyota, Unilever, and IKEA have pioneered practices such as green procurement, carbon-neutral logistics, and closed-loop product recovery systems. Such practices are driven by growing consumer awareness and regulatory pressures in developed economies. Research indicates that firms adopting GSCM experience a dual advantage—reducing operational costs while building stronger reputations as sustainability leaders (Ninlawan et al., 2022). The global diffusion of these practices has also influenced supply chain structures in developing countries, where local suppliers must adapt to green requirements to remain competitive in international markets.

An important dimension of GSCM is its intersection with digital technologies. The integration of blockchain, big data analytics, and the Internet of Things (IoT) allows industries to track resource flows, monitor carbon footprints, and ensure transparency across complex global supply chains. For instance, blockchain-based traceability systems are being applied in food and pharmaceutical supply chains to verify sustainable sourcing and reduce waste. These technological enablers ensure that environmental responsibility is not only a policy-driven obligation but also an operationally embedded practice.

4.4 Implications for Managers

For managers, these cases underline a familiar but often underutilised principle: green technologies deliver the strongest results when treated as core productivity investments rather than compliance expenses. Waste-heat recovery, clinker substitution, and RE–efficiency hybrids all reward disciplined capital planning, careful integration with existing processes, and reliable measurement systems. Managers must therefore prioritise technologies with proven TRLs, stable cash flows, and minimal disruption to throughput. Early wins from efficiency-led decarbonisation also build internal credibility, making it easier to justify subsequent investments in deeper transition technologies.

Equally important is organisational alignment. Successful adoption requires coordination between operations, finance, procurement, and sustainability teams. When emissions reduction is embedded into routine performance metrics—energy cost per tonne, yield loss, downtime, and maintenance cycles—green technologies reinforce, rather than compete with, traditional operational objectives.

4.5 Implications for Policymakers

For policymakers, these mini-cases demonstrate that India's industrial decarbonisation can progress rapidly through scale-up of mature technologies, provided the policy environment reduces execution risk. The greatest constraint is not technological readiness, but uncertainty around tariffs, open access, by-product logistics, and financing costs. Stable, long-horizon policies that reward verified abatement—rather than one-time capital subsidies alone—are more effective in sustaining adoption.

Policy design should therefore focus on enabling conditions: predictable energy regulations, demand-side procurement signals (especially for low-carbon cement and textiles), and cluster-based infrastructure support for MSMEs. By strengthening measurement, reporting, and verification (MRV) frameworks, policymakers can also unlock private capital and carbon-finance flows, reducing reliance on public expenditure.

5. EMERGING GREEN TECHNOLOGIES

5.1 Policy and Regulatory Frameworks for Green Technologies

The adoption and diffusion of green technologies in industrial systems are deeply influenced by the policy and regulatory environment in which firms operate. Policies play a dual role: they provide incentives for the adoption of sustainable practices and establish compliance mechanisms to prevent environmental degradation. Globally, governments are increasingly leveraging policy instruments such as subsidies, tax incentives, green certifications, and stringent emission standards to push industries toward greener practices. These frameworks not only stimulate technological innovation but also encourage firms to invest in renewable energy, eco-friendly materials, and waste management systems (Zhou et al., 2023).

In India, the policy ecosystem for green technologies has evolved significantly over the past two decades, particularly under the umbrella of climate change commitments and the need for sustainable industrial growth. Programs such as the National Action Plan on Climate Change (NAPCC) and its eight missions—including the National Solar Mission and the National Mission on Enhanced Energy Efficiency—have laid the foundation for widespread adoption of green technologies. Additionally, the introduction of schemes such as the Perform, Achieve and Trade (PAT) mechanism under the Bureau of Energy Efficiency (BEE) incentivizes industries to adopt energy-efficient technologies by allowing tradable energy efficiency certificates. These initiatives not only help industries reduce their carbon footprint but also integrate sustainability into economic growth models (Sharma & Thomas, 2022).

Globally, regulatory frameworks such as the European Union's Green Deal and the United States' Inflation Reduction Act are shaping industrial practices by mandating cleaner production processes, enforcing lifecycle assessments, and providing large-scale investments in renewable technologies. The EU, for instance, has introduced the Carbon Border Adjustment Mechanism (CBAM), which places tariffs on imports based on their carbon content, thereby compelling exporting countries—including India—to align with greener technologies if they want to maintain global market access (Kamal & Awasthi, 2023). Such policies demonstrate how domestic regulations often intersect with global trade, pushing industries toward accelerated adoption of green technologies.

Another emerging trend is the role of voluntary certifications and standards such as ISO 14001 (Environmental Management Systems) and the Global Reporting Initiative (GRI), which influence industries by setting benchmarks for sustainable practices. In India, several export-oriented firms have adopted these standards not only to comply with international regulations but also to enhance competitiveness in global markets. Furthermore, regulatory frameworks are increasingly being complemented by financial mechanisms such as green bonds, carbon credit markets, and public-private partnerships that lower the cost of adopting environmentally friendly technologies.

5.2 Incentives and Financial Mechanisms for Green Technology Adoption

The successful implementation of green technologies in industries depends not only on regulatory enforcement but also on the availability of robust financial incentives and mechanisms that reduce the cost burden of transition. Since green technologies often involve higher upfront costs compared to conventional methods, industries—especially in developing economies—require strong financial support systems to adopt them. Globally, financial instruments such as subsidies, tax credits, feed-in tariffs, and preferential loans have proven to be critical enablers in scaling up renewable energy and environmentally friendly industrial practices (Chen et al., 2022).

In India, a wide array of financial incentives has been designed to encourage industries to transition towards sustainable practices. For example, the government promotes renewable energy adoption through the Accelerated Depreciation (AD) benefit and the Generation-Based Incentive (GBI) schemes, which lower the cost of renewable installations. Additionally, the *Perform, Achieve and Trade (PAT)* scheme provides tradable energy efficiency credits that industries can monetize if they exceed their energy-saving targets. The expansion of *Renewable Energy Certificates (RECs)* has further enabled companies to meet compliance obligations and corporate social responsibility goals in a cost-effective manner (Ghosh & Gulia, 2023).

At the global level, mechanisms such as the Clean Development Mechanism (CDM) under the Kyoto Protocol and the Paris Agreement's Article 6 carbon markets have incentivized industries in developing nations, including India, to adopt cleaner technologies by earning carbon credits tradable in international markets. Similarly, innovative financing mechanisms such as green bonds and climate funds are gaining momentum. India, in particular, has emerged as one of the fastest-growing green bond markets, with issuances aimed at financing projects in renewable energy, sustainable agriculture, and low-carbon urban infrastructure (Sangwan & Bansal, 2022).

Another important driver is the increasing role of multilateral development banks (MDBs) and international financial institutions in supporting green industrial projects. Organizations such as the World Bank, Asian Development Bank, and International Finance Corporation provide low-interest loans, blended finance, and guarantees that reduce risks for investors and industries transitioning to green technologies. Domestically, Indian financial institutions like the Indian Renewable Energy Development Agency (IREDA) and State Bank of India (SBI) have introduced dedicated green financing products to accelerate the adoption of clean technologies in industrial systems (Yadav & Kumar, 2023).

In conclusion, financial mechanisms and incentives act as the backbone of green technology adoption. While global carbon markets, green bonds, and multilateral funding open up new opportunities, India's domestic schemes like PAT, RECs, and concessional loans are creating a favorable financial ecosystem for sustainable industrial transformation. For long-term success, industries need a blend of both domestic and international financial mechanisms that provide affordable access to capital, minimize risks, and reward sustainability performance.

5.3 Policy Support and Regulatory Frameworks

The adoption of green technologies in industrial systems is closely linked to the strength of policy frameworks and regulatory mechanisms established by governments and international organizations. Policy support creates an enabling environment by reducing uncertainty, setting clear sustainability targets, and providing industries with the guidance and incentives required for transitioning toward eco-friendly practices. Without robust regulatory frameworks, industries often hesitate to invest in technologies that have long payback periods or require large upfront investments (Mikayilov et al., 2022).

Globally, governments have adopted multiple policy instruments to promote green technology adoption. The European Union's *Green Deal* and *Fit for 55* package, for instance, have created legally binding commitments to reduce carbon emissions, mandating industries to adopt energy-efficient and low-carbon technologies. Similarly, the United States' *Inflation Reduction Act (2022)* provides tax credits and subsidies for renewable energy and industrial decarbonization initiatives, which directly influence corporate investments in sustainable innovation (Lehr et al., 2023). In East Asia, China's regulatory framework emphasizes mandatory renewable energy targets and provides policy-backed incentives for cleaner production, contributing significantly to its rapid green technology diffusion.

In the Indian context, policy support has evolved significantly over the past decade. The *National Action Plan on Climate Change (NAPCC)* and its eight national missions, including the *National Solar Mission* and *National Mission on Enhanced Energy Efficiency (NMEEE)*, have created strong regulatory pathways for industries to integrate green technologies. The *Energy Conservation Act (2001)* and subsequent amendments mandate energy audits and efficiency targets, compelling industries to comply with sustainability standards. Moreover, India's commitment to achieving net-zero emissions by 2070, announced at COP26, has further accelerated the development of green policy instruments, including renewable energy obligations, carbon pricing pilots, and state-level green energy transition policies (Kumar & Mohanty, 2023). Another crucial element of regulatory frameworks is the role of environmental compliance mechanisms. Pollution control norms under the *Environment Protection Act (1986)* and sector-specific emission standards enforced by the Central Pollution Control Board (CPCB) push industries toward adopting cleaner production methods. Additionally, state-level policies such as *Renewable Purchase Obligations (RPOs)* mandate industries to source a fixed percentage of their energy from renewable sources. These regulatory interventions, coupled with monitoring and enforcement, are gradually aligning industrial operations with global sustainability benchmarks (Shrimali et al., 2022).

In conclusion, policy support and regulatory frameworks act as a foundation for mainstreaming green technologies in industrial ecosystems. While global frameworks provide strong inspiration, India's integrated mix of national climate missions, energy efficiency regulations, and renewable energy mandates is fostering a structured transition toward sustainable industrial practices. However, the challenge lies in ensuring strict enforcement, reducing policy fragmentation across states, and enhancing industry-government collaboration to drive large-scale green technology adoption.

6. POLICY, GOVERNANCE, AND INDUSTRY STANDARDS

6.1 Economic Benefits of Green Technologies

The integration of green technologies in industrial systems provides significant economic benefits by enhancing energy efficiency, reducing resource dependency, and improving long-term competitiveness. Unlike traditional practices that rely heavily on fossil fuels and resource-intensive methods, green technologies promote circularity, waste minimization, and renewable energy use. This transition not only reduces operational costs but also fosters sustainable economic growth and innovation opportunities (Kalkuhl et al., 2022). One of the most direct economic benefits is cost savings from energy efficiency. Industries that invest in advanced energy management systems, renewable energy sources, and cleaner production technologies often experience reductions in energy bills and operating costs. For example, the International Energy Agency (IEA) reports that energy-efficient industrial technologies can lower production costs by up to 30% in heavy industries such as steel, cement, and chemicals (IEA, 2022). In India, initiatives such as the *Perform, Achieve, and Trade (PAT) scheme* under the National Mission on Enhanced Energy Efficiency have enabled large industries to cut down energy consumption while trading efficiency certificates, thereby generating both cost savings and additional revenue streams (Singh & Garg, 2023).

Beyond operational savings, green technologies also contribute to job creation and industrial growth. The renewable energy sector, particularly solar and wind energy, has become a major source of employment globally. In India, the renewable energy industry employed over 863,000 people in 2022, with projections indicating steady growth as the country advances toward its 500 GW renewable capacity target by 2030 (IRENA, 2023). These jobs span manufacturing, installation, operations, and maintenance, creating a positive multiplier effect on the economy.

Furthermore, companies adopting green technologies often gain access to green financing instruments such as carbon credits, green bonds, and concessional loans. These financial incentives lower the cost of capital and improve the return on investment in sustainability projects. For instance, India's issuance of sovereign green bonds in 2023 has attracted significant international investment, highlighting the financial attractiveness of green industrial projects (Gupta & Bansal, 2023). Similarly, global markets reward environmentally responsible companies with higher brand value and consumer preference, leading to stronger market competitiveness.

In the long run, the adoption of green technologies reduces risks associated with environmental compliance and volatile fossil fuel markets. By transitioning toward renewable energy and sustainable industrial practices, companies protect themselves against regulatory penalties, carbon taxes, and rising fuel prices. Thus, green technology adoption is not merely an environmental obligation but an economic strategy that ensures resilience, competitiveness, and profitability in a rapidly evolving global marketplace.

6.2 Environmental Certification and Compliance Mechanisms

Environmental certification and compliance mechanisms play a pivotal role in ensuring that industries adopting green technologies meet globally recognized sustainability standards. Certifications such as ISO 14001 (Environmental Management Systems), LEED (Leadership in Energy and Environmental Design), and India's Green Rating for Integrated Habitat Assessment (GRIHA) serve as benchmarks for environmentally responsible industrial operations. These certifications not only validate sustainable practices but also enhance corporate credibility and global competitiveness (Fonseca & Domingues, 2018).

Compliance mechanisms, including environmental audits, carbon footprint assessments, and adherence to regulatory frameworks such as the Paris Agreement targets, ensure that industries maintain transparency and accountability in their sustainability journey. In India, mechanisms such as the *Perform, Achieve, and Trade (PAT)* scheme and *Renewable Energy Certificates (REC)* framework provide both regulatory oversight and market-based incentives for compliance (Garg & Tiwari, 2022). Globally, mechanisms like the EU Emissions Trading System (EU-ETS) demonstrate how compliance markets can reduce industrial emissions while driving innovation.

Environmental certifications also open access to international trade opportunities, as many global supply chains now mandate adherence to sustainability standards. Companies that demonstrate compliance through credible certification not only avoid penalties but also gain preferential access to environmentally conscious consumers and investors. Thus, certification and compliance systems are no longer regulatory obligations but strategic enablers of sustainable industrial growth.

6.3 Incentives, Subsidies, and Industry Partnerships

The successful adoption of green technologies in industries often depends on supportive financial and institutional mechanisms, including incentives, subsidies, and collaborative partnerships. Governments across the globe offer a range of financial incentives such as tax benefits, concessional loans, feed-in tariffs, and direct subsidies to encourage industries to invest in renewable energy and energy-efficient technologies. For instance, India's Production-Linked Incentive (PLI) scheme for high-efficiency solar PV modules and battery storage has accelerated investments in clean energy manufacturing (Sharma & Jain, 2023).

Subsidies for renewable energy adoption reduce the upfront cost burden for industries, making the transition more economically feasible. In addition, carbon credit mechanisms and green financing options create new revenue streams for companies adopting environmentally responsible practices. Globally, countries like Germany and Japan have demonstrated how long-term subsidies for solar and wind energy catalyzed the growth of renewable industries and strengthened global leadership in clean technology innovation (Mundaca et al., 2022).

Industry partnerships, particularly public–private partnerships (PPPs), play an equally critical role in scaling green technologies. Collaborative initiatives between government agencies, research institutions, and industries help in knowledge transfer, R&D funding, and infrastructure development. In India, partnerships like the *National Solar Mission* and collaborations under the *FAME-II* scheme for electric mobility showcase how joint efforts can accelerate sustainable industrial transformation. Together, incentives, subsidies, and partnerships create a supportive ecosystem that reduces financial risks, boosts innovation, and promotes large-scale adoption of green technologies, thereby aligning industrial growth with environmental sustainability goals.

7. CASE STUDIES AND APPLICATIONS

7.1 Indian Industrial Case Studies on Green Technologies

India has witnessed a growing number of industrial case studies that highlight the adoption of green technologies to balance economic growth with environmental sustainability. The Tata Group, for instance, has integrated renewable energy sources such as solar and wind into its steel and automobile divisions, significantly reducing carbon intensity per unit of production (Kumar & Singh, 2022). Similarly, Infosys became the first major Indian IT company to achieve carbon neutrality by implementing energy-efficient data centers, green buildings, and large-scale afforestation programs (Raghavan & Menon, 2021).

In the manufacturing sector, the adoption of energy-efficient motors and waste heat recovery systems by Bharat Heavy Electricals Limited (BHEL) has showcased how legacy industries can transition toward sustainability. Another case is ITC Limited's paperboards and specialty papers division, which has successfully implemented water recycling and renewable biomass energy systems, achieving near-zero water discharge. These examples reflect how Indian industries are aligning with national missions such as *Make in India* and *National Action Plan on Climate Change* while strengthening their global competitiveness.

Mini-Case 1: Waste-Heat Recovery in Integrated Steel Manufacturing (India)

Context: Large Indian integrated steel plants generate substantial high-temperature waste heat from coke ovens and blast furnaces. Instead of venting this heat, firms such as Tata Steel have adopted waste-heat recovery (WHR)–based captive power generation.

Green Technology Deployed:

- Coke Dry Quenching (CDQ)
- Blast Furnace Top Gas Recovery Turbines (TRT)
- Heat Recovery Steam Generators (HRSG)

Indicative Capital Cost: 85–100 crore per 25–30 MW WHR unit

Payback Period: 3.5 to 4.5 years (driven by avoided grid electricity and stable base-load generation)

Environmental Impact:

- ~140,000–160,000 tCO_2e avoided annually
- Reduction of 8–10% in plant-level Scope-1 emissions

Strategic Insight: WHR aligns with the steel sector's long-standing practice of efficiency-first modernization, offering predictable returns without disrupting core metallurgical processes.

Mini-Case 2: Clinker Substitution in the Indian Cement Industry

Context: Cement manufacturing is carbon-intensive, with clinker production accounting for nearly 90% of process emissions. Indian producers such as UltraTech Cement have systematically reduced clinker content by using industrial by-products.

Green Technology Deployed:

- Fly ash–based Portland Pozzolana Cement (PPC)
- Ground Granulated Blast Furnace Slag (GGBS) blending
- Optimised grinding and quality control systems

Indicative Capital Cost: 40–60 crore per plant (blending, storage, and grinding retrofits)

Payback Period: 2 to 3 years (lower fuel use, reduced limestone consumption)

Environmental Impact:

- ~0.15–0.20 tCO_2e avoided per tonne of cement
- 18–22% reduction in plant-level carbon intensity

Strategic Insight: Clinker substitution demonstrates how incremental process innovation—rooted in industrial symbiosis—can deliver large emissions cuts without radical technology shifts.

Mini-Case 3: Renewable Energy + Efficiency Hybrid in the Textile Sector (Tiruppur Cluster)

Context: India's textile MSMEs face rising power tariffs and export-market pressure for low-carbon production. Units in the Tiruppur knitwear cluster have adopted hybrid clean-energy and efficiency solutions.

Green Technology Deployed:

- Rooftop solar PV (1–2 MW per unit)
- High-efficiency motors and variable frequency drives (VFDs)
- Waste-heat recovery from dyeing and drying processes

Indicative Capital Cost: 6–8 crore per medium-sized unit

Payback Period: 3 to 4 years (energy cost savings + net-metering benefits)

Environmental Impact:

- ~2,800–3,200 tCO_2e avoided annually per unit
- 25–30% reduction in electricity-related emissions

Strategic Insight: This hybrid model reflects India's traditional frugal innovation ethos—combining renewables with efficiency to remain competitive in global supply chains.

7.2 Global Best Practices and Industry Models

Globally, industries have set benchmarks in adopting green technologies through innovative business models and large-scale integration of sustainable practices. Tesla, for example, has revolutionized electric mobility by developing energy-efficient supply chains, integrating renewable power storage solutions, and advancing circular economy models through battery recycling (Bohnsack et al., 2020). In Germany,

Siemens has adopted digital twin technology to optimize industrial processes, minimize energy consumption, and support the transition toward Industry 4.0 sustainability (Luthra et al., 2022).

Japan offers another exemplary model where companies such as Toyota have successfully mainstreamed hybrid and hydrogen fuel cell vehicles, backed by government-industry collaboration and sustained R&D investments. Additionally, multinational corporations like Unilever and IKEA have embedded sustainability in their operations by committing to 100% renewable energy usage, sustainable sourcing of raw materials, and zero waste-to-landfill practices. These global best practices highlight that combining innovation, policy support, and corporate responsibility accelerates the adoption of green technologies.

7.3 Lessons Learned and Transferable Strategies

A comparative analysis of Indian and global case studies reveals several transferable strategies that can guide industries in scaling green technology adoption. First, *policy alignment and financial incentives* are critical enablers—India's Renewable Energy Development Agency initiatives mirror Germany's feed-in tariff models, showing that long-term policy stability fosters industrial participation (Mundaca et al., 2022). Second, *public–private partnerships (PPPs)* enhance knowledge sharing and infrastructure development, as seen in Japan's hydrogen roadmap and India's National Solar Mission.

Third, *digital innovation and data-driven optimization* play a crucial role in both contexts. Technologies such as AI-driven predictive maintenance and digital twins, when integrated with sustainability frameworks, enhance efficiency and reduce waste across sectors. Finally, *cultural and organizational shifts toward sustainability*—where companies treat green initiatives not merely as compliance requirements but as strategic growth opportunities—remain a cornerstone of successful adoption. For India, adapting global best practices requires contextualization to local challenges such as affordability, infrastructure readiness, and skill development. By combining global insights with indigenous innovations, Indian industries can leapfrog to sustainable industrialization while contributing to global climate goals.

8. CHALLENGES AND OPPORTUNITIES

8.1 Economic and Technological Barriers

The adoption of green technologies in industrial systems often encounters significant economic and technological barriers. High initial investment costs, uncertain return on investment, and limited access to green financing prevent small and medium enterprises (SMEs) from transitioning to sustainable solutions. Many industries face challenges in retrofitting older infrastructure with modern, energy-efficient technologies, as it requires both capital and operational adjustments. Furthermore, the lack of standardization and interoperability of green technologies creates difficulties in scaling solutions across sectors. Technological immaturity in areas such as large-scale energy storage, green hydrogen, and advanced carbon capture also restricts rapid industrial transformation. These barriers highlight the need for stronger financial incentives, government-backed risk mitigation schemes, and more robust innovation ecosystems to bridge the economic–technological divide.

8.2 Workforce Skills and Knowledge Gaps

A critical challenge in the implementation of green technologies is the shortage of a skilled and knowledgeable workforce. Many industries lack employees trained in renewable energy integration, energy-efficient process management, and advanced monitoring systems. The transition toward smart manufacturing and digital-driven sustainability requires expertise in data analytics, IoT, and artificial intelligence applications, which remain underdeveloped in many industrial clusters, particularly in emerging economies like India. Additionally, traditional training programs often focus on conventional production methods, leaving a gap in vocational education related to green processes. Bridging this gap requires structured capacity-building initiatives, updated curricula in technical institutions, industry-academia collaboration, and continuous reskilling of employees to adapt to rapidly evolving green technologies.

8.3 Future Opportunities in Green Industrial Growth

Despite existing challenges, the future of green industrial growth presents promising opportunities. Rising global awareness of climate change, stricter environmental regulations, and consumer preference for eco-friendly products are driving industries to innovate. In India, initiatives such as the Production Linked Incentive (PLI) scheme for renewable energy equipment and the National Green Hydrogen Mission highlight government support for sustainable industrialization. Globally, opportunities exist in areas like circular economy practices, green supply chain networks, sustainable materials innovation, and AI-driven environmental monitoring. The integration of Industry 4.0 with green technologies is set to create new business models, enhance competitiveness, and attract green investments. By leveraging these opportunities, industries can achieve both environmental sustainability and economic growth, positioning themselves as leaders in the global transition toward low-carbon economies.

9. FUTURE DIRECTIONS

9.1 Innovations in Sustainable Materials and Processes

The advancement of sustainable materials and industrial processes is revolutionizing the way industries reduce environmental footprints. Innovations such as bio-based polymers, biodegradable composites, green cement, and low-carbon steel are replacing conventional resource-intensive materials. For example, green cement reduces emissions by incorporating fly ash, slag, and other industrial by-products, while bio-based plastics derived from corn or algae reduce dependence on fossil fuels. Similarly, 3D printing with recycled feedstocks and nano-materials for energy efficiency are reshaping production systems with greater precision and minimal waste.

On the process side, closed-loop manufacturing, additive manufacturing, and digital twins are helping industries achieve higher efficiency, reduced energy use, and lower carbon intensity. For instance, advanced coatings that minimize heat absorption can lower cooling needs, while waterless dyeing technologies in textiles have drastically cut water consumption. These innovations not only contribute to sustainability but also improve product differentiation, cost competitiveness, and compliance with global sustainability standards, making them vital drivers for green industrial transformation.

9.2 Role of Green Technologies in Achieving Net-Zero Goals

Green technologies play a central role in guiding industries toward net-zero emissions pathways. Achieving net-zero requires systematic integration of renewable energy systems, electrification of processes, carbon capture utilization and storage (CCUS), green hydrogen, and energy-efficient manufacturing practices. For instance, electrification of steelmaking using hydrogen-based direct reduced iron (H2-DRI) is emerging as a key alternative to coal-based methods.

Globally, industries are aligning with climate agreements such as the Paris Agreement and India's commitment to achieve net-zero by 2070, which necessitate rapid decarbonization. Green technologies not only mitigate direct emissions but also enable scope 3 reductions by influencing supply chain sustainability. Digital solutions such as IoT-enabled monitoring, AI-driven energy optimization, and blockchain for carbon tracking are also enhancing industries' ability to measure, verify, and achieve climate goals. Thus, green technologies are not merely compliance tools but strategic enablers of industrial competitiveness in a low-carbon economy.

9.3 Multi-Stakeholder Collaboration for Green Industrial Ecosystems

The successful deployment of green technologies requires collaborative action across multiple stakeholders. Governments must establish clear policy frameworks, tax incentives, and subsidies to lower adoption barriers, while industries need to invest in research, infrastructure, and training. Academia and research institutions play a crucial role in developing localized technologies and building a skilled workforce, while civil society organizations contribute by raising awareness and ensuring accountability.

In India, platforms like the National Green Hydrogen Mission and Perform, Achieve, and Trade (PAT) scheme illustrate how public–private partnerships can accelerate the green transition. Globally, initiatives such as the European Green Deal and the UN's Sustainable Development Goals (SDGs) emphasize shared responsibility across nations and sectors. By fostering industrial symbiosis, knowledge sharing, and joint investment mechanisms, multi-stakeholder collaboration can create self-sustaining green industrial ecosystems that balance economic growth with ecological integrity.

10. CONCLUSION

Green technologies have emerged as transformative tools in reshaping industrial practices by reducing resource consumption, improving efficiency, and minimizing environmental impact. Their advancement across areas such as renewable energy integration, sustainable materials, waste valorization, and digital-enabled monitoring highlights their potential to redefine industrial sustainability. For industries, the implications extend beyond compliance, offering opportunities for cost savings, innovation, enhanced competitiveness, and alignment with global climate commitments. However, their widespread adoption depends on overcoming challenges related to financing, workforce readiness, and technological maturity. A clear roadmap for the future requires multi-stakeholder collaboration, supportive policy frameworks, scalable pilot projects, and stronger industry–academia partnerships that can drive innovation at both local and global levels. By systematically addressing these areas, green technologies can not only accelerate the transition to sustainable industrial systems but also contribute meaningfully to achieving long-term net-zero and circular economy goals.

REFERENCES

Al-Badwawi, R., Abusara, M., & Mallick, T. K. (2025). A review of hybrid solar PV and wind energy system. *Smart Science*, *3*(3), 127–138. DOI: 10.1080/23080477.2015.11665647

Allevi, E., De Giuli, M. E., Domínguez, R., & Oggioni, G. (2023). Evaluating the role of waste-to-energy and cogeneration units in district heating and electricity markets. *Computational Management Science*, *20*(1), 5. DOI: 10.1007/s10287-023-00437-3

Bai, H., Chen, Y., Bai, H., Liu, M., & Fan, Y. (2025). Energy optimization and efficiency improvement model for enterprise production process based on deep learning under the background of carbon peak and carbon neutrality. *International Journal of Computational Intelligence Systems*, *18*(3), 211–226. DOI: 10.1007/s44196-025-00901-9

Bin Amin, M., & Islam, K. M. A. (2024). Advancing sustainability: A comprehensive review of green technology innovations and their environmental impact. *Journal of Infrastructure. Policy and Development*, *8*(15), 9481. Advance online publication. DOI: 10.57125/JIPD.2024.9481

Bohnsack, R., Kolk, A., & Pinkse, J. (2020). Business models for sustainable technologies: Exploring electric vehicle industry practices. *Technological Forecasting and Social Change*, *160*, 120223. DOI: 10.1016/j.techfore.2020.120223

Chen, Y., Liu, H., & Wang, J. (2022). Financial incentives for promoting green technology adoption: Evidence from global policy frameworks. *Energy Policy*, *167*, 113081. DOI: 10.1016/j.enpol.2022.113081

Deng, Y., Zhang, X., & Zhou, D. (2023). Eco-design practices and digital innovations for sustainable product development. *Journal of Cleaner Production*, *412*, 137288. DOI: 10.1016/j.jclepro.2023.137288

Energy Conversion and Management. (2020). A comprehensive review on renewable energy integration for combined heat and power production. *Energy Conversion and Management, 224*, 113454. DOI: 10.1016/j.enconman.2020.113454

Fonseca, L. M., & Domingues, J. P. (2018). ISO 14001:2015: An improved tool for sustainability. *Journal of Cleaner Production*, *149*, 127–135. DOI: 10.1016/j.jclepro.2017.02.111

Fresner, J., Krenn, C., Haddad, J., Abdallah, R., Matalqa, B., & Abu Sada, A. (2025). A new approach to motivate micro and small enterprises for resource-efficient and cleaner production in Jordan. *Sustainability (Basel), 17*(6), 2404. DOI: 10.3390/su17062404

Garg, S., & Tiwari, A. (2022). Environmental compliance mechanisms and market-based instruments in India's industrial sector. *Environmental Science & Policy, 136*, 395–404. DOI: 10.1016/j.envsci.2022.07.012

Ghosh, A., & Gulia, S. (2023). Green financing mechanisms in India: Catalyzing sustainable industrial transformation. *Journal of Sustainable Finance & Investment, 13*(2), 325–342. DOI: 10.1080/20430795.2022.2101359

Gómez-Rubio, M.. (2025). New insights into how green innovation, renewable energy, and institutional quality affect green energy adoption in emerging economies. *Frontiers in Environmental Science*. Advance online publication. DOI: 10.3389/fenvs.2025.1525281

Gupta, A., & Bansal, R. (2023). Biomass-based cogeneration in Indian industries: A pathway towards cleaner energy transitions. *Renewable Energy Focus, 46*, 52–61. DOI: 10.1016/j.ref.2023.06.004

Gupta, A., & Sharma, P. (2023). Energy efficiency practices in India's cement industry: Progress and challenges. *Energy Policy, 174*, 113446. DOI: 10.1016/j.enpol.2022.113446

Gupta, R., & Bansal, P. (2023). Financing the green transition in India: The role of green bonds and sustainable investments. *Journal of Sustainable Finance & Investment, 13*(2), 301–317. DOI: 10.1080/20430795.2022.2157699

Hechelmann, R.-H., Seevers, J.-P., Otte, A., Sponer, J., & Stark, M. (2020). Renewable energy integration for steam supply of industrial processes—A food processing case study. *Energies, 13*(10), 2532. DOI: 10.3390/en13102532

Huang, J., Li, Z., & Xu, M. (2024). Circular product design strategies for electronic industries: A life cycle perspective. *Resources, Conservation and Recycling, 198*, 107229. DOI: 10.1016/j.resconrec.2023.107229

IEA. (2022). *Energy efficiency 2022: Analysis and outlooks to 2030*. International Energy Agency. https://doi.org/DOI: 10.1787/efficiency-2022-en

IREDA. (2024). *Renewable capacity statistics 2024*. International Renewable Energy Agency.

IRENA. (2023). *Renewable Energy and Jobs: Annual Review 2023*. International Renewable Energy Agency., DOI: 10.1787/renjobs-2023-

Jerome, A., & Ljunggren, M. (2025). Product lifetime in life cycle assessments of circular economy strategies—A review and consolidation of methodology. *The International Journal of Life Cycle Assessment, 30*(10), 2162–2189. Advance online publication. DOI: 10.1007/s11367-025-02470-7

Kalkuhl, M., Steckel, J. C., & Marschinski, R. (2022). Economic impacts of green technology adoption: Evidence from industry and energy transitions. *Energy Economics, 109*, 105995. DOI: 10.1016/j.eneco.2022.105995

Kamal, R., & Awasthi, A. (2023). Global trade policies and their impact on green technology adoption in emerging economies. *Journal of Cleaner Production, 409*, 137280. DOI: 10.1016/j.jclepro.2023.137280

Kumar, A., & Anbanandam, R. (2022). Evaluating the drivers of green supply chain management implementation in Indian manufacturing industries. *Journal of Cleaner Production, 355*, 131751. DOI: 10.1016/j.jclepro.2022.131751

Kumar, A., & Mohanty, P. (2023). Regulatory frameworks for sustainable energy transition in India: Policy pathways and industrial impacts. *Energy Research & Social Science, 101*, 103108. DOI: 10.1016/j.erss.2023.103108

Kumar, A., & Raj, R. (2023). Eco-design adoption and export competitiveness in Indian manufacturing industries. *International Journal of Sustainable Engineering, 16*(4), 733–745. DOI: 10.1080/19397038.2023.2173458

Kumar, A., & Singh, R. (2022). Adoption of green technologies in Indian industries: Case evidence from Tata Steel and auto sector. *Journal of Cleaner Production, 356*, 131870. DOI: 10.1016/j.jclepro.2022.131870

Kumar, R., Sharma, P., & Singh, S. (2024). Integrated waste-to-energy solutions for municipal solid waste management: A case study of Delhi's Ghazipur landfill. *Journal of Cleaner Production, 430*, 139876. DOI: 10.1016/j.jclepro.2023.139876

Lehr, U., Lutz, C., & Edler, D. (2023). Policy-driven green technology diffusion: Evidence from U.S. industrial decarbonization policies. *Journal of Cleaner Production, 406*, 136835. DOI: 10.1016/j.jclepro.2023.136835

Li, M., Zhao, L., & Chen, Y. (2023). Industry 4.0-enabled smart energy management in manufacturing systems. *Journal of Manufacturing Systems, 67*, 150–163. DOI: 10.1016/j.jmsy.2023.03.007

Li, X., Zhang, Y., & Wang, J. (2025). The impact of pilot policy for innovative industrial clusters on green innovation: Evidence from urban clusters. *Scientific Reports*. Advance online publication. DOI: 10.1038/s41598-025-07771-3

Luthra, S., Mangla, S. K., & Xu, L. (2022). Sustainability in Industry 4.0: Case insights from Siemens' digital twin adoption. *Resources, Conservation and Recycling, 180*, 106190. DOI: 10.1016/j.resconrec.2022.106190

Mavi, R. K., Goh, M., & Zarbakhshnia, N. (2023). Sustainable and green supply chain management: A bibliometric and content analysis. *Resources, Conservation and Recycling, 191*, 106918. DOI: 10.1016/j.resconrec.2023.106918

Mikayilov, J., Hasanov, F., & Mahmudlu, C. (2022). Regulatory frameworks and energy efficiency adoption in industries: Global policy insights. *Energy Policy, 166*, 112951. DOI: 10.1016/j.enpol.2022.112951

Mundaca, L., Busch, H., & Schwer, S. (2022). Renewable energy policies for industry decarbonization: Lessons from Germany and Japan. *Energy Policy, 164*, 112907. DOI: 10.1016/j.enpol.2022.112907

National Mission on Enhanced Energy Efficiency (NMEEE). (n.d.). *National Action Plan for Climate Change: Perform, Achieve, and Trade scheme*. Government of India., DOI: 10.1016/j.enpol.2011.05.017

Ninlawan, C., Seksan, P., Tossapol, K., & Pilada, W. (2022). The adoption of green supply chain management practices and its impact on performance. *Sustainable Production and Consumption, 31*, 550–563. DOI: 10.1016/j.spc.2022.01.019

Patil, A. A., Arora, R., Arora, R., & Sridhara, S. N. (2024). Techno-economic analysis of solar, wind, and biomass hybrid renewable energy systems to meet electricity demand of a small village in Bihar State of India. *NEPT, 23*(3), 003. Advance online publication. DOI: 10.46488/NEPT.2024.v23i03.003

Prabhu, V. S., Shrivastava, S., & Mukhopadhyay, K. (2021). Life cycle assessment of solar photovoltaic in India: A circular economy approach. *Circular Economy and Sustainability, 2*(2), 507–534. DOI: 10.1007/s43615-021-00101-5 PMID: 34888577

Raghavan, N., & Menon, V. (2021). Corporate sustainability in IT services: The case of Infosys carbon neutrality. *Business Strategy and the Environment, 30*(8), 3640–3651. DOI: 10.1002/bse.2831

Sangwan, S., & Bansal, S. (2022). Growth of green bonds in India: A step towards sustainable development. *Journal of Cleaner Production, 338*, 130505. DOI: 10.1016/j.jclepro.2022.130505

Schulze, M., Schipper, F., & Andersen, J. (2025). Advances in the social construction of energy management and industrial emissions reduction. *Nature Communications, 16*(1), 1122. DOI: 10.1038/s41467-025-59284-2 PMID: 39875403

Sharif, A., Bashir, U., Mehmood, S., Cheong, C. W., & Bashir, M. F. (2024). Exploring the impact of green technology, renewable energy and globalization towards environmental sustainability in the top ecological impacted countries. *Geoscience Frontiers, 15*(6), 101895. DOI: 10.1016/j.gsf.2024.101895

Sharma, H., & Padhi, B. (2025). Transiting towards green productivity: Evidence from Indian organized manufacturing industries. *Journal of Environmental Management, 373*(2), 123662. DOI: 10.1016/j.jenvman.2024.123662 PMID: 39693990

Sharma, M., & Thomas, A. (2022). Policy frameworks and industrial adoption of green technologies in India: Opportunities and challenges. *Environmental Policy and Governance, 32*(6), 529–543. DOI: 10.1002/eet.2008

Sharma, P., & Jain, R. (2023). Policy incentives and industrial adoption of renewable technologies: Evidence from India's solar PV sector. *Renewable & Sustainable Energy Reviews, 174*, 113065. DOI: 10.1016/j.rser.2022.113065

Sharma, R., Gunasekaran, A., Sharma, M., Luthra, S., & Joshi, S. (2025). Building a sustainable future: The role of green innovation and green premium in transforming Indian SMEs. *Environment, Development and Sustainability*. Advance online publication. DOI: 10.1007/s10668-025-06515-7

Shrimali, G., Nelson, D., & Goel, S. (2022). Enforcing industrial energy efficiency policies in India: Lessons from compliance mechanisms. *Environmental Science & Policy, 132*, 134–145. DOI: 10.1016/j.envsci.2022.03.019

Singh, A., & Garg, S. (2023). Industrial energy efficiency and economic co-benefits of the PAT scheme in India. *Energy Policy, 172*, 113308. DOI: 10.1016/j.enpol.2022.113308

Singh, R., & Bansal, S. (2023). Adoption of cleaner production technologies among Indian MSMEs: Drivers, barriers, and performance outcomes. *Journal of Environmental Management, 336*, 117642. DOI: 10.1016/j.jenvman.2023.117642

Singhania, D. S. (2024). Hydropower integration into renewable energy systems: Technological advances and economic considerations. *Indian Journal of Renewable Energy, 1*(1), 11–15. DOI: 10.36676/energy.v1.i1.3

Soni, A., Gupta, S. K., Rajamohan, N., & Yusuf, M. (2025). Waste-to-energy technologies: A sustainable pathway for resource recovery and materials management. *Materials Advances, 6*(14), 4598–4622. DOI: 10.1039/D5MA00449G

Wang, H., Liu, Y., & Zhao, X. (2023). Green product design for sustainable manufacturing: Life cycle assessment and innovation approaches. *Sustainable Production and Consumption*, *38*, 582–593. DOI: 10.1016/j.spc.2023.05.017

Wang, R., & Hynes, W. (2023). Smart environmental management practices in textile supply chains: Water and energy efficiency in South Asian mills. *Journal of Cleaner Production*, *312*, 127927. DOI: 10.1016/j.jclepro.2023.127927

Wu, J., & Strezov, V. (2024). Green technologies and sustainability: A new trend. *Green Technologies and Sustainability*, *1*(1), 1–12. DOI: 10.1016/j.gretsu.2024.100001

Yacooub, A. (2024). Cleaner production: Preventive, company-specific environmental strategies for minimizing waste and maximizing output. *Journal of Cleaner Production*, *370*, 132456. DOI: 10.1016/j.jclepro.2024.132456

Yadav, R., & Kumar, P. (2023). Financing the green transition in emerging economies: Case of India. *Environmental Economics and Policy Studies*, *25*(1), 87–106. DOI: 10.1007/s10018-022-00374-4

Zhang, Q., Liu, J., & He, W. (2023). Advanced energy-efficient technologies for sustainable manufacturing: A review. *Renewable & Sustainable Energy Reviews*, *174*, 113145. DOI: 10.1016/j.rser.2023.113145

Zhang, Y., Li, H., & Wang, X. (2023). Green technological progress and its externalities on sustainable economic growth: Evidence from emerging economies. *Journal of the Knowledge Economy*, *15*(2), 678–695. DOI: 10.1007/s13132-023-01268-y

Zhou, L., Sun, Y., & Li, Y. (2023). Regulatory drivers of green technology innovation: Evidence from international policy frameworks. *Technological Forecasting and Social Change*, *190*, 122395. DOI: 10.1016/j.techfore.2023.122395

Chapter 11
The U.S. and the Evolution of Globalization With Climate Change Implications

Frank Billingsley
https://orcid.org/0009-0000-5363-1568
Gerogetown College, USA

Hayden W. Davis
https://orcid.org/0009-0000-9385-9305
The University of Edinburgh, USA

Hana Dixon
https://orcid.org/0009-0006-3708-1870
University of Kentucky, USA

Nicole Ghirlanda
Southern Methodist University, USA

ABSTRACT

Neoliberal globalization, driven by U.S. economic policies, trade liberalization, and technology, has transformed production, consumption, and environmental outcomes. Deregulation and weakened environmental oversight enabled resource-intensive industries and offshoring of polluting production to countries such as China, Vietnam, and Indonesia. Trade agreements like NAFTA and WTO frameworks reinforced this trend, externalizing carbon emissions and social impacts. Industries such as fast

DOI: 10.4018/979-8-3373-6435-3.ch011

fashion contribute roughly four percent of global emissions. Addressing these structural harms requires enforceable supply chain standards, sustainable production, and transformative consumption to protect both the planet and vulnerable populations.

INTRODUCTION

Globalization has been one of the defining forces of the modern era, reshaping economies, societies, and political structures across the world. At its most basic level, globalization refers to the integration of national economies and cultures through flows of trade, finance, technology, migration, and ideas. Yet beyond this abstract definition, globalization has a particular historical trajectory shaped by power relations and institutional arrangements. The United States, as the preeminent global power following the Second World War, has been central in shaping the rules, norms, and material flows of globalization. Its economic leadership, cultural influence, and strategic policies have not only spurred unprecedented growth in international commerce but have also locked the global economy into fossil fuel dependency, embedding environmental consequences within the very fabric of global trade.

The postwar era of the 1950s was pivotal in establishing the United States as a global leader. Through initiatives such as the Marshall Plan, the Bretton Woods system, and the promotion of liberal trade regimes, the U.S. consolidated its economic and political dominance. This was also a period of rapid domestic industrial growth, suburbanization, and the rise of consumer culture, all of which fueled surging energy demand. Coal and oil became the backbone of production and mobility, reinforcing a carbon-intensive development model. The Cold War further cemented this trajectory, as economic integration and industrial growth were used as tools of ideological competition against the Soviet Union. In this sense, U.S. leadership did not merely encourage global trade but also entrenched fossil fuel dependency as a cornerstone of global prosperity.

The neoliberal turn of the 1980s marked the second major phase of U.S.-led globalization. Under President Ronald Reagan and other global leaders, deregulation, privatization, and trade liberalization became guiding principles of economic policy. These reforms dismantled many protections and oversight mechanisms, opening space for multinational corporations to expand globally with fewer constraints. Outsourcing production to middle- and low-income countries shifted industrial emissions abroad, while U.S. consumption continued to rise. This created a paradox in which developed countries could stabilize or reduce their domestic emissions while remaining deeply responsible for emissions embedded in imports. During this period, globalization became inseparable from financialization and corporate expansion, both of which fueled the growth of fossil fuel industries and carbon-intensive supply chains.

In the contemporary era, globalization has entered a digital and financial phase. Technology companies, e-commerce platforms, and multinational supply chains have created a more tightly connected global economy than ever before. At the same time, awareness of climate change has grown, and initiatives such as the Paris Agreement have sought to realign globalization with sustainability. Yet progress remains uneven. While renewable energy and clean technologies have expanded rapidly, fossil fuels continue to dominate the global energy mix. Outsourced production still masks the actual carbon footprint of wealthy nations, and financial institutions continue to underwrite the expansion of fossil fuel industries. Globalization thus remains a double-edged sword: it has enabled innovation, connectivity, and prosperity, but it has also amplified ecological degradation and climate risks.

This paper examines the historical and structural relationship between U.S.-led globalization and rising CO_2 emissions, focusing on three key periods, the postwar expansion of the 1950s, the neoliberal acceleration of the 1980s, and the digital globalization of the contemporary era. It argues that globalization, as shaped by U.S. policies and corporations, has entrenched systemic carbon dependency in global trade and production networks. Addressing climate change in this context requires moving beyond territorial emissions targets to confront the embedded emissions in global supply chains, the financing of fossil fuel industries, and the structural features of trade and consumer culture. The central question guiding this analysis is: how has U.S.-led globalization since the 1950s contributed to rising CO_2 emissions, and what does this mean for designing climate policies that are both effective and equitable in an interconnected world?

PART 1: KEY PERIODS OF GLOBALIZATION AND EMISSIONS GROWTH

The 1950s marked the emergence of the U.S. as a global economic leader following World War II. The U.S. promoted global economic integration rooted in capitalist principles through programs like the Marshall Plan and institutions such as the Bretton Woods system. The Cold War accelerated these efforts as economic influence became a proxy for ideological dominance over the Soviet Union.

This era witnessed rapid industrial expansion in the U.S. and Western Europe, fueled by coal and oil. The construction of suburban communities, highways, and consumer goods manufacturing led to a surge in domestic emissions. Globally, American corporate expansion began to shape early global supply chains and increase energy demands abroad (Yergin, 2009). While globalization was in its early stages, the foundations of a carbon-intensive global economy were laid.

The 1980s marked a more aggressive phase of globalization, characterized by neoliberal policies such as deregulation, privatization, and trade liberalization. Under President Reagan in the U.S. and Prime Minister Thatcher in the United Kingdom, market-driven reforms were prioritized over state-led planning (Harvey, 2005). These changes expanded U.S. corporations' global reach and facilitated the outsourcing of production to emerging markets with cheaper labor and fewer environmental regulations. As shown in Figure 1: Structural Drivers of Carbon Emissions in U.S.-Led Globalization, these policy shifts entrenched carbon dependency through three interlinked mechanisms: trade liberalization shifted production to high-emission regions, embedding emissions in U.S. imports; deregulation weakened oversight and expanded extractive industries; and consumer culture fueled demand for carbon-intensive goods, amplifying emissions across global supply chains. Together, these structural drivers explain how globalization, as shaped by U.S. economic leadership, embedded environmental degradation into the fabric of international trade.

As a result, CO_2 emissions rose sharply in the U.S. and in developing countries that supply goods to the Global North. High-emission industries, such as steel, textiles, and electronics, have shifted to countries like China, India, and those in Southeast Asia. Although the U.S. began stabilizing its domestic emissions, the global carbon footprint associated with American consumption increased through the import of goods (Davis & Caldeira, 2010). This demonstrates that while domestic mitigation efforts may appear effective, the actual environmental impact of consumption is distributed globally, highlighting the need for coordinated international climate policies and supply chain accountability.

Table 1. Structural drivers of carbon emissions in U.S.-led globalization

Driver	**Mechanism**	**CO_2 Impact**
Trade Liberalization	Shifting production to low-cost, high-emission regions	Emissions embedded in imports
Deregulation	Weakening of environmental oversight	Increased domestic and foreign extraction
Consumer Culture	Demand for carbon-intensive goods	Global supply chain emissions

This table outlines three core drivers that link U.S.-led globalization to rising CO_2 emissions. Trade liberalization shifts production to countries with weaker environmental standards, embedding emissions in imported goods. Deregulation reduces oversight, allowing for the extraction of resources and the release of pollution, both domestically and abroad. Meanwhile, consumer culture sustains demand for carbon-intensive products, driving emissions throughout global supply chains. Together, these mechanisms illustrate how systemic features of globalization perpetuate environmental harm (Hubacek et al., 2017).

Globalization today is deeply influenced by digital transformation, financial integration, and a growing focus on sustainability. Digital technologies, including the Internet, e-commerce, and digital services, have revolutionized the way communication, trade, and business operate, enabling markets worldwide to connect faster and more deeply than ever. These technologies break down traditional barriers, allowing goods, services, and information to flow rapidly across borders, creating enormous opportunities for both businesses and consumers. However, this digital progress comes with environmental costs, notably through increased carbon emissions. As climate change becomes a critical global challenge, initiatives such as the Paris Agreement have pushed the world toward renewable energy, green technologies, and sustainable business models. Yet, despite these efforts, fossil fuels still dominate energy consumption, and emissions embedded in complex global supply chains obscure the actual environmental impact of consumption, especially in developed economies where production is often outsourced to regions with less stringent environmental policies (George & Schillebeeckx, 2022). As a result, achieving meaningful global climate goals requires addressing the structural factors within global trade and production systems that perpetuate carbon dependency, necessitating a more coordinated and comprehensive approach to sustainability at the international level.

The globalized economy's energy-intensive production and supply chains continue to pose a challenge to climate mitigation efforts. Research shows that CO_2 emissions are deeply embedded in global supply networks, often escaping capture by national climate policies, which focus primarily on territorial emissions. This makes it challenging to fully account for and regulate the emissions resulting from consumption patterns in wealthier nations. Additionally, recent shocks, such as the COVID-19 pandemic and geopolitical tensions, have disrupted trade and supply chains, leading some countries to consider reshoring production. However, these disruptions have not fundamentally altered the fossil fuel-dependent model of globalization. Multinational corporations and their sprawling supply chains remain central to these emissions, complicating efforts to design effective climate policies that align economic growth with environmental sustainability (Zhang et al., 2020). As global interconnectedness persists, a more integrated approach to climate policy that addresses both local and global emissions is crucial to achieving long-term environmental and economic goals.

These insights suggest that addressing climate change in a globalized world requires international cooperation that integrates digital and financial innovations with sustainability goals. Policies must consider emissions across entire supply chains and incentivize greener production alongside digital transformation to reduce carbon footprints without stalling economic integration.

Several key issues emerge when examining the intersection of globalization and climate change. One primary concern is carbon leakage and outsourcing, where developed nations appear to reduce their emissions domestically but continue to rely on imported goods produced in high-emission countries. This practice, known as carbon leakage, obscures the actual environmental cost of consumption (Davis & Caldeira, 2010). Additionally, there is the trade versus environment dilemma: while trade liberalization can stimulate economic growth, it often encourages deforestation, resource depletion, and increased fossil fuel use in developing nations that supply global markets (Barlow et al., 2016). Another challenge is policy fragmentation. Climate policies are typically implemented at the national level, yet emissions arise from a globally interconnected system. This disconnect complicates enforcement, accountability, and the establishment of fair emissions targets (Victor, 2001). Finally, there is an ongoing debate about equity and responsibility in addressing emissions. Should the burden lie with the nations that produce carbon-intensive goods, often developing countries, or with the wealthier countries that consume them? This question remains central to discussions on global climate justice (Peters et al., 2011). Collectively, these challenges underscore that addressing climate change in a globalized economy requires coordinated international policies, shared responsibility, and an integrated approach that considers both production and consumption patterns.

Thus, the historical trajectory of U.S.-led globalization has created a carbon-intensive global economy in which emissions are deeply embedded in international trade and production systems. While technological innovation, digitalization, and international climate agreements offer pathways to mitigate environmental harm, structural drivers such as trade liberalization, deregulation, and consumption patterns continue to perpetuate carbon dependency. Addressing climate change effectively thus requires policies that transcend national borders, account for emissions throughout global supply chains, and balance economic growth with environmental stewardship. Only through coordinated international action, equitable responsibility, and integration of sustainable practices into global commerce can meaningful reductions in CO_2 emissions be achieved, ensuring long-term climate resilience.

THE HIDDEN CARBON COST OF U.S. CONSUMPTION

A striking example of carbon outsourcing is evident in the composition of the U.S. carbon footprint. In 2021, nearly 40% of the emissions associated with American consumption were embedded in imported goods, particularly from manufacturing hubs in China and Southeast Asia. These emissions, while not counted in the U.S.

territorial totals, represent a significant environmental impact driven by domestic demand. Studies by Peters et al. (2011) and Davis and Caldeira (2010) highlight how international trade masks the accurate scale of carbon emissions, allowing developed countries to appear cleaner while continuing to rely heavily on high-emission production abroad.

Ultimately, understanding the connections between globalization and CO_2 emissions is crucial for developing climate strategies that are both scientifically effective and politically and economically viable. Policies must go beyond national borders and address the structural features of global trade, finance, and production that perpetuate environmental harm. Bridging this gap will advance a sustainable and equitable global economy. Furthermore, increasing transparency in supply chains, incentivizing low-carbon production in exporting countries, and promoting consumer awareness in importing nations are essential steps to ensure that reductions in domestic emissions reflect genuine global climate progress rather than a displacement of environmental concerns.

THE FOUNDATIONS OF U.S. GLOBALIZATION: THE 1950S

Globalization refers to the interconnection of different parts and eras of the world through trade, economies, cultures, and political systems across nations. In the United States, following World War II, the 1950s saw significant GDP growth driven by increases in government spending, household consumption, and industrial production after years of wartime rationing. In the 1980s, the rise of neoliberalism under President Ronald Reagan introduced deregulation and tax cuts that stimulated the job market, increased productivity, and expanded economic growth (Prasad, 2012). Although these developments appeared beneficial, they also had long-term consequences, particularly the rise in CO_2 emissions. The modern world continues to grapple with the environmental implications of decisions made in the mid-twentieth century. Moreover, the globalization processes that fueled economic expansion embedded carbon-intensive production into global supply chains, linking domestic consumption in wealthy nations to emissions abroad and highlighting the ongoing challenge of reconciling economic growth with environmental sustainability.

DEFINING GLOBALIZATION AND U.S. SIGNIFICANCE

Globalization involves the movement of goods, services, capital, people, and ideas. After World War II, the establishment of institutions such as the International Monetary Fund and the World Bank, along with technological advancements in communication and transportation, sought to prevent the instability of the interwar period while fostering international connections. These measures promoted economic stability and global development. However, challenges also emerged, such as fluctuating exchange rates and growing national debts, which complicated efforts to maintain stability. Globalism expanded U.S. economic and political influence and fueled the spread of consumerism and media worldwide.

THE 1950S: POST-WAR EXPANSION AND COLD WAR INFLUENCE

The 1950s marked the beginning of what economists call the "Golden Age of Capitalism," a period characterized by rapid economic growth, low unemployment, and a striking rise in consumer spending, particularly in the United States. During this era, GDP grew steadily, joblessness declined, and American households embraced unprecedented prosperity and consumerism, fueling the expansion of families and the enduring appeal of the "American Dream" (Marglin, 1988). On the global stage, U.S. policy was heavily shaped by Cold War dynamics. The United States actively promoted democracy and free-market capitalism to counter Soviet influence, with the creation of NATO in 1949 serving as a cornerstone for a security alliance among Western democracies in response to perceived threats from the Soviet Union (Kaplan, 2014). Alongside economic and political strategies, cultural influence became a powerful instrument of U.S. soft power. American art, music, sports, educational models, and consumer products gained international popularity, intertwining cultural exports with diplomacy and global relations (Immerman, 2003; Iriye, 1979). Collectively, the economic expansion, geopolitical strategies, and cultural outreach of the 1950s established the United States as a dominant force in shaping both domestic prosperity and the early contours of postwar globalization, setting patterns that continue to influence global economic and cultural dynamics today.

THE 1980S: NEOLIBERALISM, DEREGULATION, AND INDUSTRIAL EXPANSION

The 1980s introduced neoliberal policies that emphasized deregulation, reduced government intervention, and the promotion of free markets. Reaganomics, championed by President Ronald Reagan, sought to stimulate economic growth by lowering taxes, curbing inflation, and encouraging competition. These policies aimed to let American families retain more of their income and reduce business constraints from government oversight, while promoting lower consumer prices through market competition. While these measures contributed to economic expansion, they also accelerated industrial production and energy consumption, which exacerbated environmental challenges (Morgan, 2008; Prasad, 2006). The economic dynamism of the decade thus came at the cost of deepening global carbon dependency, leaving a legacy of environmental strain that continues to shape the debate on globalization and sustainability. Furthermore, the outsourcing of production to emerging markets with fewer ecological regulations amplified the global carbon footprint, linking domestic consumption in the United States to environmental degradation abroad and underscoring the enduring tension between economic growth and ecological responsibility.

MODERN-DAY GLOBALIZATION

Digital technology, multinational corporations, and interconnected supply chains increasingly drive today's globalization. Tech giants and digital trade have reshaped global commerce, influencing the future of work, infrastructure, and international relations. Artificial intelligence, digital trade, and cybersecurity threats have further transformed how nations and businesses interact. Global supply chains also define the modern economy: components are designed in one country, manufactured in another, and assembled in yet another, creating efficiencies and access to broader markets while expanding consumer choice. This interconnectedness has accelerated the pace of innovation. Still, it has also increased the speed and scale of environmental impacts, as production and consumption patterns now extend far beyond national borders. The rapid industrialization of the 1950s and 1980s escalated the use of fossil fuels, particularly coal and oil, resulting in significant CO_2 emissions. Industrial activity and trade openness contributed heavily to atmospheric pollution. In the present day, globalization continues to fuel emissions, especially in emerging economies such as China. While some argue that globalization supports the spread of cleaner technologies, others point to its "pollution effect," as industries relocate to countries with fewer environmental restrictions. The relationship remains highly complex, reflecting the paradox of globalization as both a driver of innovation and an amplifier of ecological harm.

NEOLIBERAL APPROACHES TO POLLUTION REDUCTION

The embrace of market-driven environmental policies is a direct outgrowth of globalization's broader economic philosophies. As global trade and investment have accelerated, so too have cross-border environmental impacts, requiring multinational solutions. Neoliberal environmentalism, with its focus on flexible, market-based mechanisms, echoes the same logic that has driven global integration: efficiency, innovation, and competition. Yet, just as globalization has brought both prosperity and new risks, the reliance on markets for environmental progress is fraught with complexities, especially in the absence of coordinated global governance. Without robust international standards and accountability frameworks, market incentives alone may fail to adequately address carbon leakage, resource depletion, and inequities in environmental burden, leaving the promise of sustainable development partially unrealized. Ultimately, the effectiveness of market-driven approaches depends on their integration with comprehensive policy, cooperative governance, and a recognition of the interconnected nature of ecological and economic systems.

The historical trajectory of U.S.-led globalization demonstrates the intricate interplay between economic ambition, political influence, and environmental consequence. From the postwar expansion of the 1950s, with its rapid growth, cultural outreach, and Cold War strategies, to the neoliberal reforms of the 1980s that intensified industrial output and global trade, economic policies have shaped not only domestic prosperity but also the environmental footprint of production and consumption worldwide. Modern globalization, driven by technology, multinational supply chains, and digital commerce, continues to amplify both opportunities and ecological challenges. While market-based environmental strategies offer mechanisms for efficiency and innovation, they are insufficient without coordinated international governance and accountability. Understanding the historical roots of these dynamics underscores the need for integrated approaches that balance economic growth with environmental sustainability, ensuring that the benefits of globalization do not come at the expense of the planet's long-term health.

The historical development of U.S.-led globalization reveals a consistent pattern in which economic expansion, geopolitical strategy, and technological transformation have intensified carbon emissions across domestic and international systems. Beginning in the 1950s, postwar industrial growth, suburban development, and Cold War economic programs laid the foundations of a carbon-intensive global economy. By the 1980s, neoliberal reforms deepened these patterns by promoting deregulation, privatization, and trade liberalization, which expanded multinational production networks and encouraged the relocation of high-emission industries to countries with weaker environmental standards. In the present era, digital technologies and globally integrated supply chains continue to accelerate economic interconnectedness while masking the carbon embedded in imported goods, particularly in developed economies that outsource production. Research consistently demonstrates that national emissions inventories fail to capture the full environmental impact of consumption, underscoring the limitations of fragmented climate governance. Across these periods, the structural drivers of globalization have reinforced carbon dependency, revealing that effective climate action must extend beyond territorial boundaries and account for emissions embedded throughout global supply chains. A sustainable path forward requires coordinated international policy, transparent production systems, and a recalibration of global economic practices to align growth with environmental stewardship.

PART 2: THE ACCELERATION OF GLOBALIZATION: THE 1980S AND NEOLIBERALISM

As the postwar economic expansion of the 1950s established the structural foundations of a carbon-intensive global economy, the decades that followed transformed these early patterns into a far more accelerated and market-driven system. From the 1980s onward, a new wave of neoliberal economic policies reshaped global markets and intensified environmental degradation. Rooted in deregulation, privatization, and free-market capitalism, these policies, most prominently championed in the United States under President Ronald Reagan, encouraged financial expansion and global outsourcing. The rapid spread of technological innovation and supply chain optimization further integrated the world economy, but at the cost of environmental oversight. Combined with the collapse of the Soviet Union and the integration of carbon-intensive economies, such as China, into the global system, these developments significantly expanded global energy consumption and resource extraction. The consequences include increased fossil fuel dependency and widespread deforestation driven partly by U.S. trade and corporate practices.

The neoliberal turn in U.S. economic policy during the Reagan administration (1981–1989) emphasized deregulation, tax cuts, and the downsizing of federal oversight, including in environmental sectors. Known as "Reaganomics," this approach promoted the privatization of public services and land, curtailed funding for regulatory agencies like the Environmental Protection Agency (EPA) and removed constraints on industries considered burdensome to economic growth (Harvey, 2005). These shifts facilitated the expansion of capital-intensive and resource-extractive sectors, including oil, gas, and mining. Simultaneously, environmental protections established during the 1970s under laws such as the Clean Air Act and Clean Water Act were weakened or poorly enforced, creating a regulatory environment that prioritized economic growth over ecological stewardship.

Expanding financial markets, a hallmark of neoliberal globalization, led to increased investment in extractive industries. With fewer restrictions on speculative capital and transnational corporate behavior, U.S.-based firms and investors funded large-scale fossil fuel infrastructure, from pipelines to offshore drilling, domestically and abroad. These financial flows enabled companies to exploit natural resources at an unprecedented scale, often in regions with weak environmental regulations. As a result, neoliberal policies not only amplified domestic carbon emissions but also facilitated the globalization of environmentally destructive production, linking Western consumption to ecological degradation across the global south and setting the stage for ongoing challenges in climate governance.

TECHNOLOGICAL ADVANCES AND GLOBAL SUPPLY CHAINS

While neoliberal policies laid the ideological foundation for economic globalization, technological advances made it logistically feasible. The rise of telecommunications, standardized container shipping, and digital logistics revolutionized supply chains. This facilitated the outsourcing of labor-intensive and polluting industries from the U.S. to emerging markets, particularly in East and Southeast Asia. Countries like China, Vietnam, and Indonesia became manufacturing hubs for U.S. and European brands, effectively shifting emissions and environmental degradation abroad (Shen & Ma, 2019). This global redistribution of production underscores how economic integration can mask environmental costs, highlighting the need for international cooperation in addressing the ecological consequences of consumption.

This outsourcing masked the actual environmental footprint of U.S. consumption. Products consumed in America, such as electronics, textiles, and automobiles, were increasingly produced in countries with lax environmental standards and high coal dependency. A 2011 study found that approximately 22% of global CO_2 emissions were embedded in exported goods, primarily destined for high-income countries such as the U.S. (Davis & Caldeira, 2010). Furthermore, the reliance on low-cost production in the Global South intensified local pollution, deforestation, and water stress, often disproportionately affecting vulnerable communities. This trend enabled wealthier nations to appear to be reducing their domestic emissions while maintaining high-carbon lifestyles through the importation of goods.

POST-COLD WAR GLOBALIZATION AND MARKET LIBERALIZATION

The end of the Cold War in the early 1990s marked a pivotal acceleration in globalization, ushering in a new era of economic interdependence. With the dissolution of the Soviet Union, formerly closed Eastern European economies opened to foreign investment and trade. At the same time, China's market reforms under Deng Xiaoping integrated millions of workers into the global economy. These transformations created unprecedented opportunities for industrial expansion and export-led growth, fueling surges in energy demand, particularly from fossil fuel-intensive industries. China's accession to the World Trade Organization (WTO) in 2001 symbolized a formal entry into the global economic order, further solidifying its role as the "world's factory." The country's rapid industrialization relied heavily on coal, driving a near doubling of CO_2 emissions between 2000 and 2010 and establishing China as the world's largest emitter (Peters et al., 2011). Yet, much of this output was not for domestic consumption but for North American and European markets, effectively outsourcing the environmental costs of consumption in wealthy nations to emerging economies. This displacement underscores the complexity of global carbon accounting, revealing that territorial emissions often mask the true ecological footprint of consumption. The globalization of carbon-intensive production also highlights broader structural issues: the unequal distribution of environmental burdens, the dependence of advanced economies on inexpensive, high-emission manufacturing abroad, and the limitations of national policies in addressing transboundary ecological impacts. These patterns suggest that without coordinated international governance, trade and industrial integration will continue to externalize environmental costs, perpetuating a cycle in which global prosperity is built on ecological strain. Addressing these challenges requires policies that link production, consumption, and environmental accountability across borders, recognizing that sustainable development is inseparable from the global economic networks that now define modern civilization.

ENVIRONMENTAL CONSEQUENCES: FOSSIL FUELS AND THE GLOBAL SOUTH

As trade liberalization and technological innovation expanded global markets, fossil fuel consumption surged, fueling industrial growth but intensifying environmental pressures. Global energy demand rose sharply, with oil, coal, and natural gas continuing to dominate the energy mix. Despite growing awareness of climate change, fossil fuels still accounted for over 80% of global primary energy consumption in 2021 (International Energy Agency, 2021). The pursuit of free trade and resource extraction incentivized widespread deforestation in the Amazon, Southeast Asia, and sub-Saharan Africa. U.S. trade policies and multinational corporations often facilitated land use changes, including the expansion of soy, palm oil, and cattle industries, which are major drivers of carbon emissions and ecological disruption (Barlow et al., 2016). These trends highlight how the expansion of global markets, driven by trade liberalization and technological innovation, has linked economic growth to profound environmental consequences, emphasizing the urgent need for sustainable and accountable approaches to global production and consumption.

The Global South bore the brunt of this extractive development model. Governments were frequently pressured to liberalize markets, reduce environmental protections, and prioritize foreign direct investment or trade agreements, often at the expense of ecological sustainability. The resulting degradation disproportionately affected marginalized communities, contributing to biodiversity loss, water pollution, soil erosion, and forced displacement, while simultaneously sustaining high levels of consumption in the Global North. Many of these communities faced intensified economic and social vulnerability, as environmental degradation undermined livelihoods dependent on forests, fisheries, and arable land, creating cycles of poverty intertwined with ecological decline.

The neoliberal era, defined by deregulation, global financial integration, and technological transformation, profoundly shaped both economic globalization and environmental decline. U.S.-led policies promoted capital mobility, resource extraction, and industrial offshoring, often weakening local ecological safeguards. The incorporation of high-emission economies, such as China and Eastern Europe, into global supply chains further accelerated fossil fuel consumption, effectively externalizing environmental costs and obscuring the actual carbon footprint of Western consumption. International trade networks increasingly normalize ecological displacement, reinforcing the perception that economic growth could occur independently of environmental consequences, while simultaneously concentrating environmental burdens in developing regions.

These patterns underscore that climate change is not merely a scientific or technological challenge but a structural one, rooted in the economic and political architecture of globalization. Addressing the ecological consequences of neoliberal globalization will require rethinking trade policies, strengthen environmental governance, and foster equitable frameworks that link production, consumption, and ecological accountability across borders, ensuring that global prosperity does not continue to come at the expense of planetary health. Only by integrating climate justice, transparent supply chain monitoring, and binding international environmental standards can the global community begin to reconcile economic development with ecological sustainability, preventing further disproportionate harm to vulnerable populations.

The evolution of globalization from the postwar period through the neoliberal era reveals how economic integration, technological transformation, and shifting geopolitical dynamics entrenched carbon-intensive systems that continue to shape the global environmental landscape. The foundations laid in the 1950s through industrial expansion and U.S. political influence became amplified in the 1980s as neoliberal reforms promoted deregulation, financial liberalization, and the offshoring of production to regions with weaker environmental safeguards. These shifts, paired with rapid advances in logistics and digital technologies, enabled complex global supply chains that masked the true ecological cost of consumption in wealthier nations. The post-Cold War incorporation of China and other emerging markets into the global economy accelerated fossil fuel use, deforestation, and resource extraction, linking Western prosperity to environmental degradation abroad. As the Global South increasingly absorbed the burdens of pollution, biodiversity loss, and displacement, patterns of ecological inequality deepened. Together, these developments illustrate that the environmental consequences of globalization are systemic and transboundary, demanding a coordinated international response that confronts the structural roots of carbon dependency, enhances supply chain accountability, and aligns global economic activity with principles of climate justice and sustainable development.

Part 3: THE ROLE OF U.S. CORPORATE EXPANSION AND FINANCIALIZATION IN CLIMATE CHANGE

As globalization intensified the scale and reach of production, its environmental consequences became inseparable from the consumption patterns that had already taken root in the United States. Modern consumerism in the United States traces its roots to the 1920s, when mass production and the rise of advertising transformed both the availability of goods and the social meanings attached to consumption (MIT Press Reader, 2021). During this era, purchasing consumer goods became closely tied to status and identity, as Americans embraced new products and advertising encouraged people to see consumption as part of their personal and social identity. This foundation marked the beginning of a culture where material goods became symbols of self-expression and social belonging.

As technology advanced, especially with the spread of television, material goods became further associated with feelings such as happiness, success, and patriotism. Television and advertising exposed viewers to new products and lifestyles, helping to create a more uniform consumer culture and promoting the idea that material possessions were central to personal fulfillment (Five able, 2025). This shift entrenched the notion that consumption was not merely economic but emotional, shaping values and aspirations across American households.

In the contemporary era, technological change has accelerated the pace and scale of consumer behavior. The rise of the internet, e-commerce platforms, and social media marketing has made consumption more accessible, immediate, and personalized than ever before. These developments have changed the traditional shopping experience and increased the influence of brands on society. While digitalization provides more access to goods and services, it has also had a significant impact on the environment and has further tied consumption to personal worth and identity (Global Media Journal, 2024). The digital marketplace has therefore deepened both the opportunities and the dilemmas of consumerism, linking convenience with pressing questions of sustainability.

The historical rise of consumer culture in the United States set the stage for the global expansion of American multinational corporations. As demand for goods increased and consumption became central to American identity, corporations expanded internationally to maximize production efficiency and profitability, often seeking lower labor costs and less regulation (Fiveable, 2025). This shift has had profound environmental consequences. By embedding consumerist values into global trade networks, U.S. corporations exported not only products but also the environmental costs of mass production.

A clear example of these corporate practices can be found in the fast fashion industry, where multinational brands drive rapid production cycles and lower costs at significant environmental expense. Fast fashion is now responsible for about 10% of global carbon emissions and is a major source of water pollution and textile waste (Earth.Org, 2025). This industry highlights the stark reality that unchecked consumer demand, fueled by corporate expansion, has created one of the most urgent environmental challenges of our time.

EXPANSION OF FAST FASHION, ELECTRONICS, INDUSTRIAL AGRICULTURE, INCREASING CARBON FOOTPRINTS

Over the past several decades, fast fashion has become a central feature of global consumer culture and a significant driver of rising carbon emissions. The industry is built on a low-cost, high-volume production model that prioritizes speed, affordability, and trend turnover, often relying on manufacturing in regions such as Asia, where environmental and labor regulations are less stringent. According to McKinsey's Fashion on Climate report, the fashion industry alone accounts for roughly 4 percent of global greenhouse gas emissions, a carbon footprint larger than that of France, Germany, and the United Kingdom combined (Tinnes et al., 2024). This reality underscores how fast fashion has evolved from a trend-driven business model into a global environmental threat with consequences that extend far beyond the industry itself, affecting ecosystems, water resources, and carbon budgets worldwide.

The report also emphasizes that approximately 70 percent of these emissions originate from the production, preparation, and processing of raw materials, including cotton, synthetic fibers, and chemical dyes. With global consumption patterns continuing to rise, annual emissions are projected to grow by 2.7 percent (Tinnes et al., 2024). The expansion of fast fashion reflects the broader consumer culture promoted by multinational corporations, one that prioritizes constant consumption, rapid obsolescence, and disposable goods over long-term sustainability. This dynamic is further exacerbated by aggressive marketing strategies, social media trends, and the normalization of cheap clothing as a daily necessity in many high-income markets.

Addressing the environmental impact of fast fashion requires coordinated action between corporations, policymakers, and consumers. McKinsey suggests that up to 60 percent of emissions reductions could be achieved in upstream operations through the adoption of renewable energy, improved supply chain efficiency, and the substitution of lower-impact raw materials. On the consumer side, meaningful change depends on embracing circular practices such as clothing rentals, secondhand purchasing, repair, and refurbishment, along with reducing the frequency of washing and drying garments. Complementary approaches include regulatory incentives for sustainable production, stricter environmental compliance in manufacturing countries, and corporate transparency regarding lifecycle emissions. Collectively, these strategies offer pathways to transform the fast fashion industry into a more sustainable model while simultaneously reshaping consumer behavior toward responsible, long-term patterns of clothing use and disposal.

INFLUENCE OF U.S. CORPORATE PRACTICES ON GLOBAL LABOR AND PRODUCTION EMISSIONS

U.S. corporate practices have significantly influenced global labor and production emissions through the outsourcing of manufacturing to overseas locations. Many American companies reduce their domestic carbon footprints by relocating production to countries with weaker environmental regulations. This outsourcing strategy allows corporations to present a more sustainable image at home, while the actual ecological costs are shifted onto vulnerable regions, where regulatory oversight is limited.

A 2023 study by the Principles for Responsible Investment (PRI) found that outsourcing emissions is a common strategy among U.S. firms seeking to remain competitive while adhering to domestic climate policies (Rui et al., 2023). Rather than achieving genuine decarbonization, this approach relocates emissions abroad, obscuring the full environmental impact of American consumption. P&G illustrates this dynamic: while the company pledged to reduce its emissions by 50% in 2023, the commitment excludes emissions generated within its global supply chain, much of which is located outside the United States (Rui et al., 2023). Such examples demonstrate how corporate sustainability pledges can create a misleading narrative of progress while emissions remain embedded in international production networks.

These practices have far-reaching implications beyond environmental degradation, affecting labor systems in supplier countries. Workers are frequently subjected to intense production demands that prioritize low costs, high output, and rapid turnaround times, often in environments that lack robust labor protections. The PRI study highlights that this combination of environmental outsourcing and labor exploitation leaves both ecosystems and workers vulnerable to regulatory gaps and insufficient enforcement (Rui, et al., 2023). The interconnected nature of emissions and labor exploitation underscores the urgent need for stronger international standards and accountability mechanisms that compel corporations to take responsibility for their environmental and social impacts across borders.

FINANCIALIZATION AND MARKET-DRIVEN GROWTH

Banks are the main creditors for fossil fuel companies; therefore, their role in supporting a fossil fuel economy is central to the climate change debate. Financial institutions provide the loans, underwriting, and investment capital that enable oil, gas, and coal companies to explore new reserves, expand infrastructure, and maintain existing operations. Not only is investing in fossil fuel companies at this rate incompatible with international climate targets, but the financing of existing companies must be phased out (Rickman et al, 2024). This makes banks not just financial actors but decisive players in determining whether global climate goals can be realistically achieved.

Global fossil fuel debt markets amount to trillions of dollars, illustrating how deeply the global economy is intertwined with carbon-intensive industries. These financing structures make decarbonization difficult without coordinated policy action and collective accountability across borders. This dynamic is not limited to oil and gas but also sustains manufacturing sectors that rely heavily on fossil fuels, such as fashion, electronics, and agriculture (Rickman et al, 2024). In 2021, to comply with the Paris Agreement, banks committed to net-zero targets and pledged to "decarbonize" their lending portfolios. However, the authors found that fossil fuel lending has shown no significant decline over the last decade (Rickman et al, 2024). The persistence of such financing highlights the gap between institutional pledges and practical outcomes, underscoring the urgency of aligning financial systems with meaningful climate action.

TRADE AGREEMENTS (NAFTA, WTO, USMCA) PROMOTING LARGE-SCALE INDUSTRIAL EXPANSION WITH WEAK ENVIRONMENTAL OVERSIGHT

Trade liberalization via preferential trade agreements has significantly reshaped production patterns and redistributed environmental impacts to middle-income countries. Kolkata, et al. (2019) analyzed 183 countries between 1987 and 2013 and found that as trade participation deepens, ecological footprints in middle-income nations increase. This trend underscores the uneven environmental burdens of globalization, in which developing economies often absorb the ecological costs of production that support wealthier nations' consumption patterns.

While agreements such as NAFTA, its successor USMCA, and WTO frameworks have facilitated economic integration, they have also allowed multinational corporations to offshore production to countries with weaker environmental oversight (Kolcava et al., 2019). The study observed that even trade agreements containing environmental provisions failed to prevent this shift consistently. These findings highlight the limitations of voluntary or weakly enforced environmental clauses, which inadequately balance the economic advantages of trade with ecological responsibility.

Such agreements provide U.S.-based MNCs opportunities to expand operations in contexts where environmental and labor protections are limited, increasing both production scale and the carbon footprint of goods consumed in the United States. In the absence of enforceable environmental regulations, trade liberalization perpetuates the ecological and social costs of global consumer culture (Kolcava et al., 2019). Consequently, consumer demand in developed countries drives unsustainable industrial growth abroad, reinforcing global inequalities in labor conditions and environmental justice.

The apparel industry illustrates how trade agreements can accelerate large-scale industrial expansion while externalizing environmental and labor impacts. Oh et al. (2003) found that NAFTA incentivized U.S.-based multinational fashion companies to relocate production facilities to Mexico. While the agreement ostensibly aimed to foster economic integration, it effectively transferred substantial carbon emissions and labor pressures abroad. This example demonstrates how economic integration policies, when not paired with strong regulatory frameworks, can externalize social and environmental harms to trading partners.

This dynamic was propelled by MNCs' pursuit of lower labor costs, faster production cycles, and more lenient environmental and labor standards in Mexico, facilitating the high-speed, low-cost production model characteristic of fast fashion. Trade agreements thus function as structural mechanisms that enable the globalization of industries like fashion, where economic gains for corporations are closely tied to the displacement of ecological and social costs to less-regulated regions (Oh et al., 2003). In effect, trade agreements became tools for accelerating the globalization of fast fashion, embedding unsustainable practices within the broader structures of international trade.

Together, these developments show that the evolution of American consumer culture, amplified by technological change and corporate expansion, has become a central driver of global environmental strain by transforming personal identity, market behavior, and international production into interconnected sources of rising ecological pressure.

Part 4. GLOBALIZATION AND CLIMATE CHANGE: THE U.S. ROLE IN RISING CO_2 EMISSIONS

The evolution of American consumer culture, reinforced by technological innovation and corporate expansion, has become a major force behind global environmental pressures by shaping personal identity, market behavior, and international production in ways that intensify ecological strain. As these consumption patterns expanded and intertwined with global trade networks, their environmental impact became increasingly evident in the steady rise of worldwide carbon emissions. Over the past seven decades, CO_2 levels have escalated due to industrial growth, trade liberalization, and the mounting global demand for energy. Although climate awareness and renewable technologies have advanced since the early twenty first century, the global economy remains firmly dependent on fossil fuels. This tension is especially visible in the United States, a historical leader in emissions and a central figure in climate diplomacy, where economic priorities continue to complicate commitments to environmental sustainability. Ongoing questions also persist about whether U.S. emissions reductions reflect substantive progress or simply the relocation of carbon intensive production to other nations, underscoring the challenges of aligning economic systems with ecological responsibility.

The post-World War II period marked a turning point in global CO_2 emissions. From the 1950s to the 1970s, industrial expansion, particularly in the United States, Western Europe, and Japan, drove a surge in energy demand. This era witnessed the proliferation of petroleum-based transportation, mass suburbanization, and oil-powered manufacturing. The global consumption of oil increased from about 10 million barrels per day in 1950 to nearly 60 million barrels by 1973 (International Energy Agency, 2019). These decades set the stage for a deeply entrenched fossil fuel economy.

The 1980s through the 2000s saw the acceleration of trade liberalization and the globalization of supply chains. As manufacturing shifted to countries like China and India, global emissions rose despite emissions stabilizing in some developed economies. The outsourcing of production allowed countries like the U.S. to maintain high consumption levels while attempting to control emissions domestically. This period also saw a significant increase in CO_2 emissions linked to international trade, with studies suggesting that nearly a quarter of global emissions in the 2000s were tied to goods exported to developed countries (Davis & Caldeira, 2010).

From the 2010s onward, the narrative shifted toward sustainability, and renewable energy technologies, primarily wind and solar, gained traction. Investment in clean energy surged, and the share of renewables in global electricity generation rose from 20% in 2010 to about 30% in 2021 (International Renewable Energy Agency, 2022). However, fossil fuels still supply over 80% of the world's primary energy (International Energy Agency, 2021), and global emissions continue to rise, albeit at a slower pace. Despite clean energy innovations, increasing energy demand from emerging economies and continued dependence on oil, gas, and coal have hindered deep decarbonization.

US CLIMATE AWARENESS AND POLICY RESPONSES

Awareness of climate change as a scientific and policy issue began growing in the 1980s. In 1988, the United Nations established the Intergovernmental Panel on Climate Change (IPCC) to assess climate science and inform policy. That same year, NASA scientist James Hansen testified before Congress, stating with "99% certainty" that global warming was underway due to greenhouse gas emissions (Hansen, 1988). This marked a turning point in public and political recognition of the climate crisis.

Despite growing awareness, U.S. policy remained inconsistent. The 1997 Kyoto Protocol marked the first significant international agreement to mandate reductions in emissions. Still, the U.S., under President George W. Bush, refused to ratify it, citing concerns about economic competitiveness and the exclusion of developing countries from binding targets (Victor, 2001). This reluctance hampered global progress and revealed the tension between environmental commitments and national economic interests.

In contrast, the U.S. played a key role in shaping the 2015 Paris Agreement, which set voluntary national targets (Nationally Determined Contributions, or NDCs) for emissions reductions. President Obama committed the U.S. to cutting emissions by 26–28% below 2005 levels by 2025. However, this progress was undermined by the Trump administration's 2017 withdrawal from the agreement, which only allowed the Biden administration to rejoin in 2021 (United Nations Framework Convention on

Climate Change, 2021). This back-and-forth illustrates the fragility of U.S. climate commitments in the absence of binding domestic legislation.

In the contemporary era, efforts to balance economic growth with sustainability have grown more prominent. The rise of companies like Tesla and the widespread adoption of Environmental, Social, and Governance (ESG) metrics reflect a shift in corporate priorities. Tech-driven solutions, electric vehicles, carbon capture, and green hydrogen, have gained momentum, along with voluntary carbon offset programs and pledges to achieve “net-zero” emissions by mid-century.

However, critics argue that these efforts often fall short of systemic transformation. Net-zero goals are sometimes undermined by weak regulation, overreliance on offsets, and continued fossil fuel investments. As the Climate Action Tracker notes, few major economies, including the U.S., have policies that limit warming to 1.5°C (Climate Action Tracker, 2023). This disconnect raises serious doubts about the credibility of net-zero pledges and highlights the urgent need for stronger, enforceable climate policies.

Moreover, the issue of carbon outsourcing continues to cloud the picture. While U.S. territorial emissions have declined since their peak in the early 2000s, a significant portion of the country’s carbon footprint lies abroad. Products consumed in the U.S., from smartphones to apparel, are frequently manufactured in countries with more carbon-intensive energy systems. A 2014 study found that U.S. consumption-based emissions were 10–15% higher than reported territorial emissions (Peters et al., 2011), raising questions about the integrity of apparent emissions reductions.

The global rise in CO_2 emissions over the last 70 years reflects the challenges of reconciling industrial development with ecological limits. While the U.S. has played a central role in advancing climate science and participating in international climate negotiations, its policy responses have been inconsistent and often shaped by political shifts. Recent efforts toward sustainability and green innovation offer hope, but significant barriers remain, particularly in reducing fossil fuel dependency, achieving net-zero accountability, and addressing the carbon displacement embedded in global trade. A genuine transition to a low-carbon economy will require technological solutions, systemic changes to consumption patterns, trade policies, and a strong political will (see Figure 1).

Figure 1. Rise of globalization and CO_2 emissions

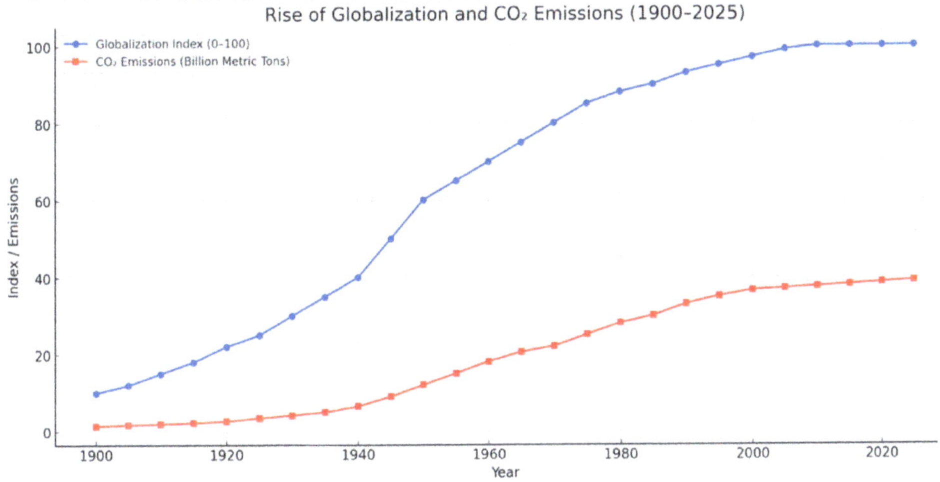

Rise of Globalization and CO_2 Emissions (1900–2025). The globalization index is a synthesized estimate representing global economic integration on a scale of 0–100, while CO_2 emissions are approximated in billion metric tons based on historical trends and projections. Data derived from Friedlingstein et al. (2022), the International Energy Agency (2021), and historical globalization research (Harvey, 2005; Friedman, 2005).

This graph illustrates the strong correlation between the rise of globalization and the growth of CO_2 emissions from 1900 to 2025. As global trade, production, and economic integration intensified, especially after World War II and during the neoliberal era in the 1980s, emissions from fossil fuels surged. While globalization brought economic growth and technological advancement, it also accelerated environmental degradation. Even amid sustainability efforts, the continued rise in CO_2 levels highlights the urgent need to decouple global economic activity from carbon emissions. If left unaddressed, this trend poses a serious threat to planetary stability and human well-being. (See Appendix A).

Taken together, these patterns reveal that rising emissions cannot be understood in isolation from the broader economic systems, cultural values, and political decisions that shaped postwar globalization, demonstrating that meaningful climate progress will require confronting the structural forces that link consumption, production, and energy use across national borders.

Part V: CHALLENGES AND FUTURE PATHWAYS FOR SUSTAINABLE GLOBALIZATION

There exists a paradox between economic expansion vs. carbon reduction. Figure 1 depicts the correlation between the rise of globalization and the growth of CO2 emissions (the economic expansion side of the paradox). Sustainability is defined as meeting the needs of today without compromising future generations (UN, 2024a). Sustainability involves reducing CO2 emissions. As globalization has increased over the years, CO2 emissions have also increased, directly related to increased travel, manufacturing, and shipping, which all require fossil fuels. So, based on what is consistent with Figure 1, globalization leads to greater levels of CO2 emissions. Now pair globalization with carbon reduction, it poses a challenge because it is globalization at the foundation that leads to increased CO2, which is the paradox we arrive at.

To decarbonize globalization, numerous changes are needed. For starters, the reduction of CO2, which will likely come from the decrease in use of fossil fuels. In the scope of logistics, this is likely to come from transportation, as logistics is made up of two main categories, transportation 80% and warehousing 20% (Tinnes et al., 2024). There has been an increase in the demand for green transportation, as the EU has required all cars produced in 2035 and later to be carbon neutral (European Parliament 2022). Ocean freight accounts for 21% of transportation emissions. Many of these freight lines have shown interest in switching to other energy sources, but where ocean freight requires such large amounts of power, most green avenues are not an option for them (Tinnes et al.). For many of these freight lines, the technology is not there yet, or it is too expensive for them to afford a switch currently. The challenges that exist and obstacles are considerable, but they are getting smaller as more people work towards bridging the gaps.

THE ROLE OF POLICY AND REGULATION

The United States faces difficulty passing environmental policies, with each administration possessing different views on climate change, fossil fuels, and ESG. This goes beyond just a federal issue in the United States; it includes divergence in states, which leaves states in control of the sustainability practices they create. California is leading the charge with two ESG laws, SB 253 and SB 261 (Tonello, 2025). As seen at the federal level, where political powers possess varying views towards environmental policies, the same is seen between states. Twenty-one states have a total of 40 laws restricting ESG efforts.

While the United States at its federal and state levels is disconnected regarding sustainability initiatives, other parts of the world are making great strides towards sustainability. The Paris Agreement is a commitment between 195 parties to reduce their emissions (UN, 2024b). The Green Deal is an agreement within the European Union to further cut emissions by 50% by 2030 (EU). China has adopted the Paris Agreement and is actively working towards carbon dioxide peaking and carbon neutrality through a 1+N policy framework (Ministry of Ecology and Environment of the People's Republic of China, 2022). These steps toward sustainability by global powers will push back against the climate challenges facing the world.

A significant barrier to resolving the paradox between globalization and carbon reduction lies in the persistent weakness and inconsistency of U.S. climate policy. As one of the world's largest historical emitters and a central architect of the global economic order, the United States has repeatedly struggled to adopt cohesive, long term climate strategies capable of aligning economic growth with environmental responsibility. Political polarization has produced abrupt policy reversals between administrations, limiting regulatory stability and undermining international credibility. Federal action has often stalled due to ideological disputes over the economic costs of environmental regulation, leading to fragmented governance in which states pursue divergent approaches and corporations face uneven expectations. Even when national climate commitments are made, they are frequently weakened by exemptions, limited enforcement, and the continued prioritization of fossil fuel interests. These dynamics reflect broader political economy constraints that impede clean energy transitions in advanced economies (Gazmararian & Tingley, 2025). These patterns have constrained the nation's ability to enact meaningful emissions reductions and have hindered global progress, since effective climate action requires leadership from major economic powers. The United States therefore exemplifies how political gridlock and fluctuating policy priorities complicate efforts to decarbonize globalization and address the structural forces that drive rising CO_2 emissions.

THE FUTURE OF ECONOMIC GROWTH IN A CARBON-CONSTRAINED WORLD

Many people associate carbon footprint with the hot topics of fossil fuels and CO2 emissions. This is an essential component; however, it is also important to understand what necessitates the extensive use of fossil fuels. The more common drivers that people have heard of are manufacturing and automobiles. There are several other drivers for CO2 emissions, such as AI and Crypto. The International

Monetary Fund (IMF) posted an article stating that crypto mining and data centers accounted for 2% of global electricity demand and approximately 0.5% of global CO2 emissions in 2022. It is estimated that by 2025, the global electricity demand for crypto mining and AI will increase to 3.5% (Hebous & Vernon-Lin, 2024). With AI growing as fast as it is, this is alarming, considering how quickly the global percentages are shifting towards crypto and AI.

Globalization has been a major contributor to growing CO2 emissions since World War II, as seen in Figure 1. Much of this growth in CO2 emissions is attributed to the dominant source of energy for much of the 20th and 21st centuries, which is fossil fuels. Fossil fuels continue to be a major contributor, with approximately 90% of CO2 emissions globally attributed to them (UN, 2024c). This doesn't mean that globalization cannot be driven by renewable energy, though, because it certainly can. Globalization has grown significantly to the point that we can talk to people around the world almost immediately. This has helped the spread of ideas and information about renewable resources. This has caused an influx in renewable energy investments as well as the diffusion of technologies around the globe (Hussein & Abdullahi, 2024). Globalization, being as widespread and growing as fast as it has, has enabled countries to have access to avenues of renewable energy that they would not have had access to otherwise.

Businesses are generally established to generate profit, and they often pursue this goal by minimizing production costs and maximizing operational efficiency. Under these conditions, firms are unlikely to adopt practices that increase expenditures unless such changes promise financial returns. External pressure from governments, investors, and consumers can, however, shift corporate priorities toward sustainability and initiate commitments associated with corporate social responsibility. As public expectations for renewable energy adoption and environmentally responsible practices have grown, corporations have increasingly responded, although their efforts frequently remain limited in scope. Many firms concentrate on reducing Scope 1 and Scope 2 emissions, which are tied to direct operations and purchased energy, yet they largely neglect Scope 3 emissions that account for the indirect environmental impacts embedded across global supply chains. This gap is significant, given that only 12 of the world's 100 most influential oil and gas companies have achieved reductions in Scope 1 and 2 emissions, and far fewer have implemented substantive measures targeting Scope 3 emissions, despite these emissions constituting the majority of their total climate footprint (WBA, 2023). Regulatory measures therefore must extend beyond operational boundaries to encompass entire value chains. While targeted policy instruments, such as a $0.047 per kilowatt hour tax on electricity consumption for crypto and AI operations, could help align energy use with global emissions goals (Hebous & Vernon-Lin, 2024), meaningful decarbonization will also require policies that address the upstream and downstream emissions that corporations continue to externalize.

Scope 3 emissions represent the most complex and expansive category of corporate greenhouse gas emissions because they include all indirect emissions associated with the production, distribution, use, and disposal of a company's goods and services (EPA, 2025). These emissions often constitute the largest share of a firm's carbon footprint, averaging roughly three quarters of total emissions and exceeding ninety percent in sectors such as agriculture, finance, and consumer goods (Hadziosmanovic et al., 2022). Despite their magnitude, Scope 3 emissions remain largely unregulated, with reporting requirements varying widely across national and international policy frameworks. This lack of standardization allows firms to highlight reductions in direct emissions while the majority of their climate impact persists within globalized supply chains. The resulting opacity contributes to carbon outsourcing, in which corporations present improved domestic emission profiles even as the environmental burden is shifted onto supplier regions. Addressing these shortcomings requires mandatory reporting standards, harmonized international accounting practices, and regulatory mechanisms capable of enforcing accountability across entire value chains so that corporate decarbonization strategies reflect comprehensive rather than selective environmental responsibility.

Taken collectively, these dynamics demonstrate that addressing the paradox between globalization and carbon reduction requires more than technological innovation or voluntary corporate efforts. It demands coordinated policy reform, meaningful regulatory enforcement, and a restructuring of global economic incentives so that growth is no longer coupled with rising emissions. Without such systemic change, the competing forces of economic expansion and environmental protection will continue to operate at cross-purposes, limiting progress toward a genuinely sustainable global future.

CONCLUSION

The trajectory of globalization since the mid-twentieth century demonstrates the deep entanglement between economic growth and environmental degradation. From the postwar foundations of the 1950s to the neoliberal acceleration of the 1980s and the digital globalization of today, U.S. leadership has shaped not only global commerce but also the carbon-intensive systems that sustain it. By embedding fossil fuel dependence into production, trade, and consumer culture, U.S.-led globalization has externalized ecological costs to the Global South while sustaining high-consumption lifestyles in the Global North. This pattern reveals that climate change is not simply the byproduct of industrial activity, but the structural outcome of a global economic system designed for expansion without adequate regard for environmental limits.

The historical record underscores that U.S.-led globalization advanced through key mechanisms that intensified carbon dependency. Trade liberalization shifted production to countries with weaker environmental standards, embedding emissions in global imports. Deregulation weakened environmental oversight and expanded extractive industries. Consumer culture normalized mass consumption and disposable goods, making material acquisition a marker of identity and success. Financialization, particularly through banking and capital markets, enabled fossil fuel companies and energy-intensive industries to access the credit and investment necessary to expand. Each of these structural drivers reveals how globalization not only facilitated economic integration but also perpetuated unsustainable production and consumption models that continue to drive climate change.

Attempts to address these problems through national emissions targets, voluntary corporate pledges, or market-based solutions have proven inadequate. Net-zero commitments, while symbolically powerful, often mask continued investments in fossil fuels and rely heavily on offsets that do little to transform structural dependencies. Trade agreements with environmental provisions, though promising on paper, have failed to prevent carbon outsourcing or to ensure accountability across borders. Meanwhile, consumers in developed nations continue to benefit from inexpensive goods produced in carbon-intensive conditions abroad. This disconnects between appearances and realities highlights the limits of fragmented, nation-centered approaches to a fundamentally global problem.

Moving forward, meaningful solutions must confront the embedded nature of carbon in global trade, finance, and production systems. Stronger international governance is required to enforce environmental provisions in trade agreements, regulate financial flows that sustain fossil fuel industries, and promote accountability for emissions across supply chains. Equitable frameworks must also address the disproportionate burden carried by developing nations, which often absorb the environmental costs of production for goods consumed in wealthier economies. Without such measures, the promises of sustainable globalization will remain rhetorical rather than real.

Yet, globalization also offers opportunities for transformation. Just as trade and technology accelerated the spread of fossil fuel dependency, they can also accelerate the diffusion of renewable energy, low-carbon technologies, and sustainable practices. Digital platforms can enhance supply chain transparency, while financial institutions can redirect capital flows toward green investments. Consumer awareness and pressure can reshape demand toward sustainable goods, while governments can enact binding regulations to hold corporations accountable. The paradox of globalization, that it both drives climate risks and provides pathways for solutions, underscores the need for systemic, coordinated action.

Ultimately, the U.S. faces a choice. As the architect and primary beneficiary of the current global economic order, it bears a particular responsibility to lead efforts toward decarbonization that extend beyond its borders. By reimagining globalization not as unchecked expansion but as a vehicle for sustainable development, the United States can help redefine the terms of global integration in the twenty-first century. The challenge is formidable, but the stakes could not be higher: the future of globalization must be aligned with planetary boundaries if it is to remain viable. Only through international cooperation, structural reform, and equitable responsibility-sharing can globalization evolve from a driver of carbon dependency into a framework for long-term climate resilience.

RERERENCES

Barlow, J., Lennox, G. D., Ferreira, J., Berenguer, E., Lees, A. C., Mac Nally, R., & Gardner, T. A. (2016). Anthropogenic disturbance in tropical forests can double biodiversity loss from deforestation. *Nature*, *535*(7610), 144–147. DOI: 10.1038/nature18326 PMID: 27362236

Climate Action Tracker. (2023). *Warming projections global update*. https://climateactiontracker.org

Davis, S. J., & Caldeira, K. (2010). Consumption-based accounting of CO_2 emissions. *Proceedings of the National Academy of Sciences of the United States of America*, *107*(12), 5687–5692. DOI: 10.1073/pnas.0906974107 PMID: 20212122

Earth.Org. (2025). Fast fashion and its detrimental effect on the environment. https://earth.org/fast-fashions-detrimental-effect-on-the-environment/

Environmental Protection Agency. (2025). Scope 3 inventory guidance. https://www.epa.gov/climateleadership/inventory-management-plan-guidanceUnited States

European Parliament. (2022). *EU ban on the sale of new petrol and diesel cars from 2035 explained*. https://www.europarl.europa.eu/topics/en/article/20221019STO44572/eu-ban-on-sale-of-new-petrol-and-diesel-cars-from-2035-explained

Fiveable. (2025). Rise of multinational corporations. https://library.fiveable.me/hs-honors-world-history/unit-11/rise-multinational-corporations/study-guide/vuhajAZdRmpceHVK

Friedlingstein, P., O'Sullivan, M., Jones, M. W., Andrew, R. M., Gregor, L., Hauck, J., Le Quéré, C., Luijkx, I. T., Olsen, A., Peters, G. P., Peters, W., Pongratz, J., Schwingshackl, C., Sitch, S., Canadell, J. G., Ciais, P., Jackson, R. B., Alin, S. R., Alkama, R., & Zheng, B. (2022). Global carbon budget 2022. *Earth System Science Data*, *14*(11), 4811–4900. DOI: 10.5194/essd-14-4811-2022

Friedman, T. L. (2005). *The world is flat: A brief history of the twenty-first century*. Farrar, Straus and Giroux.

Gazmararian, A. F., & Tingley, D. (2025). The political economy of the clean energy transition. University of Michigan and Harvard Kennedy School. https://www.alexgaz.org/files/ARPS.pdf

George, G., & Schillebeeckx, S. J. D. (2022). Digital transformation, sustainability, and purpose in the multinational enterprise. *Journal of World Business*. https://www.sciencedirect.com/science/article/pii/S1090951622000190

Global Media Journal. (2024). The rise of digital media: Transforming communication, culture, and commerce. https://www.globalmediajournal.com/open-access/the-rise-of-digital-media-transforming-communication-culture-and-commerce.pdf

Hadziosmanovic, M., Rahimi, K., & Bhatia, P. (2022). Trends Show Companies Are Ready for Scope 3 Reporting with US Climate Disclosure Rule.

Hansen, J. (1988). *Testimony to the U.S. Senate Committee on Energy and Natural Resources.*

Harvey, D. (2005). *A brief history of Neoliberalism.* Oxford University Press. DOI: 10.1093/oso/9780199283262.001.0001

Hebous, S., & Vernon-Lin, N. (2024). *Carbon emissions from AI and crypto are surging and tax policy can help.* https://www.imf.org/en/Blogs/Articles/2024/08/15/carbon-emissions-from-ai-and-crypto-are-surging-and-tax-policy-can-help

Hubacek, K., Baiocchi, G., Feng, K., & Patwardhan, A. (2017). Poverty eradication in a carbon constrained world. *Nature Communications*, *8*(1), 912. DOI: 10.1038/s41467-017-00919-4 PMID: 29066777

Hussein, O. A., & Abdullahi, A. M. (2024). Dynamic impacts of globalization and environmental performance on renewable energy. *Discover Sustainability*, *5*(1), 365. https://link.springer.com/article/10.1007/s43621-024-00604-9. DOI: 10.1007/s43621-024-00604-9

Immerman, R. H. (2003). *Culture, ideology, and U.S. foreign policy.* Pearson Longman.

International Energy Agency. (2019). *World energy balances overview.* https://www.iea.org

International Energy Agency. (2021). *World energy outlook 2021.* https://www.iea.org/reports/world-energy-outlook-2021

International Renewable Energy Agency. (2022). *Renewable capacity statistics 2022.* https://www.irena.org/publications

Iriye, A. (1979). Culture and international history. *Diplomatic History*, *3*(2), 115–128. DOI: 10.1111/j.1467-7709.1979.tb00305.x

Kaplan, L. S. (2014). *NATO 1948: The birth of the transatlantic alliance.* Rowman & Littlefield.

Kolcava, D., Nguyen, Q., & Bernauer, T. (2019). Does trade liberalization lead to environmental burden shifting in the global economy? *Ecological Economics*, *163*, 98–112. DOI: 10.1016/j.ecolecon.2019.05.006

Marglin, S. A. (1988). Lessons of the Golden Age: An overview. In Marglin, S. A., & Schor, J. B. (Eds.), *The Golden Age of capitalism: Reinterpreting the postwar experience* (pp. 1–38). Oxford University Press.

Ministry of Ecology and Environment of the People's Republic of China. (2022). *Annual report on China's policies and actions for addressing climate change (2022)*. https://english.mee.gov.cn/Resources/Reports/reports/202211/P020221110605466439270.pdf

MIT Press Reader. (2021). A brief history of consumer culture. https://thereader.mitpress.mit.edu/a-brief-history-of-consumer-culture/

Morgan, I. (2008). Reaganomics and its legacy. In Morgan, I. (Ed.), *Ronald Reagan and the 1980s: Perceptions, policies, legacies* (pp. 135–154). Palgrave Macmillan., DOI: 10.1057/9780230616196_7

Peters, G. P., Minx, J. C., Weber, C. L., & Edenhofer, O. (2011). Growth in emission transfers via international trade from 1990 to 2008. *Proceedings of the National Academy of Sciences of the United States of America*, *108*(21), 8903–8908. DOI: 10.1073/pnas.1006388108 PMID: 21518879

Prasad, M. (2006). *The politics of free markets: The rise of neoliberal economic policies in Britain, France, Germany, and the United States*. University of Chicago Press. https://books.google.com/books?hl=en&lr=&id=7OzkOJURFJIC

Prasad, M. (2012). The popular origins of neoliberalism in the Reagan tax cut of 1981. *Journal of Policy History*, *24*(3), 351–383. DOI: 10.1017/S0898030612000103

Rui, D., Duan, L., & Liang, N. (2023). *Outsourcing climate change: How firms relocate carbon emissions abroad to meet domestic policy goals. SSRN*. https://papers.ssrn.com/sol3/papers.cfm?abstract_id=3765485

Shen, L., & Ma, X. (2019). Globalization, economic growth, and environmental degradation: Evidence from 22 OECD countries. *Sustainable Development*, *27*(3), 590–600. DOI: 10.1002/sd.1913

Tinnes, E., Perez, F., Kandel, M., & Probst, T. (2024). *Decarbonizing logistics: Charting the path ahead*. McKinsey & Company. https://www.mckinsey.com/capabilities/operations/our-insights/decarbonizing-logistics-charting-the-path-ahead

Tonello, M. (2025). *Regulatory shifts in ESG: What comes next for companies?* Harvard Law School Forum on Corporate Governance. https://corpgov.law.harvard.edu/2025/04/12/regulatory-shifts-in-esg-what-comes-next-for-companies/

United Nations. (2024a). *Causes and effects of climate change*. https://www.un.org/en/climatechange/science/causes-effects-climate-change

United Nations. (2024b). *Sustainability*. United Nations Academic Impact. https://www.un.org/en/academic-impact/sustainability

United Nations. (2024c). *The Paris Agreement*. https://www.un.org/en/climatechange/paris-agreement

United Nations Framework Convention on Climate Change. (2021). The *U.S. officially rejoins the Paris Agreement*. https://unfccc.int/news/us-officially-rejoins-paris-agreement

Victor, D. G. (2001). *The collapse of the Kyoto Protocol and the struggle to slow global warming*. Princeton University Press.

World Benchmarking Alliance. (2023). *Corporate accountability: Closing the gap in pursuit of sustainable development*. https://assets.worldbenchmarkingalliance.org/app/uploads/2023/09/WBA2023_whitepaper_corporate_accountability_lr.pdf

Yergin, D. (2009). *The prize: The epic quest for oil, money & power*. Free Press.

Zhang, Z., Guan, D., Wang, R., Meng, J., Zheng, H., & Associates. (2020). Embodied carbon emissions in the supply chains of multinational enterprises. *Nature Climate Change*. https://www.nature.com/articles/s41558-020-0895-9

KEY TERMS AND DEFINITIONS

Carbon Footprint: The total amount of greenhouse gas emissions, especially CO_2, generated directly or indirectly by an individual, organization, product, or country.

Environmental Governance: The system of laws, regulations, policies, and institutions that guide human interactions with the environment, including mechanisms to enforce sustainability and mitigate ecological harm.

Fast Fashion: A business model in the apparel industry characterized by rapid production cycles, low-cost manufacturing, and frequent introduction of new styles, often contributing to environmental degradation.

Fossil Fuels: Natural carbon-based energy sources such as coal, oil, and natural gas, whose extraction and combustion are major contributors to greenhouse gas emissions.

Globalization: The increasing interconnectedness and interdependence of economies, cultures, and political systems through trade, investment, technology, and migration.

Multinational Corporation (MNC): A company that operates in multiple countries, often leveraging global labor and production networks to optimize profits and minimize regulatory constraints.

Neoliberalism: An economic and political philosophy emphasizing free markets, deregulation, privatization, and reduced government intervention in the economy, often associated with policies from the 1980s onward.

Offshoring: The practice of relocating business processes or production to another country, typically to reduce costs or bypass stricter environmental or labor regulations.

Supply Chain: The network of production, transportation, and distribution processes involved in bringing a product from raw materials to the end consumer.

Trade Liberalization: The removal or reduction of barriers to trade, such as tariffs, quotas, and regulations, intended to encourage international commerce and economic integration.

APPENDIX GLOBALIZATION AND CO_2 EMISSIONS (1900–2025): A CENTURY OF GROWTH AND ENVIRONMENTAL IMPACT

Globalization_and_CO__Emissions_Data

Year	Globalization Index (0–100)	CO_2 Emissions (Billion Metric Tons)
1900	10	1.5
1905	12	1.8
1910	15	2.0
1915	18	2.3
1920	22	2.7
1925	25	3.5
1930	30	4.2
1935	35	5.0
1940	40	6.5
1945	50	9.0
1950	60	12.0
1955	65	15.0
1960	70	18.0
1965	75	20.5
1970	80	22.0
1975	85	25.0
1980	88	28.0
1985	90	30.0
1990	93	33.0
1995	95	35.0
2000	97	36.5
2005	99	37.0
2010	100	37.5
2015	100	38.0
2020	100	38.5
2025	100	39.0

This chart presents data showing the parallel rise of globalization and carbon dioxide (CO_2) emissions from 1900 to 2025. Emissions surged as the world became increasingly interconnected through trade, technology, and economic policy, particularly after 1950 and again in the 1980s. The data highlights a critical trend: the expansion of the global economy has depended mainly on fossil fuels, contributing significantly to climate change and underscoring the urgent need for sustainable development strategies.

Chapter 12

Sampalok Lake Assessment and Monitoring Strategies Toward Sustainable Lake Ecosystems: A Qualitative Inquiry

Jherwin Pagkaliwagan Hermosa
https://orcid.org/0000-0001-8562-3028
Laguna State Polytechnic University, Philippines

Julie Fe D. Panoy
Laguna State Polytechnic University, Philippines

Michelle T. Flores
Laguna State Polytechnic University, Philippines

ABSTRACT

The study investigates assessment and monitoring strategies for Sampalok Lake in San Pablo City to promote sustainable practices. Qualitative research involving key informants reveals knowledge disparities and fragmented monitoring systems, as well as emerging threats such as microplastic contamination. Community initiatives like clean-up campaigns and youth involvement show potential in addressing these challenges. The research highlights the need for a collaborative, multidisciplinary approach incorporating scientific advancements, indigenous knowledge, and policy coordination. The study discusses various suggested measures to safeguard the eco-

DOI: 10.4018/979-8-3373-6435-3.ch012

logical health of lakes in the Philippines. Among these are setting up a collaborative lake monitoring group, promoting community-based scientific ventures, launching a university-driven extension program, utilizing real-time monitoring technology, and strengthening public outreach campaigns. These approaches underscore the significance of employing diverse and adaptable methods to guarantee the protection of lakes across the nation.

INTRODUCTION

Lakes are vital inland water bodies that play a crucial role in maintaining ecological balance and supporting socio-economic development. Globally, they serve as sources of freshwater, regulate local climates, provide habitats for diverse species, and offer cultural and recreational value. In the Philippine context, lakes are integral to local livelihoods, especially through fisheries, ecotourism, and agriculture (DENR-EMB, 2021). Among these is Sampalok Lake, the largest and most accessible of the Seven Lakes of San Pablo City in Laguna province. It is a volcanic maar lake that has long served the city's residents for aquaculture, recreation, and aesthetic appreciation (Philippine Institute for Volcanology and Seismology [PHIVOLCS], 2015).

Despite its ecological and cultural importance, Sampalok Lake is facing a growing number of environmental challenges. Factors such as urbanization, insufficient waste management practices, unregulated aquaculture activities, and encroachment on its shoreline have all played a role in the decline of water quality, eutrophication, and occasional fish die-offs (DENR-EMB, 2021). Furthermore, the rise in tourism-related activities, while economically advantageous, has placed additional strain on the lake's capacity and threatens its long-term sustainability. These issues underscore the urgent need for comprehensive monitoring and evaluation strategies that are both scientifically rigorous and inclusive of all stakeholders. Efficient lake monitoring requires the regular observation of various water quality parameters such as pH, dissolved oxygen, turbidity, and nutrient levels. While modern technologies like remote sensing, automated sensors, and GIS mapping provide useful tools for assessing ecological conditions (Dekker & Peters, 1993), constraints on resources within local governments and rural areas often limit their consistent use. Therefore, the adoption of community-based monitoring approaches, where local residents are trained to contribute to environmental observations, has emerged as a practical and empowering solution (Danielsen et al., 2005). In the case of Sampalok Lake, community participation has historically been fostered through initiatives like the "*Yakap sa Lawa*" movement, which embodies the community's dedication to environmental stewardship.

This qualitative study seeks to examine the current strategies for assessing and monitoring Sampalok Lake and determine their effectiveness in promoting sustainability. By involving various stakeholders such as city environmental officers, barangay leaders, fisherfolk, and civil society representatives, this research aims to capture their experiences and perspectives on lake governance. The research also delves into the institutional, technical, and cultural factors that impact local monitoring practices.

In line with global sustainability frameworks like Sustainable Development Goal (SDG) 6 and SDG 15, this study emphasizes the relationship between ecological monitoring, participatory governance, and environmental education. The findings from this localized study will contribute to broader discussions on freshwater conservation, adaptive management, and environmental policy in developing countries.

OBJECTIVES OF THE STUDY

The main objective of this qualitative study is to investigate and evaluate the assessment and monitoring methods currently used to manage Sampalok Lake in San Pablo City, Laguna, in order to enhance its ecological sustainability.

More specifically, the research aims to:

1. Assess and describe the current lake assessment and monitoring protocols utilized by local government entities, environmental organizations, and community participants in place.
2. Analyze the viewpoints, encounters, and understandings of key players such as fishermen, locals, environmental authorities, and advocacy groups regarding the efficiency of these monitoring methods.
3. Assess the hindrances and constraints encountered during the execution of lake monitoring and assessment activities.
4. Emphasize exemplary methods and community-driven ventures that aid in the sustainable oversight of Sampalok Lake.
5. Propose comprehensive and engaging approaches to enhance monitoring mechanisms and endorse long-lasting environmental balance and policy formulation for the lake.

Review of the Literature

Ecological, Environmental and Historical Context of Sampalok Lake

The ecological and environmental importance of Sampalok Lake, the largest of the Seven Lakes of San Pablo City, is widely acknowledged for being a maar-type volcanic crater lake. This geomorphological structure is known for its ecological sensitivity and closed-basin hydrology (Zlotnicki et al., 2015; DENR–EMB, 2021). Lakes of this type are particularly at risk of nutrient buildup and ecological imbalance due to limited water circulation, making them highly vulnerable to both natural and human-induced disruptions. Located within a rapidly expanding city, Sampalok Lake has faced continuous environmental pressures from factors such as population growth, tourism development, and increased land-use changes in its surrounding area. Ongoing research confirms that the decline in water quality in Sampalok Lake is predominantly attributed to the impact of urban runoff, domestic wastewater discharge, encroachment along the shoreline, and the expansion of aquaculture activities (Brillo, 2017; PCAARRD, 2022; DENR-EMB, 2021). Runoff from nearby residential and commercial areas carries sediments, nutrients, oils, and solid waste into the lake, resulting in increased turbidity and nutrient levels. Further exacerbating the situation is the absence of a comprehensive sewerage system in neighboring communities, leading to the direct or indirect discharge of untreated household waste into the lake and hastening the process of eutrophication.

Throughout history, aquaculture has greatly influenced the trophic state of Sampalok Lake. Early studies conducted by Lara et al. (1997) revealed concerning increases in nutrient levels due to the expansion of fish cages and unsustainable feeding practices. The accumulation of excess feed and fish waste resulted in elevated levels of phosphorus and nitrogen, leading to algal blooms and fish mortality events. These issues align with global research on cage aquaculture in closed freshwater systems, highlighting the potential for nutrient overloading to cause hypoxia, loss of biodiversity, and ecosystem instability (Beveridge, 2004; Troell et al., 2014). The lake's transition to a eutrophic to hypereutrophic state has been observed through long-term monitoring, revealing a decline in dissolved oxygen levels, an increase in chlorophyll-a concentrations, and a decrease in water transparency (DENR-EMB, 2021; PCAARRD, 2022). These environmental changes have had a detrimental impact on aquatic biodiversity and have jeopardized the lake's ability to support ecosystem services such as fisheries, recreation, and aesthetic value. Native fish species and plankton diversity have gradually been replaced by more resilient and often invasive species that thrive in low-oxygen and nutrient-rich conditions.

In terms of historical governance, the management of Sampalok Lake illustrates the enduring conflict between economic development and environmental preservation. The rapid growth of aquaculture in the 1980s and 1990s led to the establishment of fish cage operations that offered valuable economic opportunities for the local community. However, the lack of proper regulatory oversight resulted in uncontrolled expansion and overstocking, surpassing the lake's ecological limits. Despite subsequent efforts to implement regulatory measures and rehabilitate the lake, such as zoning fish cages, conducting periodic clean-ups, and monitoring water quality, the effectiveness of these initiatives has been hindered by enforcement gaps and overlapping institutional responsibilities (Brillo, 2017). Recent efforts to rehabilitate the ecosystem of the lake have led to some positive changes, such as a slight decrease in nutrient levels and an improvement in water clarity. These improvements were observed after reducing fish cage operations and clearing shorelines. However, it is important to note that the presence of aquaculture structures, informal settlements near the lake, and inadequate wastewater management systems pose ongoing threats to the lake's sustainability. The issue is further compounded by climate change, which leads to more intense rainfall, increased surface runoff, and higher water temperatures. These factors contribute to the worsening of eutrophication and oxygen depletion in the lake (DENR–EMB, 2021).

The combined influences of ecological, environmental, and historical factors on Sampalok Lake reveal a intricate interplay between natural susceptibility and ongoing human impact. Although policy changes and restoration initiatives provide avenues for improvement, the continued presence of foundational challenges—such as aquaculture, urban development, and governance constraints—underscores the critical necessity for a comprehensive, evidence-based, and community-supported approach to lake management. This is vital to preserving the enduring ecological health of this valuable freshwater ecosystem, which holds significant cultural and economic importance.

Governance, Policy and Institutional Challenges

Assessments of governance in the Seven Lakes of San Pablo City utilizing the Integrated Lake Basin Management (ILBM) framework consistently highlight significant institutional and policy challenges, especially in Sampalok Lake. Brillo (2017) identified key governance weaknesses, such as fragmented and overlapping institutional responsibilities, a reactive rather than proactive management approach, inadequate inter-agency coordination, and limited utilization of science and technology at the local government and community levels. These challenges, viewed through the ILBM lens, point to broader systemic deficiencies across fundamental governance pillars, including institutions, policies, participation, information, and technology. Despite the presence of national environmental policies like the Philippine Clean Water Act of 2004 (RA 9275), the Ecological Solid Waste Management

Act (RA 9003), and the Philippine Environmental Impact Statement (EIS) System (PD 1586) providing a solid legal foundation for water resource protection, the implementation of these policies in Sampalok Lake reveals inconsistencies and limitations (DENR–EMB, 2021; Brillo, 2017). Persistent gaps in enforcement are attributed to deficient technical expertise, inadequate environmental monitoring tools, limited financial resources, and heavy reliance on temporary, project-specific funding. Local government units (LGUs), tasked with the primary responsibility for environmental stewardship under the Local Government Code of 1991 (RA 7160), often struggle due to a lack of trained personnel in limnology, water quality assessment, and environmental engineering, thereby diminishing the efficacy of regulatory measures.

Governance is further complicated by institutional fragmentation, with various agencies such as the Department of Environment and Natural Resources (DENR), Laguna Lake Development Authority (LLDA), Bureau of Fisheries and Aquatic Resources (BFAR), and the city government of San Pablo all having jurisdiction over lake management functions. This leads to overlapping and sometimes conflicting policies, duplication of efforts, and slow decision-making, particularly in areas like aquaculture regulation, shoreline development, and wastewater discharge (Brillo, 2017). The lack of a fully empowered and dedicated lake basin authority for the Seven Lakes system hinders long-term planning and accountability. Governance challenges are heightened at the community level due to limited stakeholder participation and low environmental awareness, particularly among fisherfolk, informal settlers, and small-scale tourism operators residing near Sampalok Lake. Research highlights that while mechanisms for participation, such as lake councils and multi-sectoral monitoring teams, are in place, they mainly serve a consultative role rather than facilitating decision-making. This lack of meaningful engagement hinders local stakeholders from actively shaping policies and enforcing regulations, leading to a diminished sense of ownership over conservation efforts and ongoing non-compliance with environmental laws, particularly regarding fish cage management and waste disposal practices (Brillo, 2017; PCAARRD, 2022).

The reliance on manual water quality sampling, sporadic monitoring schedules, and insufficient data-sharing platforms at Sampalok Lake creates a significant institutional barrier due to technological constraints. The lack of real-time water quality monitoring systems, GIS-based land-use tracking, and integrated environmental databases hinders authorities from implementing effective evidence-based and adaptive management strategies. This technological gap accentuates the reactive approach to governance, as interventions are typically initiated only in response to observable ecological emergencies like fish kills and algal blooms (DENR-EMB, 2021; World Bank, 2023). Additionally, governance struggles are exacerbated by socio-political and economic pressures, notably the dependence of local populations on aquaculture and tourism for their livelihoods. Measures like reducing fish cages and clearing shorelines often face opposition due to concerns about their impact on income security, highlighting the ongoing struggle to balance environmental sustainability with local economic needs (Brillo, 2017). This ongoing dilemma frequently leads to compromises that prioritize immediate economic benefits over the long-term restoration of ecological balance.

Community-Led Monitoring Strategies

Given the constraints of government resources, community-led approaches to lake management have proven to be effective and sustainable. Danielsen et al. (2005) highlight the importance of involving local communities in monitoring biodiversity and water quality, citing benefits such as consistent data collection, cost-efficiency, community ownership, and lasting conservation impact. Other studies support this claim, showing that community-based monitoring systems not only improve adaptive management through relevant data collection but also promote environmental responsibility among stakeholders (Conrad & Hilchey, 2011; Bonney et al., 2016). In the Philippine setting, participatory environmental governance is embodied in community-based natural resource management (CBNRM) efforts, where local stakeholders are viewed as active partners rather than mere recipients of benefits (DENR–EMB, 2018). Specifically, in the vicinity of Sampalok Lake and the broader Seven Lakes system, community members have been crucial participants in conservation projects like the "Yakap sa Lawa" initiative in San Pablo City. This grassroots campaign involved the mobilization of residents, students, civic groups, and local authorities to collaborate on regular clean-up efforts, educational outreach, and advocacy for responsible lake utilization. The project not only enhanced environmental consciousness but also fostered collective ownership of the region's natural resources (Save the Lakes Movement, 2018; Brillo, 2017).

Community participation in monitoring and enforcing regulations goes beyond simply cleaning up environmental hazards. In addition to organized clean-up efforts, local communities have taken on the role of informally surveilling illegal activities such as unregulated fish cage expansion, improper waste disposal, and encroachment on shorelines. Research on co-management strategies in freshwater systems in the Philippines has revealed that when community members are actively engaged in monitoring and reporting violations, there is an increase in regulatory compliance and a decrease in environmental degradation (Pomeroy & Rivera-Guieb, 2006). Furthermore, participatory monitoring promotes social learning, enabling community members to continually improve their knowledge of ecological processes and recognize the harmful impacts of unsustainable practices. While community-led initiatives like "Yakap sa Lawa" showcase the capabilities of grassroots environmental action, the sustainability of these efforts relies heavily on ongoing institutional backing, capacity building, and policy alignment. Scholars emphasize the importance of formal recognition, technical training, and stable funding for community monitoring programs to avoid being merely symbolic or short-lived (Danielsen et al., 2005; World Bank, 2023). Therefore, the institutionalization of community-based monitoring within established lake governance structures for Sampalok Lake, supported by local regulations, technical expertise, and academic collaboration, is essential for maximizing its lasting ecological and governance benefits.

Governance and Policy Frameworks

The legal and policy framework governing inland waters in the Philippines is influenced by various national laws that set regulatory standards and institutional responsibilities. One key legislation is the Philippine Clean Water Act of 2004 (RA 9275), which requires systematic water quality monitoring, establishment of water quality management areas, regulation of wastewater discharges, and inter-agency coordination for water pollution control. Additionally, the Local Government Code of 1991 (RA 7160) grants significant authority to local government units (LGUs) to uphold environmental laws, manage local water bodies, oversee land use, and implement pollution control measures. Together, these laws establish a strong governance structure for protecting inland water resources. Despite the robust legal framework in place, policy implementation continues to face challenges due to institutional and operational inadequacies. The Asian Development Bank (ADB, 2018) has identified ongoing governance deficiencies, such as poor coordination between national agencies and local government units (LGUs), deficiencies in technical environmental personnel, ineffective monitoring and enforcement mechanisms, and inadequate funding for environmental initiatives. The World Bank (2023) has also documented similar shortcomings, noting the lack of laboratory facilities for water testing, real-time monitoring systems, and properly trained environmental officers in many LGUs. These limitations severely hamper the successful implementation of RA 9275 at the local level.

The presence of overlapping mandates among national agencies such as the Department of Environment and Natural Resources (DENR), Bureau of Fisheries and Aquatic Resources (BFAR), Laguna Lake Development Authority (LLDA), and Local Government Units (LGUs) poses a challenge to inland water governance. These institutional overlaps often result in regulatory gaps, duplicated functions, and inconsistent enforcement, particularly in the areas of aquaculture, wastewater discharge, and shoreline development (Brillo, 2017). Environmental compliance is often compromised due to fragmented accountability and ambiguous jurisdictional boundaries. Current literature emphasizes the significance of multi-stakeholder collaboration in addressing governance constraints. Lasco et al. (2017) illustrate that effective sustainable lake management in the Philippines requires coordination among government agencies, local communities, academic institutions, non-government organizations, and the private sector. Collaborative efforts promote knowledge-sharing, resource pooling, conflict resolution, and collective compliance, particularly in environmentally delicate lake ecosystems. Pomeroy and Rivera-Guieb (2006) also support these findings, suggesting that co-management arrangements enhance environmental outcomes and social equity in small-scale fisheries and inland water systems.

Additionally, it is becoming more widely recognized that adaptive and collaborative governance is crucial for effectively addressing current challenges such as urbanization and climate change. The impact of climate-related stressors like increased rainfall, flooding, and rising temperatures is exacerbating pollution levels and eutrophication in closed inland lakes, putting further pressure on already fragile local institutions (DENR–EMB, 2021). This situation highlights the growing necessity for integrated, multi-level, and participatory governance strategies that align national policy goals with local capabilities and community-driven initiatives. In general, the Philippines has a thorough legal and policy framework for managing inland waters. However, its effectiveness is hampered by issues such as coordination problems, technical and financial limitations, and institutional fragmentation. Improving implementation will necessitate greater investment in local technical expertise and monitoring infrastructure, as well as the establishment of collaborative, multi-stakeholder governance mechanisms to facilitate sustainable and adaptable lake management.

Promoting Sustainability through SDGs

The United Nations Sustainable Development Goals (SDGs) offer an internationally recognized framework for promoting environmental sustainability, inclusive governance, and the long-term protection of resources. Regarding the management of inland water and lakes, SDG 6 (Clean Water and Sanitation) and SDG 15 (Life on Land) are particularly pertinent. These goals focus on safeguarding freshwater resources, reducing water pollution, managing ecosystems sustainably, and conserving biodiversity (UNDP, 2015; United Nations, 2016). SDG 6 specifically calls for enhancing water quality by reducing pollution, minimizing untreated wastewater, and improving water-related ecosystems such as lakes and wetlands. In contrast, SDG 15 highlights the sustainable utilization of terrestrial and inland freshwater ecosystems and aims to halt biodiversity loss. It is crucial to harmonize community lake management efforts with the Sustainable Development Goals (SDGs) to ensure that sustainability endeavors are not only environmentally responsible but also socially inclusive and institutionally integrated. The United Nations Development Programme (UNDP, 2017) emphasizes the importance of localizing the SDGs to turn global commitments into tangible actions at the grassroots and municipal levels. For lakes like Sampalok Lake, this entails incorporating SDG targets into local development strategies, environmental regulations, water quality monitoring initiatives, and community-driven conservation plans. By embedding the SDGs within local governance frameworks, they become fundamental principles for future planning instead of stand-alone environmental ventures.

Furthermore, the SDG framework promotes a holistic approach to sustainability, where environmental conservation is intricately linked to poverty alleviation, healthcare, education, and economic progress. The degradation of freshwater ecosystems has direct implications for food security, public health, and the livelihoods of local communities, aligning with various SDG goals such as SDG 1 (No Poverty), SDG 3 (Good Health and Well-being), and SDG 8 (Decent Work and Economic Growth) (UN-Water, 2018). Sustainable management of lakes not only preserves ecological balance but also contributes to broader human development outcomes. For example, safeguarding Sampalok Lake supports industries like fisheries, tourism, climate resilience, and disaster risk reduction, all of which are interconnected with multiple SDG targets. The SDGs highlight the importance of engaging multiple stakeholders in governance, particularly under SDG 17 (Partnerships for the Goals), which is essential for effectively managing complex ecological systems like lakes. Research indicates that sustainability efforts that prioritize strong partnerships between government entities, civil society, the private sector, and local communities tend to be more enduring and impactful in the long run (Sachs et al., 2019; UNDP, 2020). This collaborative approach mirrors the community-based monitoring initiatives and cooperative governance models already being adopted in the management of the Seven Lakes of San Pablo City.

Despite the comprehensive nature of the Sustainable Development Goals (SDGs), scholars have noted significant implementation gaps at the local level, especially in developing countries. Factors such as limited financial resources, weak institutional capacity, lack of localized indicators, and insufficient data for monitoring can impede the achievement of SDG targets in inland water governance (ADB, 2018; UNDESA, 2021). To effectively realize the SDGs in lake ecosystems like Sampalok Lake, a context-sensitive approach that includes localization, sustained political commitment, adequate financing, and strong community engagement is essential. While the SDGs offer a valuable normative and strategic framework, their successful implementation requires tailored efforts at the local level. In essence, the SDGs function not just as aspirational objectives but as a comprehensive sustainability blueprint that can direct efforts to safeguard and restore Sampalok Lake. By purposely harmonizing local lake management initiatives with SDG goals and metrics, there is a bolstered policy consistency, global significance, and enduring focus on sustainability, which bolsters both environmental preservation and human welfare.

Methodology

Research Design

This study utilized a qualitative descriptive research design to investigate and comprehend the viewpoints and firsthand experiences of key stakeholders engaged in the oversight and maintenance of Sampalok Lake in San Pablo City, Laguna. The qualitative methodology was chosen for its ability to delve deeply into intricate insights and multifaceted occurrences that cannot be easily quantified, such as stakeholder perceptions, governance challenges, and community engagement (Creswell & Poth, 2016).

Research Locale

The research was carried out at Sampalok Lake, the largest and most easily accessible of the Seven Crater Lakes situated in San Pablo City, Laguna. The lake is enveloped by a blend of residential, commercial, and tourism-related activities and has historically been affected by aquaculture practices, solid waste disposal, and encroachment on its land.

Participants and Sampling Technique

Participants were chosen using purposive sampling to ensure that individuals with pertinent knowledge, involvement, or authority in lake management were included. The participants were as follows:

- Officials from the City Environment and Natural Resources Office (CENRO)
- Representatives from the Department of Environment and Natural Resources (DENR)
- Local government units at the barangay and city-level
- Fisherfolk and fish pen operators
- Environmental NGOs and youth volunteers
- Long-term residents residing along the lakeshore
- A total of 15 participants were interviewed until data saturation was achieved.

Methods of Data Collection

The data gathering process involved the utilization of various qualitative research techniques:

- Semi-structured Interviews: Utilized open-ended questions to allow participants to elaborate on their perspectives and experiences.
- Focus Group Discussions (FGDs): Conducted to delve into group dynamics and shared perceptions.
- Field Observations: Included observations of the lake environment, monitoring stations, and aquaculture structures.
- Document Analysis: Involved the examination of local ordinances, monitoring reports, and environmental assessments pertaining to Sampalok Lake.

All interviews and FGDs were carried out in either Filipino or English, with the participants' consent for audio recording and subsequent transcription for analysis purposes.

Data Analysis

Thematic analysis was utilized for the qualitative data examination. Following Braun and Clarke's (2006) six-phase framework, the transcripts underwent the following steps:

- Familiarization: The transcripts were read and re-read for familiarization.
- Coding: Systematic coding was applied.
- Theme and Subtheme Grouping: Transcripts were grouped into themes and subthemes.
- Review and Refinement: Themes were reviewed and refined.
- Definition and Naming: Themes were defined and named.
- Reporting: Illustrative quotations were included in the report.

Ethical Considerations

The study followed ethical research standards. Participants were informed about the research goals and voluntarily signed informed consent forms. Privacy and confidentiality were maintained throughout the research. Approval from the relevant Ethics Review Committee was obtained before collecting data.

RESULTS AND DISCUSSION

Understanding Stakeholder Awareness of Lake Monitoring

Conversations with important stakeholders unveiled differing degrees of understanding and awareness concerning the environmental monitoring activities conducted at Sampalok Lake. Officials from the Department of Environment and Natural Resources (DENR) and the Community Environment and Natural Resources Office (CENRO) underscored the significance of scientifically recognized criteria such as biochemical oxygen demand (BOD), total dissolved solids (TDS), and nutrient loading as principal markers of the lake's ecological status. These markers play a crucial role in identifying initial signs of pollution, eutrophication, and overall strain on the ecosystem. Alternatively, nearby inhabitants and fishermen primarily depended on observable and experiential signs to evaluate water quality, such as turbid water, the existence of algae blooms, unpleasant smells, and a visible decrease in the number of fish. This perception-driven evaluation demonstrates the way communities instinctively supervise environmental shifts through daily engagement with the lake. This discovery aligns with Brillo's (2017) research, which highlights that individuals within communities frequently assess the well-being of lakes based on firsthand encounters rather than relying on technical or laboratory data. The importance of public education in enhancing the understanding and appreciation of water quality indicators, especially in lake-dependent communities, has been underscored by the Philippine Council for Agriculture, Aquatic, and Natural Resources Research and Development (PCAARRD, 2022). The disparity between scientific monitoring conducted by institutions and visual assessments carried out by communities highlights a significant gap in knowledge between technical data and public comprehension. This gap hinders the community's ability to participate effectively in evidence-based decision-making and proactive environmental measures. Therefore, the results suggest a necessary implementation of systematic strategies for translating knowledge effectively to bridge the gap between scientific monitoring and community understanding. These strategies could encompass simplified reporting tools, user-friendly water quality scorecards, participatory monitoring initiatives, and consistent environmental education programs. Closing this divide is crucial not only for enhancing community engagement but also for promoting joint accountability and informed stewardship in the sustainable oversight of Sampalok Lake.

Evaluation of Current Monitoring Methods

The current lake monitoring systems for Sampalok Lake face notable restrictions in terms of consistency, collaboration, technological incorporation, and openness, as per research findings. Presently, environmental monitoring predominantly relies on intermittent water sampling carried out by personnel from the Department of Environment and Natural Resources (DENR) and the Local Government Units (LGUs). Although these initiatives yield fundamental water quality data, they are marked by irregular sampling frequencies, limited geographical reach, and delayed reporting. Consequently, the rapid identification of sudden ecological disruptions such as algal blooms and fish kill incidents is hindered. Furthermore, the current monitoring strategy heavily relies on manual data collection, lacking integration of automated sensors, real-time monitoring systems, or digital data transmission platforms. This technological gap hinders adaptive management and swift response mechanisms. These shortcomings align with the conclusions of Maravilla et al. (2019), who highlighted the advantages of utilizing unmanned aerial vehicles (UAVs), geographic information systems (GIS), and remote sensing technologies to improve the spatial and temporal resolution of lake monitoring in Philippine freshwater systems. These technologies enable continuous observation of surface water conditions, identification of algal blooms, shoreline changes, and assessment of land-use impacts within the lake's watershed—capabilities that are currently underutilized in Sampalok Lake. One major drawback is the lack of a centralized environmental database that collates monitoring data from various sources such as government agencies, academic institutions, and community initiatives. Currently, data is dispersed among different entities, leading to inconsistencies in reporting and a lack of transparency. This fragmentation hinders evidence-based decision-making, reduces coordination between agencies, and restricts the ability of stakeholders—including local communities, researchers, and civil society groups—to evaluate the status and trends of the lake independently. The absence of a comprehensive data management system also limits stakeholder engagement and public confidence, as community members face restricted access to reliable scientific information regarding water quality patterns. In the absence of transparent and timely data sharing, decision-making tends to be centralized and reactive, rather than inclusive and proactive. As a result, stakeholders frequently resort to anecdotal observations instead of validated scientific data when addressing environmental threats. In general, the monitoring structure of Sampalok Lake is considered to be technologically outdated and fragmented, hindering its ability to effectively manage the environmental challenges the lake is facing. To enhance the monitoring system, it is imperative to implement institutionalized inter-agency data sharing, utilize automated and GIS-based monitoring technologies, and establish a centralized, publicly accessible environmental database. These enhancements are crucial for improving transparency, accountability, and the overall efficiency of lake governance and rehabilitation initiatives.

Governance and Institutional Challenges

The assessment uncovered systemic governance challenges that greatly impede the efficient management of Sampalok Lake. A key issue identified is the overlapping of institutional responsibilities between the Department of Environment and Natural Resources (DENR), the Community Environment and Natural Resources Office (CENRO), and the Local Government Units (LGUs). This overlap leads to fragmented policy responses, duplicated tasks, and unclear lines of accountability. Such institutional fragmentation hinders coordinated efforts, particularly in crucial areas like water quality monitoring, aquaculture regulation, and shoreline management. Additionally, several barangay officials highlighted the challenges of enforcing environmental regulations due to vague jurisdictional boundaries and persistent funding constraints, limiting their ability to sustain monitoring and enforcement activities. The governance challenges align with Brillo's (2017) research, which highlighted ongoing issues with institutional coordination and overlapping authority in overseeing the crater lakes of San Pablo City. The absence of a centralized and empowered lake management authority worsens policy inconsistencies and hampers timely decision-making, particularly in times of environmental crises like fish kills and severe algal blooms. Furthermore, bureaucratic obstacles in enforcing fisheries regulations and issuing environmental permits perpetuate the ongoing depletion of resources in the lake. Delays in approval procedures, a shortage of inspection staff, and inconsistent monitoring result in the persistence of noncompliant activities, such as unregulated expansion of fish cages and improper disposal of waste. Despite the existence of formal policies and regulatory frameworks, their practical enforcement is patchy, reactive, and occasionally influenced by politics. This politicized approach to enforcement erodes the credibility of regulations, weakens deterrence measures, and fosters a culture of selective compliance among users of the lake's resources. Overall, the persistence of overlapping mandates, weak enforcement capacity, fiscal constraints, and political interference underscores the urgent need for streamlined institutional arrangements, clarified jurisdictional authority, and insulated enforcement mechanisms to strengthen the governance of Sampalok Lake and ensure the long-term sustainability of its ecological resources.

Community Engagement and Participatory Monitoring

The study uncovered thriving community-driven environmental projects in and around Sampalok Lake, despite ongoing institutional constraints. Various local groups, such as youth organizations, church-based communities, fisherfolk associations, and barangay councils, regularly organize clean-up campaigns, informal assessments of the lake's condition, and initiatives to raise environmental awareness. These grassroots initiatives reflect a strong commitment to environmental stewardship and shared accountability at the community level. These findings align with Danielsen et al.'s (2005) perspective that community-led monitoring promotes environmental responsibility, strengthens local engagement in conservation efforts, and encourages inclusive and participatory management of natural resources. Despite the energy and innovation of these grassroots efforts, participants lamented the lack of acknowledgment, technical assistance, and institutional support from official government bodies. Many community projects continue to operate on a voluntary basis with limited funding and without formal structures, hindering their long-term viability and preventing them from becoming fully integrated into official monitoring and governance processes. The lack of consistent communication between community-driven data and government-sponsored scientific monitoring further diminishes the effectiveness of these initiatives. The gap between hierarchical governance systems and grassroots community initiatives leads to missed opportunities for synergistic collaboration, resource allocation, and joint decision-making. The absence of formal mechanisms to acknowledge and integrate community contributions in lake management strategies means that valuable local insights are often overlooked, potentially leading to a decline in community engagement over time. To address this divide, establishing formal co-management agreements, implementing institutionalized community monitoring programs, and providing ongoing capacity-building assistance are critical in harnessing widespread civic participation to achieve effective and sustainable lake governance.

Emerging Environmental Issues

Recent studies have highlighted a critical environmental issue in Sampalok Lake: the widespread presence of microplastics (Natuel, 2023). Research has shown that substantial amounts of microplastics have built up in the lake's surface waters, primarily originating from disposable plastics used by tourists, businesses near the lake, and improper household waste disposal practices. The ecological consequences of microplastic pollution are significant, as these pollutants can disrupt the food chains of aquatic ecosystems, accumulate in fish and other organisms, and potentially pose health risks to humans who consume contaminated seafood (Cole et al., 2011; Sharma & Chatterjee, 2017). Despite the increasing scientific understanding of these hazards, the current monitoring and management strategies for Sampalok Lake do not currently include an evaluation of microplastic pollution, resulting in a lack of attention to this emerging threat. The situation is further worsened by uncontrolled tourism, rapid lakeside development, and insufficient waste management procedures, all of which contribute to the lake's environmental fragility and jeopardize its ecological strength. To effectively address this challenge, it is imperative to promptly incorporate microplastic monitoring into existing water quality assessment programs and implement preventive and regulatory measures. Sustainable tourism strategies play a key role in alleviating anthropogenic pressures, including measures such as minimizing the use of single-use plastics, promoting environmentally friendly alternatives, and introducing visitor education programs. In conjunction with this, stricter land-use regulations, enhanced waste collection systems, and community-based environmental education efforts can help reduce microplastic pollution and support broader lake conservation objectives. Failure to take immediate and coordinated action poses a threat to Sampalok Lake's ecological health and its socio-economic contributions to fisheries, recreation, and tourism.

Towards an Integrated Monitoring Framework

Based on the study's results, it is suggested to establish a collaborative lake monitoring system that integrates advanced technology with community efforts. This system would utilize remote sensing, automated data collection devices, and GIS mapping in conjunction with citizen observations to assess changes in lake conditions over time and space comprehensively. According to PCAARRD (2022), implementing automated monitoring tools and real-time data dashboards is crucial for increasing transparency, enhancing information accessibility, and enabling data-driven decision-making. One crucial aspect of this strategy involves equipping local volunteers and community members with the skills to analyze technical measurements such as BOD, TDS, nutrient levels, and emerging pollutants such as microplastics. By promoting a mutual comprehension of environmental data, these initiatives have the potential to connect scientific monitoring with community perspectives, establishing a unified platform for cooperative environmental stewardship. The effectiveness of this integrated framework relies on key factors such as secured and sustained funding, multi-sectoral collaboration among government agencies, academic institutions, civil society, and private stakeholders, and a supportive policy environment that acknowledges the importance of both specialized scientific knowledge and localized experiential knowledge. By incorporating participatory monitoring alongside technological advancements, Sampalok Lake can achieve adaptive management strategies, increased ecological resilience, and stronger community involvement, aligning local conservation efforts with broader sustainability goals.

Conclusions

Based on stakeholder interviews, thematic analyses, and a review of practices at Sampalok Lake, this section outlines the main findings of the study. The conclusions address the key themes and implications of the qualitative research, highlighting strengths and weaknesses in the lake's assessment and monitoring approaches.

To enhance lake monitoring efforts, it is crucial to recognize that stakeholders possess varying levels of knowledge in this area. While government officials and environmental organizations are typically well-informed, many community members lack access to technical information. Despite the existence of assessment systems, they are inconsistent and underfunded, limiting their ability to facilitate prompt ecological interventions. Although community-led initiatives, such as clean-up campaigns and youth engagement, show promise, they require stronger institutional support to ensure sustainability and impact. Furthermore, emerging environmental threats like microplastic contamination are not adequately addressed within current monitoring frameworks. These observations underscore the critical need for a collaborative, multisectoral approach that integrates scientific research, policy-making, and local knowledge to effectively and sustainably address lake monitoring challenges. In light of these considerations, it is recommended to encourage student-led initiatives in lake monitoring and conduct longitudinal and perception-based studies in the future.

RECOMMENDATIONS

Recommendations are proposed to guide policy-making, institutional changes, community involvement, and future research for a more sustainable lake ecosystem.

To enhance the sustainability of lake ecosystems, it is recommended that a multi-agency task force be established to oversee lake monitoring efforts and enforce compliance with buffer zone and waste management regulations. Monitoring capabilities should be strengthened through the integration of real-time sensors and satellite technologies in collaboration with academic institutions for research-based assessments. At the grassroots level, the creation of Barangay Lake Watch Councils and citizen science programs can encourage active community participation in lake conservation. These initiatives should be supported by community education campaigns and university-led extension programs to increase environmental awareness. Additionally, promoting student-led monitoring projects and conducting long-term studies will help evaluate public perceptions and inform future strategies for the sustainable management of Sampalok Lake.

REFERENCES

ADB. (2018). Philippines: Water sector assessment, strategy, and roadmap. Asian Development Bank.

Beveridge, M. C. M. (2004). *Cage aquaculture*. Blackwell Publishing. DOI: 10.1002/9780470995761

Bonney, R., Phillips, T. B., Ballard, H. L., & Enck, J. W. (2016). Can citizen science enhance public understanding of science? *Public Understanding of Science (Bristol, England)*, *25*(1), 2–16. DOI: 10.1177/0963662515607406 PMID: 26445860

Brillo, B. B. C. (2017). The governance of the seven crater lakes, San Pablo City, the Philippines. *Asian Journal of Water, Environment and Pollution*, *14*(2), 13–25. DOI: 10.3233/AJW-170012

Cole, M., Lindeque, P., Halsband, C., & Galloway, T. S. (2011). Microplastics as contaminants in the marine environment: A review. *Marine Pollution Bulletin*, *62*(12), 2588–2597. DOI: 10.1016/j.marpolbul.2011.09.025 PMID: 22001295

Conrad, C. C., & Hilchey, K. G. (2011). A review of citizen science and community-based environmental monitoring: Issues and opportunities. *Environmental Monitoring and Assessment*, *176*(1), 273–291. DOI: 10.1007/s10661-010-1582-5 PMID: 20640506

Creswell, J. W., & Poth, C. N. (2016). *Qualitative inquiry and research design: Choosing among five approaches*. Sage publications.

Danielsen, F., Burgess, N. D., & Balmford, A. (2005). Monitoring matters: Examining the potential of locally-based approaches. *Biodiversity and Conservation*, *14*(11), 2507–2542. DOI: 10.1007/s10531-005-8375-0

DENR–EMB. (2021). *National Water Quality Status Report (2016–2021)*. Department of Environment and Natural Resources – Environmental Management Bureau. https://emb.gov.ph

DOST–PCAARRD. (2022). *Real-time monitoring system for aquaculture and inland water quality*. Philippine Council for Agriculture, Aquatic and Natural Resources Research and Development. https://www.pcaarrd.dost.gov.ph

Lara, M. E., Leader, J., & Klein, D. N. (1997). The association between social support and course of depression: Is it confounded with personality? *Journal of Abnormal Psychology*, *106*(3), 478–482. DOI: 10.1037/0021-843X.106.3.478 PMID: 9241950

Natuel, D. T. (2023). Microplastic pollution in Philippine freshwater lakes: An emerging environmental concern. *Philippine Journal of Environmental Science, 45*(2), 112–126.

Pomeroy, R. S., & Rivera-Guieb, R. (2006). *Fishery co-management: a practical handbook*. CABI.

Sachs, J., Schmidt-Traub, G., Kroll, C., Lafortune, G., Fuller, G., & Woelm, F. (2019). *Sustainable development report 2019: Transformations for a sustainable world*. Bertelsmann Stiftung and Sustainable Development Solutions Network.

Save the Lakes Movement. (2018). *Community-based lake conservation initiatives in San Pablo City*. Save the Lakes Movement.

Sharma, S., & Chatterjee, S. (2017). Microplastic pollution, a threat to marine ecosystem and human health: A short review. *Environmental Science and Pollution Research International, 24*(27), 21530–21547. DOI: 10.1007/s11356-017-9910-8 PMID: 28815367

Troell, M., Naylor, R. L., Metian, M., Beveridge, M., Tyedmers, P., Folke, C., & Yarish, C. (2014). Aquaculture and ecosystem services. *Ecosystem Services, 7*, 1–8. DOI: 10.1016/j.ecoser.2013.11.9

UN-Water. (2018). *SDG 6 Synthesis Report 2018 on Water and Sanitation*. United Nations Water.

UNDP. (2015). *Philippines and the Sustainable Development Goals: Advancing water and environmental sustainability*. United Nations Development Programme.

UNDP. (2017). *Localizing the SDGs: A guide for policymakers in Asia-Pacific*. United Nations Development Programme.

UNDP. (2020). *Partnerships for the SDGs: Enabling multi-stakeholder collaboration*. United Nations Development Programme.

United Nations. (2016). *The Sustainable Development Goals report 2016*. United Nations.

World Bank. (2023). *Philippine Water Supply and Sanitation Master Plan: Investing in water for inclusive growth*. https://www.worldbank.org/en/country/philippines/publication/water-supply-sanitation

Zlotnicki, J., Lagabrielle, Y., & Coudrain-Ribstein, A. (2015). Maar-type volcanic lakes: Geomorphology, formation, and ecological sensitivity. *Journal of Volcanology and Geothermal Research, 304*, 64–75. DOI: 10.1016/j.jvolgeores.2015.07.012

Chapter 13
Sustainability-Driven Marketing Strategies: A Cross-Sectoral Study of Leading Innovators

Souvik Howlader
Jadavpur University, India

Samanway Mridha
Jadavpur University, India

Rahul Narayan Singha
Jadavpur University, India

Pratap Chandra Mandal
https://orcid.org/0000-0002-0300-4002
Indian Institute of Management, Shillong, India

ABSTRACT

Companies require formulating marketing strategies which can promote sustainable manufacturing and green products. In today's world, growing environmental concerns and consumer expectations are important. The objective of the study is to analyze how companies can sort their marketing efforts with sustainability goals. A qualitative methodology using secondary data and case studies is adopted. The analysis focuses on key elements such as transparency, green branding, consumer education, and digital engagement. Case studies of prominent companies viz. Unilever, Apple, and Tesla demonstrate how companies incorporate sustainability into their operations and brand image. The study also identifies challenges such as the attitude-behaviour gap and greenwashing. A strategic model is suggested to

DOI: 10.4018/979-8-3373-6435-3.ch013

connect marketing with sustainability. Practitioners and companies may apply the analysis to adopt sustainability initiatives. The findings provide practical understandings for businesses and how companies should incorporate sustainability in their business actions.

1. INTRODUCTION

Businesses worldwide focus on sustainable manufacturing. Both businesses and all major stakeholders are concerned about growing environmental worries, climate change, and reducing resources. Consumers also care more about these issues these days. Besides making more products, companies must ensure that their operations are sustainable for the environment, fair to individuals, while making money in the long run (UNEP, 2021; Niinimäki & Karell, 2023). This new situation requires new ideas for making products and for selling them. Companies need to create a market for sustainable products to make individuals aware, build trust, and encourage different groups of consumers to buy such products (Luchs et al., 2020).

Sustainable manufacturing means manufacturing products in ways that cause less harm to the environment. It also means saving energy and materials. These methods help individuals and still allow companies to make money (EPA, 2019). The practices help companies deal with climate change, reduce their impact on the environment, and help to reach goals for a circular economy. Even with their benefits, sustainable products often face issues in the market. They can cost more, and individuals may not know about the sustainability initiatives adopted by companies. Consumers also might not trust them, especially because of greenwashing (White et al., 2019; Carlson et al., 2021). Marketing connects sustainable products with what consumers buy. The primary focus of traditional marketing is generation of revenues. However, at present, marketing must change. It needs to show how products are made openly, exhibit their benefits over their lifetime, and whether the sources are sustainable or not. Marketing also needs to establish the emotional value of these products (Leonidou et al., 2020). Modern marketing has to deal with the attitude-behaviour gap. Sometimes, consumers say that they care about the environment. However, they do not buy eco-friendly products. This happens because of prices of products or lack of trust (Peattie & Peattie, 2022).

The study focuses on cleaner ways of making things. The study focuses on green products. Marketing can influence consumer choices. It can make sustainable buying common. This study aims to fill that gap. The study aims to analyze how businesses can match their marketing efforts with their sustainability goals. They can do this by building trust, aligning values, and using purpose-led branding.

The study uses a conceptual approach. It only uses information from existing sources. These include academic papers, market information, and examples of companies focused on sustainability. This study helps the growing field of sustainability marketing. It gives a full plan that combines smart marketing with sustainability goals. It offers useful ideas for companies that want to build a strong brand for a long time. These companies also want to help the environment and the society (Grewal et al., 2021). The new idea here is that marketing is not just for promoting products. It is a key tool for making sustainable changes in today's economy, where values matter.

The study is organized as follows: Section 2 discusses the philosophy and practice of sustainable production. Section 3 discusses the important components of sustainability-led marketing strategy. Section 4 discusses consumer attitudes and behaviour toward green products. Section 5 provides case studies of firms that have been successful in sustainable marketing and in promoting sustainable products. These include Unilever, Apple, and Tesla. Section 6 outlines a strategic model for promoting sustainable products. Section 7 discusses the implications of the research for theory and practice. Section 8 summarizes the salient points of the study by enumerating its contributions, limitations, and future research directions.

2. PHILOSOPHY AND PRACTICE OF SUSTAINABLE PRODUCTION

Sustainable production is a key part of sustainability thinking. It aims to meet the needs of both present and future generations (Ghosh et al., 2023). In manufacturing, this means reducing harm to the environment. It also means protecting the rights of workers and keeping the business profitable over time (Niinimäki & Karell, 2023). Cleaner technologies should be used. Application of such technologies help businesses save energy and materials. They also ensure fair labour conditions (Dangelico & Pujari, 2010). In real practice, sustainable manufacturing focuses on reducing waste. It also aims to cut down emissions and use more renewable materials (IEA, 2022). Products should last longer. They should be easy to repair, recycle, or break down naturally (Song et al., 2022). This is known as the lifecycle approach. The *cradle-to-cradle* idea supports this. It means repeated usage of materials. The approach helps solve environmental problems and supports a circular economy (Bocken & Ritala, 2021). At present, companies are serious about sustainability (Kiron et al., 2023). They view it as part of their main plan, and not merely as a side activity. Companies abide by the global standards which are set for the purpose viz. ISO 14001. Circular models also guide businesses. Social factors are also very important. Ethical sourcing, fair pay, and open supply chains are becoming standard practices (Carlson et al., 2022). Also, when companies act responsibly, they gain public trust. They also build stronger brands and long-term stability (Peattie & Peattie, 2022). However, some issues remain. High starting costs, poor infrastructure, and unwillingness to change are still major issues, especially in developing countries (OECD, 2021). To solve these, companies need innovation. They also need help from both the public and private sectors. Training workers with new skills is also important (Grewal et al., 2021). Sustainable production is not just about new machines or methods. It is an elaborate approach. It connects the economic, social, and environmental goals of a business. But to succeed, these values must also be shared clearly with the market. Good marketing is as important as good production methods. There is still a gap between sustainable business actions and what the market understands. Smart marketing can close this gap. It can inform, inspire, and motivate individuals to act. The next section will explore how marketing and corporate responsibility can help companies.

3. KEY COMPONENTS OF SUSTAINABLE MARKETING STRATEGY

Sustainability marketing uses ideas from sustainable production. The interactions which companies have with customers are also taken into account. Sustainable marketing is not only about making a profit but it also focuses on creating long-term value. It is based on being responsible and ethical (Grewal et al., 2021). When companies become sustainable, their marketing communications must change. They need to incorporate these values into their communication messages. They must also incorporate the values in their brands. The values need to be visible in the manner in which they interact with customers (Kumar & Christodoulopoulou, 2014). Sustainability marketing is based on being transparent and honest. Now-a-days, customers expect that companies will communicate clear information. They want to know the environmental impact created by any product. They also want to know where it comes from and its full life cycle (White et al., 2019). Companies gain trust of consumers when companies are transparent in their actions. It makes customers less sceptical. This is very important because of the issue of greenwashing (Carlson et al., 2021). When brands share information clearly, they become more believable and trustworthy. They show that they have strong values. Green branding is another important aspect. It connects the identity of a company to environmental and social causes. These brands truly live their sustainability values. Such companies are better at attracting and keeping customers who care about the environment (Leonidou et al., 2020). Education is also key. It is not enough to just be aware. Companies must run campaigns to sensitize customers about sustainable behaviour. Marketing messages should teach customers about the benefits provided by products. These could be energy efficiency, recyclability, or being carbon neutral. Digital stories, labels about a product's life cycle, and information campaigns help customers understand these ideas (Niinimäki & Karell, 2023). Third-party certifications and eco-labels are important. They prove that a product is sustainable. This gives customers peace of mind. It also makes their decisions easier (UNEP, 2021). Digital tools are helpful. Brands can use social media and blogs. They can share emotional stories about sustainability on the internet. This builds stronger bonds with customers (Jiang et al., 2022). Finally, sustainability should be in all aspects of marketing. This includes product design, pricing, and how products are sold and promoted. This makes sure that everything is consistent and real to the customer (Belch & Belch, 2004). Companies that do this meet environmental and social needs. They also improve their brand image. They build customer loyalty. They gain a competitive advantage in a market that values responsibility.

4. CONSUMER BEHAVIOUR TOWARDS GREEN PRODUCTS

Companies that manufacture and sell green products have to understand the behaviour of their customers. This includes the beliefs, perception, and feelings customers have about these products. Many individuals are now interested in green products. They are aware of environmental issues like global warming and pollution (White et al., 2019). Most consumers think green products are beneficial for them for better health and fewer harmful materials (Grewal et al., 2021). This feeling is stronger when green products work as well as or better than regular ones (Niinimäki & Karell, 2023). Trust is also an important factor. Customers are more loyal to brands that are truly eco-friendly and open about their actions (Carlson et al., 2021). There is a difference between what customers say and what they actually do. This is because of real-world issues like high costs, lack of availability, and inconvenience (UNEP, 2021). Green products often cost more because of fair sourcing and green production methods. This higher price can discourage individuals who are focused on cost, from buying them (OECD, 2020). Also, these products are usually not sold in rural or lower-income community areas. It is difficult for individuals staying in such areas to afford these products. Greenwashing happens when companies make false claims to appear to be sustainable. This makes customers doubtful and breaches their trust (Jiang et al., 2022). Individuals have different levels of commitment to the environment. There are *true greens* who will pay more for sustainable products, *light greens* who care about the idea but are more focused on price and convenience, and *non-greens* who really do not care (White et al., 2019). Other aspects, like personal values, what friends think, and belief of individuals that their actions matter, also play roles (Peattie & Peattie, 2022). Some individuals even adopt green consumption as an entire lifestyle, e.g. being a vegan or a minimalist (Niinimäki & Karell, 2023). To fix these issues, marketers need to use smart strategies. They can use educational materials to show benefits to individuals. This can motivate individuals. Third-party certifications and trustworthy eco-labels also build confidence (UNEP, 2021). Deals and loyalty programs taken for green products can help with high prices. Companies can also promote green products by showing that they are good for the environment and the planet. Also, these are innovative, reliable, and affordable. In short, if more consumers want sustainable products, then these companies have to solve issues with access, price, and trust. A marketing approach that focuses on the customer is key for both business success and environmental progress.

5. CASE STUDIES ON SUSTAINABILITY DRIVEN MARKETING

Real-world examples offer important lessons on how marketers apply profitability frameworks with social and environmental goals. The analysis discusses several successful Indian and foreign companies that combine sustainable manufacturing with effective marketing. These cases offer useful comparisons by showcasing different business practices, communication strategies, and marketing responses. These companies are operating in various global markets. Also, these companies are recognized for their strong marketing and sustainability efforts. The three companies selected are from different sectors, including electric vehicles, fast-moving consumer products, and industrial manufacturing. These three companies are distributed all around the world. They are Unilever, Apple, and Tesla. The choice of Unilever, Apple, and Tesla, three businesses from different lines of business, for a case study on sustainability provides a larger and more perceptive viewpoint to the study. These firms are from the FMCG (fast moving consumer goods), technology, and automotive/energy industries. Each industry has its own sustainability issues, challenges, and opportunities. This segmentation provides a clearer view of how sustainable processes are affected by factors specific to each industry. As an example, it can be said that product life cycles, supply chain complexity, energy use, and consumer behaviour are linked with each other. Comparing these different strategies can reveal new ideas and shared sustainable factors used across sectors.

5.1. Unilever

5.1.1. Company Overview

Unilever is a big company. It is a British-Dutch multinational consumer products company. It started in 1929. Unilever has its presence in more than 190 countries. It has over 400 brands. These brands are in food, drinks, personal care, and home care. About 2.5 billion individuals use Unilever products daily. The company has over 140,000 employees worldwide. In 2023, Unilever made about €66 billion in revenue. This is around $72 billion USD. Unilever has a large market share in personal care and ice cream. It holds almost 40 percent of the soap and detergent market. Some of its famous brands are Dove, Lipton, Hellmann's, Ben & Jerry's, Axe, Lifebuoy, Knorr, and Rexona. Unilever's mission is to make sustainable living common. It wants to improve health and well-being. It also aims to reduce harm to the environment. It uses renewable energy. It sources materials sustainably. It works to reduce waste. It also focuses on social responsibility. This makes Unilever a global leader.

5.1.2. Sustainability Initiatives in Manufacturing

Unilever is working on sustainability. It has big goals for the planet. These goals are about climate, waste, and helping individuals. Unilever wants to have zero emissions. This means that Unilever focuses on zero carbon emission. The goal is to reach this by 2039. It also wants to cut its Scope 1 and 2 emissions to zero. This will happen by the year 2030. These goals follow science-based targets (SBTi, 2022; Ghosh et al., 2023). Unilever is a leader in fighting climate change. The company is using new ways to save energy. It uses things like heat pumps (Niinimäki & Karell, 2023). This helps the environment. The company is also working on waste. It uses a circular economy plan. This means it reuses things. It recycles, composts, and uses packaging that can break down. This shows good use of materials (Ellen MacArthur Foundation, 2021; Song et al., 2022). Unilever wants all of its packaging to be recyclable by 2030. Customers want products that are good for the planet (White et al., 2020). Unilever also helps nature. It supports farming that rebuilds the soil. It gets its crops from sustainable places. This helps animals and plants (Bocken & Ritala, 2021; WWF, 2023). It pays fair wages. It helps small farmers. This shows that sustainability is also about helping individuals (Carlson et al., 2022). Unilever builds green factories. For example, its Sumerpur factory has a green rating (Kumar & Christodoulopoulou, 2014). This shows that sustainability is part of how Unilever designs its buildings. Unilever's plan for sustainability is complete. It is based on research. It shows a good way to make and sell products. It uses ideas from many smart individuals.

5.1.3. Marketing Strategies for Sustainable Products

Unilever has a plan for marketing. The plan focuses on being good to the environment. It also focuses on being good to people. This helps people trust Unilever's brands for a long time. Brands like Ben & Jerry's and Hellmann's are called *Sustain-*

able Living Brands. People want products that are good for the planet. They want products that are also good for people. These brands meet that demand. This helps Unilever keep customers (Kotler & Keller, 2022; Niinimäki & Karell, 2023). These brands tell stories. Dove and Lifebuoy are examples. They do more than just work well. Dove talks about self-esteem. Lifebuoy teaches about health. This makes people trust the brands more (Porter & Kramer, 2011; Chen & Chang, 2013). Unilever uses the internet to teach people. They use it to tell stories. They also run campaigns. Unilever makes new products. They use packaging that breaks down easily. They also use less water to make things. They get third-party certifications like RSPO and Fairtrade. This proves that their claims about sustainability are real (Thøgersen et al., 2010; Carlson et al., 2022). Unilever sees sustainability as a way to grow. They do not see it as a cost. Their eco-friendly products sell better. They also get more customers than regular products. This helps them attract investors who care about these issues (Ghosh et al., 2023; Grewal et al., 2021; Bocken & Ritala, 2021).

5.1.4. Consumer Response and Market Performance

Unilever is committed to sustainability. This appeals to consumers today. Consumers now prefer ethical and eco-friendly products (Ghosh et al., 2023). Most global buyers will pay more for brands which show a clear environmental and social impact (KPMG, 2022). This trend has helped Unilever. Unilever has Sustainable Living Brands which include Dove, Lifebuoy, and Hellmann's. These brands are doing better than others. Their growth is nearly 69 percent in the last ten years. Non-sustainable brands grew by only 10 percent (Unilever, 2023). Unilever's marketing helps this success. The marketing is transparent. It focuses on values. Unilever uses clear labels and digital stories. It links sustainability with personal values. These values include health, fairness, and responsibility. This creates strong emotional bonds and builds brand loyalty (Carlson et al., 2022). Unilever's strategy puts sustainability first. It is not an add-on. This leads to consumer engagement. It gives Unilever a competitive edge (Bocken & Ritala, 2021). Unilever also does well in ESG indices. Its sustainability reports are clear. This appeals to investors who want long-term value. They want responsible governance (BlackRock Investment Institute, 2023). The company has built stakeholder trust. It has also gained market and investor support. Unilever shows how sustainability can drive business. It is a key part of branding and operations. This leads to market success and to long-term growth (Kiron et al., 2023).

5.2. *Apple*

5.2.1. *Company Overview*

Apple Inc. is a major American technology company. Its headquarters are in Cupertino, California. It was founded in 1976. The founders were Steve Jobs, Steve Wozniak, and Ronald Wayne. Apple is known for its new ideas. It makes electronics, software, and online services. Apple has many products which include the iPhone, iPad, Mac, Apple Watch, and AirPods. Apple TV is another product. The company also offers services like iCloud, Apple Music, and the App Store. Apple has built a system where everything works well together. This system is focused on the user. It is safe and easy to use. Apple operates in more than 100 countries. It has around 164,000 employees. The company made $383 billion in revenue. More than half of that came from the iPhone. Its cloud and media services also grew. These services made $85.2 billion. Apple also cares about being eco-friendly and protecting user privacy. These things help people trust the brand. They also make Apple more respected around the world. Apple's mission is to give the best user experience through great products and services. This shows that the company values quality and smart design. Apple also includes accessibility and sustainability in its main goals.

5.2.2. *Sustainability Initiatives in Manufacturing*

Apple has a plan for being sustainable. This plan is based on research. Apple wants to be carbon neutral by 2030. This goal is in line with what the world expects. People want companies to act on climate change (Kumar & Christodoulopoulou, 2014). Apple has cut its carbon pollution by more than 55 percent since 2015. Over 320 of its suppliers now use 100 percent clean energy (CDP, 2023). Apple's plan is not just about carbon. It also uses new materials. The MacBook Air uses more than 50 percent recycled stuff. Apple has removed almost all plastic from its packaging. This supports a *circular economy* (Ellen MacArthur Foundation, 2021; Niinimäki & Karell, 2023). Apple makes its products last longer. This is part of the *Longevity by Design* plan. It offers software updates for a long time and makes devices easier to fix. This creates less electronic trash (Gartner, 2023; Bocken & Ritala, 2021). Apple also uses smart technology to manage its supply chain. It buys materials from local places. It also gets materials from many different countries. This helps Apple to be strong in a changing world (McKinsey & Company, 2022; Ghosh et al., 2023). Apple's sustainability plan follows new ideas which are about looking at a product's whole life. They are about making things in a fair way. They are also about making environmental and social goals part of a business plan (Peattie &

Peattie, 2022; Grewal et al., 2021). This makes Apple a good example for other technology companies.

5.2.3. *Marketing Strategies for Sustainable Products*

Apple's marketing plan uses good design. It tells stories that make people feel things. It also talks about being good for the planet. Again, Apple's products are expensive and well-made. This is called value-based marketing (Kotler & Keller, 2022). Apple sells high-quality products which makes the brand feel special. Apple does not compete on price. The advertisements of Apple are simple. They tell stories that make individuals feel emotions. This is a type of marketing that tries to get individuals involved (Belch & Belch, 2004). Apple's products work well together. For example, iCloud and AirDrop make things easy. This helps people feel a part of the brand. This is called *value co-creation* (Kumar & Christodoulopoulou, 2014). Apple is also focused on the environment. They use recycled materials. They want to be carbon neutral by 2030. This makes them more competitive (Niinimäki & Karell, 2023; Bocken & Ritala, 2021). Apple's stores are important. The stores give customers a special experience. This makes people more loyal to the brand. This is part of the experiential economy. Apple also focuses on privacy. They use content from real customers. They do not use paid influencers as much. This is a way to build trust (Carlson et al, 2022; Jiang et al, 2022). In short, Apple's marketing plan works. It brings together good design, emotions, and sustainability. This helps them build a strong brand.

5.2.4. *Consumer Response and Market Performance*

Apple has gained consumer trust over the year. The company is able to achieve this trust because of superior products, new technology, and its focus on the environment. Research shows Apple is one of the most trusted brands today. Its products are reliable. It also protects user privacy well (Whelan & Krivkovich, 2021). The company gives a steady and smooth experience to its users. Apple is also improving its work on sustainability. This is helping its image as an eco-friendly brand. All its offices, data centres, and stores now run on 100 percent renewable energy (Peattie & Peattie, 2022). Apple uses recycled materials in its products. For example, the MacBook Air uses only recycled aluminium. Its batteries use more than 50 percent recycled cobalt (Ghosh et al., 2023). Apple has also reduced plastic in packaging to less than one percent. The goal is to remove all plastic usage by 2025. These steps attract both investors and buyers who care about the planet (Bocken & Ritala, 2021). Still, some individuals criticize Apple. They say that it is hard to repair Apple devices. They also say that Apple creates too much e-waste. These are real concerns. In response, Apple started the Self-Service Repair Program. It now lets people fix their iPhones and MacBooks in some countries. Apple also made a plan called *Longevity by Design*. It gives software updates for up to seven years. This helps individuals use their devices longer (Niinimäki & Karell, 2023). By the end of 2023, Apple had over two billion active devices around the world. The iPhone made up more than half of the company's yearly income. Apple's service business made $85 billion in the same year. This includes Apple Music, iCloud, and the App Store. The service business grew by nine percent in one year. Apple's actions have helped stop over 28 million metric tons of CO_2 from going into the air. These efforts make Apple more popular with Gen Z and millennials who care deeply about sustainability more than ever.

5.3. Tesla

5.3.1. Company Overview

Tesla Inc. which started in 2003, focuses on electric vehicles and clean energy. The head office is in Austin, Texas. Tesla makes electric cars, solar panels, solar roof tiles, and batteries for energy storage. It works in more than 35 countries and has over 140,000 employees. In 2023, Tesla earned $96.8 billion and sold about 1.81 million vehicles around the world. It is a top company in the electric vehicle market and holds a 23 percent share of the global market. In 2024, the company was valued at around $800 billion. Its main products include vehicles such as Model S, Model 3, Model X, Model Y, Cybertruck, Semi, and Roadster, and Energy Solutions like Powerwall, Megapack, and Solar Panels. Tesla's goal is to help the world switch to sustainable energy. Sustainability is a key part of Tesla's works. The company focuses on zero-emission transport, renewable energy, and reusing resources like recycled batteries.

5.3.2. Sustainability Initiatives in Manufacturing

Tesla works as a sustainable company in the industry of manufacturing. The company uses its resources wisely with a vision to protect the planet. The company has a focus on sustainability and wants to have zero emissions by the year 2030. This includes their entire supply chain. They have already cut emissions from manufacturing cars by 95 percent since 2015. They do this by using renewable energy at their factories (Ghosh et al., 2023; Niinimäki & Karell, 2023). Tesla uses clean energy. They use technology with AI. This fits with new ideas about sustainability. They also focus on being good to the environment (Kumar & Christodoulopoulou, 2014; Bocken & Ritala, 2021). Tesla reuses and recycles materials. They recycle

96 percent of their factory waste. They work with other companies like Redwood Materials which helps them recycle old batteries. This means they do not have to use new materials as much. This supports sustainable buying and using (Song et al., 2022; Jiang et al., 2022). Tesla's factories also save water. They reuse water in a closed system. The system uses 80 percent less water than other companies. They procure their materials from responsible sources which do not have an adverse effect on the environment or individuals. These efforts follow new rules for being ethical (Carlson et al., 2022; White et al., 2020). Tesla also plants trees. They have planted over 300,000 trees in Berlin. This shows that they care about the environment in a big way (Leonidou et al., 2020; Luchs et al.,2020). In short, Tesla's efforts are about more than just working well. They show a plan for a sustainable future. This follows what smart people say about new technology. It also follows what they say about long-term effects on the planet.

5.3.3. Marketing Strategies for Sustainable Products

Tesla has a marketing plan which is about being good to the environment. It shows that the company is a modern and responsible brand. Tesla's campaigns talk about using clean energy. The brand connects itself to global climate goals. This makes Tesla look like a leader in new and clean ideas. Tesla does not use normal car dealers. It sells cars directly to individuals online. This saves money and also helps build a stronger relationship with customers (Jiang et al., 2022; Kumar & Christodoulopoulou, 2014). The company has an open culture and shares detailed reports. It also gives customers real-time information. This makes people trust the brand. It attracts buyers who care about the environment (Carlson et al., 2022; Niinimäki & Karell, 2023). When Tesla launches a new product, it shows off new ideas. It talks about its new batteries and designs. This shows that Tesla is a leader in green technology (Luchs et al., 2020; Ghosh et al., 2023). Tesla also works with other companies towards ensuring sustainability. For example, it lets others use its supercharger network which gives them a bigger presence in the car world (Whelan & Krivkovich, 2021; Bocken & Ritala, 2021). These plans show that Tesla uses green marketing. It also uses digital tools. Tesla serves as a role model for other car companies.

5.3.4. Consumer Response and Market Performance

Tesla is able to retain a significant share of their customers. It manages to maintain a strong market position and has maintained good investor confidence. The firm stands as a world leader in sustainable innovation. Its new products consistently create excitement. The Cybertruck is a perfect example that has over 1.5 million preorders and has received two unveilings in 2019 and 2023. Buyers will have to wait until late 2026 to receive their vehicles (InsideEVs, 2024). Such enormous demand reveals that customers trust the brand and their vehicles. A 2023 Bloomberg survey revealed that 98 percent of Tesla owners were satisfied with their vehicles and overwhelmingly preferred the charging points and software (Bloomberg, 2023). In 2023, it had a 23 percent market share in the global electric vehicle market. This makes the company the global leading EV manufacturer (Counterpoint Research, 2024). At the same time, its Energy Division reported revenue of $6.2 billion in 2023. It is a 54 percent increase from the previous year (Tesla, 2024a). This growth shows that Tesla is able to earn customer trust on sustainability. Tesla has also won its trust of investors. Its strong focus on ESG (Environment, Social, and Governance) concerns makes it a darling of ethical investors. As of early 2024, ESG funds held around 15 percent of Tesla's equity shares. This demonstrates that at least some institutional investors have faith in Tesla's climate impact (Morningstar, 2024). The company has mitigated more than 28 million metric tons of CO_2 emissions and is improving battery recycling while pushing for zero-emission transport. Tesla is a frontrunner in finance and ethics in the green industry.

6. STRATEGIC MODEL FOR PROMOTING SUSTAINABLE PRODUCTS

Companies require a detailed plan to market sustainable products. This plan should integrate environmental responsibility with business growth. It also needs to connect with customers. First, companies need to divide customers into groups. These groups depend on how much they care about the environment. For example, there are *Eco-Actives*, *Eco-Considerers*, and *Dismissers* (Peattie & Peattie, 2022; White et al., 2019). This helps companies create messages that fit each group. The message should talk about both practical and emotional benefits. Practical benefits include long-lasting products or items that can be reused or recycled. Emotional benefits include feeling proud or feeling like part of a bigger cause (Luchs et al., 2020; Hart, 1997). Companies should ensure that they do not *greenwash* their products. Companies must be truly sustainable and should showcase that sustainability is a real part of their brands. Brands can build trust by being honest. The brand story should be communicated clearly. Companies can use digital content or labels. QR codes may be used to explain how the products are made. Companies should partner with known influencers, get certifications from trusted organizations (Carlson et al., 2021; Dangelico & Pujari, 2010). These steps increase trust. Brands should use different ways to reach customers. This makes the brand more visible and trusted. Products should be easy to find and buy. They should not be too expensive. You can offer refill stations. Companies can also use reward programs. Some companies use product-as-a-*service* models where individuals use the product without owning it (Niinimäki & Karell, 2023; Shrivastava, 1995). Working with non-profit groups also helps. Following ESG rules builds trust among investors and business partners (Whelan & Krivkovich, 2021). Companies should also track their performance. They should use clear indicators for measuring performance and should check the product's life cycle. Continuous feedback from customers should be taken and analysed. This helps companies to improve their plan (Grewal et al., 2021; Thøgersen et al., 2010). Companies should follow global goals like the UN Sustainable Development Goals. Using tools like artificial intelligence helps companies send better personal messages to customers. This approach makes sustainability a main part of their brands. It builds long-term success and helps your companies stay strong in a market where individuals care about values.

7. IMPLICATIONS FOR THEORY AND PRACTICE

The study discusses about various aspects of sustainability and initiatives adopted by leading innovators towards ensuring sustainability. The study has both theoretical and managerial implications.

7.1. Theoretical Implications

Sustainability and marketing are coming together which means that companies need to change how they think about marketing. The old 4Ps model (Product, Price, Place, Promotion) is now getting converted into new models. These new models mix environmental and social value creation (Kumar & Christodoulopoulou, 2014). This change shows that modern marketing must look at more than just money. It also needs to consider what is good for the environment and society. Old ideas about why individuals buy things are also changing. Companies often see a gap between what consumers say and what they do. For example, individuals may say that they want sustainable products. However, they do not always buy them. New models help explain this. They include things like a person's ethical identity. Individuals think about how much impact they feel they can make, and their intentions. Trust and social norms also shape these intentions (Peattie & Peattie, 2022; Luchs et al., 2020). The job of marketers has grown. It is not just about convincing people to buy. It now includes being responsible and making a positive impact for the environment. This big change in thinking means that sustainability is not a small topic. It is the main part of marketing's purpose in today's world, where everyone is more aware of social and environmental issues.

7.2. Practical Implications

Now-a-days, companies use sustainable methods. Considering sustainable methods has become a necessary step in betterment of products. They might use carbon labels, green packaging, or fair-trade sourcing (Grewal et al., 2021; Hsu et al., 2023). This helps them meet rules. It also makes them stand out in crowded markets full of opponents. Real campaigns focus on truth, transparency, and storytelling. This reaches to customers and builds trust with customers, especially when there are some who doubt about *green* claims. Businesses use online tools more and more to communicate purpose driven narratives. They activate consumer participation in sustainability efforts (Jiang et al., 2022). To do this well, all functions of a company have to work together. Functions including marketing, supply chain, compliance, and research towards achieving the goal of sustainability. This ensures that efforts are good and have a systemic impact. Stakeholders and others now care about ESG (Environmental, Social, Governance) considerations, making marketing important for corporate accountability (Whelan & Krivkovich, 2021). Sustainability is now a key part of a brand's plan. Marketing helps businesses create new products and build customer loyalty. This also gives them a sustainable competitive advantage.

8. CONCLUSIONS

The study shows how sustainability has changed from a minor issue to a key business strategy. Unilever, Apple, and Tesla all lead their respective industries. They show how sustainable marketing changes for each industry. This includes how they use energy, design products, and manage supply chains. Unilever, in consumer products, links sustainability to daily choices. They do this through their brand's purpose. Apple, a technology company, quietly adds eco-friendliness to its high-end products. Tesla, an automaker, makes sustainability central to its brand and work. These different strategies indicate that effective sustainability marketing must go beyond surface level claims. It should be authentic, measurable, and included throughout the entire value chain. Then sustainability will become a strong tool for innovation, consumer trust, and long-term competitive edge. Sustainability initiatives are applicable not only for these three companies. It is imperative for companies and businesses in general to have a focus on sustainability and work towards achieving sustainability goals. Overall, this study shows that sustainable marketing can affect consumer behaviour. This also defines success in today's environmentally aware and resource limited global market.

8.1. *Limitations*

The study is mostly qualitative in nature. It uses only secondary data. Surveys or interviews were not conducted and empirical analysis was not performed. This limits the ability to measure what consumers actually think. It also limits comparing perceptions across different groups of individuals. The case studies are helpful. However, they focus on big multinational companies. They do not fully show the issues that small and medium-sized businesses face. The study also did not explore industry-specific challenges. Regional differences in regulations were not looked at closely.

8.2. *Avenues of Future Research*

Researchers may conduct studies which will explore the viewpoints of customers about sustainability. It may be investigated whether green marketing and green products motivates consumers in their purchasing decisions. The study may be extended to companies in different countries. Comparisons may be performed about how companies do marketing to other businesses and to regular individuals. Sustainable marketing in different countries may be investigated. The analysis may provide ideas about how to change marketing for different groups. Researchers may also investigate how to use computers and new technology with green marketing.

REFERENCES

Belch, G. E., & Belch, M. A. (2004). *Advertising and promotion: An integrated marketing communications perspective 6th*. McGraw-Hil l.

BlackRock Investment Institute (2023). *ESG and sustainable finance outlook 2023*.

Bloomberg (2023). *Tesla owners show the highest brand loyalty*.

Bocken, N. M. P., & Ritala, P. (2021). Six ways to build circular business models. *The Journal of Business Strategy*, *42*(4), 251–260.

Carlson, J. P., Rahman, M. M., & Rahman, M. H. (2022). Influencing green purchase behavior: Role of green brand equity. *Journal of Cleaner Production, 330*, 129847.

Carlson, J. P., Rosenberger, P. J.III, & Rahman, M. M. (2021). Sustainable brand equity and consumer trust: A green perspective. *Journal of Brand Management*, *28*(1), 1–12.

CDP (2023). *CDP Global Supply Chain Report*.

Chen, Y. S., & Chang, C. H. (2013). Green trust: The influences of green perceived quality and green perceived risk. *Management Decision*, *51*(1), 63–82. DOI: 10.1108/00251741311291319

Counterpoint Research (2024). *Global electric vehicle market share report*.

Dangelico, R. M., & Pujari, D. (2010). Mainstreaming green product innovation. *Journal of Business Ethics*, *95*(3), 471–486. DOI: 10.1007/s10551-010-0434-0

Ellen MacArthur Foundation (2021). *Circular economy in business strategy*.

EPA (2019). *Sustainable Manufacturing Indicators Repository*.

Gartner (2023). *Top sustainability trends in tech 2023*.

Ghosh, S., Mandal, P. C., & Roy, A. (2023). Eco-innovation strategies in global manufacturing. *Journal of Sustainable Business*, *18*(2), 145–168.

Grewal, D., Roggeveen, A. L., & Nordfält, J. (2021). The future of retailing. *Journal of Retailing*, *97*(1), 1–6.

Hart, S. L. (1997). Beyond greening: Strategies for a sustainable world. *Harvard Business Review*, *75*(1), 66–77.

Hsu, C. L., Chang, K. C., & Yansritakul, C. (2023). The role of brand love in consumer behavior for green products. *Journal of Retailing and Consumer Services, 61*, 102540.

InsideEVs (2024). *Tesla Cybertruck demand update.*

International Energy Agency (IEA) (2022). *Energy Technology Perspectives.*

Jiang, Y., Wang, X., & Gao, J. (2022). Understanding sustainable consumption behavior. *Sustainability, 14*(3), 1347.

Kiron, D., Unruh, G., & Kruschwitz, N. (2023). Corporate sustainability at a crossroads. *MIT Sloan Management Review, 64*(2), 45–52.

Kotler, P., & Keller, K. L. (2022). *Marketing Management* (16th ed.). Pearson Education.

KPMG (2022). *Global CEO Outlook: Sustainability edition.*

Kumar, V., & Christodoulopoulou, A. (2014). Sustainability and branding: An integrated perspective. *Industrial Marketing Management, 43*(1), 6–15. DOI: 10.1016/j.indmarman.2013.06.008

Leonidou, C. N., Katsikeas, C. S., & Morgan, N. A. (2020). Greening the marketing mix: Do greeners lead to cleaner performance? *Journal of Business Ethics, 164*(3), 529–552.

Luchs, M. G., Naylor, R. W., Irwin, J. R., & Raghunathan, R. (2020). The sustainability liability. *Journal of Marketing, 84*(5), 32–48.

McKinsey & Company (2022). *The State of Fashion 2022: Sustainability Focus.*

Morningstar (2024). *ESG investing report: Tesla in focus.*

Niinimäki, K., & Karell, E. (2023). Circularity and value in sustainable fashion systems. *Sustainable Production and Consumption, 36*, 150–165.

OECD (2020). *Sustainable Consumption and Production.*

OECD (2021). *Green Growth Indicators 2021.*

Peattie, K., & Peattie, S. (2022). Sustainability branding: Beyond greenwashing. *Journal of Strategic Marketing, 30*(5), 423–436.

Porter, M. E., & Kramer, M. R. (2011). Creating shared value. *Harvard Business Review, 89*(1/2), 62–77.

Science-Based Targets initiative (SBTi) (2022). *Corporate Net-Zero Standard.*

Shrivastava, P. (1995). The role of corporations in achieving ecological sustainability. *Academy of Management Review, 20*(4), 936–960. DOI: 10.2307/258961

Song, J., Yu, H., & Yu, Y. (2022). Rethinking circular economy indicators: Evidence from material loops. *Resources, Conservation and Recycling, 182*, 106290.

Tesla (2023a). *Impact Report 2023.*

Tesla (2024a). *Quarterly Shareholder Update Q1 2024.*

Thøgersen, J., Jørgensen, A. K., & Sandager, S. (2010). Consumer decision making regarding a "green" everyday product. *Psychology and Marketing, 27*(3), 445–464.

Unilever (2023). *Sustainable Living Plan Progress Report.*

United Nations Environment Programme (UNEP) (2021). *Sustainability and Circular Economy.*

Whelan, T., & Krivkovich, A. (2021). *How purpose drives financial performance.* McKinsey & Co.

White, K., Habib, R., & Hardisty, D. J. (2019). How to SHIFT consumer behaviors to be more sustainable: A literature review and guiding framework. *Journal of Marketing, 83*(3), 22–49. DOI: 10.1177/0022242919825649

White, K., MacDonnell, R., & Dahl, D. W. (2020). It's not easy being green: The effects of attribute tradeoffs on green product preference. *Journal of the Academy of Marketing Science, 48*(3), 449–470.

WWF (2023). *Business and Nature 2030 Report.*

Chapter 14
Corporate Social Responsibilities

Namita Vyas Joshi
https://orcid.org/0000-0001-9429-9629
Noida International University, India

P. Selvakumar
https://orcid.org/0000-0002-3650-4548
Department of Science and Humanities, Nehru Institute of Technology, Coimbatore, India

Kajal Maheshwari
https://orcid.org/0000-0003-3597-8001
S.B. Patil Institute of Management, Pune, India

Pradip Sanatkumar Padhye
https://orcid.org/0000-0002-1444-840X
Symbiosis International University, Pune, India

Swapnil Gorde
https://orcid.org/0009-0000-1973-3456
Symbiosis International University, India

S. Guru Basava Aradhya
https://orcid.org/0000-0002-3335-2681
Dr. D.Y. Patil B-School, Pune, India

ABSTRACT

CSR, known as dimensions. At its core, CSR is about operating. This concept has evolved significantly over the years, moving from being a peripheral issue to one that is integral to business strategies. The modern understanding of CSR recognizes that businesses operate closely with companies. The foundation of CSR lies in the belief that corporations have a responsibility that goes beyond generating profit for shareholders construct society by integrating. The traditional view sole responsibility a business to increase profits, as famously advocated by economist Milton Friedman, has gradually shifted towards a broader view where corporations are seen as key actors in ethical labor practices. CSR is therefore not just philanthropy or charity—it is a mission with values while also fulfilling societal expectations. There are several key pillars commonly associated with CSR: responsible involves minimizing waste, improving biodiversity. Ethical labor emphasizes respect for workers' rights.

DOI: 10.4018/979-8-3373-6435-3.ch014

The Concept of Corporate Social Responsibility (CSR)

CSR, known as dimensions. At its core, CSR is about operating. This concept has evolved significantly over the years, moving from being a peripheral issue to one that is integral to business strategies. The modern understanding of CSR recognizes that businesses operate closely with companies. The foundation of CSR lies in the belief that corporations have a responsibility that goes beyond generating profit for shareholders construct society by integrating. The traditional view sole responsibility a business to increase profits, as famously advocated by economist Milton Friedman, has gradually shifted towards a broader view where corporations are seen as key actors in ethical labor practices. CSR is therefore not just philanthropy or charity—it is a mission with values while also fulfilling societal expectations. There are several key pillars commonly associated with CSR: responsible involves minimizing waste, improving biodiversity. Ethical labor emphasizes respect for workers' rights. Community engagement includes supporting education, health care, economic development, and other social programs that benefit local communities. Responsible governance involves transparency, accountability, anti-corruption measures, and ethical decision-making processes. Together, these elements help establish foster consumers, investors, and regulators. CSR can be implemented in different ways, industry, geographic location of a company. Large multinational corporations often develop comprehensive CSR strategies that include sustainability reporting, codes of conduct, supplier assessments, and impact measurements. These companies may publish annual CSR or Environmental, Social, and Governance (ESG) reports detailing their commitments, actions, and performance. On the other hand, small and medium-sized enterprises (SMEs) may focus on more localized initiatives such as supporting local schools, reducing energy use, or sourcing materials from ethical suppliers. Regardless of scale, the common thread is the intentional integration of social responsibility into the company's business model. Despite its benefits, CSR also faces several criticisms and challenges.

Unlike financial performance, social and environmental outcomes are often difficult to quantify, making it challenging to assess effectiveness. Additionally, CSR initiatives may face internal resistance, especially if they are perceived as costly or not aligned with core business objectives. To address these challenges, CSR must be embedded in corporate culture and supported by leadership at the highest levels. It requires clear goals, measurable targets, adequate resources, and regular monitoring. Companies must move from reactive, ad-hoc initiatives to proactive, integrated strategies that are aligned with their long-term vision. This involves not only doing less harm but actively contributing to positive change. Companies that succeed in embedding CSR into their operations often find that it drives innovation, improves risk management, and creates business. In conclusion, transformed voluntary initiative vital component of corporate awareness, crucial and environmental about balancing profitability with purpose, success for people or the planet. As stakeholders increasingly demand greater accountability, transparency, and ethical behavior, CSR will continue to be a key driver of corporate reputation, resilience, and relevance in the global economy. It is not merely about what companies do with their profits, but how they make their profits—with integrity, respect, and a commitment to shared value creation.

The Importance of CSR in Modern Business

(CSR) defining strategy, evolving from a voluntary philanthropic initiative to a critical determinant of corporate identity, sustainability, and competitive advantage. In an era characterized by increasing environmental concerns, social inequality, and a demand for ethical transparency, businesses are expected to act not just as economic entities but also as responsible corporate citizens. The importance of CSR in contemporary business cannot be overstated, as it influences everything from brand reputation and stakeholder trust to employee engagement, investor confidence, and long-term profitability. Modern stakeholders—consumers, employees, regulators, and investors alike—expect companies to operate responsibly, ethically, and sustainably, making CSR a non-negotiable aspect of doing business. One of the most important roles of CSR in modern business is its ability to foster trust and credibility among stakeholders (Wirba A V., 2024) . In today's hyperconnected and socially conscious world, reputation is one of a company's most valuable assets. Companies that demonstrate genuine tend to enjoy communities, partners often promote sustainability, fair labor practices, and community development. Transparency, accountability, and integrity efforts the public not just profit-driven but are committed to the greater good. This perception not only strengthens customer loyalty but also enhances a company's brand equity.The importance of CSR extends to risk mitigation and regulatory compliance. Governments and international bodies are enforcing stricter laws and regulations regarding environmental protection, labor rights, corporate transparency, and anti-corruption. CSR helps companies stay ahead of regulatory requirements by embedding responsible practices into their operations, reducing exposure to financial risks (Gupta S., et al, 2021) . For example, chains practice child labor and unsafe working conditions. Similarly, businesses that invest in clean technologies and carbon reduction strategies are better positioned to adapt to emerging climate-related regulations and market shifts. CSR is also instrumental in enhancing community relationships and fostering inclusive economic development, addressing their needs, building mutual trust, and goodwill. This can take many forms, including educational programs, healthcare initiatives, support for small businesses, and infrastructure development. Such efforts were companies, creating a more stable and prosperous environment that benefits both the business and the community. In turn, strong community relations can act as a buffer against public opposition, enhance brand image, and create loyal customer bases (Kandpal V., et al, 2024) . In global supply chains, CSR plays a crucial role in promoting ethical sourcing, fair labor conditions, and environmental stewardship. Multinational companies, in particular, face complex challenges in managing their global operations ethically and sustainably. CSR provides a framework to ensure that suppliers adhere to international standards and best practices. Implementing supplier codes of conduct, auditing practices, and capacity-building programs are common CSR tools to maintain integrity across the value chain. This is especially important in industries prone to human rights violations, such as textiles, mining, and agriculture. Ethical supply chain management not only minimizes reputational risks but also contributes to building a more equitable global economy. The global challenges of climate change, inequality, public health, and social justice have further heightened the relevance of CSR in modern business. Corporations are now seen as key partners in achieving global development goals, such as the United Nations Sustainable Development Goals (SDGs) (Hongxin W., et al, 2022) . These 17 goals provide a universal framework for addressing the world's most pressing issues, from eradicating poverty to ensuring access to clean energy and promoting peace and justice. Businesses contribute systemic change while also unlocking new opportunities for growth and collaboration. Such alignment demonstrates leadership, responsibility, and commitment to global progress. In conclusion, marketing gimmick is an essential pillar of modern business success. It enables companies to operate ethically, create shared value, and build sustainable relationships with all stakeholders, integrating economic goals with social and environmental considerations. As public expectations continue to evolve, businesses must embrace CSR not just as a strategic tool, but as a core aspect of their identity and purpose. Ultimately, companies that prioritize responsible practices will be better equipped to thrive in the modern economy, where no longer but intrinsically connected.

Environmental Sustainability and CSR

In today's globalized economy, where environmental degradation, climate change, resource depletion, and pollution present growing increasingly positively environmental well-being. CSR-driven environmental sustainability embodies a strategic approach that embeds consciousness in all facets of operations, value creation processes. Environmental organizations demonstrate responsibility not only to shareholders but also to future generations and the planet. At its core, environmental sustainability in CSR uses natural resources theirs including limiting generation, transitioning to renewable energy sources, preserving biodiversity, and supporting ecosystem regeneration (Mostepaniuk A., et al, 2022) . These actions are not only vital from a moral and ecological standpoint, but they also make sound business sense, enhance reputation, and the mitigation of long-term environmental and regulatory risks. In this context, sustainability is viewed not as a constraint, but as a source of innovation, resilience, and strategic advantage. A significant component of environmental sustainability in CSR is climate change mitigation. With global temperatures rising due to excessive carbon emissions and deforestation, businesses are under increasing pressure to reduce their carbon footprints and support climate action. CSR programs often include the implementation of sustainable energy policies, such as switching to renewable sources like solar and wind, investing in energy-efficient technologies, and offsetting emissions through carbon credits or reforestation projects (Sánchez-Teba E M., et al, 2021) . Companies are also developing climate adaptation strategies, particularly in industries vulnerable to extreme weather events, such as agriculture, manufacturing, and logistics. By incorporating climate resilience into their business models, companies not only protect their assets but also demonstrate leadership in the global transition toward a low-carbon economy. Resource efficiency is another critical element of environmental sustainability under CSR. The concept emphasizes doing more with less—maximizing output while minimizing input and waste. Companies are adopting circular economy principles to minimize dependency on finite minimize. This includes durability, reparability, as well as reusing materials and minimizing waste throughout the product lifecycle. For instance, many corporations are redesigning packaging to be biodegradable or eliminating single-use plastics. Others are optimizing manufacturing processes to conserve water and energy, reduce emissions, minimize by-products, reduce environmental harm, boost innovation, and improve overall competitiveness. Waste management and pollution control are also central to environmentally sustainable CSR initiatives (Rehman S U., et al, 2022) . Improper waste disposal, toxic emissions, and chemical contamination have significant environmental and public health impacts. Responsible companies are adopting stringent waste reduction strategies such as zero-waste goals, recycling initiatives, composting programs, and hazardous waste treatment. In addition, air and water pollution control technologies—like scrubbers, filters, and eco-friendly detergents—are being integrated into industrial processes to minimize harmful outputs. Sustainable supply chain management also plays a role in reducing environmental impact, as companies select suppliers based on their adherence to eco-friendly practices and environmental standards.

Biodiversity conservation is another important area where CSR and environmental sustainability intersect. Habitat destruction, species extinction, and ecosystem degradation threaten not only the planet's natural heritage but also the long-term viability of human societies and economies. Forward-thinking businesses are engaging in biodiversity-friendly practices such as supporting wildlife conservation projects, protecting natural habitats near their operational sites, reducing pesticide use, and promoting sustainable agriculture and forestry. Some companies also participate in land restoration programs or fund research initiatives to support biodiversity conservation (Li H., et al, 2022). These efforts help companies build trust within the communities they operate, meet environmental compliance requirements, and position themselves as responsible stewards of the environment. Environmental sustainability in CSR is closely linked to corporate governance and leadership. Top-level commitment is essential for fostering a sustainability-oriented culture within the organization. This involves setting ambitious environmental goals, integrating sustainability into corporate mission and vision statements, and establishing dedicated sustainability committees or departments. Furthermore, companies are tying executive compensation and employee performance incentives to the achievement of environmental targets. Leadership plays a pivotal role in embedding sustainability across organizational levels of logistics. Without a clear and sustained commitment from leadership, environmental initiatives often remain fragmented or superficial. Education and consumer engagement are other dimensions of environmental sustainability in CSR. Companies are leveraging their platforms to raise awareness about environmental issues and promote sustainable consumer behavior (Banala S., et al, 2025). This can include eco-labeling, green marketing campaigns, educational outreach, and sustainability-themed social media content. By empowering consumers with information and sustainable choices, companies can drive behavioral change and build stronger emotional connections with their audiences. Moreover, businesses that promote sustainable lifestyles contribute to broader environmental goals and position themselves as purpose-driven brands. Collaboration and partnerships are essential for scaling environmental sustainability through CSR. Businesses cannot address environmental challenges alone; collaboration with other corporations is critical. These partnerships can lead to shared innovation, knowledge exchange, joint ventures, and collective action platforms. For instance, industry coalitions may work together to establish sector-wide sustainability standards, invest in clean technologies, or support community-based conservation projects. Through multi-stakeholder engagement, companies can amplify their impact, share the burden of sustainability investments, and foster systemic change.

Social Impact and Community Development

Social impact and community development are foundational dimensions role in addressing societal challenges, improving quality of life, and promoting inclusive development. As modern businesses operate within diverse social contexts, they hold accountable contributions to social Social impact refers to the effect on individuals, communities, and broader society. Community development, a core component of this impact, involves strategic engagement with local populations to enhance living standards, empower marginalized groups, and create sustainable socio-economic opportunities. Together, these elements of CSR help bridge the gap between corporate success and social progress, creating mutual benefits for both businesses and the communities in which they operate (Alamineh G A., et al, 2023) . In a rapidly globalizing world marked by persistent inequality, poverty, and social fragmentation, companies have a responsibility to implement positive Social impact initiatives under CSR encompassing a broad spectrum of activities, including education, healthcare, skills training, infrastructure development, women's empowerment, disaster relief, and cultural preservation. These efforts are not limited to corporate philanthropy or ad hoc donations; instead, they represent strategic, long-term commitments that align with business goals and stakeholder expectations. By actively promoting well-being, ethical obligations social capital, strengthen stakeholder trust, and enhance their license to operate. Healthcare is another critical area of focus within CSR social impact programs. Companies often invest in healthcare initiatives that address pressing community needs such as maternal and child health, disease prevention, nutrition, mental health, and sanitation. These projects may include setting up clinics, conducting medical camps, distributing essential medicines, or running awareness campaigns on issues like hygiene, HIV/AIDS, vaccination, and reproductive health. In regions where public health infrastructure is weak, corporate interventions improve access to healthcare service outcomes (Siemieniako D., et al, 2021) . Moreover, health-focused CSR programs contribute to productivity, reduce absenteeism, and promote the overall resilience of the communities companies depend on. By helping individuals gain the skills and resources needed to become economically self-sufficient, businesses foster inclusive growth and contribute to poverty reduction. Furthermore, these initiatives often strengthen the local supply chain, create a more stable operating environment, and generate customer and employee loyalty. Infrastructure development is another dimension of community-oriented CSR. This includes constructing roads, schools, healthcare centers, sanitation facilities, and clean water systems in underserved regions. While infrastructure projects often require significant investment, their long-term benefits are substantial. They enhance access to essential services, improve mobility, attract investment, and elevate the overall standard of living. In many developing economies, companies are stepping in to fill infrastructural gaps where government resources are limited. These projects are often undertaken in collaboration with local authorities, NGOs, and community organizations to ensure alignment with local needs and sustainable impact.

In disaster-prone or conflict-affected areas, companies can play an essential role in relief and rehabilitation through timely CSR interventions. Disaster response efforts may include providing emergency supplies, funding rebuilding efforts, supporting trauma counseling, and offering employment to affected individuals. In the aftermath of crises such as floods, earthquakes, pandemics, or armed conflicts, businesses can help restore stability and accelerate recovery. Long-term resilience building—such as climate adaptation, early warning systems, and disaster risk reduction programs—is also increasingly being integrated into CSR strategies. Cultural preservation and social inclusion are additional areas where businesses can contribute meaningfully (Brooks C., et al, 2023) . In a world of increasing homogenization, many companies are supporting indigenous communities, local artisans, and cultural traditions through CSR. This may involve funding heritage projects, supporting local crafts, promoting indigenous languages, or sponsoring cultural festivals. Inclusive practices also extend to persons with disabilities, LGBTQ+ individuals, and other historically marginalized groups. By promoting diversity, equity, and inclusion, companies foster more cohesive societies and create workplaces and markets that reflect the richness of human experiences. Community participation is vital to the success of any CSR initiative aimed at social impact. Top-down approaches often fail to address real needs and can lead to resentment or project failure. Engaging communities in the planning, execution, and evaluation of CSR programs ensures that initiatives are relevant, inclusive, and sustainable. Participatory approaches empower beneficiaries, build local ownership, and enable more effective outcomes (Lokhande M., et al, 2025). This participatory ethos aligns with the broader principle of shared value—where businesses and communities co-create solutions that deliver both social and economic benefits. In conclusion, social impact and community development are indispensable components. They reflect a business's goals beyond the traditional goal of profit maximization. By addressing social issues, empowering communities, and fostering inclusive development, companies help build more resilient societies and sustainable economies. In turn, these actions reinforce business success by creating stable operating environments, loyal stakeholders, and long-term value. As societal expectations evolve, impact will remain a hallmark of responsible and future-ready enterprises. CSR, when genuinely committed to social progress, becomes a vehicle for transformational corporate community spheres.

Ethical Business Practices and CSR

Ethical business practices form the moral and operational backbone of integrity, fairness, accountability, and respect for all stakeholders. CSR encompasses various domains, and its foundation lies in conducting business in a manner that is ethically sound. Ethical business conduct ensures that companies adhere not only to legal obligations but also to broader moral and ethical behavior. Today's complex and interconnected global marketplace, where transparency and accountability are increasingly demanded by stakeholders, ethical practices are more than just a compliance requirement—they are central to building trust, reputation, and long-term success (Upadhyaya A N., et al, 2025). At its core, ethical business conduct involves making decisions that are not only legal but morally justifiable. This includes fairness in dealings with customers, suppliers, employees, investors, and society at large. It entails being honest in marketing, ensuring product safety and quality, paying fair wages, upholding workers' rights, preventing corruption, and respecting human rights. Companies that practice ethical governance integrate these values into their corporate culture, codes of conduct, and everyday operations. In doing so, they create a transparent, accountable, and principled business environment that promotes sustainability and stakeholder confidence. Fair treatment of employees is another central ethical obligation for businesses. They promote diversity and inclusion, offer opportunities for career development, and foster respectful workplaces free of harassment or exploitation. Ethical labor practices also involve respecting freedom of association and ensuring that employee rights are protected, particularly in global supply chains (Lee S., et al, 2021) . When companies invest in the well-being of their workforce, they not only fulfill ethical responsibilities but also improve productivity, loyalty, and organizational morale. Ethical sourcing and supply chain management are increasingly important aspects of CSR. In an era of global trade, many companies rely on complex networks of suppliers, some of which may operate in regions with weak labor laws or environmental regulations. This may include conducting regular audits, requiring compliance with codes of conduct, offering training and capacity-building, and ceasing business with suppliers who violate ethical standards. By extending ethical practices across their value chain, companies demonstrate a holistic commitment to responsible business. Anti-corruption and compliance are non-negotiable elements of ethical business conduct. Corruption not only distorts markets and increases the cost of doing business but also undermines public trust and institutional integrity. Ethical businesses implement robust anti-bribery and anti-corruption policies, provide employee training on compliance, conduct due diligence on partners and contractors, and encourage whistleblower protections (Albitar K., et al, 2021) . Many companies adopt international frameworks and OECD Guidelines for Multinational Enterprises to guide their compliance efforts. Ethical businesses also recognize the importance of tax responsibility and avoid aggressive tax avoidance schemes that deprive governments of critical revenue for public services. Ethical business also involves environmental responsibility. While environmental sustainability is often treated as a distinct CSR category, it is inherently ethical. Companies have an ethical duty to minimize harm to the planet, conserve resources, reduce emissions, and preserve biodiversity (Khuong M N., et al, 2021) . Ethical environmental practices include compliance with environmental laws, investing in green technologies, reducing waste and pollution, and supporting climate action initiatives. Ethical businesses understand that environmental degradation disproportionately affects the most vulnerable communities and that protecting the planet is an intergenerational responsibility. Stakeholder engagement is also enhanced by ethical practices. Whether interacting with employees, customers, communities, investors, or regulators, ethical companies prioritize open communication, mutual respect, and shared value. This creates stronger, more resilient relationships and facilitates collaboration, problem-solving, and innovation. Ethical companies listen to stakeholder concerns, involve them in decision-making processes, and demonstrate a willingness to evolve and improve. This participatory and inclusive approach is essential for sustaining trust and legitimacy in a dynamic business environment (Halkos & Nomikos, 2021). In conclusion, ethical business practices are not only the moral foundation of CSR but also a strategic asset in modern business. They define how a company engages with stakeholders, makes decisions, manages risks, and contributes to society. In a world where reputational capital, stakeholder trust, and social legitimacy are paramount, companies that prioritize ethics stand out as leaders. Ethical practices help build resilient organizations, foster inclusive and just societies, and support sustainable development. As CSR continues to evolve, ethics will remain at the heart of responsible business—guiding companies to do the right thing, even when it is not the easiest or most profitable choice. Ultimately, ethical business is good business: it creates value, inspires trust, and contributes to a better world for all.

Developing a CSR Strategy

Developing a strategy critical process for organizations to systematically align their social, environmental, and ethical commitments with their core business objectives. A well-crafted CSR strategy ensures that social responsibility is not an isolated activity or a marketing gimmick, but rather an integrated, strategic component of a company's identity and long-term success. In today's increasingly transparent and stakeholder-driven business environment, companies that implement thoughtful, impactful CSR strategies are more likely to build trust, drive innovation, mitigate risk, and contribute meaningfully to sustainable development (Anzola-Román P., et al, 2024). Developing such a strategy requires deliberate planning, inclusive engagement, and continuous evaluation to ensure it delivers real value to both business and society. This involves identifying and analyzing how social responsibility can be embedded within them. The CSR strategy should be a natural extension of the organization's ethos, enhancing its reputation and strengthening its technology may focus on digital education access, while a food company may emphasize sustainable agriculture and nutrition. The alignment of CSR initiatives with business objectives helps ensure that social and environmental goals support overall corporate performance rather than being perceived as separate or secondary endeavors. Once the foundation is established, the company must assess its internal and external environment to identify key areas where CSR can have the greatest impact. This includes conducting a materiality assessment—a process used to determine the most important environmental, social, and governance (ESG) issues that affect the business and its stakeholders. Materiality assessments involve engaging internal stakeholders (e.g., employees, management) and external stakeholders (e.g., customers, investors, communities) to identify shared priorities. Climate change, community development, ethical supply chains, and transparency often emerge as key concerns. These findings help shape the strategic priorities and scope of the CSR program. Stakeholder engagement is fundamental throughout the CSR strategy development process (Hayat K., et al, 2023). Effective CSR strategies are not created in a vacuum—they are shaped through dialogue and partnership with those who are affected by or have influence over the company's operations. This includes customers, employees, shareholders, regulators, suppliers, local communities, and civil society organizations. Engaging stakeholders ensures that the strategy reflects real needs, garners support, and builds mutual trust. Methods of engagement may include surveys, focus groups, interviews, public consultations, and advisory panels. After defining goals, companies must design initiatives and programs that translate the strategy into concrete actions. These programs should align with core business operations, leverage the company's strengths, and address the most pressing challenges retail might create sustainable fashion line, reduce packaging waste, and launch a global sourcing transparency campaign. A financial services firm may invest in financial literacy education, support underbanked communities, and implement ethical investment principles. Programs can be structured at local, national, or global levels depending on the company's scale and reach. Strategic partnerships with NGOs, academic institutions, governments, and other companies can also amplify impact and resource efficiency.

Figure 1. Importance of corporate social responsibility

Resource allocation is another critical component in the development of a CSR strategy. Successful implementation requires adequate financial investment, dedicated personnel, and access to tools and technologies. CSR should be integrated into budgeting processes, with funding allocated based on strategic priorities and anticipated outcomes. Companies often create dedicated CSR departments or embed CSR responsibilities across various compliance departments. Training programs, internal awareness campaigns, and employee engagement initiatives help embed a culture of responsibility throughout the organization. Encouraging employee volunteerism, for instance, strengthens internal engagement and reinforces the company's values in action. Monitoring, evaluation, and reporting are essential to ensure accountability and continuous improvement. A robust CSR strategy includes key performance indicators (KPIs) and metrics to track progress toward goals (Lu H., et al, 2022). These metrics may include environmental indicators (e.g., carbon footprint, water use), social indicators (e.g., community investments, employee diversity), and governance indicators (e.g., board diversity, anti-corruption compliance). Regular internal audits and external evaluation programs are areas of adjustment. Public sustainability reports, aligned with recognized reporting standards, offer transparency to stakeholders and help build credibility. Continuous feedback loops support learning, innovation, and the ability to respond to changing stakeholder expectations and global trends. Internally, this involves engaging employees at all levels, integrating CSR into performance management systems, and recognizing contributions to CSR goals. Externally, communication builds brand reputation, attracts ethically-minded customers and investors, and fosters trust. This can be achieved through sustainability reports, digital media, press releases, and participation in sustainability indices or rankings. Honest and transparent communication, including acknowledgment of challenges and setbacks, enhances the credibility of CSR efforts and positions the company as a responsible and ethical actor. Adaptability is another key characteristic of a resilient CSR strategy. Social, environmental, and economic conditions evolve, and companies must be prepared to adjust their approaches accordingly. Whether responding to new regulations, emerging global crises, technological advancements, or shifting consumer values, a flexible CSR strategy allows companies to remain relevant and impactful. Scenario planning, trend analysis, and stakeholder consultations can support proactive adaptation and strategic foresight (Nisha Angeline M., et al, 2025). For instance, during the COVID-19 pandemic, many companies pivoted their CSR efforts to focus on public health, remote education, and employee welfare—demonstrating responsiveness to urgent societal needs. In conclusion, developing a CSR strategy is a comprehensive and dynamic process that requires clarity of purpose, stakeholder collaboration, strategic focus, and ongoing commitment. A well-developed CSR strategy aligns business objectives with ethics, creating a roadmap for sustainable growth and positive impact. It is not merely about fulfilling obligations but about embracing the opportunity to be a force for good. As expectations from consumers, investors, and communities continue to rise, companies that invest in developing robust, authentic, and effective CSR strategies will not only enhance their reputation and profitability but also contribute meaningfully to building a more equitable, sustainable, and inclusive world.

Implementing CSR Initiatives in Business Operations

Implementing initiatives and operations is the critical phase where strategic intentions are translated into tangible actions. Socially, developing a CSR strategy sets the foundation; its success ultimately depends on effective implementation and functions, CSR daily, decision-making processes, and corporate culture rather than being treated as a separate or ancillary activity. Through thoughtful planning, stakeholder engagement, employee participation, performance monitoring, and transparent communication, companies can embed CSR into their operational DNA and ensure that their activities contribute meaningfully to sustainable development and societal well-being. The first and most crucial step in implementation is internal alignment and integration (Coates J M., et al, 2025). CSR initiatives must be communicated across all departments and hierarchies to ensure shared understanding and collective commitment. This often begins with leadership demonstrating active support for CSR through policies, public statements, and resource allocation. Leadership commitment sets the tone from the top, signaling that CSR is a priority embedded in the organization's core values. Department heads and team leaders must then cascade these priorities into operational plans, ensuring that each unit identifies and executes its role in advancing CSR objectives. For example, the procurement department might prioritize ethical sourcing, while human resources may focus on diversity and employee well-being. One of the core components of implementation is aligning CSR with core business operations and competencies. Instead of treating CSR as a separate philanthropic activity, companies should embed socially responsible practices directly into how they conduct business. Manufacturing redesigns its processes, energy use, and emissions; a logistics firm may optimize delivery routes to minimize fuel consumption and environmental impact (Yousaf Z., et al, 2021). A financial institution could integrate environmental, social, and governance (ESG) criteria into lending and investment decisions, while a technology company might develop products that enhance digital inclusion or promote environmental awareness. Operationalizing CSR in this way ensures that positive impact is generated as a direct outcome of core business activities. Partnering with external stakeholders plays a vital role in CSR implementation. Companies cannot operate in isolation—they must work alongside non-governmental organizations (NGOs), governments, academic institutions, community leaders, suppliers, and customers to address complex social and environmental challenges. Partnerships help extend reach, pool resources, enhance credibility, and gain insights into local contexts. For instance, an apparel brand may collaborate with an NGO to audit its supply chain for labor rights compliance, or a food manufacturer may partner with local farmers to promote sustainable agriculture practices. These collaborative efforts are more likely to create long-term, systemic change than standalone company-driven programs.

Transparency and communication are equally important during the implementation phase. Stakeholders—ranging from investors and customers to regulators and communities—expect companies to communicate their CSR actions, results, and challenges openly. This includes publishing annual sustainability or CSR reports, maintaining interactive dashboards, and using social media to highlight key initiatives and outcomes. Transparent communication builds credibility and trust while inviting dialogue and feedback (Wentzel L., et al, 2022). Companies should also be willing to acknowledge setbacks and explain how they plan to address them. Authenticity is key—stakeholders are increasingly adept at identifying superficial or insincere CSR efforts and demand meaningful, measurable impact. Technology plays an enabling role in implementing CSR initiatives effectively. Digital tools and platforms can enhance data collection, streamline reporting, automate sustainability processes, and improve stakeholder engagement. For example, environmental sensors can monitor resource use in real time, employee apps can promote volunteer opportunities, and artificial intelligence can help identify ESG risks in supplier databases. Additionally, companies can use technology to crowdsource ideas from employees or customers, manage donations and grantmaking, or track the carbon footprint of products and services. Technology not only increases efficiency but also expands the scale and reach of CSR initiatives. Adopting recognized CSR standards and certifications can also support implementation. Frameworks like ISO 26000 (social responsibility), SA8000 (labor standards), or ISO 14001 (environmental management systems) provide structured guidance for embedding CSR into operational procedures. Certifications signal to stakeholders that a company adheres to best practices and is committed to continual improvement (Issa A., et al, 2023). Additionally, adhering to global frameworks such as or aligning with them helps ensure consistency and comparability across sectors and geographies. In conclusion, implementing CSR initiatives into business operations is both a strategic imperative and a moral responsibility. It requires aligning CSR with core operations, engaging employees and stakeholders, managing supply chains ethically, measuring impact, embracing transparency, and leveraging technology. In today's interconnected global economy, companies that embed CSR deeply into their operations are better positioned to thrive while making a lasting difference.

Measuring and Reporting CSR Performance

Performance critical component, any comprehensive CSR strategy organization's initiatives demonstrate accountability to stakeholders, and identify areas for continuous improvement. As CSR becomes increasingly embedded in core business practices, regulators demand clear, accurate, and transparent data on corporate. Without robust systems for measurement and transparent reporting, even the most well-intentioned CSR programs can appear superficial, unaccountable, or insincere. Therefore, CSR performance measurement and reporting not only reinforce credibility but also enhance strategic decision-making, stakeholder trust, and long-term organizational sustainability (Damtoft N F., et al, 2025). The first step in measuring CSR performance is establishing clear and relevant KPIs. The organization's values indicators serve as benchmarks for success and can vary depending on industry, geography, stakeholder expectations, and the company's specific sustainability priorities. Common CSR diversion rates, diversity inclusion ratios, employee training hours, community investment, supply chain compliance, and employee engagement levels. Quantitative metrics provide objective data, while qualitative indicators—like stakeholder feedback or case studies—offer context and insight into the lived impact of initiatives. Employee training and cross-functional coordination are also vital in ensuring high-quality data collection. Employees across departments must understand how their activities contribute to CSR metrics and be equipped to report relevant data accurately. Creating internal protocols and standard operating procedures for CSR data collection minimizes inconsistencies and enhances comparability across business units and reporting periods. In addition, establishing internal controls, such as audits and verification procedures, strengthens data integrity and reduces the risk of greenwashing or misinformation (Wang Q., et al, 2023). Transparency and credibility are essential elements of CSR reporting. Stakeholders must trust that reported data is accurate, complete, and reflective of actual performance. To enhance assurance, external auditors review data collection processes, validate metrics, and provide assurance statements that increase stakeholder confidence. Third-party certifications, such as B Corporation status or Fair Trade certification, also bolster trust by signaling compliance with rigorous social and environmental standards.

Stakeholder engagement is not only a source of input for materiality assessments but also a feedback mechanism for CSR reporting. Reports should be accessible, easy to understand, and tailored to the needs of different stakeholder groups. For instance, investors may prioritize financial materiality and risk disclosures, while customers might be more interested in product sustainability and ethical sourcing (Chen X., et al, 2023). Digital reporting formats, such as interactive websites, dashboards, infographics, and videos, enhance engagement and facilitate data exploration. Incorporating storytelling elements—such as employee spotlights, community testimonials, or project case studies—can humanize the data and convey the real-world impact of CSR efforts. In addition to annual CSR or sustainability reports, many companies now issue periodic updates and use real-time reporting platforms. This shift toward continuous reporting reflects the dynamic nature of CSR and the increasing demand for timely, up-to-date information. For example, real-time dashboards may track key environmental metrics like carbon intensity or waste reduction, while quarterly updates can highlight progress on social initiatives (Stojanović A., et al, 2021). Integrating CSR data into financial reports and investor briefings further reflects the mainstreaming of sustainability in corporate strategy. Despite its benefits, CSR measurement and reporting also face several challenges. These include data availability and quality, lack of standardization across frameworks, the complexity of impact assessment, resource constraints, and transparency relevance, so that reports are detailed without being overwhelming or unfocused. They must also navigate tensions between aspirational goals and actual performance, avoiding the pitfalls of greenwashing or excessive self-promotion. Continuous improvement, stakeholder dialogue, and adherence to emerging best practices are essential in overcoming these challenges. In conclusion, measuring and reporting CSR performance is a cornerstone of responsible and effective corporate sustainability (Chowdhary S., et al, 2025). It enables companies to assess progress, demonstrate accountability, engage stakeholders, and make informed decisions. By selecting relevant KPIs, establishing robust data systems, aligning with global standards, and embracing transparency and stakeholder engagement, organizations can achieve credible, long-term business success. As the global emphasis on sustainability grows, those companies that excel in CSR measurement and reporting will be better positioned to lead with integrity, attract investment, and create meaningful value for both society and shareholders.

Benefits of CSR for Businesses and Society

As global awareness of environmental, social, and ethical issues increases, voluntary philanthropic gestures are becoming a vital strategic tool. Businesses are no longer evaluated solely on financial performance but also on their contributions to sustainable development, social equity, and ethical conduct. When effectively designed and implemented, CSR delivers a wide array of benefits—not only enhancing a company's image and profitability but also driving meaningful change in society at large (Limbore N V., et al, 2025). From bolstering brand reputation and employee engagement to addressing pressing social issues, CSR for both businesses and society is extensive and interdependent. One of the most immediate and visible benefits of CSR for businesses is the enhancement of brand reputation and customer loyalty. Conscientious individuals increasingly prefer to support social causes, demonstrate environmental responsibility. CSR initiatives—such as sustainable sourcing, fair trade practices, and charitable giving—help cultivate a positive public image and distinguish a company from competitors can leading to increased customer trust and brand loyalty, ultimately influencing purchasing decisions and fostering long-term relationships. CSR also enhances investor appeal and access to capital with many institutional screens of CSR performance. Strong CSR practices are perceived as indicators of effective risk management, long-term viability, and ethical governance. Companies with robust CSR frameworks often enjoy lower capital costs, better credit ratings, and improved shareholder relations (Singh K., et al, 2021). Socially responsible investment funds and ESG-focused indices are growing in popularity, creating new opportunities for companies that can demonstrate positive societal and environmental impact. From an operational standpoint, CSR can drive efficiency, innovation, and cost savings. Sustainability-focused conservation, efficient use, and reducing operational expenses while minimizing environmental impact. Green sustainable practices not only mitigate ecological footprints but also encourage innovation and process improvement. For example, adopting renewable energy sources or reducing packaging materials can simultaneously cut costs and align with environmental goals.

For society, the benefits of CSR are equally compelling and far-reaching. One of the most significant societal impacts of CSR is the advancement of social equity and community development. Corporate initiatives focused on education, healthcare, poverty alleviation, women's empowerment, and rural development are tangible. By investing in local communities, businesses help build infrastructure. For example, a company that establishes vocational training centers in underdeveloped regions not only enhances its local labor pool but also contributes to sustainable livelihoods and economic mobility. CSR also plays a crucial role in environmental protection and climate action (Akbari M., et al, 2022). Companies that commit to reducing emissions, conserving biodiversity, eliminating single-use plastics, or supporting reforestation efforts directly contribute to the health of the planet. These efforts are particularly important given the scale and urgency of global environmental challenges. When businesses act as environmental stewards, they complement governmental and non-governmental efforts, amplify collective impact, and drive industry-wide transformation. Additionally, CSR-related environmental innovation—such as the development of biodegradable products or carbon-neutral technologies—has the potential to influence consumer behavior. CSR can also foster greater civic engagement and democratic participation. When businesses promote transparency, human rights, and ethical governance, they help strengthen the social fabric and reinforce democratic values. Organizations can raise awareness of social issues, influence public policy, and empower marginalized voices. For example, CSR campaigns that address domestic violence, mental health, or voter education can bring critical issues into public discourse and mobilize collective action. In conclusion, businesses are both extensive and interconnected (Yan X., et al, 2022). For businesses, CSR enhances reputation, strengthens employee and customer relationships, opens new markets, drives innovation, ensures regulatory compliance, and attracts investment. For society, CSR contributes to community development, environmental protection, economic inclusion, public health, and democratic values. When executed authentically and strategically, CSR is not a cost, but an investment—a path to sustainable growth, stakeholder success. Global shift toward greater accountability and ethical responsibility, CSR will remain a vital pillar of resilient, future-ready businesses and thriving, equitable societies.

Challenges in Implementing CSR Initiatives

While its implementation is fraught with a range of challenges that can hinder effectiveness and sustainability. Despite the growing global emphasis on responsible business conduct, organizations often struggle to integrate CSR into core operations in a meaningful, consistent, and measurable way. The complexity of stakeholder expectations, resource limitations, regulatory ambiguities, and internal organizational dynamics contributes to the multifaceted nature of these challenges (Dal Mas F., et al, 2022). As businesses are profit-making, understanding and addressing the obstacles to successful CSR implementation becomes essential for achieving long-term impact and credibility. Resource constraints—both financial and human—are another significant hurdle. Implementing effective CSR programs often requires substantial investments in research, training, infrastructure, and technology. For small and medium-sized enterprises (SMEs), in particular, the costs associated with CSR can be prohibitive. Limited budgets may force companies to prioritize short-term financial survival over long-term social or environmental investments. Furthermore, a shortage of skilled personnel to plan, manage, and evaluate CSR initiatives can hamper execution (Suska M., 2021). Without dedicated CSR teams or cross-functional collaboration, programs may lack direction, continuity, and impact. Overcoming these limitations requires innovation in resource allocation, partnerships with NGOs or academic institutions, and leveraging external funding sources such as grants or impact investments. Internal resistance to change is another critical barrier. Implementing CSR often requires shifts in corporate culture, leadership priorities, and operational practices. Employees or executives may perceive CSR as a distraction from core business goals or view it as a compliance burden rather than an opportunity for innovation and growth. In hierarchical or conservative corporate environments, securing buy-in for CSR initiatives can be especially difficult. Overcoming resistance necessitates strong leadership, clear communication inclusive engagement strategies that involve employees at all levels. By demonstrating how CSR can create value—through improved brand loyalty, employee retention, or operational efficiencies—companies can build internal momentum and foster a culture of social responsibility (Khalil N., et al, 2024). Where laws do exist, they may be vague, inconsistent, or inadequately enforced, leading to confusion and uneven application. Companies operating in such environments may find it challenging to navigate legal obligations, assess risks, or benchmark performance. Moreover, weak governance and corruption in certain regions can undermine CSR efforts, particularly when businesses face pressure to conform to unethical practices or encounter bureaucratic barriers. Advocating for stronger regulations, engaging in industry self-regulation, and partnering with transparent local institutions can help mitigate these risks and promote a more enabling environment for CSR.

Stakeholder management is another persistent challenge. CSR inherently involves multiple stakeholders—including customers, employees, investors, suppliers, governments, NGOs, and local communities with distinct priorities, expectations, and power dynamics. Balancing these interests and fostering genuine dialogue can be resource-intensive and politically sensitive. Misalignment or miscommunication can lead to skepticism, conflict, or backlash. For example, communities may feel marginalized if CSR projects are imposed without consultation, or employees may disengage if internal initiatives appear tokenistic. Effective stakeholder engagement requires early and ongoing involvement, transparent communication, and mechanisms for feedback and grievance redressal (Yong R Y M., et al, 2024). A further complication is the risk of reputational damage from CSR failures or inconsistencies. Companies that are caught violating their stated values—whether through environmental damage, labor abuses, or unethical sourcing—face significant public backlash, legal consequences, and loss of consumer trust. The rise of social media has amplified this risk, as negative incidents can quickly go viral and attract global scrutiny. Maintaining consistency between corporate messaging and actual practices is therefore crucial. This requires rigorous internal audits, supply chain transparency, ethical sourcing policies, and crisis management protocols. Furthermore, companies must be prepared to admit mistakes, take corrective action, and learn from failures to rebuild trust and credibility. Finally, communicating CSR efforts effectively presents its own set of challenges. Many companies struggle to articulate their CSR vision in a way that is authentic, compelling, and accessible to diverse audiences. Overly technical reports, generic messaging, or a lack of storytelling can fail to engage stakeholders or differentiate the company's efforts. Worse, overhyping achievements without evidence invites skepticism and accusations of greenwashing or purpose-washing. Effective CSR communication requires honesty, humility, and a focus on both successes and areas for improvement. Leveraging multiple platforms—such as sustainability reports, social media, employee testimonials, and community events—can help humanize the company's impact and build trust through transparency (Mehra J., et al, 2025). In conclusion, implementing CSR initiatives is a complex endeavor that involves navigating strategic, operational, financial, cultural, and regulatory challenges. From resource constraints and internal resistance to stakeholder complexity and reputational risk, the obstacles are diverse and interrelated. Yet these challenges are not insurmountable. With strong leadership, strategic alignment, stakeholder engagement, and a commitment to transparency, businesses can overcome barriers and embed CSR as a core driver of sustainable success. In doing so, they create ethical, resilient, trusted, and future-ready organizations in an increasingly responsible global economy.

CONCLUSION

In today's increasingly conscious and connected global economy, Corporate Social Responsibility (CSR) can no longer exist on the periphery of business operations. Forward-thinking organizations are embracing CSR not just as a supplementary effort, but as an essential component of their core strategic framework. Integrating CSR into a company's primary mission, vision, values, and business model merely advantages fuels innovation, strengthens stakeholder trust, enhances brand reputation, ensures regulatory compliance, mitigates risk, and opens new market opportunities. Corporate strategy transcends philanthropy and becomes a driving force for sustainable growth and competitive differentiation. This integration ensures that businesses remain resilient, responsible, and relevant in a world where long-term success is increasingly measured not only by financial performance but also by social and environmental impact. Strategic Alignment with Corporate Vision and Mission: The integration of CSR begins with aligning social and environmental objectives with Companies that aspire to be purpose-driven need to reflect this purpose in the way they define success. A CSR-aligned mission might emphasize commitment to sustainability, inclusivity, and ethical growth, while the vision may project long-term goals for community development, environmental stewardship, and responsible innovation. For example, Patagonia's mission—"We're in business to save our home planet"—demonstrates how deeply CSR can be rooted in a brand's identity. Such alignment sends a strong signal to stakeholders that CSR is not an afterthought but a foundational value. Embedding CSR into Corporate Governance: CSR integration requires top-level commitment, especially from boards and C-suite executives. Incorporating CSR into corporate governance structures ensures that sustainability and responsibility become strategic priorities at the highest level. This includes: Establishing CSR or ESG (Environmental, Social, Governance) committees on the board. Linking executive compensation to sustainability performance. Requiring regular CSR performance reporting and risk assessments. When leadership champions CSR, it becomes embedded in the organizational culture, fostering alignment across departments and geographies. Clear accountability frameworks are necessary to move beyond rhetoric and achieve meaningful results. Human Capital and Organizational Culture: Strategic CSR integration influences organizational culture, talent acquisition, and employee retention. Millennials and Gen Z—who form the majority of today's workforce—expect employers to have a strong sense of purpose and societal contribution. Integrate ESG criteria into capital budgeting and investment analysis. Use sustainability ratings to guide mergers, acquisitions, or divestments. This financial integration sends a powerful message: CSR is not a cost center but a value creator that enhances shareholder and stakeholder returns over time. Transparent Reporting and Communication: Strategically integrated CSR is visible, credible, and consistently communicated. Businesses must share their goals, progress, and challenges openly through Integrated annual reports combining financial and CSR performance. Standalone sustainability reports. Real-time digital dashboards or CSR microsites. Transparent reporting builds stakeholder trust and allows companies to demonstrate integrity, align with regulatory expectations, and invite constructive feedback. Greenwashing—exaggerating CSR efforts—can backfire, so honest and evidence-based reporting is essential. Adapting to Changing Global Contexts: A dynamic CSR strategy must evolve with changing societal expectations, geopolitical shifts, technological innovations, and environmental urgency. Integration means embedding agility into the company's strategic DNA so that CSR efforts remain relevant. For instance, Climate adaptation strategies may require updated infrastructure investment. Social unrest may call for stronger community engagement and DEI commitments. Technological disruptions may enable new CSR tools or demand data ethics policies. By viewing CSR as a living component of the corporate strategy, businesses stay resilient and responsive in a volatile world. The future belongs to companies that understand that profit and purpose into core business strategy transforms corporate responsibility from a checkbox exercise into a powerful lever for innovation, resilience, trust, and growth. It fosters a culture of accountability, promotes inclusive development, and positions the business as a catalyst for systemic change. This integration requires intentional design, leadership commitment, stakeholder engagement, and robust measurement. When done right, CSR not only improves societal outcomes but enhances business success in ways that are enduring, ethical, and economically sound. In a world where stakeholders are increasingly values-driven, CSR is not just good business—it is essential business.

REFERENCES

Akbari, M., Nguyen, L. L. H., & Nguyen, T. Q. (2022). Corporate social responsibility in Vietnam: Views from corporate and NGO executives. *Social Responsibility Journal*, *18*(5), 1027–1044. DOI: 10.1108/SRJ-06-2020-0259

Alamineh, G. A., Hussein, J. W., Endaweke, Y., & Taddesse, B. (2023). The local communities' perceptions on the social impact of tourism and its implication for sustainable development in Amhara regional state. *Heliyon*, *9*(2), e14692. DOI: 10.1016/j.heliyon.2023.e17088 PMID: 37484348

Albitar, K., Al-Shaer, H., & Elmarzouky, M. (2021). Do assurance and assurance providers enhance COVID-related disclosures in CSR reports? An examination in the UK context. *International Journal of Accounting & Information Management*, *29*(2), 295–313. DOI: 10.1108/IJAIM-01-2021-0020

Anzola-Román, P., Garcia-Marco, T., & Zouaghi, F. (2024). The influence of CSR orientation on innovative performance: Is the effect conditioned to the implementation of organizational practices? *Journal of Business Ethics*, *196*(3), 877–898. DOI: 10.1007/s10551-023-05406-z

Banala, S., Panyaram, S., & Selvakumar, P. (2025). Artificial intelligence in software testing. In Chelliah, P., Venkatesh, R., Natraj, N., & Jeyaraj, R. (Eds.), *Artificial intelligence for cloud-native software engineering* (pp. 237–262). IGI Global Scientific Publishing., DOI: 10.4018/979-8-3693-9356-7.ch009

Brooks, C., Waterton, E., Saul, H., & Renzaho, A. (2023). Exploring the relationships between heritage tourism, sustainable community development and host communities' health and wellbeing: A systematic review. *PLoS One*, *18*(3), e0278487. DOI: 10.1371/journal.pone.0282319 PMID: 36989275

Chen, X., Liang, X., & Wu, H. (2023). Cross-border mergers and acquisitions and CSR performance: Evidence from China. *Journal of Business Ethics*, *190*(1), 1111–1134. DOI: 10.1007/s10551-021-05025-6

Chowdhary, S., Selvakumar, P., Sangeetha, B., Sarvaani, V. V., Anute, N., & Sharma, M. (2025). The role of social capital in entrepreneurial development. In Martínez-Falcó, J., Sánchez-García, E., & Marco-Lajara, B. (Eds.), *Support networks for entrepreneurs: Social capital and strategic collaborations* (pp. 311–338). IGI Global Scientific Publishing., DOI: 10.4018/979-8-3693-9836-4.ch013

Coates, J. M., Santhanalakshmi, K., Dinesh, N., Gandhimathi, S., & Chandel, P. S. Mohit, Selvakumar P, (2025). "Assessment and evaluation in transnational staff development." In Morris G, Ye J (Eds.), Training and development in transnational higher education (pp. 387-406). IGI Global Scientific Publishing. DOI: DOI: 10.4018/979-8-3373-1897-4.ch019

Dal Mas, F., Tucker, W., Massaro, M., & Bagnoli, C. (2022). Corporate social responsibility in the retail business: A case study. *Corporate Social Responsibility and Environmental Management*, *29*(5), 1447–1457. DOI: 10.1002/csr.2198

Damtoft, N. F., van Liempd, D., & Lueg, R. (2025). Sustainability performance measurement–a framework for context-specific applications. *Journal of Global Responsibility*, *16*(2), 267–290. DOI: 10.1108/JGR-05-2023-0082

Gupta, S., Nawaz, N., Alfalah, A. A., Naveed, R. T., Muneer, S., & Ahmad, N. (2021). The relationship of CSR communication on social media with consumer purchase intention and brand admiration. *Journal of Theoretical and Applied Electronic Commerce Research*, *16*(5), 1228–1248. DOI: 10.3390/jtaer16050068

Halkos, G. E., & Nomikos, S. N. (2021). Reviewing the status of corporate social responsibility (CSR) legal framework. *Management of Environmental Quality*, *32*(6), 1280–1299. DOI: 10.1108/MEQ-04-2021-0073

Hayat, K., Jianjun, Z., Ali, S., & Ageli, M. M. (2023). Eco-advertising and ban-on-plastic: The influence of CSR green practices on green impulse behavior. *Journal of the Knowledge Economy*, *14*(4), 3664–3682. DOI: 10.1007/s13132-022-01014-w

Hongxin, W., Khan, M. A., Zhenqiang, J., Cisma , L. M., Ali, M. A., Saleem, U., & Negru , L. (2022). Unleashing the role of CSR and employees' pro-environmental behavior for organizational success: The role of connectedness to nature. *Sustainability (Basel)*, *14*(6), 3466. DOI: 10.3390/su14063191

Issa, A., & Zaid, M A A. (2023). Firm's biodiversity initiatives disclosure and board gender diversity: A multi-country analysis of corporations operating in Europe. *Business Strategy and the Environment*, *32*(1), 318–335. DOI: 10.1002/bse.3404

Kandpal, V., Jaswal, A., & Pathania, P. (2024). Corporate social responsibility (CSR) and ESG reporting: redefining business in the twenty-first century. In *Circular Economy and Sustainability* (pp. 557–575). Springer., DOI: 10.1007/978-3-031-32461-3_25

Khalil, N., Che Abdullah, S. N., Haron, S. N., & Hamid, M. Y. (2024). A review of green practices and initiatives from stakeholder's perspectives towards sustainable hotel operations and performance impact. *Journal of Facilities Management*, *22*(2), 233–250. DOI: 10.1108/JFM-03-2022-0025

Khuong, M. N., Truong An, N. K., & Thanh Hang, T. T. (2021). Stakeholders and Corporate Social Responsibility (CSR) programme as key sustainable development strategies to promote corporate reputation—Evidence from Vietnam. *Cogent Business & Management*, *8*(1), 1984172. DOI: 10.1080/23311975.2021.1917333

Lee, S., Zhai, X., Lee, M., & Luo, Q. (2021). Current status of CSR practices in the casino industry: A comparison between the US and Macau. *Journal of Hospitality and Tourism Management*, *46*, 112–123. DOI: 10.1016/j.jhtm.2021.07.011

Li, H., Kuo, Y. K., Mir, M. M., & Omar, M. (2022). Corporate social responsibility and environmental sustainability: Achieving firms sustainable performance supported by plant capability. *Ekonomska Istrazivanja*, *35*(1), 1692–1716. DOI: 10.1080/1331677X.2021.2015612

Limbore, N. V., Anute, N., Ahluwalia, G. K., Chordiya, P. B., Kalshetti, P., & Selvakumar, P. (2025). Ethical leadership and corporate responsibility. In Harriott, S. (Ed.), *Innovative approaches to managing conflict and change in diverse work environments* (pp. 193–218). IGI Global Scientific Publishing., DOI: 10.4018/979-8-3693-9556-1.ch009

Lokhande, M., John Deva Prasanna, D. S., Selvakumar, P., Ragupathi, T., Vaissnave, V., & Manjunath, T. C. (2025). "Novel ML algorithms for healthcare." In Ajibesin A, N G, Thangavel S (Eds.), Synergizing data envelopment analysis and machine learning for performance optimization in healthcare (pp. 113-134). IGI Global Scientific Publishing. DOI: DOI: 10.4018/979 8-3373-0081-8.ch005

Lu, H., Liu, X., & Falkenberg, L. (2022). Investigating the impact of corporate social responsibility (CSR) on risk management practices. *Business & Society*, *61*(3), 585–617. DOI: 10.1177/0007650320928981

Mehra, J., Madaan, S., & Balu, N. Mohit, Selvakumar P, Manjunath T C, (2025). "The ubiquitous internet and smart computation available 24/7 for anyone and anywhere." In Babulak E (Ed.), Educational AI humanoid computing devices for cyber nomads (pp. 247-270). IGI Global Scientific Publishing. DOI: DOI: 10.4018/979-8-3693-8985-0.ch010

Mostepaniuk, A., Nasr, E., Awwad, R. I., Hamdan, S., & Aljuhmani, H. Y. (2022). Managing a relationship between corporate social responsibility and sustainability: A systematic review. *Sustainability (Basel)*, *14*(4), 2201. DOI: 10.3390/su141811203

Nisha Angeline, M., Selvakumar, P., Gupta, M. K., Nazar, N., Elumalai, D., & Bhattacharya, S. (2025). Emerging materials for energy storage and environmental applications. In Muduli, K., Moharana, B., Ales, S., & Biswal, D. (Eds.), *Using computational intelligence for sustainable manufacturing of advanced materials* (pp. 53–72). IGI Global Scientific Publishing., DOI: 10.4018/979-8-3693-7974-5.ch003

Rehman, S. U., Bresciani, S., Yahiaoui, D., & Giacosa, E. (2022). Environmental sustainability orientation and corporate social responsibility influence on environmental performance of small and medium enterprises: The mediating role of green innovation. *Corporate Social Responsibility and Environmental Management*, *29*(3), 806–818. DOI: 10.1002/csr.2293

Sánchez-Teba, E. M., Benítez-Márquez, M. D., Bermúdez-González, G., & Luna-Pereira, M. M. (2021). Mapping the knowledge of CSR and sustainability. *Sustainability (Basel)*, *13*(6), 3402. DOI: 10.3390/su131810106

Siemieniako, D., Kubacki, K., & Mitręga, M. (2021). Inter-organisational relationships for social impact: A systematic literature review. *Journal of Business Research*, *134*, 437–449. DOI: 10.1016/j.jbusres.2021.04.026

Singh, K., Misra, M., & Yadav, J. (2021). Corporate social responsibility and financial inclusion: Evaluating the moderating effect of income. *MDE. Managerial and Decision Economics*, *42*(7), 1783–1798. DOI: 10.1002/mde.3306

Stojanović, A., Mihajlović, I., Safronova, N. B., Kunev, S., & Schulte, P. (2021). The multi-criteria analysis of corporate social responsibility: A comparative study of Russia, Bulgaria and Serbia. *Journal of Management & Organization*, *27*(6), 1327–1343. DOI: 10.1017/jmo.2020.40

Suska, M. (2021). Environmental corporate social responsibility (ECSR) on the example of Polish champion oil, gas and mining companies. *Sustainability (Basel)*, *13*(9), 4740. DOI: 10.3390/su13116179

Upadhyaya, A. N., Debnath, P., Rani, P., Selvakumar, P., Manjunath, T. C., & Bhattacharya, S. (2025). AI-based disaster prediction and early warning systems. In Mobo, F. (Ed.), *AI and emerging technologies for emergency response and smart cities* (pp. 55–74). IGI Global Scientific Publishing., DOI: 10.4018/979-8-3693-9770-1.ch003

Wang, Q., & Yan, W. (2023). Your behaviour reflects your risk attitude: The influence of CEOs' insurance behaviours on corporate social responsibility. *British Journal of Management*, *34*(3), 1172–1191. DOI: 10.1111/1467-8551.12649

Wentzel, L., Fapohunda, J. A., & Haldenwang, R. (2022). The relationship between the integration of CSR and sustainable business performance: Perceptions of SMEs in the South African construction industry. *Sustainability (Basel), 14*(19), 12516. DOI: 10.3390/su14031049

Wirba, A. V. (2024). Corporate social responsibility (CSR): The role of government in promoting CSR. *Journal of the Knowledge Economy, 15*(3), 1505–1522. DOI: 10.1007/s13132-023-01201-6 PMID: 40479158

Yan, X., Espinosa-Cristia, J. F., Kumari, K., & Cioca, L. I. (2022). Relationship between corporate social responsibility, organizational trust, and corporate reputation for sustainable performance. *Sustainability (Basel), 14*(24), 16996. DOI: 10.3390/su14148737

Yong, R. Y. M., Chua, B. L., Fakfare, P., & Han, H. (2024). Sustainability à la carte: A systematic review of green restaurant research (2010-2023). *Journal of Travel & Tourism Marketing, 41*(2), 159–179. DOI: 10.1080/10548408.2023.2293014

Yousaf, Z., Radulescu, M., Nassani, A., & Aldakhil, A. M. (2021). Environmental management system towards environmental performance of hotel industry: Does corporate social responsibility authenticity really matter? *The Engineering Economist, 32*(4), 364–376. DOI: 10.5755/j01.ee.32.4.28878

Chapter 15
Sustainable Development Goals

Pamarthi Satyanarayana
https://orcid.org/0009-0008-4041-3674
Vel Tech University, India

Pullagora Chiranjeevi
Saida Moulana Law College, India

S. T. Naidu
https://orcid.org/0000-0002-6523-2027
Vel Tech Rangarajan Dr. Sagunthala R&D Institute of Science and Technology, India

G. Saraswathy
https://orcid.org/0009-0004-1464-3688
Dr. MGR Educational and Research Institute, Dr. MGR University, India

Bensha C. Shaji
https://orcid.org/0000-0002-8298-0908
VIT-AP School of Law, VIT-AP University, Amaravati, India

P. Selvakumar
https://orcid.org/0000-0002-3650-4548
Department of Science and Humanities, Nehru Institute of Technology, Coimbatore, India

ABSTRACT

Launched specific programs aimed at addressing a broad spectrum of issues designed (2000–2015) while correcting their limitations by adopting a more inclusive, comprehensive, and universally applicable approach. The Origin and Evolution of the SDGs: The SDGs were born out of decades of international development dialogue. They were shaped through extensive consultations across the globe. After the conclusion of the MDGs, which had notable successes in areas like poverty reduction and primary education but were limited in scope and geographical focus, world leaders recognized the need for a broader, more integrated development agenda. The result was unanimously all centered around a rallying cry to "leave no one

DOI: 10.4018/979-8-3373-6435-3.ch015

behind. Key Principles Guiding the SDGs Several core principles shape the SDG framework Special attention is given to marginalized and disadvantaged populations. Integration: Economic, social, and environmental dimensions are addressed together.

UNDERSTANDING THE SUSTAINABLE DEVELOPMENT GOALS (SDGS)

Launched specific programs aimed at addressing a broad spectrum of issues designed (2000–2015) while correcting their limitations by adopting a more inclusive, comprehensive, and universally applicable approach. The Origin and Evolution of the SDGs: The SDGs were born out of decades of international development dialogue. They were shaped through extensive consultations across the globe. After the conclusion of the MDGs, which had notable successes in areas like poverty reduction and primary education but were limited in scope and geographical focus, world leaders recognized the need for a broader, more integrated development agenda. The result was unanimously all centered around a rallying cry to "leave no one behind. Key Principles Guiding the SDGs: Several core principles shape the SDG framework: Inclusivity: Special attention is given to marginalized and disadvantaged populations. Integration: Economic, social, and environmental dimensions are addressed together. Accountability: Countries are encouraged to develop transparent reporting mechanisms. Participation: The active engagement of citizens, communities, and stakeholders is vital. These principles underscore the vision of a just, equitable, and inclusive global society, in which the rights and dignity of every person are upheld. The Role of Stakeholders: Governments: Establish policy frameworks, allocate budgets, and provide public services aligned with SDGs. Civil Society: Monitor implementation, promote accountability, and advocate for marginalized voices. Academia: Provide research, data analysis, and evidence-based policy recommendations. Youth and Communities: Engage in advocacy, entrepreneurship, and innovation at the grassroots level. True success depends on collaborative action, where all stakeholders work toward shared goals with mutual accountability and shared responsibility. A Call to Action: The Decade of Action (2020–2030): Recognizing the urgency of accelerating SDG progress, launched in 2020, galvanizes transformative change at global, local, and individual levels. This involves: Mobilizing greater political leadership and funding. Unlocking science, technology, and innovation. Enhancing citizen engagement and accountability mechanisms. Citizens are encouraged to make sustainable lifestyle choices, participate in community initiatives, and hold leaders accountable. An unprecedented planet, prosperity, peace, and partnerships. They embody a holistic vision for a world that leaves no one behind, ensuring that development is inclusive, equitable, and enduring. While the path to 2030 is complex and challenging, it is also filled with hope, innovation, and opportunity. Understanding the SDGs is the first step toward engaging with them meaningfully—whether as a policymaker, business leader, educator, student, or active citizen.

The Importance of SDGs for Global Development

serve as more equitable, peaceful, and 169 associated targets most urgent health, education. These goals represent an ambitious global agenda aimed at transforming our world by 2030, uniting a shared commitment to sustainable development. Not only a comprehensive scope but also in their capacity to foster long-term systemic change across nations and societies. They have become a guiding framework for global development—encouraging coordinated action, fostering innovation, promoting inclusivity, and ensuring accountability on a planetary scale. A Holistic and Integrated Vision for Development: One of the core strengths of the SDGs is their integrated and indivisible nature, addressing development not as a collection of isolated issues but as a web of interconnected priorities (Elalfy A., et al, 2021). This systems-oriented approach acknowledges that eradicating poverty (Goal 1) cannot happen without addressing hunger (Goal 2), improving, among others. By promoting synergies and reducing trade-offs, the SDGs enable a more strategic and effective approach to development. They encourage countries to adopt cross-sectoral policies and foster cooperation between stakeholders who might otherwise work in silos. Universal Relevance and Global Solidarity: Unlike their predecessor (MDGs) largely targeted at nations SDGs are universal. They apply equally to developed and developing countries, recognizing that challenges like inequality, environmental degradation, and unsustainable consumption are global. This universality fosters among nations, promotes global, encouraging cooperation and resource-sharing across borders. The SDGs thus create a common language for international dialogue, helping align national priorities with global needs and facilitating cross-border collaboration, knowledge transfer, and mutual accountability. Eradicating Poverty and Reducing Inequality: The SDGs place a strong emphasis on the world and its populations (Kleespies M W., et al, 2022). Poverty eradication (Goal 1) remains the cornerstone of the agenda, but the goals go beyond income poverty to address multidimensional deprivation, including education, decent housing. Similarly, tackling systemic barriers to inclusion, such as discrimination, unequal opportunities, and income disparities. By promoting inclusive growth, equitable service delivery, and social protection, the SDGs lay the foundation for more just and resilient societies.

Promoting Peace, Justice, and Strong Institutions: Sustainable development is impossible in the absence of peace. SDG 16 highlights the importance of building and reducing corruption, promoting access to justice. These elements are essential for maintaining social cohesion, enabling economic development, and protecting human rights. Countries with strong institutions are better equipped to respond to crises, manage resources, and deliver public services that make development efforts both sustainable and equitable. The SDGs also promote inclusive decision-making, encouraging citizen participation and civic engagement at all levels of governance (Mishra M., et al, 2024). Catalyzing Innovation and Economic Transformation: The SDGs recognize the transformative power of innovation, infrastructure, and industry (Goal 9) in driving science, technology, and research. Countries can develop new solutions to persistent development problems, renewable energy technologies to climate-smart agriculture, and e-health platforms. The SDGs also promote decent work and inclusive economic growth (Goal 8), supporting job creation, entrepreneurship, and social enterprise. These initiatives are particularly critical in the face of rising automation, digital disruption, and the future of work. Driving Accountability and Transparency: The SDGs have redefined how global development is monitored and evaluated, with an emphasis on evidence-based policy, robust data systems, and transparent reporting mechanisms. Countries voluntarily submit their showcases and challenges in implementing the goals. These reviews promote accountability, highlight successful practices, and stimulate healthy competition among nations. The focus on data disaggregation also ensures that no group—whether based on gender, age, disability, geography, or income—is left behind or rendered invisible in national statistics. Inspiring Local and Global Action: The SDGs are not just a government agenda—they are a people-centered movement (Leal Filho W., et al, 2021). Local governments, cities, schools, businesses, and grassroots organizations across the world are adapting the goals to their unique contexts. This localization of the SDGs ensures that the global agenda becomes a reality in communities everywhere. Furthermore, the SDGs offer a framework for citizens to act, from choosing sustainable products and reducing waste to advocating for social and environmental justice. This bottom-up momentum is essential for driving the transformative change needed to meet the 2030 targets. The SDGs are more than aspirational goals—they are a roadmap for transforming our world. They provide a unifying vision that connects global development efforts with national priorities and local realities. Their importance lies in promoting inclusive growth, environmental stewardship, gender equality, peace, and partnerships that benefit everyone, especially the most vulnerable of need to accelerate progress has never been more urgent. The SDGs are building resilient, equitable, prosperous societies. Embracing the SDGs means committing to a shared future—one where no one is left behind and every person has the opportunity to thrive.

No Poverty (SDG 1): Eradicating Poverty Worldwide

It is the first and foundational goal that it seeks, and the broader dimensions. The importance stems from the fact that poverty is the root cause of numerous global challenges, including hunger, poor health, inequality, conflict, and environmental degradation. SDG 1 promotes inclusive economic growth, increasing protection, and ensuring equal opportunities. With nearly a million more experiencing multidimensional poverty, the urgency of SDG 1 cannot be overstated. Its success is fundamental to achieving all other SDGs. Understanding Poverty: A Multidimensional Challenge is solely defined as a complex and that includes lack of healthcare, housing, employment, and social participation (Hoy C., et al, 2021). People in poverty often experience vulnerability to disasters, displacement, violence, discrimination, and systemic exclusion. SDG 1 acknowledges this complexity by promoting policies that address all aspects of deprivation, not just financial. (MPI), developed by UNDP and the Oxford Poverty and Human Development Initiative, plays a vital role in highlighting how poverty affects people in overlapping ways and underscores the importance of targeted interventions across sectors. Social Protection as a Cornerstone Strategy: Target 1.3 of SDG 1 focuses on expanding social protection systems, which are critical tools in mitigating poverty and inequality. Social protection includes cash transfers, pensions, health insurance, unemployment benefits, child support grants, and food subsidies. These mechanisms not only provide a safety net during times of crisis but also promote economic inclusion and human development. Countries like Brazil (through Bolsa Família) and India (via Pradhan Mantri Jan Dhan Yojana and PM-KISAN) have demonstrated that targeted cash transfers can significantly reduce poverty, improve health and education outcomes, and stimulate local economies (Li Y., et al, 2021). Gender Equality and Poverty Eradication: Poverty affects women and girls disproportionately due to gender discrimination, unpaid care responsibilities, limited access to education and financial resources, and gender-based violence. Addressing gender-based poverty is essential for achieving SDG 1. Empowering women through education, land rights, access to credit, and reproductive healthcare can have a multiplier effect, improving entire households and communities. Integrating gender-sensitive policies into poverty reduction strategies is vital, as is strengthening women's voices in decision-making processes.

Climate Resilience and Disaster Risk Reduction: Target 1.5 of SDG 1 highlights the importance of resilience against shocks, especially climate-related events. Poor populations are the most vulnerable to floods, droughts, cyclones, rising sea levels, and environmental degradation. These hazards can destroy homes, livelihoods, infrastructure, and crops, plunging millions back into poverty. Strengthening climate resilience involves early warning systems, climate-smart agriculture, insurance schemes, resilient infrastructure, and ecosystem restoration. Investing in disaster risk reduction (DRR) not only saves lives but reduces long-term economic losses, especially in low-income communities (Arora N K., et al, 2022). Innovative Approaches and Digital Solutions: Emerging technologies offer new ways to tackle poverty. Digital payments, mobile banking, blockchain for aid transparency, big data for targeting beneficiaries, and AI for needs assessment are transforming poverty reduction strategies. For example, have sent remittances, grown micro-businesses. Digital identification systems also help ensure that social benefits reach the intended recipients. Innovation must be harnessed responsibly and inclusively to bridge the digital divide and promote equity. Urgency and the Way Forward: As we move toward the 2030 deadline, the world must accelerate action on SDG 1. The current pace of progress is insufficient, and rising inequality, geopolitical instability, pandemics, and climate change continue to threaten hard-won gains. Governments must strengthen political will, increase funding for poverty alleviation, and scale up effective interventions. International cooperation, especially with least developed and conflict-affected countries, is vital (Chib S., et al, 2025). There must also be a renewed focus on long-term solutions that empower people economically, socially, and politically. SDG 1 — No Poverty — is not just a goal; it is a moral, social, and economic imperative. Poverty robs individuals of dignity and opportunity while weakening societies and economies. Its eradication is essential for sustainable development, human rights, and global stability. Through inclusive growth, social protection, climate resilience, equitable access to resources, and participatory governance, the goal of a poverty-free world is both achievable and transformative. The time to act decisively is now—to ensure that every person, regardless of where they are born or live, chance security and opportunity.

Zero Hunger (SDG 2): Achieving Food Security

Hunger is urgent in our time. Despite global efforts, more than 735 million people worldwide were chronically undernourished in 2023, with millions more experiencing food insecurity and malnutrition. SDG 2 addresses not only the need for sufficient caloric intake but also the broader issues of nutritional quality, equitable access to food, and sustainable food systems. Ending hunger is central to the realization of including economic growth, climate action, and inequality. To achieve this goal by 2030, the global community must address the root causes of hunger, empower small-scale farmers, invest in rural development, and transform global food systems to be inclusive, resilient, and sustainable (Avalekar U A., et al, 2025). Understanding the Dimensions of Hunger and Food Insecurity: In development terms, it is more accurately reflected by chronic undernourishment, and Food insecurity can range from anxiety about food availability to skipping meals and going days without eating. Additionally, hidden hunger — or micronutrient deficiencies — affects billions who consume enough calories but not enough vitamins and minerals. SDG 2 seeks to eliminate these problems by ensuring nutritional adequacy, food safety, and the reliability of supply chains. Current Global Hunger Landscape: The global hunger crisis has been exacerbated by a series of overlapping crises — including conflict, economic shocks, climate change, and biodiversity loss, for example, disrupted food production, distribution, and access, worsening food insecurity worldwide. Additionally, ongoing conflicts in countries like Yemen, Syria, South Sudan, and Ukraine have led to massive displacements, destroyed agricultural infrastructure, and triggered food shortages. According to the FAO, global hunger has risen since 2015, and without immediate action, SDG 2 is unlikely to be met by 2030, and those in rural or conflict-affected areas are at greatest risk (Chen X., et al, 2023). Agriculture and Rural Development: SDG 2 emphasizes the importance of supporting small-scale farmers, including women, Indigenous peoples, and pastoralists, who often lack access to land, inputs, markets, finance, and technology. Increasing their productivity and incomes is critical to poverty reduction and food security. Sustainable agriculture practices — such as agroecology, organic farming, and water-efficient irrigation- increase yields while conserving the environment. Investments in rural infrastructure, storage, transportation, and market access are equally important.

Food Waste and Loss: An estimated 1.3 billion tons of food — roughly one-third of all food produced — is lost or wasted annually. This wastage occurs throughout the supply chain, from harvest and post-harvest handling to retail and consumption. Food loss not only contributes to hunger but also squanders water, energy, land, and labor. Reducing food loss and waste is critical for increasing food availability, lowering prices, and reducing environmental impact. Solutions include improved storage facilities, cold chains, packaging, efficient logistics, consumer awareness campaigns, and policies to redistribute surplus food (Sporchia F., et al, 2024). Innovations and Digital Solutions: Technology plays a pivotal role in enhancing Precision drones, remote sensing, mobile advisory services, e-markets, and blockchain for supply chain transparency are revolutionizing farming practices. Farmers can now access weather forecasts, market prices, pest alerts, and best practices via mobile apps, reducing risk and improving efficiency. Digital platforms also connect farmers directly to buyers, reducing intermediaries and increasing profit margins. Innovations must be affordable, accessible, and appropriate for small-scale farmers, especially in resource-constrained settings. Policy and Governance: Effective governance is central to achieving food security. Governments must adopt rights-based approaches to food, enshrine and implement policies that ensure universal access to affordable, nutritious food. Social protection programs — such as school feeding, food vouchers, public distribution systems, and nutrition subsidies — help cushion the impact of food price volatility and protect the most vulnerable. Integrating food security into national development plans and aligning agriculture policies with environmental and social objectives are crucial for long-term sustainability (Atukunda P., et al, 2021). SDG 2 — Zero Hunger — is not only about feeding people; it is about ensuring human dignity, health, and opportunity for all. It requires a multi-dimensional approach that combines agricultural innovation, climate resilience, gender equity, nutritional quality, poverty reduction, and sustainable development. Ending hunger is a moral imperative and a test of our collective resolve to leave no one behind. As the world races against the 2030 deadline, bold actions, strong political will, and inclusive partnerships are essential to make Zero Hunger a reality.

Good Health and Well-being (SDG 3): Promoting Health

(SDG 3) aims central pillar Without good individuals cannot their communities from and to safety, substance abuse, environmental health risks. The overarching goal is to improve life expectancy, reduce mortality rates, and build resilient health systems capable of delivering equitable, high-quality care. While significant progress has been made globally, persistent disparities, pandemics, emerging diseases, and systemic barriers continue to challenge the realization of universal health coverage (Macassa G., 2021). Progress and Challenges in Global Health: Since 2000, there has been notable progress in global health outcomes: Global maternal mortality has declined by nearly 34%, and child mortality has fallen by more than 50%. New HIV infections have decreased, and more people now receive antiretroviral therapy than ever before. Malaria deaths have declined due to vector control and rapid diagnostic tests. However, many regions, particularly low-income and conflict-affected countries, continue to experience inadequate healthcare access, underfunded systems, and high disease burdens. The COVID-19 pandemic further reversed years of gains, straining health infrastructure, disrupting immunization and maternal health services, and deepening inequalities. Mental health, often overlooked, has also emerged as a global crisis. Universal Health Coverage (UHC): At the heart of SDG 3 is health coverage, which ensures that everyone can access the health services they need — from prevention to treatment to rehabilitation — without suffering financial hardship (Rad D., et al, 2022). Currently, at least half the world's population lacks full coverage of essential health services, and 930 million people are pushed into poverty due to out-of-pocket health expenditures. Achieving UHC involves: Expanding primary healthcare systems, Recruiting and training health workers, Investing in health infrastructure and supply chains, Implementing financial protection mechanisms, such as public insurance and subsidies. Health Systems Strengthening: Strong and sustainable backbone for health improvements. The COVID-19 pandemic revealed weaknesses in health governance, data collection, and emergency preparedness. Strengthening systems involves: Reliable surveillance and early warning systems, well-functioning referral mechanisms, Digital health technologies, Supply chain and logistics for vaccines and medications, and Equitable allocation of health resources. Community health workers, nurses, and midwives must be empowered and compensated fairly as frontline responders (Zickafoose A., et al, 2024). Road Safety and Injury Prevention: with the suffering from injuries and disabilities road by 2030. Solutions include: Stricter enforcement of speed limits and drunk driving laws, Use of helmets, seatbelts, and child restraints, Improved road design and pedestrian zones, Driver education, and traffic law reform. Sexual and Reproductive Health Rights, including contraception, safe abortion, STI prevention, and comprehensive sexuality education, are fundamental to individual autonomy and public health. SDG 3.7 supports universal access to such services, especially for adolescents, marginalized groups, and those in humanitarian crises. Educating young people, eliminating harmful practices like child marriage, and ensuring confidential and affordable care are essential for reducing unintended pregnancies and promoting reproductive choice. Global Partnerships and Innovation: Global collaboration is vital in achieving SDG 3. Organizations like GAVI, UNICEF, the Fund, and CEPI play key roles in financing, coordinating, and delivering health services worldwide (Upadhyaya A N., et al, 2025). Multilateral partnerships facilitate vaccine development, health technology transfer, and emergency response. Innovative tools — such as telemedicine, mHealth apps, AI-powered diagnostics, and community-based surveillance systems — are transforming health delivery in remote and underserved areas. Sustainable Development Goal 3 envisions a world where every person — regardless of age, gender, geography, or socioeconomic status — enjoys a long and healthy life. Promoting good health and economic development. While significant strides have been made, persistent inequities, rising health threats, and systemic gaps demand renewed commitment. By prioritizing prevention, care, fostering global solidarity, and realizing ambitious goals.

Figure 1. Sustainable development goals

Quality Education (SDG 4): Ensuring Inclusive Education

Quality Education cornerstone "ensure inclusive and equitable quality education and promote lifelong learning opportunities for all." Education catalyst individual empowerment, poverty eradication, gender equality, people of all ages, backgrounds, and abilities, SDG 4 seeks to break the cycle of disadvantage and Despite progress, education landscape remains marked by deep inequalities, learning deficits, and systemic barriers—challenges that SDG 4 seeks to overcome through ambitious, inclusive, and transformative targets (Nisha Angeline M., et al, 2025). The Importance of Inclusive and Equitable Education: Inclusion and equity are core to SDG 4. Inclusive education ensures that all learners—regardless of gender, socioeconomic status, ethnicity, language, disability, refugee status, or geographical location quality learning. Equity ensures that those who are disadvantaged or marginalized receive additional support to achieve comparable outcomes. Global data reveal that children from the poorest households are five times more likely to be out of school than those from the richest. Girls, especially in conflict zones and rural regions, often face barriers such as child marriage, domestic labor, or cultural bias. Ensuring inclusivity means addressing both access and participation, and removing physical, social, and financial barriers. Universal Primary and Secondary Education: One of the most fundamental goals of SDG 4 is to ensure that every child completes While enrollment rates have improved, completion of learning outcomes remains a major concern. More than adolescents, and even those in school, often lack basic skills. In many low-income countries, fewer than half of children complete lower secondary education. Challenges include: Lack of trained teachers, Overcrowded classrooms, Gender-based violence in schools, Conflict and displacement, Poverty, and child labor (Roy C K., et al, 2022). Strategies must focus on curriculum reform, inclusive teaching methods, remedial learning, and gender-sensitive policies to improve both access and quality. Lifelong Learning and Adult Education: SDG 4 promotes lifelong learning, recognizing that education should extend beyond childhood and be accessible at all stages of life. Adult education includes literacy programs, continuing education, and upskilling initiatives. Yet globally, over overengage in civic life and lead healthier lives. Governments must prioritize adult education in national education policies and make use of community centers, mobile learning, and non-formal learning channels.

Quality of Education and Learning Outcomes: Access alone is not enough—education must lead to meaningful learning. Many students complete years of schooling without acquiring foundational literacy and numeracy. The concept of "learning poverty" is alarming many. Improving quality involves: Qualified, motivated, and supported teachers, Student-centered pedagogy, culturally relevant and inclusive curricula, and Assessment systems that promote learning, not rote memorization (Monaco S., 2024). Technology, when appropriately used, can support differentiated learning and bridge educational gaps. Infrastructure and Learning Environments: Safe, inclusive, and effective learning environments are essential infrastructure, such as toilets, clean water, electricity, or digital connectivity in schools. Schools should be child-friendly, gender-sensitive, and disaster-resilient. Investments in digital infrastructure and climate-smart facilities are necessary to ensure that every child can learn in a safe and supportive space. Global Cooperation and Financing: Achieving SDG 4 requires strong political will, adequate financing, and global partnerships. Yet many countries spend less than the recommended 4-6% of GDP or 15-20% of public expenditure on education. The Global Partnership for Education (GPE), Education Cannot Wait, and other multilateral mechanisms support low-income countries in building resilient education systems (Tanui–Too P., 2021). Private sector engagement, innovation, and civil society advocacy also contribute to expanding quality education. Sustainable Development Goal 4 aspires to transform education into a powerful force for equality, sustainability, and human dignity. It calls for a radical rethinking of how we educate, who we include, and what outcomes we prioritize. From early childhood through adulthood, from formal schooling to informal skills development, education must be inclusive, equitable, and learner-centered. Despite challenges, the commitment to SDG 4 remains a beacon of hope. By investing in quality education for all, the world will be full every individuals laying a just, prosperous foundation.

Gender Equality (SDG 5): Empowering Women and Girls

One of the transformative and ambitious goals within the United Nations' inequalities in nearly every aspect of life: education, healthcare, and freedom from violence. SDG 5 seeks to eliminate discrimination and structural barriers, ensure economic rights, and and girls across all spheres of society (Louis N., et al, 2021). The Persistence of Discrimination and Inequality: In nearly all societies, patriarchal norms, stereotypes, and systemic discrimination continue to disadvantage women and girls. Legal and cultural biases often restrict mobility and decision-making power. In 2024, only 55% of women aged 15–49 globally were empowered to make their own decisions regarding sexual and reproductive health and rights. In over 100 countries, laws still exist that limit women's participation in the workforce, deny them inheritance rights, or grant unequal custody rights. Addressing these structural inequalities requires comprehensive legal reforms, robust enforcement, and transformative shifts in cultural attitudes. Ending Harmful Traditional Practices: Target 5.3 focuses on ending child marriage and female genital mutilation (FGM), both deeply rooted in cultural traditions and gender norms. These practices have devastating health, educational, and economic consequences. Efforts to eliminate them must involve: Community engagement and education, Enforcement of legal bans, Support for girls at risk, and Involving men and boys in the conversation. Programs led by grassroots women's organizations and survivor-led movements have shown a strong impact in reducing these practices (Aman H., et al, 2024). Gender Equality in the Digital Age: Target 5.b emphasizes leveraging technology and innovation to promote gender empowerment. Yet, a global digital gender divide persists: Women are 18% less likely to own a smartphone. Girls often have less access to STEM education and digital skills training. Closing the digital gap can open new opportunities for employment, entrepreneurship, health information, civic participation, and education. Gender-responsive digital policies, inclusive tech design, and digital literacy campaigns are critical.

Intersectionality and Inclusive Gender Policies: Gender inequality is not experienced in isolation. Women and girls may also face marginalization due to disability, race, ethnicity, sexual orientation, age, caste, religion, or displacement status. For instance, Indigenous women or LGBTQ+ persons often face compounded barriers to education, healthcare, and justice (Küfeoğlu S., 2022). An intersectional approach ensures that SDG 5 addresses the needs of all women, especially the most marginalized. The Role of Men and Boys in Achieving Gender Equality: Engaging allies and advocates is critical to transforming power dynamics and dismantling patriarchal norms. Programs like "HeForShe" and men's engagement in fatherhood, caregiving, and anti-violence initiatives show the importance of shared responsibility. Linkages Between SDG 5 and Other Goals: Education (SDG 4): Educated girls marry later, have healthier children, and earn higher incomes. Poverty Reduction (SDG 1): Empowering women economically lifts entire families out of poverty. Health (SDG 3): Reproductive rights reduce maternal mortality and improve child health. Peace (SDG 16): Women's participation in peace processes leads to more durable peace. Progress on SDG 5 accelerates progress across the 2030 Agenda. Achieving an empowering, just moral imperative—it is a strategic and societal one (Mehra J., et al, 2025). SDG 5 offers a bold vision of a world where every woman and girl can live free from violence, realize her full potential, and participate fully in shaping her future and that of her community. But this vision can only be realized through collective action, political will, adequate resources, and cultural transformation. Must contribute to dismantling the structural barriers that continue to hold women back. Only when gender equality becomes a lived reality will all be fulfilled.

Clean Water and Sanitation (SDG 6): Ensuring Access

One of the most fundamental human rights and critical necessities for life is agriculture, industry, and environmental sustainability. Yet, worldwide, basic facilities are lacking. Water scarcity, pollution, climate change, poor infrastructure, and inequitable distribution continue to undermine progress, particularly in low-income and rural communities. Achieving SDG 6 by 2030: water security, hygiene, sustainable management, and inclusive governance. The Importance of Water, Sanitation, and Hygiene (WASH): Ensures protection from such as pathogens, heavy metals, and industrial waste. Promotes sanitation systems that safely separate human waste from contact with people and the environment (Coates J M., et al, 2025). Emphasizes hand hygiene and menstrual hygiene management to prevent disease and enable school attendance, especially for girls. Investments in WASH have a multiplier effect: every $1 invested in clean water and sanitation returns up to $5–$28 in increased productivity, reduced healthcare costs, and improved well-being. Gender and Social Equity in Access to Water: Water scarcity inadequate impact on women and girls disproportionately. In many regions, Poor sanitation also restricts their ability during menstruation and privacy. Gender-sensitive water programs must: Ensure privacy, dignity, and safety in WASH infrastructure. Include women's voices in water governance. Provide menstrual hygiene education and supplies. Promote equal access and reduce time burdens (Rajapakse J., et al, 2023). Empowering women in WASH management improves outcomes and strengthens community resilience. Improving Water Quality and Ecosystem Protection: Targets 6.3 and 6.6 highlight the urgent need to address pollution and ecosystem degradation. Industrial waste, untreated sewage, agricultural runoff, and plastic pollution severely impact freshwater systems. Effective measures include: Expanding wastewater treatment facilities, Regulating chemical use and industrial discharges, Protecting and restoring wetlands, rivers, aquifers, and watersheds, Encouraging nature-based solutions such as reforestation and constructed wetlands. Healthy water ecosystems are vital for biodiversity, carbon sequestration, flood control, and sustainable livelihoods (Bhattacharya R., et al, 2022). Community Participation and Behavioral Change: The success of water and sanitation programs depends on community engagement and behavioral change. Local communities must be involved in designing, managing, and maintaining WASH services to ensure cultural appropriateness and ownership. Hygiene education campaigns—such as handwashing with soap—reduce disease spread and are especially vital in schools, health centers, and emergency settings. Youth, women, and marginalized groups must have a voice in water governance and decision-making. Social mobilization, capacity building, and local leadership are crucial for long-term success and sustainability. Monitoring Progress and Addressing Data Gaps: Accurate and disaggregated data are essential to track progress on SDG 6. However, many countries lack reliable data on water quality, service coverage, water withdrawals, and wastewater treatment. Strengthening data systems requires: Investment in monitoring infrastructure, Standardized methodologies, Community-based data collection, and Open access to water data platforms (Zehri C., et al, 2024). Transparent reporting and data-driven planning help identify gaps, target interventions, and ensure accountability. Water is life, yet its safe and sustainable provision remains one of humanity's greatest challenges. Sustainable Development Goal 6 is not only about building water pipes and toilets—it's about building systems that ensure dignity, health, equity, and resilience. To meet the 2030 targets, we must scale up investment, prioritize marginalized populations, promote integrated and community-led solutions, and build capacity at all levels. As urbanization intensifies strain on water resources, urgent and coordinated global action is essential distant aspiration critical imperative that will define the future of human and planetary well-being.

Decent Work and Economic Growth (SDG 8): Fostering Economic Opportunities

(SDG 8) "promote sustained, inclusive and sustainable economic growth, full and productive employment, and decent work for all." It is strongly driven by innovation, productivity, and employment, and is long-term. Yet, this growth must be inclusive and environmentally sustainable to benefit all sections of society. The goal encompasses a broad spectrum of targets that emphasize youth employment, equal pay, labor rights, entrepreneurship, financial inclusion, and safe working environments. In a global context marked by rising inequality, technological disruptions, and climate change, SDG 8 provides a framework for advancing environmental stewardship. Understanding the Pillars of SDG 8:SDG 8 rests on the following key pillars: Sustained and inclusive economic growth: Encouraging GDP growth that benefits all, especially in least developed countries (LDCs). Productive employment and decent work: Ensuring jobs that are fairly paid, secure, and uphold labor rights (Chigbu B I., et al, 2023). Entrepreneurship and innovation: Promoting micro, small, and medium enterprises (MSMEs) through support and access to finance. Youth and vulnerable groups: Tackling unemployment and underemployment among young people and marginalized populations. Labor protection and rights: Safeguarding workers' rights and ensuring workplace safety and dignity. Sustainable tourism and resource efficiency: Encouraging job-rich growth while managing natural resources responsibly. These elements are reflected in SDG 8's 12 targets, including measurable indicators that track progress in areas such as unemployment rates, child labor, access to banking, and labor productivity. Global Employment Trends and Challenges: As of 2025, several challenges threaten the achievement of SDG 8: Global unemployment remains high, especially among youth (ages 15–24), with over 400 million people lacking adequate employment opportunities. Informal employment is prevalent, especially in developing regions where more than 60% of workers operate outside of legal and regulatory frameworks. The rise of platform work and the gig economy has brought flexibility but also insecurity and a lack of benefits for many workers. Automation and digitalization are displacing certain job sectors, especially among low-skilled workers (Kreinin H., et al, 2022). The COVID-19 pandemic caused massive job losses and widened disparities, and recovery remains uneven globally. Despite these challenges, targeted interventions and sustainable policy frameworks can reverse these trends and stimulate inclusive labor market growth. Inclusive Economic Growth: A Catalyst for Development: Inclusive Economic Growth—meaning it reduces inequality and benefits marginalized populations— is crucial to sustainable development. According to SDG 8, Least developed countries (LDCs) should aim for at least 7% GDP growth per year. Growth should be accompanied by investments in infrastructure, industry, education, and healthcare to broaden its impact. Fiscal policies must ensure redistribution and social safety nets to protect vulnerable communities. However, growth must be decoupled from environmental degradation. Green growth strategies, investments allow economies to grow while respecting planetary boundaries.

Entrepreneurship and MSME Development: Micro, Small, and Medium Enterprises (MSMEs) are the backbone of most economies, accounting for over 90% of businesses and more than 50% of employment worldwide. SDG 8 recognizes their role in innovation, local development, and job creation. However, MSMEs face: Limited access to affordable credit. Regulatory burdens and red tape. Lack of business support services and infrastructure. To promote entrepreneurship, Governments and financial institutions must offer inclusive financing mechanisms. Digital platforms and e-commerce should be leveraged to expand market reach. Business training and incubation services can improve survival rates. Special support should be given to women-, youth-, and minority-led enterprises (UN ESCAP., 2021). Creating an enabling ecosystem for MSMEs leads to resilient, diversified, and inclusive economies. Financial Inclusion and Economic Empowerment: Yet, nearly 1.4 billion people remain unbanked, particularly in rural areas and among women. SDG 8 promotes: Universal access to banking, insurance, and financial services. Support for digital finance tools like mobile banking and fintech. Financial literacy programs to empower users. Policies to ensure consumer protection and ethical lending. Financial inclusion boosts savings, encourages entrepreneurship, reduces vulnerability, and strengthens economies from the ground up. Digital Transformation and Future of Work: Technological disruption is reshaping the world of work. While AI, automation, and digital platforms create new jobs and efficiencies, they also raise challenges such as Job displacement in manufacturing and services. Growing digital divides. Erosion of traditional employer-employee relationships. SDG 8 policies must: Prepare workers through lifelong learning and reskilling. Ensure digital infrastructure is available in underserved regions. Update labor laws to include gig and platform workers. Support remote work while addressing its mental health impacts and work-life balance. Technology must be harnessed not to replace humans, but to augment human potential and create new pathways (Limbore N V., et al, 2025). Sustainable Development Goal 8 envisions a world where economic growth leads to widespread opportunity, human dignity, and environmental stewardship. It challenges outdated models of development that rely on exploitation, inequality, and ecological harm. Instead, SDG 8 calls for growth that is inclusive, rights-based, job-rich, and future-oriented. Achieving this vision requires the collaboration of workers. With the right policies, investments, and innovations, we can build economies that offer decent work for all—fueling prosperity while safeguarding the planet and advancing global equity.

Climate Action (SDG 13): Combating Climate Change

Critical global priorities, escalating effects on global extremes, and disrupted ecosystems are all symptomatic of anthropogenic climate change. These environmental shifts are already undermining food security, health, economic stability, and geopolitical peace. SDG 13 addresses this global emergency by urging nations, communities, and individuals to adopt ambitious climate policies, economies and build resilience against related hazards. The goal recognizes that both imperative and strategic necessity with other such as SDG 11, Failure to act on change jeopardizes progress across all development areas (Chowdhary S., et al, 2025). Conversely, effective climate action can yield co-benefits for health, employment, innovation, and environmental sustainability. Adaptation: Strengthening Climate Resilience: While mitigation tackles the root cause, adaptation measures are already unfolding: Climate-resilient agriculture uses drought-resistant crops, efficient irrigation, and crop diversification. Disaster risk reduction (DRR) improves early warning systems, evacuation plans, and infrastructure to withstand floods and storms. Urban resilience includes green roofs, permeable surfaces, and cool corridors to reduce urban heat islands. Coastal adaptation strategies include sea walls, mangrove restoration, and managed retreat from high-risk zones. Adaptation is particularly critical for vulnerable regions where livelihoods depend on natural resources and adaptive capacity is limited. Financing, technology transfer, and knowledge sharing are essential to support effective adaptation efforts in the Global South (Owusu-Sekyere E., et al, 2024). Climate Finance: Bridging the Gap: Measures According to the UNFCCC: Trillions of dollars are needed annually to align economies with climate goals. The \$100 billion per year commitment from developed to developing countries remains unfulfilled. Climate finance is needed for: Renewable energy infrastructure and grid upgrades. Adaptation projects in agriculture, water management, and health. Capacity-building and climate education programs. Technology innovation and deployment. Institutions help channel finance to climate-vulnerable countries. However, better coordination, transparency, and effective access mechanisms.

Youth and Civil Society: Driving Momentum: Young people and grassroots organizations are at the forefront of climate action: Movements such as Fridays for Future, Extinction Rebellion, and Climate Strike have mobilized millions globally. Youth activists like Greta Thunberg, Vanessa Nakate, and Licypriya Kangujam have become voices of urgency and intergenerational equity. Indigenous communities advocate for climate justice by protecting biodiversity, practicing sustainable resource management, and resisting extractive industries. SDG 13 underscores the importance of inclusive participation, ensuring that women, youth, and marginalized groups have a voice in climate decision-making and benefit from climate solutions (Arora N K., et al, 2023). Technological Innovation and Data for Climate Action: Science and technology are vital tools in combating climate change: Earth observation satellites, remote sensing, and AI-powered modeling enhance climate forecasting and disaster preparedness. Smart grids, electric vehicles, and green hydrogen are enabling decarbonization. Blockchain and digital platforms support transparency in carbon markets and supply chains. Collaboration between academic institutions, governments, and industry is essential for driving climate innovation. Challenges and Opportunities Ahead: Despite progress, numerous challenges remain: Many countries' current policies could lead to over 2.5°C of warming. Fossil fuel subsidies persist, and carbon pricing mechanisms remain underutilized. Vulnerable countries face challenges in accessing climate finance and implementing adaptation strategies. However, climate action also offers immense opportunities: The green economy can create millions of jobs. Nature-based solutions restore ecosystems while providing climate benefits. Resilient communities are more prosperous, equitable, and secure (He X., et al, 2023). To seize these opportunities, climate action must be embedded in every facet of development planning and governance. Challenge SDG 13 calls for urgent, ambitious, and coordinated action for the future. It demands political will, public engagement, and international solidarity transition — not just in economic terms, but in lives lost, livelihoods destroyed, and ecosystems degraded. Achieving SDG 13 is not merely about reducing emissions. It is about transforming the way we live, produce, consume, and relate to the Earth. The path to a climate-resilient world is also a path to justice, peace, and prosperity for all. The time to act is now — before climate change becomes irreversible, and the opportunity to safeguard our future slips away.

CONCLUSION

The SDGs, a comprehensive blueprint for achieving inclusive, global development. As we approach the final stretch of the timeline, it becomes essential to reflect on the progress made, acknowledge the challenges that remain, and reaffirm the collective commitment to realizing the full potential of this transformative global agenda. The SDGs have had a profound impact on national policies, international partnerships, civil society efforts, private sector strategies, and academic research. Countries have integrated the goals into development frameworks, and businesses are aligning operations with sustainability benchmarks. Civil society organizations and youth movements are raising awareness and advocating for action. Innovations in data science, technology, and financing mechanisms have emerged to support implementation and monitoring. According to the United Nations' latest reports, the COVID-19 pandemic, conflicts, climate crises, and economic instability have reversed or stalled progress across multiple goals. Millions have been pushed back into extreme poverty, gender disparities have widened, and access to healthcare, education, and nutrition has deteriorated for vulnerable populations. The climate emergency, coupled with biodiversity loss and environmental degradation, threatens to undermine decades of progress. Furthermore, funding gaps, weak governance structures, lack of accountability, and insufficient data continue to impede effective implementation and evaluation. Nevertheless, the SDGs have proven to be more than a checklist of development indicators—they serve as a moral compass, guiding humanity toward dignity, justice. They have encouraged a shift from siloed interventions to long-term, systemic thinking. By emphasizing "leave no one behind," the SDGs have brought equity and inclusion to the forefront of development discourse, advocating for the disadvantaged. Another critical element is data and monitoring. Robust data collection, real-time tracking, disaggregated statistics, and evidence-based policymaking are indispensable for understanding progress and course-correcting where necessary. Advances in digital technology, satellite imagery, big data, and AI can aid in mapping, analyzing, and forecasting trends. At the national and local levels, governments must prioritize SDG-aligned policies, decentralize implementation efforts, and foster inclusive governance. Localizing the SDGs—adapting them to suit regional and cultural contexts—ensures relevance and impact on the ground. In this regard, local authorities, community leaders, and grassroots organizations become crucial actors in delivering the goals. The role of education and youth cannot be overstated. Education systems must integrate sustainability, global citizenship, and critical thinking to prepare future generations for the challenges and opportunities of the 21st century. Young people are not only beneficiaries of the SDGs but also active contributors—driving change through innovation, advocacy, entrepreneurship, and social movements. Furthermore, gender equality and social inclusion are foundational to the SDG agenda. Empowering women and girls, supporting indigenous communities, addressing systemic discrimination, and ensuring equal access to opportunities for all are not just ethical imperatives—they are essential for sustainable development. Societies that prioritize inclusivity tend to be more resilient, innovative, and economically dynamic. Importantly, achieving the SDGs also requires rethinking consumption and production patterns. Overconsumption, waste, and unsustainable lifestyles contribute to environmental degradation and social inequity. As we move toward 2030, SDGs are still achievable—but only through a radical acceleration of efforts, reimagining of systems, and reinvigoration of commitments. The decade ahead must be one of bold leadership, innovative thinking, and collective resolve. Governments must act decisively, businesses must lead responsibly, communities must engage meaningfully, and individuals must contribute consciously. In conclusion, the Sustainable Development Goals offer the most ambitious, inclusive, and integrated framework ever conceived for global progress. They reflect humanity's highest aspirations—to end poverty, heal the planet, and ensure peace and prosperity for all. Their success hinges not merely on the actions of institutions but on the will of people to imagine and create a better world. The SDGs challenge us to transcend borders, ideologies, and interests in pursuit of a future where every person can thrive with dignity, in harmony with nature. As such, they are not only a roadmap—they are a covenant of hope, action, and shared destiny.

REFERENCES

Aman, H., Doost, Z. H., Hejran, A. W., Mehr, A. D., Szczepanek, R., & Gilja, G. (2024). Survey on the challenges for achieving SDG 6: clean water and sanitation: a global insight. *Knowledge, Innovation and Sustainability*, *4*(2), 98–117. DOI: 10.51526/kbes.2024.5.3.21-42

Arora, N. K., & Mishra, I. (2022). Current scenario and future directions for Sustainable Development Goal 2: A roadmap to zero hunger. *Environmental Sustainability*, *5*(1), 25–34. DOI: 10.1007/s42398-022-00235-8

Arora, N. K., & Mishra, I. (2023). Sustainable development goal 13: Recent progress and challenges to climate action. *Environmental Sustainability*, *6*(2), 201–210. DOI: 10.1007/s42398-023-00287-4

Atukunda, P., Eide, W. B., Kardel, K. R., Iversen, P. O., & Westerberg, A. C. (2021). Unlocking the potential for achievement of the UN Sustainable Development Goal 2–'Zero Hunger'–in Africa: Targets, strategies, synergies and challenges. *Food & Nutrition Research*, *65*. Advance online publication. DOI: 10.29219/fnr.v65.7686

Avalekar, U. A., Mehta, V., Mishra, B. R., Prasanna, L., Rajlakshmi, P. V., Manjunath, T. C., & Selvakumar, P. (2025). The current advances in AI-driven education. In Babulak, E. (Ed.), *Educational AI humanoid computing devices for cyber nomads* (pp. 173–196). IGI Global Scientific Publishing., DOI: 10.4018/979-8-3693-8985-0.ch008

Bhattacharya, R., Kumari, A., & Bose, D. (2022). Impact of Covid-19 on SDG 6 and integrated approaches for clean water access and sanitation. *Sustainability and Climate Change*, *15*(2), 75–87. DOI: 10.1089/scc.2022.0040

Chen, X., Shuai, C., & Wu, Y. (2023). Global food stability and its socio-economic determinants towards sustainable development goal 2 (Zero Hunger). *Sustainable Development (Bradford)*, *31*(6), 2611–2626. DOI: 10.1002/sd.2482

Chib, S., Mehta, A. K., Selvakumar, P., Saravanan, A., Mishra, B. R., & Manjunath, T. C. (2025). Mental health challenges and solutions in high-pressure work environments. In Harriott, S. (Ed.), *Innovative approaches to managing conflict and change in diverse work environments* (pp. 247–272). IGI Global Scientific Publishing., DOI: 10.4018/979-8-3693-9556-1.ch011

Chigbu, B. I., & Nekhwevha, F. (2023). Exploring the concepts of decent work through the lens of SDG 8: Addressing challenges and inadequacies. *Frontiers in Sociology*, *8*, 1181984. DOI: 10.3389/fsoc.2023.1266141

Chowdhary, S., Selvakumar, P., Sangeetha, B., Sarvaani, V. V., Anute, N., & Sharma, M. (2025). The role of social capital in entrepreneurial development. In Martínez-Falcó, J., Sánchez-García, E., & Marco-Lajara, B. (Eds.), *Support networks for entrepreneurs: Social capital and strategic collaborations* (pp. 311–338). IGI Global Scientific Publishing., DOI: 10.4018/979-8-3693-9836-4.ch013

Coates, J. M., Santhanalakshmi, K., Dinesh, N., Gandhimathi, S., & Chandel, P. S. Mohit, Selvakumar P, (2025). "Assessment and evaluation in transnational staff development." In Morris G, Ye J (Eds.), Training and development in transnational higher education (pp. 387-406). IGI Global Scientific Publishing. DOI: DOI: 10.4018/979-8-3373-1897-4.ch019

Elalfy, A., Weber, O., & Geobey, S. (2021). The Sustainable Development Goals (SDGs): A rising tide lifts all boats? Global reporting implications in a post SDGs world. *Journal of Applied Accounting Research*, *22*(3), 521–540. DOI: 10.1108/JAAR-06-2020-0116

He, X., Khan, S., Ozturk, I., & Murshed, M. (2023). The role of renewable energy investment in tackling climate change concerns: Environmental policies for achieving SDG-13. *Sustainable Development (Bradford)*, *31*(3), 1081–1094. DOI: 10.1002/sd.2491

Hoy, C., & Sumner, A. (2021). The End of Global Poverty: Is the UN sustainable development goal 1 (Still) achievable? *Global Policy*, *12*(4, S6), 101–114. DOI: 10.1111/1758-5899.12992

Kleespies, M. W., & Dierkes, P. W. (2022). The importance of the Sustainable Development Goals to students of environmental and sustainability studies—A global survey in 41 countries. *Humanities & Social Sciences Communications*, *9*(1), 162. Advance online publication. DOI: 10.1057/s41599-022-01242-0

Kreinin, H., & Aigner, E. (2022). From 'Decent work and economic growth' to 'Sustainable work and economic degrowth': A new framework for SDG 8. *Empirica*, *49*(2), 983–1006. DOI: 10.1007/s10663-021-09526-5

Küfeoğlu, S. (2022). SDG-6 clean water and sanitation. In *Emerging Technologies: Value Creation for Sustainable Development* (pp. 201–219). Springer., DOI: 10.1007/978-3-031-07127-0_9

Leal Filho, W., Lovren, V. O., Will, M., Salvia, A. L., & Frankenberger, F. (2021). Poverty: A central barrier to the implementation of the UN Sustainable Development Goals. *Environmental Science & Policy*, *120*, 116–129. DOI: 10.1016/j.envsci.2021.08.020

Li, Y., Wu, W., & Wang, Y. (2021). Global poverty dynamics and resilience building for sustainable poverty reduction. *Journal of Geographical Sciences*, *31*(5), 681–702. DOI: 10.1007/s11442-021-1890-4

Limbore, N. V., Anute, N., Ahluwalia, G. K., Chordiya, P. B., Kalshetti, P., & Selvakumar, P. (2025). Ethical leadership and corporate responsibility. In Harriott, S. (Ed.), *Innovative approaches to managing conflict and change in diverse work environments* (pp. 193–218). IGI Global Scientific Publishing., DOI: 10.4018/979-8-3693-9556-1.ch009

Louis, N., Shyleen, C., & Tirivangasi, H. M. (2021). Barriers towards sustainable development goal (SDG 5) on achieving gender equality and empowerment of women and girls in Nyanga rural district of Manicaland. *African Journal of Gender, Society and Development*, *10*(1), 33–46. DOI: 10.31920/2634-3622/2021/v10n3a2

Macassa, G. (2021). Can sustainable health behaviour contribute to ensure healthy lives and wellbeing for all at all ages (SDG 3)? A viewpoint. *Journal of Public Health Research*, *10*(2), 2215. DOI: 10.4081/jphr.2021.2051

Mehra, J., Madaan, S., & Balu, N. Mohit, Selvakumar P, Manjunath T C, (2025). "The ubiquitous internet and smart computation available 24/7 for anyone and anywhere." In Babulak E (Ed.), Educational AI humanoid computing devices for cyber nomads (pp. 247-270). IGI Global Scientific Publishing. DOI: DOI: 10.4018/979-8-3693-8985-0.ch010

Mishra, M., Desul, S., Santos, C A G., & Mishra, S. K. (2024). A bibliometric analysis of sustainable development goals (SDGs): A review of progress, challenges, and opportunities. *Environment, Development and Sustainability*, *26*, 7567–7588. DOI: 10.1007/s10668-023-03855-5

Monaco, S. (2024). SDG 5. Achieve gender equality and empower all women and girls. In *A Guide for Achieving the UN Sustainable Development Goals* (pp. 89–102). Emerald Publishing., DOI: 10.1108/978-1-83797-549-520241006

Nisha Angeline, M., Selvakumar, P., Gupta, M. K., Nazar, N., Elumalai, D., & Bhattacharya, S. (2025). Emerging materials for energy storage and environmental applications. In Muduli, K., Moharana, B., Ales, S., & Biswal, D. (Eds.), *Using computational intelligence for sustainable manufacturing of advanced materials* (pp. 53–72). IGI Global Scientific Publishing., DOI: 10.4018/979-8-3693-7974-5.ch003

Owusu-Sekyere, E., Nyam, Y. S., & Selelo, O. T. (2024). "Sustainable Development Goal 13: Urgent action to combat climate change and its impacts." In: Handbook on Public Policy and Climate Action, ElgarOnline, 201–218. DOI: DOI: 10.4337/9781839105449.00035

Rad, D., Redeş, A., Roman, A., Ignat, S., & Lile, R. (2022). Pathways to inclusive and equitable quality early childhood education for achieving SDG4 goal—A scoping review. *Frontiers in Education*, *7*, 969341. Advance online publication. DOI: 10.3389/feduc.2022.969341

Rajapakse, J., Otoo, M., & Danso, G. (2023). Progress in delivering SDG6: Safe water and sanitation. *Cambridge Prisms: Water*, *1*, e18. DOI: 10.1017/wat.2023.5

Roy, C. K., & Xiaoling, H. (2022). Achieving SDG 5, gender equality and empower all women and girls, in developing countries: How aid for trade can help? *International Journal of Social Economics*, *49*(5), 691–705. DOI: 10.1108/IJSE-12-2020-0813

Sporchia, F., Antonelli, M., Aguilar-Martínez, A., Bach-Faig, A., Caro, D., Davis, K. F., Sonnino, R., & Galli, A. (2024). Zero hunger: Future challenges and the way forward towards the achievement of sustainable development goal 2. *Sustainable Earth*, *7*(1), 22. Advance online publication. DOI: 10.1186/s42055-024-00078-7

Tanui–Too, P. (2021). Gender equality and empowerment of women and girls. *Africa Habitat Review*, *15*(2), 47–55. DOI: 10.37284/ahr.15.2.821

UN ESCAP. (2021). *SDG 13: Climate action: take urgent action to combat climate change and its impacts*. United Nations Economic and Social Commission for Asia and the Pacific., DOI: 10.18356/e-escwa-rfsd-2021-inf13-

Upadhyaya, A. N., Debnath, P., Rani, P., Selvakumar, P., Manjunath, T. C., & Bhattacharya, S. (2025). AI-based disaster prediction and early warning systems. In Mobo, F. (Ed.), *AI and emerging technologies for emergency response and smart cities* (pp. 55–74). IGI Global Scientific Publishing., DOI: 10.4018/979-8-3693-9770-1.ch003

Zehri, C., El Amin, B. M., Inaam, Z., & Sekrafi, H. (2024). Exploring the nexus of decent work, financial inclusion, and economic growth: A study aligned with SDG 8. *Sustainable Futures : An Applied Journal of Technology, Environment and Society*, *6*, 100299. DOI: 10.1016/j.sftr.2024.100213

Zickafoose, A., Ilesanmi, O., Diaz-Manrique, M., Adeyemi, A. E., Walumbe, B., Strong, R., Wingenbach, G., Rodriguez, M. T., & Dooley, K. (2024). Barriers and challenges affecting quality education (Sustainable Development Goal# 4) in sub-Saharan Africa by 2030. *Sustainability (Basel)*, *16*(4), 1636. DOI: 10.3390/su16072657

Chapter 16
Waste Management and Resource Optimization in Local Industries

Sandeep Gupta
https://orcid.org/0000-0002-3734-3723
Graphic Era University, Dehradun, India

Anjali Singh
https://orcid.org/0009-0009-9132-1072
Amity University, Lucknow, India

Naghma Ali
United Institute of Management, Prayagraj, India

Shagufta Ali
https://orcid.org/0000-0002-1839-6109
United Institute of Management, Prayagraj, India

ABSTRACT

This chapter investigates different approaches, methods, and technologies implemented in waste management and resource utilization, emphasizing local industrial systems. As sustainable practices become the norm and green technologies are enforced by stricter environmental regulations, we in Malaysia have started to see local industries adopting intelligent waste management, which not only reduces wastage but also becomes a source of revenue. This chapter on Waste-to-Resource practices also covers the modern waste management scenario and principles of the Circular economy. It approaches resource efficiency measures, with a couple of case studies demonstrating how these approaches have been integrated into local industries. The intention is to give researchers and relevant industrial practitioners

DOI: 10.4018/979-8-3373-6435-3.ch016

a complete view of the holistic resource management systems that are applicable to different local industrial contexts. The chapter highlights the technology, policy framework, and stakeholder engagement required to ingrain sustainable practices in different sectors.

INTRODUCTION

One such new-age problem is making local industries more sustainable amid environmental degradation, resource depletion, and climate change. Industries play a crucial role in economic development, but they also create a mess by generating waste, consuming energy, and depleting resources. Allocation of waste management practices and recovery of resources at the local level has therefore become critical, especially for industrial systems. Small industries at the local level do not have the infrastructure and support systems (backward and forward linkages) that large industries have (Jones et al., 2012).

The emphasis in this section will be on how local industries may implement and include waste management practices to mitigate the impact on the environment and improve operational efficiency, on the other hand, strict operational regulations. Waste management, an evil that was necessary to do business, is changing. It is not just about disposing of things at the end-of-pipe anymore. It is undergoing an integrated and strategic agenda in which wasteful practices are eliminated by embedding sustainable design into the industrial process itself.

Resource efficiency means using less material, energy, and water in production, while minimising waste and emissions, and optimising the lifecycle of resources. Organizations can make huge environmental and economic gains through waste reduction as well as optimizing resource use (Roberge et al., 1994). Furthermore, these projects are frequently linked to larger international goals, such as reducing carbon footprints and achieving sustainability standards (lean manufacturing) or corporate social responsibility objectives (CSR).

This section will contribute the necessary background to appreciate the importance of waste and resource optimisation in local industries while enhancing their type of engagement to limit environmental impact with a view to sustainable industrialisation. This chapter provides an applied account of the principles, strategies, and challenges of waste-to-resource management that will enable the industrial stakeholders (e.g., SMEs & Local Industry leaders), policy-makers, and researchers to see through industrial technological enablement demonstrated in practices.

The Urgency of Addressing Waste and Resource Management in Local Industries

The area-based economy typically relies on local industries that generate a significant amount of waste during their operation. In essence, these sectors produce raw materials and by-products that cause environmental pollution and depletion of resources. Thus, their sustainable optimisation must become the need of the hour (Anwar et al., 2018). In other areas, waste is having an even bigger impact as limited resources are stretched further, while the consequences of waste impact both the environment and public health.

Economic Value: Most monocultures support some local industry that provides essential goods or services to the region. In pursuit of economic advancement, the damage inflicted on nature can sometimes be bypassed or overlooked. The chapter emphasizes the balancing of trade-offs between economic drive and sustainable waste management and resource optimization.

Environmental Implications: Water pollution, air pollution, and habitat loss are contaminating the environment at an alarming rate; they are releasing greenhouse gases into the environment. Subsequently, enhanced waste management and efficient use of resources can help combat climate change and other challenges.

Regulatory Pressures: As global governments encourage harsher environmental regulations, local industries will be made to comply with them. By deploying best-in-class waste management systems and resource optimization practices, the local industries are able to comply with regulatory requirements and emerge as sustainability leaders.

The Transition to Sustainability in Local Industrial Systems

In recent decades, much of the industry from all over the world has been made sustainable. Earlier designed industrial systems did not take into account waste generation and resource consumption. Technological advancements, growing international environmental awareness, and dissatisfaction with operational processes in countries are forcing domestic industries to reassess their relevance.

Industrial Revolution: The phrase " an issue is explained as the stage at which the problem interrupts the flow of the present system. The single-use cycle promotes the linear model, which is wasteful and accumulates in the environment without effective use of resources. An important aspect of sustainability is a shift towards a circular economy where resources are not just used but are re-used, re-purposed, or recycled, where loss is minimised, and maximum value is extracted from what we have (Fotis et al., 2025).

Technology Integration: Technological advancement in Waste Management & Resource Utilization would be an inevitable and critical enabler for the transition to a sustainable waste economy. By adopting smart manufacturing, automation, energy efficiency, and advanced waste treatment technologies, local industries will be able to improve both the ecological footprint and economic performance (Awasthi et al., 2024).

Planetary Trends: The global focus on sustainability, manifested in frameworks such as the UN Sustainable Development Goals (SDGs), is increasing attention to waste reduction, resource efficiency, and environmental stewardship. Local industries are anticipated to develop the economy, but within the framework of these broader objectives.

The Challenges and Opportunities of Waste Management in Local Industries

There exists a huge opportunity for both waste management and resource optimization, although there are several on-the-ground challenges that will be difficult to solve. These challenges vary from constraints on resources and funding to limited access to high-tech equipment. However, they also provide a plethora of opportunities for innovation and scaling.

Challenges:

Resource Constraints: Small and medium enterprises, or SMEs, are typically small in scale and organically grow at most times; thus, they tend to be economically strapped for cash that is needed to put in place sophisticated waste management systems or acquire capital-intensive resource optimization technologies.

Lack of Expertise: Many local industries lack the technical know-how for developing effective waste management strategies. The transition to sustainable technologies can also be inhibited by this.

Poor Infrastructure: Many areas do not have infrastructure to segregate, recycle, and dispose of waste. As a result, it makes it difficult for local businesses to install a proper waste management mechanism, especially at places where the waste gets dumped directly at the dumping ground or burnt without being treated.

Opportunities:

Cost Savings: As waste management and resource optimization can lead to cost savings, reduction in operating costs, and improvement in the bottom line, through reduced waste, minimizing energy use, raw material use, etc., in Industries.

Innovation and Competitive Advantage: The importance of innovation and competitive advantage should not be underestimated. Local industries that adopt green technologies and sustainable practices are likely to enjoy the benefits of innovation and competitive advantage (Shukla et al., 2010). When consumers have a positive perception of a brand, it helps improve the reputation of that brand. People who care about the environment would help those companies that have the checks to ensure that they do not leave any adverse impacts on the environment.

Creation of Jobs and Local Development: The creation of jobs and local development can happen when initiatives are implemented that focus on waste disposal management and waste recycling for progressive development logic and optimization of resource recovery.

This chapter offers a review of waste management–what can we do to minimize waste, improve resource use, recycle, and develop waste-to-energy technology? This paper explores the role of technologies in better waste management, waste treatment, resource recovery, and circular economy models. There are plenty of case studies that showcase sustainable Waste Management in real-life applications. The chapter talks about issues obtaining and opportunities for the local industry, with actionable recommendations to improve the sustainability of industrial systems. The aim is to enable readers to understand how their local industry is able to include waste management and resource optimisation into their operations.

Figure 1. Waste hierarchy diagram

PRINCIPLES OF WASTE MANAGEMENT IN LOCAL INDUSTRIES

The essential feature is a waste management system to protect the local industrial systems in the long run. It involves collecting, transporting, treating, further disposing of, and recycling of industrial waste. The effective management of waste helps to minimise the environmental impact of industrial activity. It will provide financial benefits as it conserves resources and further reduces costs. Complying with these principles requires local industries to enhance the use of resources and apply appropriate waste management techniques essential for sustainable practices. Through the following section, we will see the basic theories and acts of waste management. It is a starter pack for small industries to reduce waste, save resources, and popularize sustainable development at large.

Waste Hierarchy: A Guiding Framework

The waste hierarchy is the conceptual framework for defining how to deal with waste. It aims at reducing and effectively dealing with waste by focusing on a few simple ideas in order of preference. A roadmap to reduce the waste stream, not make mountains of it. This hierarchy places a bias on action to prevent anything from becoming waste before actions involving disposal are considered as a last resort. Applying the waste hierarchy can provide a substantial contribution to local industries in environmental conservation and efficiencies arising from its operations (Mostaghimi et al., 2023). These key stages are known as the waste hierarchy, as shown in Figure 1.

Prevention: The best waste management solution is preventing waste from ever being created. Waste prevention can greatly benefit the environment through the development of processes and products that reduce the amount of raw materials necessary. E.g., reduced material wastage during manufacturing, or an optimized production schedule to prevent excessive ordered materials, reduces the amount of waste that is created. Preventive approaches also consist of designing products to last longer, use fewer resources, and have less toxicity.

Minimization: This is the process of reducing the volume and toxicity of the waste generated. Thus, it would require new, cleaner production techniques, process improvements, and resource-efficient technologies. For example, applying precision manufacturing processes to cut down the Number of losses in materials or enhancing packaging configurations to prevent excessive over-packaging can largely reduce waste generation.

Reuse: One of the best ways to reduce reliance on new resources is to reuse materials within the industrial process or use by-products from one process as raw materials for another. For instance, in the textiles or plastics sector, waste products from production can be repurposed for use at a later stage of the manufacturing process, which altogether reduces the need and consumption of new raw materials.

Recycling: This is the process of converting waste materials into new products, and it aims to reduce the consumption of fresh raw materials. Local industries can also set up recycling facilities that refine and recover valuable metal, plastic, paper, and other recyclable materials to return the resource into the supply chain. In this regard, recycling was an attractive alternative not only because it reduces waste but also because the impact is lower compared to producing new materials from raw resources (Johri et al., 2025).

Figure 2. Circular economy model applied to local industry

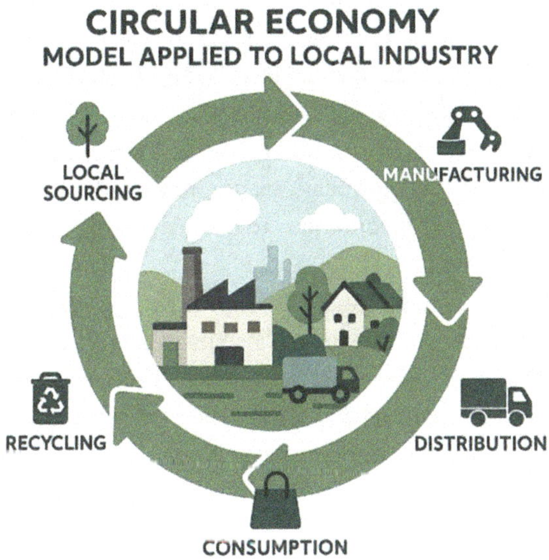

Energy recovery: making use of waste that cannot be reused or recycled to convert it into energy. Waste-to-energy technologies, such as incineration, anaerobic digestion, and gasification, capture the energy content of waste materials and convert it into heat, electricity, or biogas. These technologies assist the industries in minimizing waste, and at the same time provide an alternative for producing Clcan/ Green energy (Shankar et al., 2023).

Disposal: Last but not least, disposal is the only other man standing in the waste hierarchy; it should always be the most important thing: disposal, final treatment, and disposal of waste without recovery (e.g., landfilling or incineration) (Liao & Wang, 2020). Although disposal cannot be eliminated, industries should attempt to cut back on the use of landfills as their top option for waste dispersal by favouring alternative waste management practices, including recycling and energy recovery.

Circular Economy in Local Industries

Circular economy in local industries is a sustainability model that starkly contrasts the traditional linear economy, in which products are made, utilized, and thrown away. In a CE, materials are maintained in use for as long as realistically possible through reusing, recycling, and remanufacturing. Local industries can reduce their environmental footprints and minimize their resource consumption by implementing circular economy principles, as shown in Figure 2. Circular economy practices aim to maximize the life cycle of products, which reduces the output of waste and eliminates the requirement for additional virgin raw materials (Fatimah et al., 2020). Some principles that can be applied in the local industries include:

Durability design: All products must have a long life cycle, designed to be more repairable and easier to recycle. Such designs mean that more products can be utilized completely before they are removed, and create less trash when they reach the end of their life. Examples of such applications would include the electronics industry; it could utilize manufacturing techniques to construct a PC that could last 50 years and be repaired by the owner.

Source critical materials: Local industries can design systems for sources of rare materials found in end-of-life products. These processes could include dismantling and sorting, and the extraction of metals, plastics, or other recyclable materials that can be reused for another product. Industries can also redesign industrial processes so that resource consumption can be reduced, waste production can be eliminated, and the team can reuse waste resources to develop other new products (Thandlam et al., 2025).

Supply chain management: Supply chain management can be characterized as the coordination of flows between a company's upstream and downstream partners. The approach is known as Supply Chain Management. This is a crucial aspect of logistics since firms are now seeking to expand supplier bases while simultaneously expanding their customer bases.

Circular business model: *Circular* business models sell services instead of products. Take, for example, the movement of industries away from selling products to employing product-as-a-service models in which consumers pay for use of a product rather than owning it. It encourages manufacturers to make long-lasting and repairable products.

Collaborative consumption & sharing: Sharing platforms and collaborative consumption models also hold THE potential of becoming important components of the circular economy. This may include machinery or equipment sharing across industries, or inter-sectoral sharing, both of which reduce the need for new resource production but also utilize current assets more effectively (Nishad et al., 2024).

For most industries, moving towards a circular economy means revisiting how they design, produce, and consume. The circular economy provides local industries with many opportunities to reduce waste, save money, and give value by efficiently using resources (Aiguobarueghian et al., 2024).

Extended Producer Responsibility (EPR) and Regulatory Compliance

Extended Producer Responsibility (EPR) is the crux of this policy philosophy, where producers have to take care of their products from "cradle to grave, a principle that holds for other consumer waste too. EPR programs are designed to create incentives for producers to design products that are more recyclable, safer in their use of hazardous materials, and have lower environmental impact at the end of life.

EPR encourages local industries to:

Provide for Recycling: Industries should design products to ensure that the end-of-life phase of their products is taken care of, which would then motivate them towards making products that are recyclable or at least easily repurposed. As a result, waste can be more affordable and easier to manage, and ensures that high-quality materials get reclaimed.

Take-Back Programs: Now, industries give the option of take-back programs, where the consumer returns used products to the producer for safe disposal or recycling. According to F, producers of electronic goods may have to retrieve and recycle old devices for themselves.

Regulatory Standards: The EPR policies have regulatory standards like recycling rates and waste diversion targets. To enforce these rules, the local industries will require developing the strength of waste management systems to recycle and reduce the waste going to landfills, and utilize these discarded materials.

Role of Stakeholders in Waste Management and Resource Optimization

Stakeholders in industries play an active role in other product waste management and resource optimization, which include:

Government and Regulatory Bodies: Governments play an important role in defining regulations, incentivising, and providing facilitating environments to encourage sustainable practices. In order to promote the required behaviours, certain central enablers such as the Extended Producer Responsibility (EPR), eco-innovative tax incentives, and waste management grants for innovation are essential.

Industry Associations: These associations can facilitate the integration of principles of energy efficiency and optimum resource management in their regional matrix with the help of industry bodies. They are usually behind the guidelines for industrial processes, techniques, and practices.

Consumers: Consumers can make responsible choices to help reduce waste and use resources efficiently. Hence, we are demonstrating that the demand for sustainable goods is very real. This leads us to solution number two: to spend money on eco-friendly products. Changing consumer behavior will only be possible through public awareness and education campaigns.

Waste management service provider: The waste management service provider essentially forms one of the most vital links in the whole industry waste management process that is engaged in collecting, sorting, recycling, and disposing of waste. Proper management of waste and recovery of resources can happen through this collaboration.

Researchers and innovators working with nuclear technologists are continuously innovating to recycle waste as well as recover resources. India is becoming mechanized rapidly, with big R&Ds happening in India, our local industries must also remain at par in adopting the latest and most efficient cost-effective solutions available.

STRATEGIES FOR WASTE MINIMIZATION

This entails lowering the quantity of waste, its toxicity, and the effect that waste could have on the environment caused by manufacturing processes. Implementation of waste minimization through source reduction is an extremely profitable process for local industries to augment their dwindling economic and environmental sustainability. Waste minimization is not only about decreasing the amount of waste, but it is also related to making processes more efficient, reducing material and energy use, and creating a sustainable environment inside companies. This section looks at the different measures local industries can take to reduce waste production for every stage of the production process.

Lean Manufacturing and Waste Reduction

The incorporation of Lean Manufacturing principles into industrial operations is one of the best approaches to waste minimization, as shown in Figure 3. Lean Manufacturing is a methodology first put into practice by Toyota to eliminate waste and maximize efficiency in the automobile industry. The objective is to produce a greater value using minimum resources by shedding the ones that do not offer anything to the product.

There are several types of waste that Lean Manufacturing recognizes as the "Seven Wastes" or "7 Mudas," which are:

Overproduction: this is making more of the product than necessary, consuming resources in excess for a wasteful cause.

Waiting: excessive time spent waiting for equipment, information, and materials, leading to idle manhours.

Transportation: Material/product movement that does not add value.

Excess Processing: Performing more work than is required to meet customer specifications, wasted labor and resources.

Inventory: Excesses and unnecessary stock or raw materials in anticipation of being used, which will incur positive costs on resources and space.

Motion: Wasted movement by individuals or equipment that results in inefficiency and potential damage to parts.

Defects: Typical imperfection types include producing a defective product requiring rework or scrapping, resulting in waste of materials and labor.

It reduces waste by streamlining processes, improving workflow, and minimizing resource consumption. A few important Lean tools useful in waste removal are:

5S: This means that Sort, Set in order, Shine, Standardize, Sustain (The practice helps to standardize work processes and improve organization to reduce waste.)

Kaizen (Continuous Improvement): The Kaizen philosophy denotes small but continuous and incremental process improvements that slowly add up to a significant reduction in waste.

Just-In-Time (JIT): On-time manufacturing: Producing goods when and only when they are needed; in the right place, at the right time, in exact amounts.

Implementing Lean practices benefits in helping industries to recognize and eliminate the inefficient parts of their organization, hence their waste generation will be less, leading to an easier operation and also a reduction in production Costs.

Figure 3. Lean manufacturing principles in industrial waste reduction

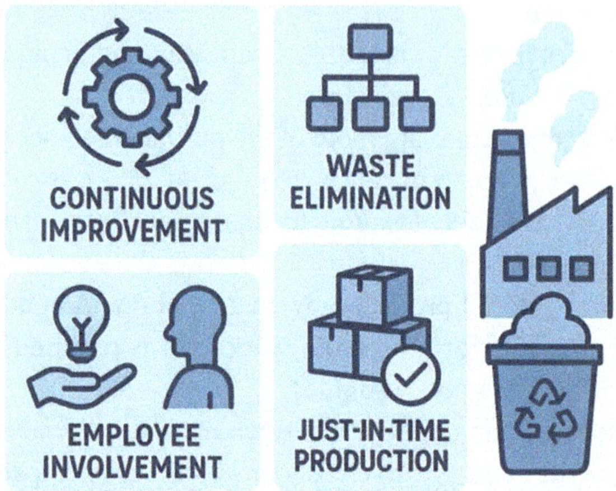

Waste Prevention through Process Redesign

Redesigning processes attacks the problem at its source, and it is a well-accepted fact that Process Redesign, even before Scale Reduction, is a very proactive strategy. Through applying cleaner production methods and process redesign, domestic industries are able to reduce demand for raw materials, thereby achieving reductions in energy consumption, which can lead to greatly reduced waste generation. Waste Prevention Process Redesign Strategies/actions that can be ineffective/ processes redesign strategies may hit the mark when we select:

Closed-Loop Systems: This is when you make products, use them, and return the materials to start all over again. For instance, companies from many sectors, not only food processors but also those representing textiles, exploit the re-utilization of water resources, solvents, and by-products in the production cycle, which creates no waste and minimizes the consumption of additional raw materials.

Process Optimization: Manufacturers can enhance their production systems or processes to ensure maximum utilization of resources by using such data. Methods like process simulation, energy audits, or optimization modeling help to locate where the waste generation can be minimized, through a lower electricity usage, for example, or even by lessening the raw material waste.

Design for Resource Efficiency: Industries must ensure that their products are designed in such a manner that they consume fewer resources per unit produced, produce less (or no) energy while producing, and create fewer wastes. A more sustainable material or a more recyclable product designed such that even at the end of life, useful discarding would be easy to recycle.

Technology Integration: Using robots and machines, automation requires digital monitoring systems, which put an accurate process into place and help avoid wastage. On one end, AI-driven predictive maintenance can help machines operate reliably and longer, thereby stopping disastrous machine failures that cause waste; on the other end, precision manufacturing prevents material waste by making parts with minimal excess.

Resource Efficiency and Material Substitution

Resource efficiency means using fewer raw materials to produce the same amount of output, by reducing waste that is generated in the process. Local industries could do this by employing nature-inspired processes to minimize the use of material resources, decrease demand for unnecessary inputs, and produce nothing more than is needed.

Materials Substitution: Similarly, in several industries, traditional raw materials can be replaced using alternative materials that are eco-friendly, cost-effective, and consume fewer resources. For example, using recycled plastics in packaging or biodegradable materials reduces the demand for virgin materials and reduces waste. Similarly, replacing hazardous chemicals with environmentally friendly alternatives contributes to preventing the generation of toxic waste.

Material Optimization: Out of all optimizations, material optimization is concerned with deriving the maximum yield out of each unit of raw material. It reduces the scrap material in the cutting, moulding, and assembly operations, hence making the manufacturing process more efficient and less wasteful. And reduced waste, for example, 3D printing permits parts to be built up layer-by-layer, directly from plastic or metal powders.

Product Redesign for Maximum Use of Resources: The overall optimization is curtailed to a large extent if the volume of raw materials used or the share of energy spent on production can be lessened by means of redesign at the product level. Packaging redesigns that use less material, or designing products for easy disassembly and recycling at the end of their life, can further reduce the costs and waste generated.

Energy Efficiency Waste Reduction

Reduced consumption of energy is often the source of waste, and trash generation is closely tied to the use of resources. The potential for energy-intensive industries (e.g., cement, steel, and chemicals) to reduce the waste arising from their use of energy through improved efficiency is substantial (Laureti et al., 2024).

Energy Audits: The first thing businesses should do is perform energy audits as a great starting point for reducing energy use and pinpointing waste areas. These evaluations examine energy usage at diverse points of production, from heating and cooling to machinery use and lighting. We then identify the most impactful points for improvement.

Energy-efficient Technologies: Energy-efficient technologies like LED lighting, high-efficiency motors, and heat recovery systems will save energy. Moreover, waste heat recovery systems can capture and utilize the waste heat generated by industrial processes as an energy source in a different part of the production process, thus reducing the requirement for additional net input energy.

Turn to clean energy (also known as renewable energy): Switching from dirty fossil fuels to clean, renewable resources such as biogas and renewables is one way of reducing waste-related problems and carbon emissions. Industrial clusters can create renewable energy facilities on-site or take power from the grid to a lesser extent than fossil fuel generation, while not entirely relying on coal for energy production.

Employee Engagement and Culture of Sustainability

In the world of waste minimization, the employee role must not be overlooked, even though technology and process improvements are necessary. A shift in the culture of the organization within can help develop and implement large-scale waste reduction and long-term sustainability strategies. When employees at all levels participate, a culture of ownership and responsibility for waste reduction is built.

Employee Training and Awareness: Employees should be regularly trained to understand the benefits of waste reduction, resource optimization, and sustainability. Training is one of the most important awareness methods. Regular workshops and campaigns create awareness in employees about how their efforts help minimise waste generation and enhance the overall sustainability of the company.

Additional sustainability rewards: A clever technique to get your employees to go the extra mile is by giving them some incentives if they give you an idea that you haven't thought of yet for reducing waste or optimizing your resources for reducing the waste you produce as a business. Recognizing and rewarding employees for their commitment to sustainability can aid the integration of sustainability into your culture.

Employees to be Encouraged: Employees from different areas should be brought together to work on common sustainability projects within the firm to develop superior waste reduction. Encouraging collaboration between local industries can yield positive results in the management of waste from various perspectives.

Collaborative Waste Reduction Initiatives

Your local business may want to partner with other organizations on efforts to reduce waste and maximize resource use. Initiatives that can help in waste reduction:

Shared Infrastructure: At a local level, industries often belong to the same region and/or the same industry cluster. Such industries can set up recycling infrastructure and use common waste disposal to save costs and enhance the recycling rate. It might be a good idea for smaller-scale businesses or SMEs to pool resources that would otherwise lack the capacity to implement an effective waste system on their own.

Industry Partnerships for Waste Exchange: In the industry partnership of waste exchange, businesses come together to exchange waste. It can also lead to a joint venture where the waste of one industry can become raw material for another. After many years of effort, an agricultural processing plant will be built. Excess organic waste will be used in biogas production. Another example of a bioenergy company could get a secure supply in this manner.

RESOURCE OPTIMIZATION TECHNIQUES

The efficient use of material, energy, and water with a focus on minimizing waste will enhance the overall sustainability of your industrial operations. This process refers to the strategic application of technologies, practices, and management systems used to lower the consumption of resources while either maintaining or improving production quality. Optimization of rendering and improvement of available resources. The content, compared to the local industry context, proves to be an economic device. Of course, optimization and improvement help reduce the emission impact on the environment. It creates benefits, mainly of a financial order, which are: Cost reduction, better operational scale efficiency, and adaptation to the legal environment. In this section, we present important waste management solutions that would help local industries to utilize resources efficiently and reduce waste generation to become more sustainable.

Energy Efficiency Improvements

One of the most significant waste producers and negative environmental impact sources is energy consumption. Implementing energy-efficient technologies and practices can help local industries reduce their energy footprint. Industries, by optimizing energy use, can lower operational costs and, in the process, reduce greenhouse gas emissions as well as reduce dependence on non-renewable resources, as shown in Figure 4.

Figure 4. Energy efficiency solutions for local industries

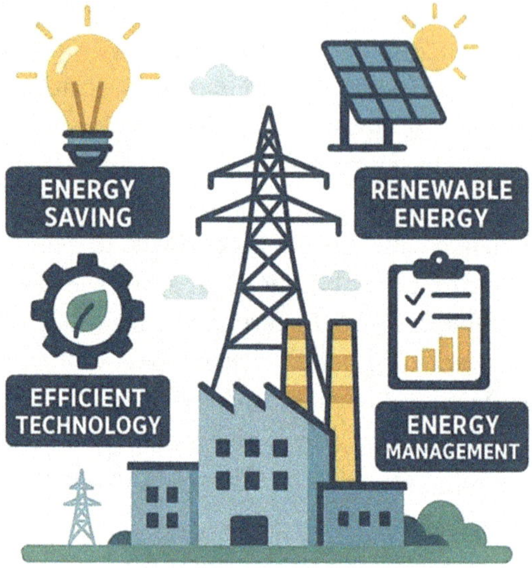

Key energy efficiency improvements include:

Energy Audits: A detailed energy audit should be the first step to improving energy efficiency. An energy audit identifies areas in which energy is being wasted and provides measures for their improvement. This audit includes all fields of industrial operation, such as heating, cooling, lighting, or machinery, and shows potential for optimization. Energy audits may show that machines are operating inefficiently, in which case replacing or upgrading them would reduce the amount of power being consumed (Gupta et al., 2010).

Energy Efficient Equipment: Replacing old and inefficient equipment with energy-efficient models is one of the best ways to increase energy efficiency. This will involve updating the machinery, gas compressors and pumps, and HVAC unit to much more efficient newer models. There are further ways for schools to cut energy costs, such as retrofitting their current equipment, which will help cut down power requirements, maybe even adding additional non-critical but costly components to save money by changing out a few components to more efficient parts.

Optimize & Automate Processes: This means not only do industrial processes need to be optimized using every possible method towards energy efficiency, but also that they consume less power in operation. Automated advanced process control systems could optimize production schedules to reduce wasted energy. Automated systems can even adjust machine operations in real-time using energy only when needed, and no more than that. Preventive maintenance subsystems will need to be a step ahead of mechanical failure, but through artificial intelligence, predictive maintenance systems can identify wasteful actions before they cause an expensive catastrophe (Alshaikh & Abdelfatah, 2024).

Waste Heat Recovery: A number of industrial processes, such as steel manufacturing, chemical production, and others, create a lot of heat as waste, which is not traditionally harnessed.

Renewable Energy Integration: Local industries can take a proactive approach to integrate renewable energy sources in their facility operations as an alternative to conventional fuel sources like fossil fuels. By installing solar panels, wind turbines, or biogas systems and addressing some of the energy requirements, it not only helps decrease environmental impact but also makes industries more resilient to fluctuating energy prices. They usually involve high upfront costs, but they pay long-term dividends and also contribute to sustainability, so it pays to invest your money in this.

Figure 5. Water conservation techniques in industrial applications

Water Conservation and Wastewater Treatment

Sectors of the food & beverages, textiles, and chemicals are highly consumer-oriented and need water to consume their product. Local sustainable enterprises need to ensure that they use water in the most efficient way, and wastewater treatment is the best practice (Chang et al., 2024). It is also cost-effective by saving costs in water usage and preventing depletion of fresh water resources, and reducing the environmental & ecological impact caused by industrial processes. Key water conservation techniques include, as shown in Figure 5:

Closed-Loop Water Systems: Closed-loop water systems can help to substantially decrease the consumption of water in contexts that require large amounts of it, such as textiles or food processing industries. A closed-loop system treats water and sends it back through the same production process, where it can be put to use elsewhere. One of the major benefits is reducing demand for fresh water, which helps cut waste and protect the valuable resource. For example, in the textile industry, water used for dyeing fabrics can be treated, and the same water reused any Number of times, such that there is minimal waste of water.

Rainwater Harvesting: Industries receive and collect rainwater to add to their water supply. Rainwater Harvesting Systems are a process of collecting rainwater from otp and storing it in tanks to use in the future. The water can be used for non-potable purposes, such as cooling, irrigation, or cleaning, reducing the demand for municipal water (Del Carmen et al., 2020).

Water-saving Technologies: Rainwater harvesting, low-flow faucets, equipment to reduce water waste, and efficient cooling technology, among others, can be used by industries to save water in their processes, such as laser hole drilling or whatever process that doesn't coming under water usage but involved with water flow, so the use of an efficient type of nozzles that can spray out in proper manners is a good solution to reduce overall quantity being used per drilling without having to compromise production quality.

Wastewater Treatment and Recycling: They can opt for state-of-the-art wastewater treatment plants that would help in treating the wastewater to make it usable in production processes. Wastewater decontamination is possible with membrane filtration, reverse osmosis, or activated sludge systems. Through this treatment, waste can be removed from wastewater. By lowering total water demand and preserving natural resources, this can enable treated water applications for cooling, watering, or sanitation.

Zero Liquid Discharge (ZLD): A ZLD is an advanced, amenable, and sustainable water management technology designed to eliminate discharge of wastewater by recovery of all usable water from the effluent stream and converting it into clean, reusable water along with other recoverable components like acids, salt, etc. ZLDs use a combination of filtration, evaporation, and crystallization processes to treat the water and form solid waste. ZLD systems have higher initial costs (in terms of capital costs), but they have substantial ecological returns in these areas.

Materials Efficiency

Resource efficient means using a lesser amount of resource materials to produce the same (or better) output. It also refers to less waste and environmental impact arising from the production process. In particular, the construction, automotive, and manufacturing sectors have a high dependence on raw materials. Strategies that boost material efficiency include:

Material Replacement: Material replacement is one of the processes that exist to replace traditional raw materials with others that are more sustainable. In building construction, substituting conventional cement with alternative constituents can help reduce the use of raw materials and the emissions of carbon – examples of alternative constituents are fly ash and slag. Using either recycled or bio-based materials in products can help reduce dependency on virgin materials.

Resource-Efficient Industrial Production: The industrial production of these products uses fewer resources and creates less waste in their manufacturing. The circular economy for industries can be obtained by making the concept of Ecol-design, a design that is guided to facilitate the dismantling, reinstallation, and recycling. Using methods like modular design to develop a product that can be repaired or upgraded means fewer raw materials have to be extracted, and fewer products will end up in the landfill at the end-of-life.

Efficient Use of Raw Material: The manufacturing industries in the region may use next-gen process efficiencies to produce lesser quantities of raw material while producing a variety of other goods. This entails improving machine precision, lowering scrap, and maximizing material use, etc., which is part of it. In metalworking, for example, using CAD software and a laser cutter will generate almost no off-cuts of metal sheets when they are cut.

Collaborating with suppliers: Forging collaboration agreements among suppliers can help prevent waste in the supply chain. By collaborating with their suppliers, companies can reduce waste and enhance circularity through supply chain materials, packaging sizing, and eco-friendly raw material sourcing. Further, the supply chain can promote waste reduction and material optimization by suppliers, which would bring about a change in the supply chain.

Technology and Innovation in Resource Optimization

Technology is the game-changer for better resource utilization in local industries. The use of modern technologies can enable industries to track, monitor, and improve resource efficiency in every field of their work. Manufacturing, automation, and digitalization technologies are also rapidly advancing to provide industries with better means of optimizing the use of materials, energy, and water. The main technological innovations of resource optimization in this case are:

Internet of Things (IoT): IoT can share instant information about its resources to have better control by the industry over energy, water, or material consumption. IoT sensors in factory equipment enable energy tracking to identify waste, correct inefficiencies, and take action promptly. This data is then analyzed to enable production schedules that can be optimized, a reduction in waste generation, and better use of resources (Lakhouit, 2025).

Smart Manufacturing & Industry 4.0: Industry 4.0 technologies such as robotics, Artificial Intelligence (AI), and big data analytics help industries streamline their production processes, reduce waste, and improve resource efficiency. For example, AI-powered systems are capable of predicting when equipment is going to fail before it actually does, optimising the consumption of energy and ensuring that production lines operate more efficiently, i.e., every operation adjusts in real-time according to both resource availability and demand.

Additive Manufacturing (3D Printing): The correct term used for 3D printing is Additive Manufacturing, which means manufacturing with precision of the segment (material) only to pay for what we use. Since the printer only uses what is necessary, there is no waste from excess material, and highly intricate or customized items can be produced without the need for molds or tooling. And 3D printers work with input materials, further recycling the consumables to be better for our planet.

Advanced Process Control Systems: Advanced process control (APC) systems allow industries to control their production processes in real-time to conserve resources. They also adjust temperature, pressure, and flow rates by utilizing mathematical models and predictive algorithms that save money, time, and resources (Selvakanmani et al., 2024).

In order to create more sustainable local industries, it is necessary to pursue resource optimisation. Incorporating energy-saving technology, water-saving technology, optimized usage of raw material, and the use of smart technology can aid different industries to reduce their negative environmental impacts, besides their positive economic impacts, in terms of cost-effectiveness and efficiency. Naturally, the optimization of resources is a lengthy and continuous process, and one that is always changing due to technology and consensus across industries and supply chains. The local industries can be sustained and successfully compete with the international growth by following these strategies.

TECHNOLOGICAL INNOVATIONS IN WASTE MANAGEMENT

With the advent of technology, waste management has changed face, particularly at the local level industries. In recent years, industries worldwide have generated enormous quantities of waste, which destroys the earth at an alarming rate. Ad$e - Waste Trading Mobile App has brought smarter solutions for better waste management. Innovations that facilitate efficient disposal of waste and recycling that help recover value from waste to build a circular economy. Various technological innovations have been implemented in local industrial systems to manage waste, from waste-to-energy technologies to advanced recycling methods and smart waste monitoring systems.

Industrial Waste-to-Energy Technologies

Waste-To-Energy (WTE) technologies convert the heat from the combustion of wastes into electricity and strategic waste material. The Waste to Energy technology is specifically designed to meet the twin challenges of waste disposal and energy generation by diverting the unwanted elements from surrounding landfills, while also reducing dependence on burning fossil fuels. WTE technologies are especially beneficial in industries that produce large quantities of waste materials, such as food processing, agriculture, and manufacturing. Key waste-to-energy technologies include:

Incineration: Probably the most well-known of the WTE technologies, incineration is when trash is burned at high temperatures to create heat that can be used to produce electricity. Based on this method, the waste volume is much lower, as it can reduce the mass of the waste to 90%. However, incineration is a careful process that can release harmful emissions and use advanced air pollution control systems in modern incinerators to prevent harmful gases from going into the atmosphere (Gupta & Tripathi, 2013).

Anaerobic Digestion: Anaerobic digestion is a combination of processes that uses microorganisms to break down biodegradable material in the absence of oxygen. The biogas is used to produce electricity and heat; the rest of the digestate can be a very good fertilizer. Anaerobic digestion is widely used in food processing, agriculture, and wastewater treatment, where the organic waste is generated.

Gasification: Gasification is a thermochemical process that can convert carbonaceous feedstock such as coal, petcoke, and biomass into syngas (i.e., synthesized gas comprising CO, H2, and CH4) by heating it to high temperatures in the presence of relatively limited oxygen. This syngas is a building block for pick up to be utilized either as a fuel source for producing electricity or heat, or can be further refined into chemicals and fuels. Gasification, because it results in less-harmful emissions and can mean better control over the energy production end of things, is generally considered gentler than incineration.

Pyrolysis: Pyrolysis is a thermochemical technique that decomposes organic waste material by pre-naturally heating it at high temperatures in the absence of oxygen to produce oil, char, and gases. Both the liquid oil and the solid char can replace petroleum-based fuels, and the solid char can be used as a soil amendment or in industrial applications that require a carbon source. This makes pyrolysis, especially for turning plastic waste into high-energy products and by-products, very useful (Hernandez et al., 2024).

Ferrous Metal: Ferrous metal, such as steel and iron, is easily extracted due to the magnetic properties it has from the plasma arc technology process. The process of gasification has proven to be both highly efficient and able to handle a wide variety of waste types, such as hazardous waste, medical waste, and others that are more complex or toxic in nature; this makes it well-suited for such industries (Prasanna et al., 2024).

Advanced Recycling Technologies

Recycling is one of the most effective ways of waste management for industries producing tons of materials like metals, plastics, and paper. The employment of advanced recycling techniques allows re-industries to extract high-value materials from waste streams and to lessen their reliance on virgin resources, thereby decreasing the environmental footprint. By using these technologies, recycling becomes more efficient, pure, and cost-effective, making it applicable for local industries to use.

Automated Sorting Systems: Traditional recycling methods commonly use manual sorting, which takes a lot of time and produces errors. Automated sorting systems are powered by the use of artificial intelligence (AI), robotics, and machine vision and are equipped with sensors and cameras to detect and separate all waste materials. These systems are able to separate materials such as plastics, metals, and paper, which would allow for a cleaner stream of recyclables being collected and fewer contaminants getting mixed within the recycling process.

Chemical-Encoding: Chemical recycling is an evolving technology that includes converting plastics and other materials into their chemical components for the production of new products using them as raw material. Since mechanical recycling is limited in the multiple plastics it can handle, chemical recycling opens a new universe of recalcitrant materials like mixed plastics that all vehicle producers have to deal with. Pyrolysis, solvolysis, and depolymerization are chemical recycling technologies.

Hydrometallurgical and Biotechnological Recycling: One of such methods is hydrometallurgical recycling, which uses aqueous solutions to recover useful metals, e.g., gold, silver, and copper, from waste electric appliances or mining by-products. Using microorganisms or enzymes to deconstruct complex organic materials like plastics or textiles into recycling stock (biotechnological recycling). This technique provides a clean and efficient option for recycling materials that are not easily recyclable.

Closed-Loop Recycling Systems: This is where materials are completely recycled and used to make the same product or in the same process for which they were designed. For instance, closed-loop recycling in the electronics industry enables manufacturers to reclaim metals from old gadgets and reuse them to create new electronics. This will drastically reduce the environmental impact of production processes and work towards achieving a circular ecology.

IoT for Smart Waste Monitoring and Management Systems

As a result, the idea of smart waste management has been coined. This is a creative use of technology to monitor the waste and dispose of it smartly. Smart waste management systems help local companies to monitor the generation of waste by their business, streamline the disposal procedure, and improve recycling efficiency.

IoT Sensors for Waste Tracking: Internet of Things (IoT) devices/sensors have opened up new avenues of waste tracking enabled by them. These tools can be placed in waste containers, bins, production lines, etc to record how much and what type of waste is created. These sensors help in understanding the patterns of the waste that is generated and identify places where money can be saved; this will consequently improve the waste management process (Sinha et al, 2023). An example of this is using sensors to monitor waste at various stages of production. By doing this, managers can notice where waste is occurring and adapt their processes accordingly until it results in zero waste.

Smart Bins and Collection Systems: Smart Bins and Collection Systems: To monitor the fill level status of waste containers, smart bins are equipped with sensors. They send an automatic alert to the waste management teams when human intervention is needed in emptying the bin(s). Waste is collected faster and more agilely because it reduces the number of wasted trips and raises the overall range of collection times. Smart bins may also be used to segregate waste by type, e.g., recycling materials separated from ordinary waste.

Waste Generation Data Analysis: A better understanding of trends in waste generation using data analytics and machine learning algorithms on waste generation data, so that the city can plan collection routes efficiently and also increase recycling rates. (Gupta, 2025). One example is the use of data by industry on the type of materials being wasted to reduce generation or pursue more targeted recycling (Akbarpour et al., 2021).

AI (Artificial Intelligence) & Robotics in Waste Sorting and Recycling: Waste sorting and recycling with the employment of AI (Artificial Intelligence) & Robotics. Automatic sorting of waste using AI & robotics can simplify waste extraction and recycling water. Some of the coolest waste-separating technologies include machine learning, robotics, and even image recognition.

Retail and hospitality are adopting smart waste systems that help in managing waste efficiently, such as source separation of recyclables and ensuring efficiency in waste collection. High-tech applications are changing the waste management game. With help from these applications, different sectors are able to reduce their overall waste and recover some useful materials, as well as optimize their functions. Waste-to-energy helps us harness renewable energy from waste, whilst advanced recycling can allow us to retrieve important resources that would otherwise be thrown away.

Furthermore, intelligent waste management solutions allow industrial facilities to keep an eye on waste generation on a real-time basis to adjust the recycling plan and enhance decision-making based on the data. As illustrated in this post, leveraging local industries mid-mile can greatly help in the transition to a circular economy, while also reducing their environmental impact and operating sustainably through these tech disruptions.

POLICY AND REGULATORY FRAMEWORKS

Usually, a combination of policy, regulation, and standards drives local industries to turn waste into a resource and play an effective role in waste management. Educational institutions need to grow frameworks for an ecosystem where sustainable practices will grow. Governments, global organizations, and industrial bodies framed guidelines based on which local industries are to manage waste and optimize the utilization of resources in a sustainable way (Brotosusilo & Naldi, 2021). As illustrated in Figure 6, these actions will not only contribute to a safe and sustainable future but also hold the industries accountable for their environmental footprints. This chapter looks at the differential policy and regulatory framework for pushcarts that makes waste management and resource optimisation more than just a local phenomenon, with the presence of noteworthy laws, incentives, and global goodies.

Figure 6. Policy and regulatory framework for industrial sustainability

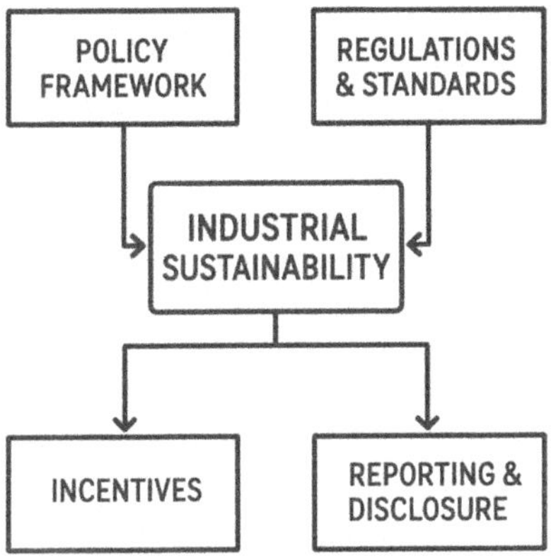

Government Regulations

Resource efficiency and waste management require national laws and regulations that all local industries in most parts of the world must comply with. The regulations outlined herein shall serve the purpose of environmental protection, resource conservation, and management of waste. Governments have instituted mandatory and voluntary standards to encourage various industrial sectors to use cleaner technologies and to better manage wastes. Regulatory frameworks for effective waste and resource management:

Extended Producer Responsibility (EPR): The principle of Extended Producer Responsibility (EPR) as a component of local industries is one of the most impactful regulations within this aspect. Under EPR, companies are liable for everything that happens to their product from cradle to grave, including the collection phase itself, recycling, over the phase, and disposal at end-of-life. It has seen success in electronics, packaging, and automotive manufacturing sectors. For instance, electronics firms must take back old devices to recycle them, ensuring the recovery of value and proper end-of-life e-waste management. As a result, product design (for recyclability) has improved, and an efficient recycling infrastructure has developed because of EPR.

National and regional laws: These usually have mandatory waste diversion, recycling rate, and landfill reduction targets (waste management laws). This regulation once again necessitates the development of strategies in the industry to manage the waste, indicating the source separation of waste, recycling, or reduction of the mass of waste in production. For instance, the European Union has recycling targets in place under its Waste Framework Directive, which require certain percentages of both municipal and industrial waste to be recycled or reused.

Environmental Impact Assessments (EIA): In several countries, industries are required to perform Environmental Impact Assessments (EIA) prior to commencing production on a larger scale. The purpose of EIAs is to assess the possibility that a proposed industrial project will be in a position to have environmental and other unfavorable side effects due to waste management practices, consumption of resources, and overall footprint. The EIA process enables industries to know the best possible way of reducing their negative environmental impacts and sustainable waste management practices from the outset.

Regulations Contamination Control: Governments far and wide have enacted contamination control laws to lessen mechanical smog discharges by preventing air and water contamination. These laws regulate the pollution of air, water, and soil and establish standards for permissible emissions, wastewater discharges, and hazardous waste disposal. Industrials are particularly vulnerable to this form of regulation because the Clean Air Act in the US and the EU Water Framework Directive both place heavy restrictions on how much waste a wastewater treatment plant can take and discharge.

Legislation Supporting the Circular Economy: Some governments are even creating legislation to support the plan of a circular economy, which is posing as one of the many requirements for sustainability goals set globally. The regulations support industries' use of practices that lengthen the life of products, extract valuable materials, and reduce waste. This includes policy, such as being a factor in the design of laws that France, Germany, and the Netherlands are implementing that force companies to remanufacture, recycle, and reuse materials when producing goods.

Enforcement and Compliance Role:

Industry Standards and Voluntary Initiatives

Aside from government requirements, lots of industries also adopt their own standards and schemes that encourage sustainability and responsible waste management. Many of these voluntary frameworks are even more stringent than regulatory requirements, incentivizing industries to implement best waste and resource reduction practices. Main Voluntary Initiatives and Industry Standards:

ISO 14001 Environmental Management System (EMS): One of the most adopted standards in the world of environmental management, ISO 14001 identifies requirements for an organisation to establish/manage/improve its EMS. It is a set of guidelines developed for organizations to implement an EMS and reduce their environmental impact, specifically waste generation and resource usage. Adhering to ISO 14001 guidelines not only supports local industries to improve their environmental performance but also helps them save costs and meet all legal requirements. ISO 14001-certified companies will have the ability to show that they are serious about the environment and get an edge on their competitors.

ISO 50001 Energy Management: ISO 50001 is a global standard for energy management systems. Offers a structured methodology for sectors to improve energy utilization, reduce power consumption, and increase the quality of energy consumed. When industries implement ISO 50001, they can reduce the use of non-renewable energy sources and lower the energy costs to improve sustainability. It also aids in waste reduction due to energy overconsumption, thereby aiding various resource optimization initiatives.

Cradle-to-Cradle (C2C) Certification: A Cradle-to-Cradle certification is a voluntary sustainability standard that helps companies earn points to meet other, more generic green rating systems and encourages designs for products and processes that result in the industry closing the loop on waste. Rowejo lowers the barriers for companies designing 100% recyclable, compostable, and biodegradable products to reduce waste and resources consumed. The C2C certification looks at a company´s potential to use eco-friendly design, optimize material flows, and carry out waste-free production.

LEED Certification: LEED (Leadership in Energy and Environmental Design) certification is the globally recognized standard for green building construction and design in the world of construction and real estate. This includes happenings like sustainable materials, energy-efficient systems, and practices that help in waste reduction. Local construction is encouraged to follow LEED standards, which results in the reduction of a great deal of construction waste, resource consumption, and reduces energy and water use at the regional and final buildings.

Benefits of Voluntary Standards:

These standards often help industries get a leg up on upcoming regulations and forge stronger ties with various stakeholder groups (Customers, Investors, Regulatory bodies, etc.). In a city, how industries recycle and optimise their resources is decided by the policy and regulatory framework. In the same way, regulatory guidance and incentives are also critical: Governments (which regulate for environmental impact) and international organizations lay important groundwork for helping industries embed sustainability in their offering. Sectors can maintain their viability and cost-competitiveness in a sustainability-oriented marketplace by aligning with government regulations, but also voluntary standards and economic incentives, and superior outcomes beyond the legal requirement. Although these issues will always play a prominent role in the establishment of a sustainable and resource-efficient industrial future, investors want regulations that are well-designed. These will effectively provide financial rewards in favour of high-performing firms.

CHALLENGES IN IMPLEMENTING WASTE MANAGEMENT PRACTICES

Even though the benefits of waste minimization and resource conservation are clear, local industries have a lot of trouble achieving them successfully after gaining a good understanding. A lack of funding, technical difficulties, insufficient infrastructure, and reluctance to change create barriers in transitioning from unsustainable practices to more sustainable ones. It is very important to understand the deficiencies in the industries and what the possible fixes could be. This region of the publication investigates the obstacles to the application of waste management practices, looks at why these are unjustifiably hindering, and how they may be overcome.

Financial Constraints

The lack of funds or resources is one of the main challenges for local industries to adopt practices around waste management and resource optimization. This is particularly the case with regard to certain industries and all SMEs generally, who may not have the spare cash to implement such technology/systems. Buying new equipment, building waste treatment plants, or introducing better recycling systems all require huge capital investment, which may not be possible for poorer businesses.

Challenges and Solutions:

- ***Capital Intensive:*** Establishing waste treatment systems or spending on energy-efficient equipment requires significant capital investment. Conversely, the costs of such nets, although they can be saving over the long-term, may put a company in a position of not being able to afford that investment.

    Solution: To meet the upfront costs of deploying sustainable technologies, governments and financial institutions can provide subsidies, tax credits, or low-interest loans. Allows the local industry to invest in various garbage disposal units, which involves low cost. It greatly contributes to operational costs in the long run.
- ***Operational Costs and Maintenance:*** The waste management sector faces high operational costs and maintenance costs. There are maintenance costs, energy costs, and labor costs involved in the process. These types of wastes

can help carry out various operations & activities at a reduced cost. This is another hit on the financials.

Solution: The waste services will be used by industries on a performance contract basis or they will finance the lease of capital items such as PURVA-related equipment. This will reduce capital expenditure and allow for higher certainty in operating costs.

- ***Payback Period Too Long:*** Many waste management technologies (especially waste-to-energy systems or closed-loop water recycling) can have a payback period of several years, and for this reason, the sector is often not in a position to invest.

  Solution: A few long-term benefits of these systems (Ex, reduced utility bills, lower waste disposal fees, regulatory compliance), when demonstrated, may convince businesses to make the same investments. At the same time, it could be a potential revenue growth opportunity for companies by catering to an increasing consumer demand for sustainable products and services!

Lack of Technical Expertise

One of the most important obstacles in creating strategies for treating waste and diminishing natural resources is the local industries' lack of know-how. Public policy initiatives that reusable liquid bulk packaging can help drive, such as waste management, resource optimization, and sustainability efforts, also often require knowledge and skills that many smaller or previously less focused on these areas companies may be lacking.

Challenges and Solutions:

- ***Lack of Awareness of Technologies:*** Many would also be unaware of the new and emerging technologies in the area of waste management, including automated sorting systems, waste-to-energy technologies, or energy-efficient machinery, especially in the case of traditional sector industries. Such unawareness prevents businesses from owning the most efficient and effective systems.

  Solution: Trade associations, government agencies, and environmental groups create training programs, workshops, and other informational resources to help businesses learn about new technologies and best practices in waste management. The technology providers or consultancy organizations can also support the local industries to reach this expertise.

- ***A dearth of Skilled Labor:*** Even if the industries are acquainted with waste management technology, they may not have abundant, proficient labor to bring in the functionality and future inspection and maintenance of these systems.

  Solution: Governments and trade associations may sponsor workforce development initiatives to train folks in each sustainability work and waste management technologies. Certification programs, apprenticeships, and other incentives can populate a workforce.
- ***Technological Change Resistance:*** Most local industries have shown skepticism when it comes to adopting new technologies, due to the fear of interrupting existing operations as well as the perceived complexity of adding any new implementation.

  Approach: Offering concrete proof of similar industry success using case studies and relevant examples enables an easy switch to embrace the change. Furthermore, introducing new technology in phases and receiving proper training and support helps transition much more smoothly, which is often less disruptive.

Insufficient Infrastructure

Time and place. In many areas of the world, particularly in developing countries, random dumps or overflowing landfills are common, and many of them return resources to where they came from with little or no waste management practices. For instance, recycling facilities are scarce, as are waste treatment plants and waste collection systems.

Challenges and Solutions:

- ***No Recycling and Waste Treatment Facilities*****:** Without local recycling or waste treatment facilities, local industries are forced to manage their waste onsite (if that), as any cost of shipment will come from their bottom line. Without these facilities, industries have no choice but to bury or burn off their waste, and that is not only a non-sustainable method of handling the waste.

  Solution: Governments invest in the building or upgrading of recycling and waste treatment infrastructure, particularly in less-developed areas. Public-private partnerships (PPPs) can help to procure government funding and combine it with the expertise of the private sector in order to build such infrastructure.

Struggles of private industries in adopting prudent waste practices and resource conservation are being faced all over the globe. But if these hurdles are known and informed solutions are developed, industries can learn to deal with the obstacles standing in their way of a more sustainable mode of operation.

The Future of Waste & Resource Optimization

As the world focuses more and more on sustainability, so too does the role of technology drive innovation and creation. Used to transform attitudes would form part of the reshaping of waste management/resource optimisation in our local industries now and into the future, with continually changing regulatory landscapes and economies of scale requiring smaller or larger, more efficient and organically sensitive means too. Use of waste and resource management will also require new elements; existing sectors will be expected to invest in them for more cost-efficient and circular outcomes. In the important field of managing local industrial waste, optimising resources, which includes new technologies, policies, and better practices, we look at the trending lines that will define the next.

CONCLUSION

Adopting sustainable industrial systems is a lengthy process that necessitates engaging local industry, government, and community stakeholders. Due to the global environmental crises that include climate change, natural resource extraction, and unmanaged waste management, effective waste management and resource generation are essential. We need to integrate local industries, which are often at the heart of the economy and society of regions, in climate change policy, not only for the short run but for greater sustainability. Through various examples, the current chapter shows how local industries can better manage wastes, optimize resources, and sustainably use resources at the business level, community level, and environmental level.

Incorporating sustainability into local industries through re-designing processes and products can reduce waste generation and enhance materials' longevity. The adoption of eco-design principles makes products easier to recycle, repair, or repurpose, thereby lowering the demand for new resources and reducing waste. The circular economy model encourages the reuse, recycling, and regeneration of resources so that they can be used continuously. Local industries can be built on a circular economy model. The success of sustainability practices depends on engaging stakeholders. In this way, stakeholder engagement fosters a sustainability culture in an organization. Further, it involves all stakeholders. Recycling and Resource Recovery, as well as technology and innovation, are being urged to help small- and medium-sized enterprises realise Industry 4.0 transformations for smart manufacturing. In addition, personal computers, internet connections, automotive technologies, and renewables, such as solar, wind, and waste-to-energy technologies, are addressed, with specific design supporting improvements in efficiency and effectiveness, as well as cost reductions, in waste minimisation, resource recovery, and cleaner production.

The case studies in this chapter show that any local industry can simultaneously combine waste management and resource optimization solutions, regardless of sector, to achieve environmental sustainability and operational efficiency. In this way, waste production is reduced, resources are saved, and worldwide market competition is increased through recycling systems, reuse, and resource-efficient strategies. In addition to showcasing the advantages of sustainable growth, the real-life cases will also serve as inspiring and instructive resources for other industries attempting to adopt best practices.

When local industries adopt sustainable waste management and resource optimization practices, they gain significant economic and environmental benefits. The benefits extend beyond regulatory compliance. They enhance financial performance, strengthen market reputation, and foster better relations with customers, employees, and investors. From this chapter, we can conclude that economic benefits cover savings, new revenues, attracting investors and customers, complying with laws, limiting risks, reducing waste generation, conserving resources, cutting down carbon footprint, and protecting biodiversity. By implementing the processes, the industries ensure compliance with the local, national, and international laws, mitigate risk from resource scarcity and environmental legislation, and remain competitive.

REFERENCES

Aiguobarueghian, I., Adanma, U. M., Ogunbiyi, E. O., & Solomon, N. O. (2024). Waste management and circular economy: A review of sustainable practices and economic benefits. *World Journal of Advanced Research and Reviews*, *22*(2), 1708–1719. DOI: 10.30574/wjarr.2024.22.2.1517

Akbarpour, N., Salehi-Amiri, A., Hajiaghaei-Keshteli, M., & Oliva, D. (2021). An innovative waste management system in a smart city under stochastic optimization using vehicle routing problem. *Soft Computing*, *25*(8), 6707–6727. DOI: 10.1007/s00500-021-05669-6

Alshaikh, R., & Abdelfatah, A. (2024). Optimization techniques in municipal solid waste management: A systematic review. *Sustainability (Basel)*, *16*(15), 6585. DOI: 10.3390/su16156585

Anwar, S., Elagroudy, S., Abdel Razik, M., Gaber, A., Bong, C. P. C., & Ho, W. S. (2018). Optimization of solid waste management in rural villages of developing countries. *Clean Technologies and Environmental Policy*, *20*(3), 489–502. DOI: 10.1007/s10098-018-1485-7

Awasthi, B., Gupta, S., Kumar, S., & Tripathi, A. (2024, January). Business management and planning using queuing model with heterogeneous customers for electrical applications. *In 2024 2nd International Conference on Intelligent Data Communication Technologies and Internet of Things (IDCIoT)* (pp. 279-283). IEEE. DOI: 10.1109/IDCIoT59759.2024.10467673

Brotosusilo, A., & Naldi, A. (2021, March). Policy on optimization of household waste and hazardous waste management based on community empowerment at the local level. [). IOP Publishing.]. *IOP Conference Series. Earth and Environmental Science*, *716*(1), 012085. DOI: 10.1088/1755-1315/716/1/012085

Chang, H. L., Liu, Y. L., Keng, C. J., Jiang, H. L., & Hu, J. (2024). Challenges of industrial wastewater treatment: Utilizing Membrane bioreactors (MBRs) in conjunction with artificial intelligence (AI) technology. *Journal of Industrial and Production Engineering*, *41*(5), 422–427. DOI: 10.1080/21681015.2024.2330401

Del Carmen Munguía-López, A., Zavala, V. M., Santibañez-Aguilar, J. E., & Ponce-Ortega, J. M. (2020). Optimization of municipal solid waste management using a coordinated framework. *Waste Management (New York, N.Y.)*, *115*, 15–24. DOI: 10.1016/j.wasman.2020.07.006

Fatimah, Y. A., Govindan, K., Murniningsih, R., & Setiawan, A. (2020). Industry 4.0 based sustainable circular economy approach for smart waste management system to achieve sustainable development goals: A case study of Indonesia. *Journal of Cleaner Production*, *269*, 122263. DOI: 10.1016/j.jclepro.2020.122263

Fotis, G., Gupta, S., Varshney, T., Dhar, S. L., & Gulzar, M. M. (Eds.). (2025). *Modern Computing Technologies for EV Efficiency and Sustainable Energy Integration*. IGI global., DOI: 10.4018/979-8-3373-2382-4

Gupta, S. (2025). *The Future of Machine Learning with Automation in Engineering*. Nova Science Publishers., DOI: 10.52305/YFIX2832

Gupta, S., & Tripathi, R. K. (2013, February). An LQR and pole placement controller for CSC based STATCOM. *In 2013 International Conference on Power, Energy and Control (ICPEC)* (pp. 115-119). IEEE. DOI: 10.1109/ICPEC.2013.6527634

Gupta, S., Tripathi, R. K., & Shukla, R. D. (2010, November). Voltage stability improvement in power systems using facts controllers: State-of-the-art review. *In 2010 International Conference on Power, Control and Embedded Systems* (pp. 1-8). IEEE. DOI: 10.1109/ICPCES.2010.5698665

Hernandez, D., Pasha, L., Yusuf, D. A., Nurfaizi, R., & Julianingsih, D. (2024). The role of artificial intelligence in sustainable agriculture and waste management: Towards a green future. *International Transactions on Artificial Intelligence*, *2*(2), 150–157. DOI: 10.33050/italic.v2i2.552

Johri, S., Selvakumar, P., Satyanarayana, P., Das, A., Gupta, S., & Bhattacharya, S. (2025). Urbanization and Energy: The Nexus of Global Challenges. In Geopolitical Landscapes of Renewable Energy and Urban Growth (pp. 453-478). *IGI Global Scientific Publishing*. DOI: 10.4018/979-8-3693-8814-3.ch016

Jones, H., & Boger, D. V. (2012). Sustainability and waste management in the resource industries. *Industrial & Engineering Chemistry Research*, *51*(30), 10057–10065. DOI: 10.1021/ie202963z

Lakhouit, A. (2025). Revolutionizing urban solid waste management with AI and IoT: A review of smart solutions for waste collection, sorting, and recycling. *Results in Engineering*, *25*, 104018. DOI: 10.1016/j.rineng.2025.104018

Laureti, L., Costantiello, A., Anobile, F., Leogrande, A., & Magazzino, C. (2024). Waste management and innovation: Insights from Europe. *Recycling*, *9*(5), 82. DOI: 10.3390/recycling9050082

Liao, B., & Wang, T. (2020). Research on industrial waste recovery network optimization: Opportunities brought by artificial intelligence. *Mathematical Problems in Engineering*, *2020*(1), 3618424. DOI: 10.1155/2020/3618424

Mostaghimi, K., & Behnamian, J. (2023). Waste minimization towards waste management and cleaner production strategies: A literature review. *Environment, Development and Sustainability*, *25*(11), 12119–12166. DOI: 10.1007/s10668-022-02599-7

Nishad, D. K., Tiwari, A. N., Khalid, S., & Gupta, S. (2024). Power quality solutions for rail transport using AI-based unified power quality conditioners. *Discover Applied Sciences*, *6*(12), 651. DOI: 10.1007/s42452-024-06372-5

Prasanna, S., Verma, P., & Bodh, S. (2024). The role of food industries in sustainability transition: A review. *Environment, Development and Sustainability*, *27*(7), 1–21. DOI: 10.1007/s10668-024-04642-1

Roberge, H. D., & Baetz, B. W. (1994). Optimization modelling for industrial waste reduction planning. *Waste Management (New York, N.Y.)*, *14*(1), 35–48. DOI: 10.1016/0956-053X(94)90019-1

Selvakanmani, S., Rajeswari, P., Krishna, B. V., & Manikandan, J. (2024). Optimizing E-waste management: Deep learning classifiers for effective planning. *Journal of Cleaner Production*, *443*, 141021. DOI: 10.1016/j.jclepro.2024.141021

Shankar, R., Chaudhary, B. P., Mehraj, H., Gupta, S., Arslan, F., Srivastava, A. K., & Goswami, C. (2023). Impact of node mobility on the DL based uplink and downlink MIMO-NOMA network. *International Journal of Information Technology : an Official Journal of Bharati Vidyapeeth's Institute of Computer Applications and Management*, *15*(6), 3391–3404. DOI: 10.1007/s41870-023-01362-z

Shukla, R. D., Tripathi, R. K., & Gupta, S. (2010, November). Power electronics applications in wind energy conversion system: A review. *In 2010 International Conference on Power, Control and Embedded Systems* (pp. 1-6). IEEE. DOI: 10.1109/ICPCES.2010.5698663

Sinha, R., Kaur, N., Gupta, S., & Thakur, P. (2023, November). Diagnosis of Parkinson's Disease using Hybrid Ensemble Technique. *In 2023 International Conference on Ambient Intelligence, Knowledge Informatics and Industrial Electronics (AIKIIE)* (pp. 1-5). IEEE. DOI: 10.1109/AIKIIE60097.2023.10390458

Thandlam, A. K., Vignesh, T., Saravanan, A., Subramani, P., Marimuthu, R., & Gupta, S. (2024, April). Next-gen vehicle safety: The futuristic approach to auto-stop in modern vehicles. In 2024 *10th International Conference on Communication and Signal Processing (ICCSP)* (pp. 265-270). IEEE. DOI: 10.1109/ICCSP60870.2024.10543666

Chapter 17

Biofortification and the SDGs:

A Pathway to Healthy and Sustainable Futures

K. Parimala Gandhi
https://orcid.org/0000-0003-4643-6734
Nehru Institute of Technology, India

ABSTRACT

Biofortification, which enhances the nutrient density of staple foods through agronomic methods, conventional breeding, or biotechnology, offers a sustainable way to combat micronutrient deficiencies or "hidden hunger." In line with the 2030 SDGs, it provides a scalable and cost-effective solution to improve diets, benefiting vulnerable communities with limited food diversity. By enriching crops at the production stage, it supports SDG 2 (Zero Hunger), SDG 3 (Good Health), and SDG 1 (No Poverty), while climate-resilient varieties advance SDGs 13 and 15. Examples include iron-rich beans in Rwanda, zinc wheat in India, and vitamin A-rich sweet potatoes in Africa. Success depends on political will, seed systems, farmer training, awareness, women's empowerment, and private sector partnerships. Though not a cure-all, biofortification is a key strategy to fight malnutrition, poverty, and ensure sustainability.

INTRODUCTION

One of the most enduring worldwide issues of the twenty-first century is malnutrition, which poses a serious threat to human health, productivity, and socioeconomic advancement. More than two billion people worldwide suffer from

DOI: 10.4018/979-8-3373-6435-3.ch017

micronutrient deficiencies also known as "hidden hunger" caused by inadequate consumption of vital vitamins and minerals like iron, zinc, iodine, and vitamin A, despite notable improvements in food production and economic growth. These deficits disproportionately impact low- and middle-income nations, where staple food-based diets predominate and where nutritional susceptibility is exacerbated by poverty, insufficient health systems, and limited dietary diversity. With irreversible cognitive damage, compromised immunity, diminished labor capability, and elevated mortality risks, children under five and women of reproductive age face the brunt of this burden. In this regard, addressing micronutrient deficiency is both a health necessity and a requirement for attaining sustainable and equitable development. As humanity's main food supply, agriculture is crucial in influencing the results of nutrition. Agricultural systems have always prioritized boosting yields and caloric availability, frequently at the price of nutritious quality. Food quantity has increased in many areas, but food quality has remained insufficient due to this "production-centric" worldview. Because of this, a lot of people eat enough calories yet don't get the vital nutrients needed for good health. One important method to close this gap is through the transition to nutrition-sensitive agriculture. Countries may address hunger at its source by rethinking agricultural policies to improve both production and nutritional value, guaranteeing that food systems provide not only more food but better food. Therefore, biofortification increasing the nutrient content of staple crops through plant breeding, agronomic techniques, or contemporary biotechnology has gained prominence as a sustainable agricultural strategy that can improve micronutrient intake at the population level without necessitating significant dietary or food consumption changes.

The significance of biofortification stems from its capacity to reach vulnerable groups that frequently lack access to a variety of diets, supplements, or processed foods that have been fortified. Biofortification is a one-time investment with long-term advantages since it incorporates nutrients directly into the crops as they develop, unlike conventional fortification, which necessitates a strong food processing sector and ongoing financial investment. Farmers can create a self-sustaining system of increased nutrition by cultivating nutrient-rich varieties season after season once they are produced and distributed. Certain crops have already shown significant gains in nutritional outcomes, such as vitamin A-rich sweet potatoes in sub-Saharan Africa, zinc-enhanced rice in Bangladesh, and iron-rich pearl millet in India. These achievements demonstrate the promise of biofortification as a scalable, economical, and eco-friendly strategy to fight hidden hunger worldwide. When considering the Sustainable Development Goals (SDGs) of the UN, the significance of biofortification becomes even more clear. By improving the nutritional value of basic foods, biofortification is a multifaceted approach that directly supports SDG 2 (Zero Hunger) and indirectly supports a number of other objectives (*Bouis H E 2003).*

By lowering illness burdens and enhancing the health of mothers and children, improved nutritional status contributes to SDG 3 (Good Health and Well-Being). SDG 1 (No Poverty) and SDG 8 (Decent Work and Economic Growth) are advanced by increased agricultural productivity and better market prospects for biofortified crops. Because women are crucial to seed selection, household nutrition, and local food value chains, biofortification also supports SDG 5 (Gender Equality). Additionally, by encouraging agricultural systems that can flourish in a variety of environmental circumstances, climate-resilient biofortified cultivars support SDG 13 (Climate Action). As a result, biofortification functions as a cross-cutting intervention that tackles interrelated issues pertaining to sustainability, equity, health, and food systems (*Arora, N.K &Mishra P, 2025*).

This study aims to present a thorough and critical examination of biofortification as a means of attaining sustainable, healthy futures that are in line with the SDGs. The study specifically looks at the prevalence of micronutrient malnutrition worldwide, evaluates the scientific and practical underpinnings of biofortification, assesses its contributions to several SDGs, examines the obstacles preventing its widespread adoption, and suggests methods for incorporating biofortification into more comprehensive changes to the food system. The study aims to describe the transformative potential of biofortification while outlining the policy, technological, and social concerns required for sustainable implementation by examining empirical data, case studies, and new inventions. The paper is divided into multiple main sections to accomplish these goals. The worldwide background of malnutrition and the shortcomings of current nutrition interventions are described in the part that follows the introduction. A thorough explanation of biofortification techniques and the underlying scientific ideas comes next. With examples from throughout the world, the next part looks at how biofortification supports several SDGs. After analyzing the difficulties and obstacles to scale, the report offers solutions for successful integration between the agricultural, health, and policy sectors. The concluding parts highlight the crucial function of biofortification in creating resilient, nutrition-sensitive food systems and examine future implications.

MALNUTRITION AND GLOBAL FOOD SECURITY

i. Current Status of Micronutrient Deficiencies

Malnutrition is still a serious public health concern worldwide, even with significant advancements in food production over the last 50 years. The Green Revolution greatly expanded the number of calories available, but it did not sufficiently address diets' nutritional value. Millions of people still experience inadequacies in vital vitamins and minerals as a result. According to estimations from the World Health Organization (WHO), over two billion individuals globally suffer from deficiencies in at least one essential micronutrient; this condition is commonly known as "hidden hunger." Contrary to overt hunger, concealed hunger is more often characterized by diets deficient in iron, zinc, iodine, vitamin A, and other micronutrients essential for normal growth and development rather than by empty stomachs.An estimated 1.6 billion individuals suffer from iron deficiency anemia, the most common nutritional condition, which primarily affects young children and women of reproductive age. Due to decreased immunity, vitamin A deficiency is a major contributor to child mortality and one of the main causes of avoidable blindness in children. While iodine insufficiency during pregnancy can cause developmental delays and poorer cognitive outcomes in newborns, zinc deficiency affects growth and immune function (*Mudadirwa D et al 2023*). These inadequacies are not limited to low-income countries; the incidence of malnutrition is increased even in middle-income and affluent countries due to a high reliance on processed foods and a lack of dietary diversification.

Dietary dependence on staple crops like rice, maize, and wheat which supply the majority of calories but are naturally poor in micronutrients is one factor contributing to the persistence of hidden hunger. Due to their heavy reliance on these staples, people in many parts of the world especially in South Asia and Sub-Saharan Africa are particularly susceptible to micronutrient deficiencies. As a result, global food systems must simultaneously ensure sufficient nutritional quality and supply enough calories to support expanding populations.

ii. Health and Economic Impacts of Malnutrition

Malnutrition has serious repercussions for both people and entire societies. From a medical standpoint, consuming insufficient amounts of micronutrients weakens the immune system, lowers resistance to infectious infections, and causes chronic illnesses. Malnutrition in children is closely linked to stunting, wasting, and delayed cognitive development. In addition to lowering living quality, these illnesses also lower educational attainment, which restricts the options available to those who are impacted in the future.Malnutrition raises the risk of pregnancy and delivery problems for women, which raises the mortality rates for both mothers and babies. From conception to the child's second birthday, inadequate nutrition throughout the first 1,000 days of life has lasting effects, including impaired brain development and long-term reductions in earning potential and productivity. Thus, malnutrition contributes to a generational cycle of poverty and injustice.

Malnutrition has a significant financial impact on countries. The Global Nutrition Report estimates that productivity losses brought on by inadequate nutrition cost nations between 2% and 3% of their GDP each year. This covers both direct medical expenses as well as indirect losses brought on by decreased economic growth, lower educational attainment, and decreased worker productivity. These financial pressures are particularly significant in low- and middle-income nations, where sizable segments of the populace are impacted. According to World Bank estimates, higher productivity, education, and health outcomes result from every dollar spent on nutrition interventions.Malnutrition's twin burden makes the world's problems even more complex. Undernutrition is still a problem in many low-income areas, but in affluent populations, obesity and non-communicable diseases linked to diet are on the rise. Undernutrition, micronutrient deficiencies, and overnutrition make up the "triple burden," which presents a special challenge for policymakers and emphasizes the urgent need for systematic, long-term solutions that address all aspects of malnutrition at once (*Pompano LM et al 2022*).

iii. Limitations of Supplementation and Food Fortification

Global health initiatives have historically focused on industrial food fortification and supplementation as important strategies to address the pervasiveness of malnutrition. Even though these strategies have produced noteworthy results, they also have serious drawbacks that limit their applicability as long-term fixes.

For high-risk groups, micronutrient supplementation programs, such giving out vitamin A capsules or iron tablets, work well in the short run. But in order to get broad coverage, they need steady financial investment, reliable supply chains, and robust healthcare delivery infrastructures. The viability of supplementation programs is hampered in many low-resource environments by logistical difficulties, inadequate infrastructure, and poor awareness. Additionally, people may forget to take supplements on a regular basis or decide not to, which can lead to variable compliance.Similarly, several deficiencies have been successfully reduced by food fortification, which is the practice of adding nutrients to regularly consumed foods like flour, salt, or cooking oil. One common example of a global public health success story is the iodization of salt. However, rural people that eat mostly home-grown or minimally processed foods might not be reached by fortification because it also depends on centralized food processing and delivery infrastructures. For instance, fortified commercial items might not be beneficial to households that mostly eat locally produced staples or home-milled grains. Fortification schemes are also susceptible to changes in legislation, difficulties with quality control, and problems with accessibility and pricing.

Customer acceptability is another constraint. In some cultural contexts, supplements and fortified foods may be regarded with suspicion, especially if they are connected to industrial food systems or outside aid initiatives. Changes in flavor, texture, or color may also make them less desirable, which would decrease uptake. These drawbacks highlight the necessity of complementing approaches that, as opposed to depending exclusively on outside interventions, include nutrition enhancements directly into food production systems.

iv. Significance of Solutions Based on Agriculture

Agriculture-based solutions are becoming more and more acknowledged as crucial for tackling malnutrition in a sustainable way, given the limitations of supplements and fortification. For billions of people, especially in low- and middle-income nations, agriculture provides both food and a living. Malnutrition can be addressed at its source by coordinating agricultural development with nutrition objectives.As a plan powered by agriculture, biofortification has a number of special benefits. First of all, it is naturally sustainable: after been created and embraced, biofortified crop varieties may provide nutritional advantages year after year without requiring ongoing outside funding. In the long term, this makes it very cost-effective. Second, rural communities who mostly depend on subsistence farming are among the populations that biofortification reaches yet are frequently left out of supplementing and fortification programs. Biofortification guarantees direct and equal access to better diets by incorporating nutrition into the same crops that these communities cultivate and eat. Agriculture based solutions have the potential to provide co-benefits for environmental sustainability and economic growth in addition to addressing nutrition. Adopting biofortified crops gives farmers access to more productive, climate-resilient cultivars, increasing food security and stability of revenue. In addition to boosting market prospects, strengthening regional value chains, extension services, and seed systems around biofortified crops boosts rural economies. Furthermore, including biofortification into climate-smart farming methods might increase resistance to shocks like pests, droughts, and warming temperatures.Globally, biofortification and other agriculture-based solutions fit in perfectly with the larger goal of sustainable food systems. They acknowledge that hunger is a complicated, multifaceted issue that calls for coordinated responses from the fields of agriculture, nutrition, trade, and policy. Biofortification may change agricultural systems into engines of sustainability, equity, and wellness by directly integrating nutritional goals into food production.

CONCEPT AND METHODS OF BIOFORTIFICATION

i.Understanding the Concept of Biofortification

A nutrition-sensitive agricultural technique called biofortification uses biological processes to increase the micronutrient content of major food crops. Biofortification concentrates on enhancing crops during the production stage, as opposed to post-harvest interventions like industrial fortification and supplementation. Biofortification offers an inherent remedy for common micronutrient deficiencies by directly incorporating vital vitamins and minerals into the seeds and edible tissues of staple crops. For low-income rural people that rely on subsistence farming and have limited access to supplements, fortified foods, or varied diets, this approach is especially beneficial.With programs like HarvestPlus, which has promoted the creation and global distribution of biofortified crops, the Consultative Group on International Agricultural Research (CGIAR) invented the idea. Conventional breeding, agronomic biofortification, and biotechnology techniques are the three main ways that make up biofortification today. Each has advantages and disadvantages of its own. When combined, these tactics offer a thorough foundation for using agriculture to combat hidden hunger.

ii.Conventional Breeding

For biofortification, conventional plant breeding is the most commonly used technique. It is based on the selection and crossbreeding of crop varieties with desirable agronomic qualities like yield stability, pest resistance, or drought tolerance with those that naturally display higher amounts of micronutrients. Plant breeders create stable varieties over many generations that combine traits that farmers and consumers like with better nutritional profiles.The general acceptance of conventional breeding is one of its main benefits. Compared to biotechnology-based methods, it is less contentious and subject to fewer regulatory obstacles because it does not entail genetic change. Additionally, traditional breeding makes use of the biodiversity that already exists, especially in crop gene banks that include landraces and wild cousins with naturally greater micronutrient contents.

Instances of Traditional Breeding in Action

Iron-Rich Pearl Millet (India): In India's dry and semi-arid regions, where diets frequently lack iron, pearl millet is a staple crop. The "Dhanashakti" cultivar, created by HarvestPlus and its collaborators, has up to 70% more iron than traditional types.

Regular consumption of iron-rich pearl millet has been shown in clinical trials to lower iron deficiency anemia in children and women of reproductive age.

Vitamin A Maize (Africa): In Sub-Saharan Africa, maize is a staple crop but naturally low in vitamin A. Through selective breeding, provitamin A carotenoids have been added to biofortified cultivars. Provitamin A maize dramatically raises children's vitamin A status, lowering their risk of illness and blindness, according to field tests and nutrition studies conducted in nations like Zambia and Nigeria.

Zinc Rice (Bangladesh): Although rice is the main staple food in Bangladesh, it has a low zinc content by nature. Zinc-biofortified rice cultivars, such "BRRI dhan62," were released by the Bangladesh Rice Research Institute in partnership with HarvestPlus. These cultivars have the production and cooking characteristics that farmers and consumers are accustomed to, but they also have increased zinc contents.

Numerous biofortified crops have been introduced throughout Asia, Africa, and Latin America, demonstrating the scalability of conventional breeding. However, not all nutrients can be efficiently increased using this strategy, and progress may be gradual because multi-season breeding, testing, and release take time.

iii.Agronomic Biofortification

In order to raise the concentration of micronutrients in crops during growth, agronomic biofortification entails using fertilizers, soil amendments, or foliar sprays. This method offers a supplementary approach to breeding and tackles soil shortages, which frequently lead to crops with low nutritional levels (*Kendal, 2020*).

Agronomic techniques can be used to efficiently provide micronutrients including iodine, zinc, and selenium. In South Asia, for instance, where soils frequently lack zinc, resulting in low dietary zinc intake and poor crop yields, zinc-enriched fertilizers are widely utilized. In areas where soil nutrient deficits are well-established and where farmers currently use fertilizers as part of crop management, agronomic biofortification works particularly well.

Agronomic Biofortification Examples in Action

Zinc-Enriched Wheat (India): Researchers have advocated for the use of zinc sulfate fertilizers in wheat production in India, where about half of the population suffers from zinc shortage. This approach raises the zinc content of harvested grains while simultaneously increasing crop production. Agronomic and nutritional advantages have been observed by farmers who use zinc-enriched wheat.

Selenium Biofortification (Finland): On a national level, Finland is a shining example of agronomic biofortification.As a public health measure to address the prevalent selenium shortage in the population, selenium has been added to fertilizers since the 1980s. By effectively raising selenium intake through cereals and other food products, this project has shown the potential of agronomic practices on a nationwide level.

Iodine Biofortification (Vegetables): Application of iodine-containing fertilizers has been shown in experiments to raise iodine concentrations in crops like spinach and lettuce, adding another dietary source of this essential vitamin.

Agronomic biofortification has the advantage of being immediate; soil and foliar sprays can raise crop nutrient levels in a single season, whereas breeding takes years. However, the method necessitates constant fertilizer input, which raises questions regarding sustainability, expense, and possible environmental effects. When combined with more comprehensive soil fertility management techniques, it works best.

iv.Biotechnology Approaches

Biotechnology provides strong alternatives to traditional breeding and agronomic techniques where they are limited. Scientists can directly alter plant genomes to increase micronutrient content or nutrient bioavailability using genetic engineering and genome editing. Inserting genes that allow plants to produce more of a certain nutrient or improve its uptake from the soil is known as genetic engineering. Golden rice, which was genetically modified to produce beta-carotene, a precursor of vitamin A, in the endosperm, is a well-known example.Golden rice has been authorized for production in a number of nations, including Bangladesh and the Philippines, despite controversy and delays brought on by biosafety and regulatory issues (*Mishra P et al 2023*).

Newer methods that enable precise editing of plant genomes without introducing foreign DNA include CRISPR-Cas and other gene-editing tools. For instance, gene editing can improve nutrient synthesis routes or eliminate genes that restrict nutrient accumulation. Compared to standard transgenic techniques, this approach is quicker and might be more acceptable to the general public because it might not be classified as genetically modified organisms (GMOs) in some jurisdictions.

Biotechnology in Biofortification Examples

Golden Rice (Asia): Golden Rice, one of the most well-known examples of biotechnology-based biofortification, was created to address vitamin A deficiency. According to clinical research, when incorporated into diets, it can supply between 30 and 50 percent of the daily required amount of vitamin A.

High-Iron Beans (Africa): To increase the iron content of beans, a staple food in East Africa, genetic engineering has been used. These initiatives seek to provide greater, more consistent amounts of iron buildup and to supplement traditional breeding methods.

CRISPR in Wheat and Rice: Recent studies have shown that by modifying genes related in the absorption and transport of nutrients from roots, CRISPR can improve the uptake of zinc and iron in cereal crops. Although these developments are currently in the experimental phase, they have potential for use in the future.

The speed and accuracy of biotechnology techniques allow researchers to target nutrients that are challenging to improve using traditional methods. But there are still a lot of obstacles to overcome, including ethical discussions, legal restrictions, and public acceptance.

v.Successful Adoption Case Studies

Orange-Fleshed Sweet Potato (Sub-Saharan Africa): The promotion of orange-fleshed sweet potatoes (OFSP) enhanced with beta-carotene is among the most well-known examples. Since its introduction, OFSP has been enthusiastically embraced by smallholder farmers in nations including Rwanda, Uganda, and Mozambique. Nutrition studies reveal notable improvements in women's and children's vitamin A intake, which results in quantifiable decreases in health problems associated with deficiencies. The OFSP's success emphasizes the value of farmer training, community involvement, and consumer awareness initiatives.

Iron Beans (Rwanda and Democratic Republic of Congo): In Rwanda and the Democratic Republic of Congo, where beans are a staple food, high-iron bean types created through conventional breeding have been made available. Because of the cultivars' agronomic benefits and consumer familiarity, adoption rates are high. Research validates the health impact of these beans by confirming improvements in iron status among youngsters who eat them.

Zinc Wheat (India and Pakistan): In South Asia, where zinc insufficiency is a major problem, zinc-biofortified wheat types have been introduced. Farmers in Pakistan have embraced "Zincol-2016" and included it into school lunch programs. Zinc wheat consumption has been associated with better growth and immunological results in children.

Multi-Crop Adoption in Latin America: Through HarvestPlus collaborations, nations like Brazil and Colombia have released biofortified beans, maize, and cassava. With assistance from government food programs, these crops are being incorporated into local diets, illustrating how biofortification may be expanded to include a variety of food staples.

These case studies demonstrate that, when paired with encouraging legislation, community involvement, and efficient extension services, biofortification is not only technically possible but also socially and economically sustainable.

BIOFORTIFICATION AND THE SUSTAINABLE DEVELOPMENT GOALS

Adopted in 2015, the Sustainable Development Goals (SDGs) of the United Nations offer a global framework for tackling interrelated issues pertaining to inequality, poverty, health, and environmental sustainability. This framework is directly supported by biofortification, which connects agricultural innovation to better nutrition and long-term development results. Biofortification is a revolutionary tool for progress across several SDGs since it incorporates sustainability and health into food production, unlike short-term nutritional treatments. Its contributions fall into two categories: direct effects on food systems and nutrition, and indirect effects that affect social, economic, and environmental development.

i. Direct Contributions

SDG 2: Zero Hunger

SDG 2 is centered on attaining food security and eradicating hunger. Global agricultural productivity has increased, but hunger and nutritional shortages still exist in large numbers. By enhancing the nutritional value of staple crops that serve as the basis for the diets of billions of people, biofortification directly tackles this issue. In Sub-Saharan Africa, for instance, the introduction of vitamin A-biofortified orange-fleshed sweet potatoes (OFSP) has greatly boosted dietary vitamin A intake among women and children, lowering the risk of blindness and sickness associated with vitamin A insufficiency. Similar to this, the spread of iron-rich beans in East Africa and zinc-biofortified wheat in South Asia guarantees that populations reliant on these staples obtain more vital elements without necessitating significant dietary changes.Crucially, biofortification improves the accessibility and availability of nutrient-dense foods. Distributing biofortified seeds allows farmers to consistently grow nutrient-rich crops year after year, which makes it an affordable option for communities with limited resources. SDG objectives 2.1 and 2.2, which aim to eradicate hunger and all forms of malnutrition by 2030, respectively, are advanced by this. Biofortification promotes dietary sufficiency at scale and fortifies food systems against future crises by directly integrating nutrients into crops (Habib et al., 2025).

SDG 3: Good Health and Well-Being

SDG 3 emphasizes ensuring healthy lives and promoting well-being for all, yet malnutrition remains a major barrier to achieving this goal. Deficiencies in iron, zinc, and vitamin A significantly increase the risk of infectious diseases, maternal complications, and child mortality. Biofortified foods help address these challenges by consistently improving micronutrient intake and reducing associated health risks. Clinical evidence demonstrates these benefits: in Rwanda, consumption of iron-rich beans improved iron status in school-aged children and reduced anemia rates. Similarly, in Mozambique, children who consumed orange-fleshed sweet potatoes showed improved vitamin A levels, lowering their risk of illness and vision problems (Mitra-Ganguli et al., 2019). By strengthening immune function and reducing healthcare burdens, biofortification contributes to long-term health resilience. It prevents malnutrition-related diseases before they occur, an especially valuable advantage in low-income countries with limited health services. In doing so, biofortification directly supports SDG targets 3.2 (ending preventable deaths of children under five) and 3.4 (reducing premature mortality through prevention).

SDG 12: Responsible Consumption and Production

Modern food systems are criticized for encouraging diets high in calories but low in nutrients, which leads to obesity on the one hand and malnutrition on the other. Sustainable patterns of production and consumption that minimize food waste, save resources, and guarantee the prudent use of agricultural inputs are called for by SDG 12.

By incorporating nutrition into crops without requiring extra steps during food processing, biofortification helps achieve these objectives. Biofortification offers inherent nutritional benefit at the manufacturing stage, in contrast to industrial fortification, which necessitates centralized processing and ongoing expenses. Food supply chains become more resilient and sustainable as a result of the decreased reliance on external fortification systems.Furthermore, biofortified crops are frequently cultivated for yield stability and climate resistance, guaranteeing that farmers can produce wholesome food even in the face of unfavorable environmental circumstances. Biofortification tackles the ecological and nutritional aspects of responsible production by encouraging nutrient-dense, environmentally friendly crops (*Mishra P et al 2025*). SDG targets 12.2 and 12.3, which focus on sustainable management of natural resources and minimizing food waste and losses along the value chain, respectively, are in line with this.

ii. Indirect Contributions

SDG 1: No Poverty

Malnutrition and poverty are closely related, creating a vicious cycle whereby low productivity due to inadequate nutrition feeds poverty. Biofortification indirectly contributes to poverty reduction by improving economic potential, educational performance, and labor productivity through improved nutritional outcomes.

Increased yields and market opportunities are frequent advantages for farmers that use biofortified crops. For instance, the introduction of iron-rich pearl millet (Dhanashakti) in India reduced the chance of crop failure by giving farmers a crop that was suited to arid circumstances while simultaneously increasing dietary iron intake. In a similar vein, Pakistan's adoption of zinc wheat cultivars raised household earnings by increasing market demand and production.Additionally, better nutrition lowers healthcare expenses for families, freeing up scarce funds for savings, livelihoods, and education. Biofortification indirectly contributes to SDG target 1.2, which seeks to eradicate poverty in all of its manifestations, by enhancing human capital and economic resilience.

SDG 4: Quality Education

Since cognitive development, academic performance, and learning ability are all strongly impacted by nutritional condition, biofortification is a crucial component of SDG 4. When compared to their starving peers, children who eat biofortified crops show improvements in their ability to focus, remember things, and attend school.

Children who eat vitamin A sweet potatoes do better in school because they have stronger immune systems and miss fewer school days owing to illness, according to evidence from initiatives in Sub-Saharan Africa. Likewise, increased cognitive function and attention spans have been associated with decreased iron insufficiency caused by iron-biofortified vegetables.

The educational underpinnings required to end intergenerational cycles of poverty are supported by biofortification, which enhances child nutrition. This addresses learning difficulties associated with hunger, which is in line with SDG objective 4.1 (ensuring completion of free, equitable, quality primary and secondary education).

SDG 5: Gender Equality

Women play a critical role in household nutrition and agriculture, yet they are disproportionately affected by malnutrition. Nearly one-third of women of reproductive age globally suffer from iron deficiency anemia, which affects their health and productivity. Biofortification helps advance gender equality by improving women's nutrition, reducing maternal health risks, and strengthening their role in agricultural value chains. Women farmers are central to cultivating, processing, and preparing biofortified foods. Self-help groups and women's cooperatives have effectively promoted awareness and adoption, as seen in Uganda where women-led initiatives accelerated the use of orange-fleshed sweet potatoes, improving both nutrition and income. With access to biofortified seeds and training, women become change agents who promote healthier diets within families and communities. By enhancing women's agency and access to resources, biofortification supports SDG targets 5.1 and 5.a.

SDG 8: Decent Work and Economic Growth

By encouraging innovation, opening up markets for agricultural products, and raising worker productivity, biofortification promotes economic growth. Because healthier people can work longer, learn better, and contribute to economic development more successfully, improved nutrition boosts labor efficiency. From an agricultural standpoint, the creation and distribution of biofortified seeds increases demand for novel inputs, opening doors for agro-dealers, seed firms, and rural business owners. Jobs in processing, marketing, and distribution are also created by value chains centered on biofortified crops. For instance, biofortified beans and maize have been integrated into regional food sectors in Latin America, assisting small businesses and farmers alike.Biofortification contributes to SDG targets 8.5 (reaching full and productive employment for all) and 8.2 (achieving better levels of economic production through diversification and innovation) by tying together agriculture, nutrition, and economic opportunity.

SDG 13: Climate Action

Global food security is seriously threatened by climate change, which worsens malnutrition by lowering agricultural yields and nutritional quality. The micronutrient content of vital staples like rice and wheat has been shown to decrease when carbon dioxide levels rise.By cultivating crops that are both nutrient-dense and stress-tolerant, biofortification offers a climate-smart adaptation method. For example, drought-resistant biofortified cassava and pearl millet cultivars have been created to provide food and nutrition security in areas susceptible to climate shocks. Incorporating biofortification into climate-smart agriculture enhances community resilience to environmental threats while also protecting nutrition.

CHALLENGES AND BARRIERS

Although biofortification has the potential to revolutionize the fight against hidden hunger and further the Sustainable Development Goals (SDGs), there are many obstacles in the way of its widespread adoption. These issues are complex, involving markets, consumer behavior, government, cultural contexts, and scientific limitations. In order to mainstream biofortification into agriculture and nutrition plans globally, it is imperative to comprehend and solve these obstacles.

i.Policy and Governance Gaps

Biofortification has not yet been completely included into national and international nutrition policies, despite its demonstrated benefits. Instead of considering biofortification as a standard agricultural and health intervention, many governments continue to view it as an auxiliary initiative. The integration of policies is one significant gap. Health policies might not fully acknowledge agriculture's role in reducing malnutrition, while national agricultural policies frequently place a higher priority on production maximization than nutritional outcomes. Because of this, biofortification lacks integrated planning and finance, placing it in a gray area between the agriculture and health sectors.Additionally, different nations continue to have different regulatory frameworks for licensing and developing biofortified crops. While some countries, like Brazil, Nigeria, and India, have included biofortified varieties into their national plans, others have done so slowly because of concern about genetically modified (GM) techniques, bureaucratic obstacles, or a lack of understanding. This delay restricts consumer reach and limits farmers' access to better seeds.

Another obstacle is funding. Long-term investments in crop development, seed distribution, and awareness-raising are necessary for biofortification. However, because they produce faster outcomes, many governments and donors choose short-term measures like fortification and supplementation. Despite its long-term viability, biofortification runs the risk of becoming marginalized in the absence of political will and financial support.Lastly, there is little international alignment. Despite the fact that international programs like HarvestPlus have led the charge on biofortification, regional coordination is still lacking. To bridge these governance gaps, a more robust multilateral strategy is required, complete with common research platforms and established procedures (*Shohael A M, et al 2025*)

ii.Consumer Acceptance

The availability of biofortified crops does not ensure consumer adoption. Perceptions of new technologies, cultural customs, and sensory preferences all have a significant influence on people's eating choices. Taste and appearance are a major obstacle. For example, some communities have opposed orange-fleshed sweet potatoes, which are high in vitamin A, since they prefer the more conventional white-fleshed types. Similarly, in some regions of Africa, yellow-colored biofortified maize was initially disapproved of since it was thought to be a nutritional enhancement rather than food aid for the underprivileged. Without strategies to demonstrate similar or superior flavor, color, and cooking skills, adoption may stall. The nutritional advantages of biofortified crops are still unknown to many consumers, and suspicion can be exacerbated by false information, particularly when it comes to biotechnology. Adoption is frequently hampered by worries about safety, possible negative effects, or "unnatural" food. Biofortification through conventional breeding may also be misinterpreted as genetic manipulation in places where GM crops encounter resistance, making acceptance even more difficult. Furthermore, compared to rural populations, urban customers frequently have different expectations. Urban consumers usually favor branding, convenience, and market image, but rural households could prefer affordability and availability. The secret to biofortified crops' acceptability in larger markets is to present them as aspirational rather than just poverty-focused.In the end, initiatives for behavioral change, nutrition education, and social marketing are necessary for successful customer participation. Long-term acceptance and trust can be increased by showcasing better health results, incorporating communities in crop choices, and matching products to regional eating customs.

iii.Market Integration

The successful adoption of biofortified crops depends on their full integration into markets and value chains, yet this remains a major challenge. Weak supply chains, limited incentives, and minimal private sector participation hinder widespread uptake. Farmers are often reluctant to adopt biofortified seeds because they lack access to reliable markets that offer favourable prices. Unlike cash crops, biofortified staples are typically grown for household consumption, limiting their commercial scalability. Without strong consumer and food industry demand, farmers perceive biofortified varieties as less profitable than conventional high-yielding cultivars. The seed industry also presents significant bottlenecks. Producing biofortified seeds requires specialized breeding, certification, and distribution systems, which are underdeveloped in many low- and middle-income countries. Smallholder farmers, who grow most staple crops, frequently struggle to access certified seeds, especially in remote rural areas. Ensuring the affordability and availability of high-quality biofortified seeds is therefore essential.

Post-harvest and processing factors further impact nutrient retention. Vitamin A in orange-fleshed crops is highly sensitive to heat and light, while iron and zinc levels may diminish during storage or milling. Without improved processing technologies and adequate farmer training, a considerable portion of the nutritional value may be lost before consumption. Commercialization challenges also persist in labeling and branding (Boliko, 2019). Consumers are more willing to pay for nutritional benefits when these are clearly communicated, yet many markets lack certification systems that distinguish biofortified products from conventional crops. The absence of standardized labeling limits premium pricing opportunities and reduces consumer awareness. Finally, private sector engagement remains limited despite its critical importance. Food companies, processors, and retailers need incentives—such as subsidies, tax benefits, or strong consumer demand—to incorporate biofortified crops into their supply chains. Without sustained private sector participation, biofortification risks remaining donor-driven rather than evolving into a viable, long-term market-based solution.

iv.Technical Limitations

Despite tremendous progress, biofortification's full potential is still limited by scientific and technological obstacles. The biological ceiling is one restriction. Due to genetic limitations, not all crops are able to accumulate large amounts of micronutrients. For example, it has been challenging to get enough iron in rice as opposed to millet or beans. It is still difficult to breed crops that are high-yielding and nutrient-rich without sacrificing agronomic qualities. Another problem is the bioavailability of nutrients. Antinutritional elements like phytates can prevent the human body from fully absorbing nutrients, even if a crop has higher nutrient levels. Advanced breeding methods, advancements in food processing, and associated dietary behaviors are necessary to guarantee that biofortified crops offer accessible nutrients (*Batool, S et al 2021*).The intricacy is increased by climate change. Increased carbon dioxide levels jeopardize the benefits of biofortification by lowering the nutritional density of many staple foods. Investing heavily in research is necessary to breed crops that are both nutrient-rich and climate-resilient.

Furthermore, lengthy breeding cycles impede advancement. Seven to ten years may pass before a biofortified variety is developed, tested, and released. This timetable postpones significant effects, particularly in areas where there is an immediate nutritional crisis. Although advanced biotechnology techniques like genome editing and genetic engineering have the potential to speed up development, they encounter obstacles related to ethics, regulations, and public acceptance.Monitoring and evaluation are also subject to technical constraints. Extensive research, biomarkers, and long-term surveillance systems are necessary for assessing nutritional impact at the population level. It is challenging to measure the effectiveness of biofortification initiatives, nevertheless, because many nations lack the infrastructure necessary to monitor micronutrient results.

v.Ethical and Cultural Considerations

Food serves as a cultural identity in addition to being a source of nutrition. To prevent opposition or unforeseen repercussions, attempts to modify staple crops through biofortification must negotiate ethical and cultural difficulties. Equity of access is one ethical issue. Biofortification may unintentionally increase inequality rather than decrease it if the poorest farmers cannot buy or obtain biofortified seeds. Similarly, underprivileged people would not benefit from nutritional advances if commercial farmers or seed firms reap the majority of the benefits. Therefore, it is morally necessary to guarantee inclusive access.Another concern is informed choice. Communities must understand what biofortified crops are, how they differ from traditional varieties, and why they are beneficial. Without transparency, introducing biofortified foods risks being seen as paternalistic or even exploitative, particularly in regions with histories of food aid dependency.Cultural food preferences also matter. In many societies, staple foods are tied to traditions, rituals, and identities. Efforts to replace or modify these staples may meet resistance if they conflict with established norms. For example, in South Asia, white rice is deeply ingrained in dietary culture, making adoption of biofortified alternatives difficult despite their nutritional advantages.

Furthermore, ethical questions of corporate control, environmental effect, and safety are brought up by biotechnology biofortification. Critics contend that local seed sovereignty may be undermined by genetic engineering, which might transfer control to global corporations. This issue emphasizes the necessity of using participatory breeding techniques that honor traditional knowledge and involve nearby farmers. Lastly, the issue of nutrition justice is brought up by biofortification: should governments promote greater nutritional diversity or should people be forced to rely on genetically modified crops for basic health? Although biofortification works well in the short and medium term, it needs to be incorporated within a comprehensive nutrition plan that honors food sovereignty and cultural diversity.

STRATEGIES FOR SCALING AND INTEGRATION

By raising the micronutrient density of staple crops, biofortification has become a financially viable and scientifically validated strategy to combat hidden hunger. However, widespread integration into agricultural systems, food markets, and nutrition policy is necessary for biofortification to succeed; it is not only about crop development. This calls for a comprehensive approach that incorporates systemic policy integration, gender inclusion, private sector involvement, farmer training, cross-sectoral alliances, and innovative research.

The following tactics show how biofortification can be institutionalized and scaled up in national and international food systems.

i. Cross-Sectoral Collaboration: Building Multi-Stakeholder Synergies

Agriculture–Nutrition–Health Nexus

Biofortification has applications in agriculture, nutrition, and public health. These historically divided industries need to work together for scaling to be successful. To guarantee that biofortified crops are acknowledged as nutritional treatments as well as agricultural products, ministries of agriculture should collaborate closely with ministries of health and education.

Global Development Agencies

Global institutions including the World Bank, FAO, WHO, and CGIAR institutes are essential to the advancement of biofortification. Increased political visibility and donor commitment can be achieved by integrating biofortification into global nutrition objectives such as the Scaling Up Nutrition (SUN) campaign.

NGOs and Civil Society

Campaigns for awareness and grassroots mobilization heavily rely on non-governmental groups. They can ensure that biofortified crops are promoted and adopted in a way that is culturally appropriate by bridging the gap between local populations and scientific organizations. NGOs also improve accountability by tracking household-level effects.

PPPs, or public-private partnerships

PPPs are necessary for the incorporation of biofortification into food systems. Governments can encourage commercial businesses to use biofortified crops in food processing, retail, and school feeding programs by offering subsidies, tax breaks, or co-finance plans. Through these collaborations, biofortification can become viable beyond donor funds and increase market demand.

ii. Farmer Training and Extension Services: Enabling Adoption at the Grassroots

Building Farmers' Capacity

Although many people are still ignorant of the possibilities of biofortification, farmers are essential to its success. Training courses ought to include the Biofortified variety management and seed selection.

Agronomic techniques to improve the retention of nutrients.

Possibilities for the biofortified crop market; Post-harvest handling to avoid nutrient loss.

Enhancing Extension Frameworks

It is necessary to revitalize agricultural extension services, which are frequently underfunded in developing nations, in order to function as knowledge channels. To illustrate the financial and nutritional advantages of biofortified crops, extension professionals ought to receive training. Participatory learning, demonstration plots, and farmer field schools can all help promote broad adoption.

Digital Platforms and ICT

Farmers can receive up-to-date information on market prices, weather patterns, crop management, and seed availability through digital innovations including community radio, SMS notifications, and mobile apps. Additionally, interactive portals can link farmers with buyers, extension agents, and nutrition specialists.

Rewards for Early Adopters

It takes momentum to scale. Early adoption can be promoted by offering incentives like premium market pricing for biofortified crops, input credits, or seed subsidies. By showcasing real advantages and encouraging peer-to-peer knowledge sharing, these early adopters can subsequently serve as advocates in their communities.

iii. Private Sector Engagement: Driving Market Demand and Value Chains

Participation of the Seed Industry

When it comes to expanding biofortification, seed firms are essential. Working together with private seed firms guarantees that biofortified seeds will be more widely available and more reasonably priced. Distribution networks can also be extended to far-flung rural areas through collaborations with cooperatives, agro-dealers, and farmer associations.

Food Processing and Labeling

Biofortified crops can be incorporated by food producers into high-value goods including cereals, snacks, and flours. Clear nutritional information on the label and branding of biofortified products can raise customer awareness and increase demand in urban and peri-urban areas*(Priya Singla &Kiran Grover2017).*

Channels of Distribution and Retail

For biofortified food, supermarkets, neighborhood markets, and online market-places can serve as multipliers. Promoting and encouraging stores to carry biofortified crops raises awareness of them and normalizes their use in consumer diets.

CSR, or corporate social responsibility

Biofortification is something that private sector actors can incorporate into their CSR plans. To combine profitability with social impact, food and agribusiness firms can, for instance, fund farmer training, awareness campaigns, or school-feeding programs that incorporate biofortified foods.

Impact Investors and Financial Institutions

It is essential to have access to investment funds and credit. Microfinance products for smallholder farmers using biofortified crops can be created by financial institutions. In the meanwhile, value-chain businesses that deal with biofortified crops, like seed companies, processors, and transporters, can be funded by impact investors.

iv. Innovation and Research: Advancing Science and Technology

Advanced Methods of Breeding

Even though traditional breeding has produced successful crops (such as vitamin A maize and iron-rich pearl millet), innovative techniques like CRISPR-Cas9, genetic engineering, and marker-assisted selection can speed up development and improve nutritional levels. To break through biological ceilings, investment in crop genomics and bioinformatics is essential (*Dhaliwal SS et al 2022*).

Studies on Bioavailability

To evaluate bioavailability, research needs to go beyond nutrient concentration. Working together with medical experts and nutrition specialists can guarantee that biofortified crops have noticeable health benefits, especially for children and women of reproductive age.

Adaptable Biofortification to Climate Change

Climate resilience and nutrient augmentation must be combined in future breeding. Crops must retain their nutrient density while withstanding heat stress, pests, and drought. Food and nutrition security is ensured by this dual approach, especially in light of climate change.

Innovations after Harvest

To minimize nutritional losses, research on processing, grinding, and storage is essential. Nutritional results can be improved, for instance, by creating inexpensive storage bags that reduce vitamin deterioration or by using grinding methods that retain iron and zinc.

Systems for Data and Monitoring

To assess nutritional impact, yield performance, and adoption rates, strong monitoring mechanisms are required. Policymakers can obtain evidence for scaling by incorporating biofortification measurements into agricultural censuses and national health surveys.

v. Improving Gender Inclusion through the Role of Women in Value Chains

Women in Agriculture and Caregiving

In homes, women are the major caregivers and food producers. Providing women farmers with training, credit, and access to biofortified seeds can greatly enhance household nutrition results. Research indicates that women are more likely than males to spend money on food and health, which increases the benefits of biofortification.

Women in Processing and Seed Systems

Local value chains can be strengthened by women's involvement in seed cultivation, processing, and retailing. In addition to fostering biofortification, teaching women's cooperatives how to process food or multiply seeds generates employment prospects.

Social Behavior Change and Nutrition Education

Food choices in the home are frequently influenced by women. Adoption at the family level can be promoted by educating women about the advantages of biofortified foods through targeted nutrition education initiatives. It is possible to mainstream the production of biofortified crops by involving women in cooking lessons or community demonstration kitchens (*Kumar, S et al 2022*).

Policy Sensitivity to Gender

Gender differences in land ownership, financial access, and extension services must be addressed by policy. Initiatives for biofortification can attain better inclusivity and long-term sustainability by guaranteeing women's equal involvement.

vi. Agricultural Policy Mainstreaming: Transitioning from Niche to Mainstream Integration into Food Systems and Policies

Targets for biofortification should be specifically included in national agricultural strategies. To guarantee farmer access and cost, governments might incorporate biofortified varieties into public procurement programs, seed certification programs, and subsidy programs.

Participation in Public Nutrition Initiatives

Programs for food assistance, school feeding, and maternal and child nutrition offer great opportunities to incorporate biofortified foods. For instance, incorporating iron beans into hospital meal programs or vitamin A maize into school lunches can directly impact vulnerable populations.

Regulation of Trade and Markets

To certify biofortified crops and products, governments must set up criteria and labeling schemes. In addition to increasing customer confidence, this opens up new markets for the global sale of biofortified foods*(Oluwole O et al 2023*).

Methods of Financing

Biofortification requires consistent investment to scale. Innovative funding tools, such as blended finance, nutrition-sensitive agricultural budgets, and public bonds devoted to food system innovation, should be created by governments and donors.

Toward Food Systems That Are Sensitive to Nutrition

Integrating and scaling biofortification requires a (*Sheoran, S et al 2021*) systemic change rather than a single endeavor. Strong policy frameworks that mainstream biofortification within national food systems, cross-sectoral partnerships to bridge agriculture and nutrition, farmer empowerment through training and extension, private-sector leadership in markets and value chains, ongoing scientific innovation, gender inclusion for equity and sustainability, and coordinated action across multiple dimensions are all necessary. Together, these approaches have the potential to transform biofortification from isolated pilot projects into a global nutrition revolution that will improve the diets of billions of people while promoting economic growth, gender equality, and food security. Although the path necessitates foresight, cooperation, and perseverance, the benefits healthier generations and more robust food systems make the work invaluable.

PROSPECTS FOR BIOFORTIFICATION IN THE FUTURE

The potential of biofortification to reduce hidden hunger and enhance nutritional security has previously been shown. However, the future of biofortification must change as food systems undergo changes due to the effects of climate change, technological advancements, and evolving sociopolitical environments. To guarantee the scalability and sustainability of biofortified crops in the ensuing decades, new strategies that incorporate digital innovation, climatic resilience, international co-operation, and moral foresight will be needed (*Bharti C et al 2024).*

i. Climate-Smart Biofortification's Function

Climate Change Adaptation

Due to increased atmospheric CO_2 levels, climate change not only lowers agricultural yields but also lowers micronutrient concentrations in key crops, posing a twofold danger to global food security. Therefore, the goal of climate-smart biofortification will be to create nutrient-dense cultivars that are also resistant to insect outbreaks, floods, salt, and drought.

Climate-Smart Agriculture (CSA) Integration

CSA frameworks that prioritize resource efficiency, soil health, and sustainable farming might incorporate biofortification. For example, maximizing nutritional and environmental results can be achieved by combining biofortified seeds with conservation agriculture, crop variety, and precision irrigation.

Food Systems That Are Resilient

In the future, biofortification is also expected to support the resilience of the food system. In line with SDG 13 (Climate Action), biofortification supports adaptation and mitigation goals by providing smallholder farmers with crops that provide greater nutrition while thriving under climate stress (*Labuschagne M 2023*).

ii. Applications of AI and Digital Agriculture

Genomics and Precision Breeding

Crop improvement is being revolutionized by developments in big data analytics and artificial intelligence (AI). Breeding cycles can be sped up by using machine learning algorithms to evaluate genomic datasets and anticipate desirable features like increased iron or zinc levels. Scientists will be able to more precisely discover genetic markers for nutrient enrichment thanks to AI-driven bioinformatics tools.

Advisory and Intelligent Farming Systems

Drones, IoT sensors, satellite imaging, and smartphone apps are examples of digital agricultural tools that can optimize farming methods for biofortified crops. For instance, remote sensing can track crop growth and identify stress early, guaranteeing higher yields of biofortified types; AI-based advising applications may offer tailored advice on planting dates, nutrition management, and pest control.

Blockchain for Tracking

Biofortified crops may be tracked from seed to plate using digital platforms like blockchain, guaranteeing their quality and authenticity. This is especially crucial for international trade in biofortified products and customer confidence.

Overcoming the Digital Gap

Investments in inclusive platforms, digital literacy, and rural connectivity are essential if smallholder farmers are to profit from digital technologies. In the absence of this, biofortification runs the risk of becoming just another invention that is out of reach for those who need it most.

iii. International Collaborations and Funding Sources

International Cooperation

Stronger international partnerships are essential for future advancement. The foundation has already been established by programs such as HarvestPlus, but scaling calls for further cooperation between governments, multilateral organizations (*Neeraja, C.N et al 2017*), academic institutions, and the commercial sector. Global adoption can be accelerated through coordinated seed policies, knowledge-sharing platforms, and cooperative research initiatives.

Models of Sustainable Financing

Although donor support has been crucial, long-term funding is also necessary. Future strategies could involve: Blended finance, which combines charitable donations, private investment, and government subsidies. Budgets for agriculture that are responsive to nutrition, in which governments set forth money expressly for biofortification. Development impact bonds that incentivize quantifiable dietary results.

Local Centers and South-South Collaboration

Biofortification hubs for seed production, training, and distribution can be established through regional cooperation, especially in Africa and Asia. Through cooperation, nations dealing with comparable issues will be able to share technologies, tactics, and success stories.

iv. Possible Dangers and Discussions about Relying Too Much on One Solution

Despite being an effective strategy, biofortification cannot take the place of a varied diet. One possible danger is that donors or governments may favor biofortified staples over more comprehensive changes to the food chain, unintentionally limiting nutrition plans.

Concerns about Equity

There are worries that if big businesses control the delivery of seeds, smallholder farmers would be left out. Public regulation and community involvement will be necessary to guarantee that biofortification continues to be available, reasonably priced, and equitable.

Considerations of Culture and Ethics

If biofortified crops are seen as artificial or imposed from without, certain people may oppose them. Overcoming skepticism requires open communication, participatory breeding, and culturally informed awareness initiatives.

GM and Biotechnology Discussions

Ethical discussions may get more heated as biofortification incorporates genetic modification (GM) and genome editing techniques more and more. Regulatory regimes must strike a balance between innovation and public trust while guaranteeing the protection of consumer rights, safety, and the environment.

Assessing Long-Term Effects

Additional longitudinal research is also required to assess the long-term health effects of biofortified crops. Policymakers can be reluctant to make large investments in the absence of solid proof.

Moving Toward a Resilient and Inclusive Future

Biofortification's future depends on its capacity for innovation, adaptation, and integration into larger changes in the food system. Crop resilience will be increased by climate-smart methods, while breeding and agricultural efficiency will be accelerated by digital and AI-driven technologies. Long-term sustainability will be guaranteed by international collaborations and funding sources, and continuous discussions will hone the implementation's ethical, cultural, and equitable aspects. In the end, biofortification has promise not only for (*Narayanan N et al 2019*) micronutrients but also for empowering communities, enhancing resilience, and rethinking food systems to provide sustainability, equity, and health in a world that is changing quickly.

CONCLUSION

At the nexus of agriculture, nutrition, and sustainable development, biofortification provides a socially revolutionary and scientifically sound solution to the world's micronutrient deficits. In particular, SDG 2 (Zero Hunger), SDG 3 (Good Health and Well-Being), SDG 12 (Responsible Consumption and Production), and SDG 13 (Climate Action) can all be advanced by biofortification, as this paper has shown. There is ample evidence that incorporating nutrient-rich staple crops into food systems can significantly lower hidden hunger among disadvantaged communities, lessen the need for supplementing programs, and enhance public health indicators on a large scale.

The difficulties mentioned governance gaps, market restrictions, technology limitations, cultural attitudes, and consumer acceptance show that biofortification cannot be successful on its own. The effectiveness of multi-level support systems, ranging from inclusive market processes and community participation to policy frameworks and extension services, determines its impact. Governments must establish clear regulatory procedures, incorporate biofortification into national nutrition missions and agricultural policies, and guarantee cross-sectoral coordination between the ministries of health, agriculture, education, and rural development in order to address these systemic obstacles.

The tactics discussed in this study highlight the fact that scaling biofortification is essentially a team effort. To maximize yields and nutrient content, farmers require specialized training, easily accessible seed systems, and evidence-based agronomic advice. The private sector has to be encouraged to make investments in consumer-focused product development, value chain improvement, and seed multiplication. Research organizations need to keep coming up with new ideas, especially in the areas of digital agriculture platforms, genomic tools, and climate-resilient breeding, which can reduce breeding cycles and increase accuracy. Crucially, women must be acknowledged as key players in biofortification value chains since they are essential to seed selection, household nutrition, and food processing. Their involvement guarantees fair benefit distribution, increases adoption rates, and speeds up acceptance at the community level. Looking ahead, how well biofortification adjusts to new obstacles will determine its destiny. Because crop yield and nutrient density are threatened by climate change, climate-smart biofortification is not only desirable but also essential. Breeding innovation, advisory delivery, and traceability systems will be driven more and more by digital agriculture and AI, enhancing efficiency and transparency. Whether biofortification remains a donor-driven endeavor or develops into a nationally-led, sustainably supported program with long-term resilience will depend on international collaborations and creative financing methods.

However, alertness is crucial. Broader food system changes including dietary variety, soil health restoration, and fair market access should be complemented, not replaced, by biofortification. The development and application of biofortified crops must be guided by ethical concerns, guaranteeing that cultural values, consumer autonomy, and safety requirements continue to be of utmost importance. Maintaining public trust will depend on open communication, harmonizing regulations, and community involvement as biotechnology disputes become more heated. One of the most scalable, economical, and scientifically proven methods for combating hidden hunger and promoting sustainable development is biofortification. Its long-term viability, however, hinges on the integration of research with social systems, policy with grassroots realities, and innovation with inclusivity. Biofortification can develop from a focused nutrition intervention into a game-changing force for resilient food systems, healthier people, and sustainable futures with concerted effort, improved governance, and a dedication to equity.

REFERENCES

Arora, N. K., & Mishra, P. (2025). Bridging nutritional and sustainability gaps: The role of biofortification in achieving the SDGs. *Environmental Sustainability*, *8*(2), 135–138. DOI: 10.1007/s42398-025-00361-z

Batool, S., Asghar, H. N., Shehzad, M. A., Yasin, S., Sohaib, M., Nawaz, F., Akhtar, G., Mubeen, K., Zahir, Z. A., & Uzair, M. (2021). Zinc-solubilizing bacteria-mediated enzymatic and physiological regulations confer zinc biofortification in Chickpea (Cicerarietinum L.). *Journal of Soil Science and Plant Nutrition*, *21*(3), 2456–2471. DOI: 10.1007/s42729-021-00537-6

Bharti, C., Fatima, T., Mishra, P., Verma, P., Bhattacharya, A., Alaylar, B., & Arora, N. K. (2024). Salt-tolerant endophytic Bacillus altitudinis NKA32 with ACC deaminase activity modulates physiochemical mechanisms in rice for adaptation in saline ecosystem. *Environmental Sustainability*, *7*(2), 231–249. DOI: 10.1007/s42398-024-00316-w

Boliko, M. C. (2019). FAO and the situation of food security and nutrition in the world. *Journal of Nutritional Science and Vitaminology*, *65*(Supplement), S4–S8. DOI: 10.3177/jnsv.65.S4 PMID: 31619643

Bouis, H. E. (2003). Micronutrient fortification of plants through plant breeding: Can it improve nutrition in man at low cost? *The Proceedings of the Nutrition Society*, *62*(2), 403–411. DOI: 10.1079/PNS2003262 PMID: 14506888

Dhaliwal, S. S., Sharma, V., Shukla, A. K., Verma, V., Kaur, M., Shivay, Y. S., Nisar, S., Gaber, A., Brestic, M., Barek, V., Skalicky, M., Ondrisik, P., & Hossain, A. (2022). Biofortification-A Frontier Novel Approach to Enrich Micronutrients in Field Crops to Encounter the Nutritional Security. *Molecules (Basel, Switzerland)*, *27*(4), 1340. DOI: 10.3390/molecules27041340 PMID: 35209127

Habib, M., Singh, S., Jan, S., Jan, K., & Bashir, K. (2025). The future of the future foods: understandings from the past towards SDG-2. *npj Science of Food*, *9*(1), 138.

Kendal, D. (2020). Genetically Modified Plants: Nutritious, Sustainable, yet Underrated. *The Journal of Nutrition*, *150*(10), 2628–2634. DOI: 10.1093/jn/nxaa220 PMID: 32725215

Kumar, S., Sharma, S. K., Thakral, S. K., Bhardwaj, K. K., Jhariya, M. K., Meena, R. S., Jangir, C. K., Bedwal, S., Jat, R. D., Gaber, A., Atta, A. A., & Hossain, A. (2022). Integrated Nutrient Management Improves the Productivity and Nutrient Use Efficiency of Lens culinaris Medik. *Sustainability (Basel)*, *14*(3), 1284. DOI: 10.3390/su14031284

Labuschagne, M. (2023). Biofortification to improve food security. *Emerging Topics in Life Sciences*, *7*(2), 219–227. DOI: 10.1042/ETLS20230066 PMID: 37962270

Mishra, P., Mishra, J., & Arora, N. K. (2023). Biofortification revisited: Addressing the role of beneficial soil microbes for enhancing trace elements concentration in staple crops. *Microbiological Research*, *275*, 127442. DOI: 10.1016/j.micres.2023.127442 PMID: 37437425

Mishra P, Mishra J, Bharti C, Arora NK (2025) Salt-Tolerant Pseudomonas taiwanensis PWR-1 mediated organic acid production for biofortification of zinc and reducing fertilizer dependency in wheat under saline conditions. *J Plant Growth Regul.* d o i: 1 0 . 1 0 0 7 / s 0 0 3 4 4 - 0 2 5 - 1 1 6 2 3 – 9

Mitra-Ganguli, T., Boyd, K., Uchitelle-Pierce, B., & Walton, J. (2019) Proceedings of the workshop 'biofortified food - Working together to get more nutritious food to the table in india'. J *Nutr Intermediary Metabolism* 18:100100. DOI: 10.1016/j.jnim.2019.100100

Mudadirwa D, Ndhlela T, Olakojo SA, Gonhi T (2023) The potential of provitamin A maize (Zeamays L.) biofortification in sub-Saharan africa: current status, opportunities, challenges and perspectives: A review. *CABI Reviews*. .DOI: 10.1079/cabireviews.2023.0041

Narayanan, N., Beyene, G., Chauhan, R. D., Gaitán-Solís, E., Gehan, J., Butts, P., Siritunga, D., Okwuonu, I., Woll, A., Jiménez-Aguilar, D. M., Boy, E., Grusak, M. A., Anderson, P., & Taylor, N. J. (2019). Biofortification of field-grown cassava by engineering expression of an iron transporter and ferritin. *Nature Biotechnology*, *37*(2), 144–151. DOI: 10.1038/s41587-018-0002 1 PMID: 30692693

Neeraja, C. N., Babu, V. R., Ram, S., Hossain, F., Hariprasanna, K., & Rajpurohit, B. S. Prabhakar; Longvah, T.; Prasad, K.S.; Sandhu, J.S.; et al (2017). Biofortification in Cereals: Progress and Prospects. *Curr. Sci.* 113, 1050–1057. https://www.jstor.org/stable/26494165

Oluwole, O., Ibidapo, O., Arowosola, T., Raji, F., Zandonadi, R. P., Alasqah, I., Lho, L. H., Han, H., & Raposo, A. (2023). Sustainable transformation agenda for enhanced global food and nutrition security: A narrative review. *Frontiers in Nutrition*, *10*, 1226538. DOI: 10.3389/fnut.2023.1226538 PMID: 37599683

Pompano, L. M., Luna, S. V., Udipi, S. A., Ghugre, P. S., Przybyszewski, E. M., & Haas, J. (2022). Iron-biofortified Pearl millet consumption increases physical activity in Indian adolescent schoolchildren after a 6-month randomised feeding trial. *British Journal of Nutrition*, *127*(7), 1018–1025. DOI: 10.1017/S000711452100180X PMID: 34078482

Sheoran, S., Kumar, S., Kumar, P., Meena, R. S., & Rakshit, S. (2021). Nitrogen fixation in maize: Breeding opportunities. *Theoretical and Applied Genetics*, *134*(5), 1263–1280. DOI: 10.1007/s00122-021-03791-5 PMID: 33677701

Shohael, A. M., Kelly, J., Venkataraman, S., & Hefferon, K. (2025). Unlocking Opportunities and Overcoming Challenges in Genetically Engineered Biofortification. *Nutrients*, *17*(3), 518. DOI: 10.3390/nu17030518 PMID: 39940376

Singla, P., & Grover, K. (2017). Biofortified Cereal Crops: A Sustainable Approach for Food and Nutritional Security. *British Journal of Applied Science and Technology*, *24*(4), 1–13. DOI: 10.9734/CJAST/2017/37865

World Health Organization. (2006) Guidelines on food fortification with micronutrients. https://www.who.int/publications/i/item/9241594012

About the Contributors

Syed A. Rehman Khan completed his postdoctoral fellowship at Tsinghua University. Prof. Khan is currently associated with Business Research and Consultancy Limited in the United Kingdom. Professor Khan has consistently received recognition as one of the world's top 2% highly cited scholar. His impressive portfolio includes nearly 200 high-level academic research articles and 11 published academic books related to business administration, circular economy, and green practices in business operations. Additionally, he has led numerous research funding projects. Professor Khan is a sought-after keynote speaker, invited speaker, and conference chair, having participated in over 70 international academic conferences. His expertise and influence in the field are unparalleled.

Zhang Yu is an accomplished academic with a PhD in logistics and Operations Management and extensive experience in teaching and research, specializing in closed-loop supply chain and green logistics. With over 150 publications in Scopus and Web of Science-indexed journals and a solid understanding of AACSB accreditation standards, She is dedicated to advancing educational excellence through innovative teaching methodologies, strategic curriculum development, and impactful research. Seeking to leverage her expertise in both academic leadership and industry collaboration to drive the institution's growth, enhance global research visibility, and foster a dynamic learning environment that bridges theory with real-world applications.

Muhammad Tanveer Nawaz is a top 2% Scientist of the World (Stanford Ranking). He is a researcher and academic at the College of Business, Imam Mohammad Ibn Saud Islamic University (IMSIU), Riyadh, Saudi Arabia. His expertise includes management, sustainability, and emerging technologies, with a focus on Green HRM, AI, Corporate Performance, and Sustainable Development Goals. He has authored high-impact publications and is committed to advancing knowledge through research, mentorship, and collaboration.

***

Rahmath Nisha is a dedicated professional with comprehensive, hands-on experience in Human Resource Management (HRM), having successfully managed the entire employee life cycle—from recruitment and onboarding to performance management and employee engagement. She holds an MBA in Human Resource Management, where she was recognized as a Gold Medalist and awarded the title of Best Outgoing Student from Saveetha University. Her academic journey began with a Bachelor's degree in Electronics and Communication Engineering, showcasing her diverse educational foundation. Currently serving as an Assistant Professor, she combines industry expertise with academic insight to mentor and inspire the next generation of HR professionals.

A. K. Subramani is a distinguished academic with a PhD in Business Administration from Manonmaniam Sundaranar University, Tirunelveli, India. With 18 years of teaching experience, he has made significant contributions to the field of human resource management and organisational behaviour. He has an impressive publication record with 80 articles in international journals, including two papers in the ABDC-A category, eight in the ABDC-B category, three in the ABDC-C category, 30 in Scopus-indexed journals, and eight in Web of Science journals. He has served as a resource person for various training workshops on SPSS and AMOS software packages

A. V. Senthil Kumar is working as Principal, Nehru Institute of Information Technology and Management, Coimbatore, India. He has worked as Professor and Director, PG and Research Department of Computer Applications, Hindusthan College of Arts & Science, Coimbatore, Tamilnadu for 15 years and as Senior Grade Lecturer in CMS College of Science and Commerce for 14 years. To his credit he has industrial experience for five years and teaching experience of 29 years. He has also received his Doctor of Science (D.Sc in Computer Science). He has to his credit 86 Book Chapters, 234 papers in International and National Journals, 85 papers in International Conferences in International and National Conferences, and edited 17 books and 3 Text books. He is as Associate Editor of IEEE Access. He is an Editor-in-Chief for many journals and Key Member for India, Machine Intelligence Research Lab (MIR Labs). He is an Editorial Board Member and Reviewer for various International Journals. He is also a Committee member for various International Conferences. He has guided 16 Ph.D scholars and guiding 3 Ph. D scholars now.

Naghma Ali is serving as an Assistant Professor in the Department of Business Administration at the United Institute of Management, Prayagraj, India. Her research interests include Stress of working professionals, burnout at the workplace, personality, organizational behaviour and human resource management.

Shagufta Ali is an Assistant Professor in the Department of Business Administration at the United Institute of Management, Prayagraj, India. Her research interests centered on professionals, work-life balance, and women's empowerment, the psychosocial dimensions of professional life, emphasizing well-being, equity, empowerment in organizational contexts, organizational behaviour, and human resource management.

Souvik Banerjee, PhD. is a faculty at Management Development Institute (MDI) Murshidabad. His research interests include Corporate Finance, Corporate Governance, Behavioral Finance, and Financial Economics. His work has been published in Scopus-indexed and ABDC-listed journals. At present, he is the Editor of the MDIM Journal of Management Review and Practice (Sage). He has developed several business cases for MBA and executive education. He has trained executives from Fortune 500 companies in Corporate Finance and Corporate Valuation.

Shikha Bhagat is an expert in International Business and Marketing. She graduated with a gold medal from Guru Nanak Dev University. Dr. Shikha earned her Ph.D. from Noida International University, specializing in international business. With over 14 years of experience, she has published numerous papers and book chapters. Her work appears in prestigious journals indexed in Scopus and Web of Science. Dr. Shikha regularly participates in academic forums and conferences on global business trends. Her research excellence has earned her Best Paper Awards at international conferences in 2023 and 2024. As an Assistant Professor at Christ University, she is committed to mentoring students. Dr. Shikha emphasizes practical applications and critical thinking in her teaching. She is a member of the International Social Marketing Association. This highlights her active role in global academic and professional communities. Her career reflects a dedication to research, education, and scholarly collaboration.

Patcha Bhujanga Rao is a distinguished professional with a stellar reputation in the realms of human resources (HR) and soft skills development. With over two decades of experience and an extensive academic background encompassing degrees such as M.Com., DCFA., M.Phil., Ph.D., MBA (HR), M.Sc (Psychology), and LL.B, he possesses a comprehensive understanding of multiple disciplines. Currently serving as a professor at Jain Deemed-to-Be University in Bengaluru, Dr. Rao is highly regarded for his expertise in HR management, which has been instrumental in shaping the careers of numerous professionals. His commitment to nurturing essential skills such as effective communication, leadership, and interpersonal abilities reflects his dedication to facilitating personal and professional growth. Dr. Rao's contributions have earned him widespread respect and admiration within the HR and soft skills domains, positioning him as a respected authority and mentor for aspiring professionals seeking to excel in their respective fields.

Frank Billingsley has had a diverse and extensive career spanning multiple domains, including education, social services, and administration. He has had extensive international experience, and qualifications in higher education administration and teaching are prominent, with notable roles as a program director, head of department, and faculty member at various institutions. He has contributed to academic publications and presented at conferences on topics related to higher education and gender studies. Beyond academia, the candidate has a social work and counseling background, demonstrating a versatile skill set. His precise community service and involvement in various organizations further underscore their commitment to positively impacting.

Ankita Chaturvedi is currently working as an Associate Professor at IIS (deemed to be University), Jaipur and has over 16+ years of teaching experience. She received her Ph.d degree(2011), Master's in Accounting and Business Statistics(2006), and Master's in Economics and Financial Management(2018) from University of Rajasthan, India. She has qualified NET in Commerce in 2015. She is an author or Co-author of more than 46 research papers published in national or international refereed journals, 11 chapters in edited books, 3 books with ISBN number and more than 50 conference contributions. She has received several (22) awards in the field of academics and research. She has supervised around 18 PG dissertations. She is supervising 6 Ph.d Thesis and 6 Ph.d degrees has been awarded under her guidance. She has been convener of various academic events like international conference, training programmes, Research workshop, Board of Studies etc. She has attended various more than 22 FDP's and training programmes.

Pullagora Chiranjeevi holds multiple degrees including B.L., LL.M., M.A. in English, and a Ph.D. in Law from Acharya Nagarjuna University, Guntur. His academic journey reflects a strong interdisciplinary foundation with additional qualifications in Political Science and a Postgraduate Diploma in Ambedkar Studies. Dr. Chiranjeevi specializes in Law and has consistently contributed to legal education and research. His expertise spans across jurisprudence, constitutional law, and socio-legal studies. Alongside his academic achievements, he is skilled in data creation and research methodologies. Dr. Chiranjeevi is deeply committed to promoting justice, social equity, and educational empowerment through legal discourse. He is the son of Late Mastan Rao and continues to carry forward a legacy of intellectual pursuit and public service. His multi-dimensional knowledge makes him an asset in the legal and academic community.

Hayden Davis is a attending the University of Edinburgh for a Masters in accounting and financial management.

Januari Dawolo, S.Si., M.Si. Lecturer, Aquatic Resources Study Program Faculty of Science and Technology, University of Nias He is a permanent lecturer in the Aquatic Resources Study Program, Faculty of Science and Technology, University of Nias. He completed his undergraduate degree in Chemistry at the State University of Medan (UNIMED) and obtained his master's degree in Chemistry at Bogor Agricultural University (IPB). The author specializes in writing and research.

Hana Dixon is currently a student at the University of Kentucky working on obtaining a Masters in Health Administration. She graduated from Georgetown College in May 2025 with a Bachelors of Science majoring in Healthcare Administration and a minor in Accounting.

Michelle T. Flores is an Assistant Professor IV at Laguna State Polytechnic University, San Pablo City Campus, where she assigned as the Program Coordinator of the Bachelor of Science in Biology program since 2013. She earned her Bachelor's degree in Biology from the University of the Philippines Los Banos with Microbiology as her field of specialization. She later completed her Master in Public Affairs major in Education Management from the same university and currently writing her dissertation to complete the Doctor of Philosophy in Education (Biology) at the University of the Philippines Open University.

Saraswathy G, is a legal scholar and academician associated with the Faculty of Law, Dr. MGR Educational and Research Institute,(Dr.MGR University). Saraswathy's interests include contemporary legal issues, constitutional principles, and the evolving role of law in social transformation.

G. Vanishree has 28 years in Academics & administration. Having worked as HOD, Principal, and vice principal in Vignan Institutions, Institute of Aeronautical Engineering, Badruka, St. Patricks, St.Joseph College Kristu Jayanti College, etc, she also worked as Director for Kempe Gowda Institute of Management Studies and Research Bangalore. Presently, she is working as a professor at IBS-B Bangalore. Prof has been a credit participator in several National seminars & conferences and published journal articles. She has won the Best Teacher, parent, and Adarsh Yuvathi awards. She makes herself suited to the position placed. She is responsible and inspiring, has a good attitude, and believes in ethical norms. Having a good record of good results, she believes in the statement, "Learning is a continuous process."

Swapnil Gorde is an Associate Professor and PhD scholar with 20+ years of experience across academia, Big-4 firms, and FinTech. An expert in taxation and finance, he has founded multiple academies, led customer success at Intuit India, and holds several patents and SCOPUS-indexed publications

Sandeep Gupta is presently working as an Associate Professor at Graphic Era (Deemed to be University), Dehradun (India). He has a total of 15 years of experience. His areas of interest in research are the application of artificial intelligence to power system control design, FACTS devices, renewable energy, power electronics, and stability of power systems with machine learning. He is a member of numerous professional organizations, including IE, IRED, IDES, IAENG, IAER, and the Society of Professional Engineers. He has been the author or co-author of more than 150 papers published in international journals and presented at different national and international conferences. He is currently guiding the research scholars under Machine Learning, Control & Power Systems, Wireless Technology, and the Internet of Things.

Jherwin P. Hermosa is an Associate Professor 3 at the Laguna State Polytechnic University in San Pablo City Campus. Dr. Hermosa earned his Doctor of Education in Educational Management in 2023 from the same university. He also holds degrees in Philosophy and Theology from Rome, Italy. Dr. Hermosa currently teaches Social Sciences and Philosophical Subjects at both the Undergraduate and Graduate Program levels. Additionally, he serves as an editorial board member for the International Review of Social Sciences Research and member of the Philosophical Association of the Philippines (PAP). Dr. Hermosa has authored books such as Foundations of Education (2022) Philosophy of Education (2023) and Advanced Social Philosophy (2024), and Ethics: A Guide to a Moral Living (2025) along with numerous scientific articles and literary works that have been published internationally. Currently, he is pursuing a Juris Doctor degree at San Pablo Colleges, College of Law. Dr. Hermosa's research interests primarily focus on the social sciences, philosophy, educational management, and law.

Namita Vyas Joshi is a distinguished academician and eminent legal scholar with over two decades of exemplary experience in legal education, advanced research supervision, and academic leadership at premier law institutions in India. She holds a Doctorate in Law, with her doctoral research making a significant scholarly contribution to the legal framework governing the registration of marriages in India. Currently, Prof. (Dr.) Joshi holds pivotal leadership positions as Director, Dean, and Chairperson of the Board of Studies, where she plays a transformative role in curriculum design, academic planning, regulatory compliance, and institutional governance. Her visionary leadership has been instrumental in strengthening academic standards and fostering a culture of excellence and innovation in legal education. A prolific author and researcher, Prof. (Dr.) Joshi has authored and edited several authoritative books on law, in addition to publishing extensively in reputed national and international journals. Her scholarly works are widely referenced and contribute meaningfully to contemporary legal discourse, particularly in the areas of family law, criminology, and socio-legal studies. She also serves on the editorial boards of reputed academic journals, further underscoring her standing in the legal academic community. Her academic affiliations include active membership in esteemed professional bodies. Widely respected for her subject expertise and academic leadership, Prof. (Dr.) Joshi is frequently invited as a resource person, keynote speaker, and organizer of national and international conferences, seminars, and workshops. Through her sustained contributions as an educator, author, mentor, and academic leader, Prof. (Dr.) Namita Vyas Joshi continues to shape legal education and scholarship while inspiring future generations of legal professionals..

Ranjithaa K C is an Assistant Professor with a Ph.D. in Finance from Saveetha School of Management, SIMATS. She holds dual specializations in Finance and Marketing through her MBA and has cleared TNSET in Management. With over six years of experience in teaching and research, she brings academic rigor and industry relevance to the classroom. She has presented papers at national and international conferences and authored a book titled "Management Concepts in Chanakya (2019)". Her expertise lies in blending traditional principles with modern financial insights to foster impactful learning.

Meera K L is a finance professional with over 20+ years of experience in industry and academia. Currently, she is working as an Associate Professor at Christ Academy Institute of Advanced Studies, Bangalore. She has held leadership roles, including IQAC Coordinator and HOD at Dayananda Sagar Business Academy. A Cambridge-trained educator, she adopts innovative, student-centered teaching methods. Her research focuses on investment avenues for middle-class investors, with publications in ABDC, Scopus, and Web of Science journals. She believes in the transformative power of education to inspire growth.

M. R. Arun completed his BE, ME and PhD from Anna University, Chennai. He completed PG Diploma (PGDEEM) from Annamalai University, M.Sc. Counselling Psychology from TNOU University, M.A. Public Administration from Madras University and B.Lib.Sc. from Alagappa University. He is in the Teaching Profession for the past 17 years, Published around 115 papers in Journals, Patents and Conferences, Active participant crossing more than 200 events and internationally networked through various Professional bodies. He is skilled and technically proficient in various simulation Softwares associated to Research in Engineering. He Organized and Coordinated around 15 events in working Institution. Authored more than 15 Books, 20 Book Chapters and he has been technically integrated in various international journals as an Editorial Board Member and Reviewer. His area of interest includes Video Processing, Remote Automation, Wireless Communication, IoT Sensors, Quantum Computing, Machine Learning & Artificial Intelligence.

Kajal Maheshwari is an Associate Professor at S. B. Patil Institute of Management, Nigdi, Pune. She is actively involved in teaching, research, and academic administration, with expertise in management education and a strong focus on student development. Dr. Maheshwari contributes to curriculum design, research initiatives, and institutional development, and is committed to nurturing future managers through a blend of academic rigor and practical insight.

Pratap Chandra Mandal is an Assistant Professor (Marketing) at Indian Institute of Management, Shillong, India. He has completed graduate degree from the reputed Indian Institute of Technology, Kharagpur (IIT Kharagpur), India (Bachelor of Technology in Mechanical Engineering), post-graduate degree from Vinod Gupta School of Management, IIT Kharagpur (Masters in Business Adminstration), PhD (Marketing) from Vinod Gupta School of Management, IIT Kharagpur. His research concerns customer relationship management, customer satisfaction, services marketing, marketing intelligence, and qualitative methods in management. He is the editor-in-chief of two international journals and is on the editorial board of journals like Journal of Global Marketing. Pratap has won several prestigious scholarships and awards throughout his academic career.

T. C. Manjunath was born on 6th February 1967. He did his B.E. from RV College of Engg. in 1989 (B'lore Univ), M.E. in Robotics & Automation from Govt.'s LD College of Engg., (Gujarat Univ., Ahmedabad) in 1992 with FCD-First Rank & PhD in Systems & Control Engg. from IIT Bombay in the year 2007, got best research scholar award for the entire IITB. He is currently working as the Dean Research (R & D) and as Professor in Dept. of Computer Science & Engg. with specialization in IoT, Cyber Security, Blockchain Technology in Rajarajeswari College of Engineering, Bengaluru, Karnataka. He has got a teaching (academic), research & administrative experience of more than 35+ years (10 years as Principal & 8 years as HoD, 1 year as Dean Research & rest as faculty) and has published more than 1500+ research papers in various national & international journals & conferences, with 175+ patents, 100+ book-chapters, 25 awards, 100s of Keynote talks and also published 25 text books in various field of science, engineering &

technology, completed projects worth Rs. 9 Crores till date & guided 10 PhD scholars in VTU. He is a member of 30 Professional Society Chapters and is a fellow of IE, IETE, AMIEE, Sr. Member of IEEE and a chartered engineer.

Ridho Nazara is a permanent lecturer in the Agrotechnology Program at the Faculty of Science and Technology, Nias University. He completed his undergraduate degree in the Agrotechnology Department at the Methodist University of Indonesia (UMI) and his master's degree in the Agrotechnology Department at the University of North Sumatra (USU). The author specializes in Writing and Research.

P. Selvakumar is currently an Associate Professor in the Department of Science and Humanities at Nehru Institute of Technology, Coimbatore, Tamil Nadu, India. Prior to joining Nehru Institute of Technology, he held positions as an Assistant Professor in the Department of Science and Humanities various engineering colleges in Coimbatore, Tamil Nadu, India, spanning from 2011 to 2023. Dr.P. Selvakumar's academic and research interests lie primarily in the field of phytochemistry. He has made significant contributions to his field, with an impressive record that includes 30 journal publications, 12 patents, and 20 conference papers. Additionally, he has organized 5 seminars and conferences, further demonstrating his commitment to academic and scientific exchange. Throughout his career, Dr. P.Selvakumar has been recognized for his dedication and excellence in teaching, having been honored with several awards, including the Best Teacher Awards. He holds a prominent position in the academic community, contributing actively to research.

Ranjitha P.K. is an experienced academician with over 13 years in education and management. Completed Ph.D. in Management from Vivekananda Global University, currently working as Assistant professor in Jain CMS deemed to be university,her research explores social media's influence on consumer preferences with respect to destination branding. She has also undergone training in Data Analytics at IIM Rohtak, specializing in tools like Power BI. Having served as an Assistant Professor at esteemed institutions, she has taught Marketing, Business Statistics, and E-Business while mentoring numerous student projects. A prolific researcher, she has published papers in reputed journals and presented at global conferences.

Pradip Sanatkumar Padhye is an Assistant Professor at the Symbiosis School of Banking and Finance, Symbiosis International (Deemed University), Pune. He is engaged in teaching, academic research, and mentoring students in the areas of banking, finance, and related disciplines. With a strong commitment to academic excellence and industry relevance, Dr. Padhye contributes to curriculum development, scholarly activities, and institutional initiatives, supporting the school's mission of developing competent and ethically grounded finance professionals.

Revati Ramrao Rautrao Associate Professor, is a scholar, HR expert, and educator with 20+ years of experience, known for her research in sustainability, AI in education, and women's empowerment.

Filiz Resuloğlu studied English Language and Literature at Ankara University between 1999-2003. She completed her Master's education on Communication Sciences at Kocaeli University in 2014 .Her thesis was on transmedia storytelling and the popular Tv series "Game of Thrones". She completed her Phd in January 2025 at Kocaeli University. She is interested in new media studies, transmedia storytelling, dataveillance,sustainability and governance.

Guru Basava Aradhya S. is a distinguished academic leader and Professor with 18 years of experience. Holding a PhD in Management and several international patents, his expertise encompasses HR analytics, AI-driven management, and strategic growth. He has published extensively in Scopus and ABDC journals and authored influential books on HR metrics and strategy . Currently, he serves as Professor in Dr DY Patil B School, Pune, Maharashtra.

S. Haashni is a dedicated student of Vimal Jyothi Convent Matriculation Higher Secondary School, located in Coimbatore, Tamil Nadu, India. Known for her commitment to academic excellence and personal growth, S.Haashni is actively involved in Academic activities and strives to make the most of her educational journey. She is passionate about reading and writing books, and enjoys playing games in her free time. S. Haashni is a curious learner who actively pursues her interests both inside and outside the classroom.

S.T. Naidu, an Associate Professor specialising in Labour and Industrial Law & Human Resource Management (HRM). I hold dual doctorates in Management and Law from Acharya Nagarjuna University and have over 23 years of experience in teaching, training, research, and consultancy. Currently, I am an Associate Professor of Law at Vel Tech Rangarajan Dr. Sagunthala R&D Institute of Science and Technology in Chennai. I have also coordinated Research & Development, served on the Board of Studies, and supervised six Ph.D. Research Scholars at the School of Law. I have extensive experience in training and consulting within educational institutions, having authored 52 articles, 5 books, 2 patents, and contributed to 8 edited volumes, many featured in UGC CARE-listed and Scopus-indexed journals. I've participated in 26 FDP Programs, 8 international and 16 national seminars, completed 14 Coursera certificates, attended 12 webinars, and held 4 professional memberships. I also serve as a reviewer, editor, and board member for respected journals.

T.P. Saravanan earned a Doctorate in Management from Periyar University and a Master of Business Administration (MBA) from Bharathiar University. He is an Indian Institute of Management, Ahmedabad alumnus who completed a four-month FDP in Management in the year 2013. With 25 years of experience in industry and academia, he has published 10 research publications in recognised journals, some of which are indexed by Scopus, IEEE, and UGC Care. He has also delivered several research papers at international conferences hosted by major institutions like as IIMs, IITs, and others. Dr. T P Saravanan is an active researcher and consultant for government, public, and commercial organisations. He has organised several Faculty Development Programmes (FDPs), workshops, and Management Development Programmes (MDPs). He has also acted as a resource person for over 40 management development programmes across the nation, including those at State Bank of India, Dena Bank, URC Constructions Pvt Ltd., Rennacon Pvt Ltd., Angel Starch Foods Pvt Ltd., etc. His favoured teaching topics include Branding, CRM, Product Management, Rural Marketing, Retail Management, Consumer Behaviour, Marketing Analytics, Digital & Social Media Marketing and Business Research Methodology. Dr. T P Saravanan is also a trained ISO Certification Trainer under the National Quality Academy. He received the 'Best Researcher Award-2021' from American College in Madurai, Tamil Nadu, as well as the Tamil Nadu Government's "Best Service to the Nation" Award.

Shilpa Sarvani, holds a PhD in Retail Marketing from GITAM University, she seamlessly integrates academic rigor with practical insights, drawing from her four-year stint as an In-Store Brand Manager at Colgate-Palmolive Limited. Dr. Shilpa boasts four years of teaching experience in prominent Bengaluru universities, employing innovative methodologies that emphasize case studies and industry collaborations. Her extensive industry connections contribute to a dynamic learning environment, providing students with the latest insights. An accomplished researcher, Dr. Shilpa has authored several papers in esteemed journals, including those indexed by ABDC and Scopus and also contributed chapters to the Scopus indexed books. Her commitment to advancing knowledge is further exemplified by her participation in national and international conferences. Notably, she received the multiple best paper awards at the international conferences. Driven by a passion for education and industry relevance, Dr. Shilpa Sarvani stands as a thought leader, inspiring students to navigate the ever-evolving marketing landscape. Her area of in interest and expertise in research lie with Retail marketing, Private label brands, organised retail, e commerce, digital marketing, Advertising and sustainable marketing.

Pamarthi Satyanarayana, specialize in Constitutional Law and Legal Order, focusing my research on the interface between individual rights and public policy, with a strong emphasis on civil liberties, social justice, and the right to privacy. Over the past 36 months, I have served as an Assistant Professor of Law at Vel Tech Rangarajan Dr. Sagunthala R&D Institute of Science and Technology in Chennai, Tamil Nadu. Here, I teach courses such as Constitutional Law, Legal Research and Writing, and both Civil and Criminal Law. My passion for teaching and in-depth subject knowledge have earned me recognition as a dedicated educator. In addition to my academic role, I actively engage with the community, partnering with legal clinics to provide pro bono services to underprivileged individuals. Looking forward, I aim to expand my research into emerging legal issues in the digital age, especially in privacy rights, while guiding aspiring law students and future legal scholars.

S. C. Vetrivel is a faculty member in the Department of Management Studies, Kongu Engineering College (Autonomous), Perundurai, Erode Dt. Having experience in Industry 20 years and Teaching 16 years. Awarded with Doctoral Degree in Management Sciences in Anna University, Chennai. He has organized various workshops and Faculty Development Programmes. He is actively involved in research and consultancy works. He acted as a resource person to FDPs & MDPs to various industries like, SPB ltd, Tamilnadu Police, DIET, Rotary school and many. His areas of interest include Entrepreneurship, Business Law, Marketing and Case writing. Articles published more than 100 International and National Journals. Presented papers in more than 30 National and International conferences including IIM Bangalore, IIM Kozhikode, IIM Kashipur and IIM Indore. He was a Chief Co-ordinator of Entrepreneurship and Management Development Centre (EMDC) of Kongu Engineering College, he was instrumental in organizing various Awareness Camps, FDP, and TEDPs to aspiring entrepreneurs which was funded by NSTEDB – DST/GoI

Bensha C. Shaji, an academician specializing in Taxation Laws, Corporate Law, and Commercial Law, currently serving as Assistant Professor (Senior Grade) at the School of Law, VIT-AP University, Amaravati. My teaching interests include tax law, corporate governance, and commercial regulations, with a focus on blending theoretical foundations with contemporary legal developments. My doctoral research critically examined the legal framework surrounding double taxation in India, particularly overlapping tax structures, providing a foundation for my ongoing work in fiscal federalism, tax policy reform, and digital compliance mechanisms. I actively engage in academic mentoring, curriculum development, and student-led research initiatives. My current research explores the intersection of law and technology, especially how AI and digitalisation affect tax compliance and corporate regulation. I remain committed to fostering critical legal scholarship and contributing meaningfully to discourse on emerging challenges in national and transnational regulatory frameworks.

Anjali Singh is a doctoral scholar at Amity Business School, Amity University, Lucknow, India. Her research interests include digital pedagogy, educational innovation, and learning experience design, organizational behaviour, and human resource management.

Sanjayan T. S. is an assistant professor and Head, postgraduate department of Education in GVMs Dr.Dada Vaidya College of Education, Goa University. Delivered talks in almost 57 countries in the world . published papers in renowned journals . Done collaborative projects in international level. speaker in UNESCO, EUROPEAN COMMISSION AND World forum of Education in different countries.

Putra Hidayat Telaumbanua, S.P., M.M., is an Assistant Lecturer in the Agrotechnology Study Program, Faculty of Science and Technology, Universitas Nias, Indonesia. The author holds a Bachelor's degree in Agriculture and a Master's degree in Management, both obtained from Universitas Nasional, Jakarta. The author's academic expertise includes Marketing Management and Agricultural Economics. The author teaches courses such as Principles of Agricultural Management, Agricultural Ecology, Agribusiness Feasibility Study, Agrotourism, and Introduction to Agricultural Science. The author is also actively involved in the Merdeka Belajar Kampus Merdeka (MBKM) program as a Field Supervisor for the student internship program during the Even Semester of the 2024/2025 academic year. The author's research interests focus on digital marketing strategies for agricultural products and the empowerment of local farmers in the Nias Islands region.

V. Sabareeshwari currently serves as Assistant Professor in Department of Soil Science, Amrita School of Agricultural Sciences, Coimbatore. Having more than 5 years of research experience and more than 2 years of teaching experience. She got 7 awards in the field of agriculture. Her field of expertise are soil genesis, soil pedological studies as well as soil fertility mapping using advanced software like Arc GIS. She had published 22 research papers and more than 10 book chapters and books in high- impact reputed journals. She has actively participated and presented her papers in more than 20 conferences and seminars. She not only restrict her contribution only in the academic and research part, she had extension experience at farm level (lab to land) with varied crop research.

V.P. Arun is a driven and accomplished professional with a diverse educational background and extensive hands-on experience across various industries. Graduating with honors, Arun earned his Master of Business Administration (M.B.A) with a specialization in Human Resources and Marketing from the renowned Sona School of Management in Salem in 2018, where he excelled academically with an impressive 8.3 Cumulative Grade Point Average (CGPA).Throughout his academic journey, Arun displayed an unwavering commitment to learning and personal growth, actively seeking opportunities to expand his knowledge and skills beyond the confines of traditional education. He sought practical experiences to complement his theoretical understanding, such as a 45-day summer internship focused on conducting a feasibility study for R-Doc Sustainability in the market. Additionally, Arun broadened his horizons through a 7-day industrial visit to Malaysia and Singapore, immersing himself in diverse cultural and professional environments. Arun's academic pursuits were further enriched by his involvement in hands-on projects, including a comprehensive study on Employee Job Satisfaction at Roots Cast Private Limited. His professional trajectory includes serving as a Growth Officer at Parle Agro Private Limited, where he played a crucial role in driving sales growth and market expansion. Arun is a dedicated professional, currently employed at JKKN Engineering College in the Management Studies Department. He holds expertise in his field and has contributed significantly to academia through the publication of two journal articles. Arun's work focuses on advancing knowledge and understanding in management studies, making valuable contributions to the academic community.

Index

S

T

www.ingramcontent.com/pod-product-compliance
Lightning Source LLC
LaVergne TN
LVHW061529290926
849162LV00004B/722